T0192054

Lecture Notes in Computer Science 13889

Advanced Research in Computing and Software Science

Subline of Lecture Notes in Computer Science

More information about this series at https://link.springer.com/bookseries/558

Sun-Yuan Hsieh · Ling-Ju Hung ·
Chia-Wei Lee
Editors

Combinatorial Algorithms

34th International Workshop, IWOCA 2023
Tainan, Taiwan, June 7–10, 2023
Proceedings

Editors
Sun-Yuan Hsieh
National Cheng Kung University
Tainan, Taiwan

Ling-Ju Hung
National Taipei University of Business
Taoyuan, Taiwan

Chia-Wei Lee
National Taitung University
Taitung, Taiwan

ISSN 0302-9743 ISSN 1611-3349 (electronic)
Lecture Notes in Computer Science
ISBN 978-3-031-34346-9 ISBN 978-3-031-34347-6 (eBook)
https://doi.org/10.1007/978-3-031-34347-6

This Springer imprint is published by the registered company Springer Nature Switzerland AG
The registered company address is: Gewerbestrasse 11, 6330 Cham, Switzerland

Preface

The 34th International Workshop on Combinatorial Algorithms (IWOCA 2023) was planned as a hybrid event, with on-site activity held at the Magic School of Green Technologies in National Cheng Kung University, Tainan, Taiwan during June 7–10, 2023.

Since its inception in 1989 as AWOCA (Australasian Workshop on Combinatorial Algorithms), IWOCA has provided an annual forum for researchers who design algorithms for the myriad combinatorial problems that underlie computer applications in science, engineering, and business. Previous IWOCA and AWOCA meetings have been held in Australia, Canada, the Czech Republic, Finland, France, Germany, India, Indonesia, Italy, Japan, Singapore, South Korea, the UK, and the USA.

The Program Committee of IWOCA 2023 received 86 submissions in response to the call for papers. All the papers were single-blind peer reviewed by at least three Program Committee members and some trusted external reviewers, and evaluated on their quality, originality, and relevance to the conference. The Program Committee selected 33 papers for presentation at the conference and inclusion in the proceedings.

The program also included 4 keynote talks, given by Ding-Zhu Du (University of Texas at Dallas, USA), Kazuo Iwama (National Ting Hua University, Taiwan), Peter Rossmanith (RWTH Aachen University, Germany), and Weili Wu (University of Texas at Dallas, USA). Abstracts of their talks are included in this volume.

We thank the Steering Committee for giving us the opportunity to serve as Program Chairs of IWOCA 2023, and for the responsibilities of selecting the Program Committee, the conference program, and publications.

The Program Committee selected two contributions for the best paper and the best student paper awards, sponsored by Springer.

The best paper award was given to:

- *Adrian Dumitrescu and Andrzej Lingas* for their paper "Finding Small Complete Subgraphs Efficiently"

The best student paper award was given to:

- *Tim A. Hartmann and Komal Muluk* for their paper "Make a Graph Singly Connected by Edge Orientations"

We gratefully acknowledge additional financial support from the following Taiwanese institutions: National Science and Technology Council, National Cheng Kung University, NCKU Research and Development Foundation, Academic Sinica, National Taipei University of Business, and Taiwan Association of Cloud Computing (TACC).

We thank everyone who made this meeting possible: the authors for submitting papers, the Program Committee members, and external reviewers for volunteering their time to review conference papers. We thank Springer for publishing the proceedings in their ARCoSS/LNCS series and for their support. We would also like to extend special thanks to the conference Organizing Committee for their work in making IWOCA 2023 a successful event.

Finally, we acknowledge the use of the EasyChair system for handling the submission of papers, managing the review process, and generating these proceedings.

June 2023

Sun-Yuan Hsieh
Ling-Ju Hung
Chia-Wei Lee

Organization

Steering Committee

Maria Chudnovsky	Princeton University, USA
Henning Fernau	Universität Trier, Germany
Costas Iliopoulos	King's College London, UK
Ralf Klasing	CNRS and University of Bordeaux, France
Wing-Kin (Ken) Sung	National University of Singapore, Singapore

Honorary Co-chairs

Richard Chia-Tong Lee	National Tsing Hua University, Taiwan
Der-Tsai Lee	Academia Sinica, Taiwan

General Co-chairs

Meng-Ru Shen	National Cheng Kung University, Taiwan
Sun-Yuan Hsieh	National Cheng Kung University, Taiwan
Sheng-Lung Peng	National Taipei University of Business, Taiwan

Program Chairs

Ling-Ju Hung	National Taipei University of Business, Taiwan
Chia-Wei Lee	National Taitung University, Taiwan

Program Committee

Matthias Bentert	University of Bergen, Norway
Hans-Joachim Böckenhauer	ETH Zurich, Switzerland
Jou-Ming Chang	National Taipei University of Business, Taiwan
Chi-Yeh Chen	National Cheng Kung University, Taiwan
Ho-Lin Chen	National Taiwan University, Taiwan
Li-Hsuan Chen	National Taipei University of Business, Taiwan
Po-An Chen	National Yang Ming Chiao Tung University, Taiwan
Eddie Cheng	Oakland University, USA
Thomas Erlebach	Durham University, UK
Henning Fernau	Universität Trier, Germany
Florent Foucaud	LIMOS - Université Clermont Auvergne, France
Wing-Kai Hon	National Tsing Hua University, Taiwan
Juraj Hromkovič	ETH Zurich, Switzerland
Mong-Jen Kao	National Yang Ming Chiao Tung University, Taiwan
Ralf Klasing	CNRS and University of Bordeaux, France

Tomasz Kociumaka	Max Planck Institute for Informatics, Germany
Christian Komusiewicz	Philipps-Universität Marburg, Germany
Dominik Köppl	Tokyo Medical and Dental University, Japan
Rastislav Královic	Comenius University, Slovakia
Van Bang Le	Universität Rostock, Germany
Thierry Lecroq	University of Rouen Normandy, France
Chung-Shou Liao	National Tsing Hua University, Taiwan
Limei Lin	Fujian Normal University, China
Hsiang-Hsuan Liu	Utrecht University, The Netherlands
Hendrik Molter	Ben-Gurion University of the Negev, Israel
Tobias Mömke	University of Augsburg, Germany
Martin Nöllenburg	TU Wien, Austria
Aris Pagourtzis	National Technical University of Athens, Greece
Tomasz Radzik	King's College London, UK
M. Sohel Rahman	Bangladesh University of Engineering and Technology, Bangladesh
Adele Rescigno	University of Salerno, Italy
Peter Rossmanith	RWTH Aachen University, Germany
Kunihiko Sadakane	University of Tokyo, Japan
Rahul Shah	Louisiana State University, USA
Paul Spirakis	University of Liverpool, UK
Meng-Tsung Tsai	Academia Sinica, Taiwan
Shi-Chun Tsai	National Yang Ming Chiao Tung University, Taiwan
Ugo Vaccaro	University of Salerno, Italy
Tomoyuki Yamakami	University of Fukui, Japan
Hsu-Chun Yen	National Taiwan University, Taiwan
Christos Zaroliagis	CTI and University of Patras, Greece
Guochuan Zhang	Zhejiang University, China
Louxin Zhang	National University of Singapore, Singapore

External Reviewers

Akchurin, Roman
Akrida, Eleni C.
Aprile, Manuel
Bruno, Roberto
Burjons, Elisabet
Caron, Pascal
Chakraborty, Dibyayan
Chakraborty, Dipayan
Chakraborty, Sankardeep
Chang, Ching-Lueh
Cordasco, Gennaro
Červený, Radovan
Dailly, Antoine
De Marco, Gianluca
Dey, Sanjana
Dobler, Alexander
Dobrev, Stefan
Fioravantes, Foivos
Francis, Mathew
Frei, Fabian
Gahlawat, Harmender
Gaikwad, Ajinkya
Ganguly, Arnab
Garvardt, Jaroslav
Gawrychowski, Pawel
Gehnen, Matthias
Georgiou, Konstantinos
Gregor, Petr
Hakanen, Anni
Hartmann, Tim A.
Im, Seonghyuk
Jia-Jie, Liu
Kaczmarczyk, Andrzej
Kalavasis, Alkis
Kaowsar, Iftekhar Hakim
Kellerhals, Leon
Klobas, Nina
Konstantopoulos, Charalampos

Kontogiannis, Spyros
Kralovic, Richard
Kunz, Pascal
Kuo, Te-Chao
Kuo, Ting-Yo
Kurita, Kazuhiro
Lampis, Michael
Lehtilä, Tuomo
Li, Meng-Hsi
Lin, Chuang-Chieh
Lin, Wei-Chen
Liu, Fu-Hong
Lotze, Henri
Majumder, Atrayee
Mann, Kevin
Mao, Yuchen
Martin, Barnaby
Melissinos, Nikolaos
Melissourgos, Themistoklis
Mock, Daniel
Morawietz, Nils
Muller, Haiko
Mütze, Torsten
Nisse, Nicolas
Nomikos, Christos
Pai, Kung-Jui
Pardubska, Dana
Pranto, Emamul Haq
Raptopoulos, Christoforos
Renken, Malte
Rodriguez Velazquez, Juan Alberto
Roshany, Aida
Ruderer, Michael
Saha, Apurba
Sarker, Najibul Haque
Siyam, Mahdi Hasnat
Skretas, George
Sommer, Frank
Sorge, Manuel
Stiglmayr, Michael
Stocker, Moritz
Strozecki, Yann
Terziadis, Soeren
Theofilatos, Michail
Toth, Csaba
Tsai, Yi-Chan
Tsakalidis, Konstantinos
Tsichlas, Kostas
Unger, Walter
Vialette, Stéphane
Wang, Guang-He
Wang, Hung-Lung
Wild, Sebastian
Wlodarczyk, Michal
Wu, Tsung-Jui
Yang, Jinn-Shyong
Yeh, Jen-Wei
Yongge, Yang
Zerovnik, Janez

Sponsors

NSTC 國家科學及技術委員會
National Science and Technology Council

NATIONAL CHENG KUNG UNIVERSITY
1931

成大研究發展基金會

中央研究院
ACADEMIA SINICA

National Taipei University of Business

Taiwan Association of Cloud Computing
TACC

Springer

Lecture Notes in
Computer Science
LNCS
LNAI
LNBI

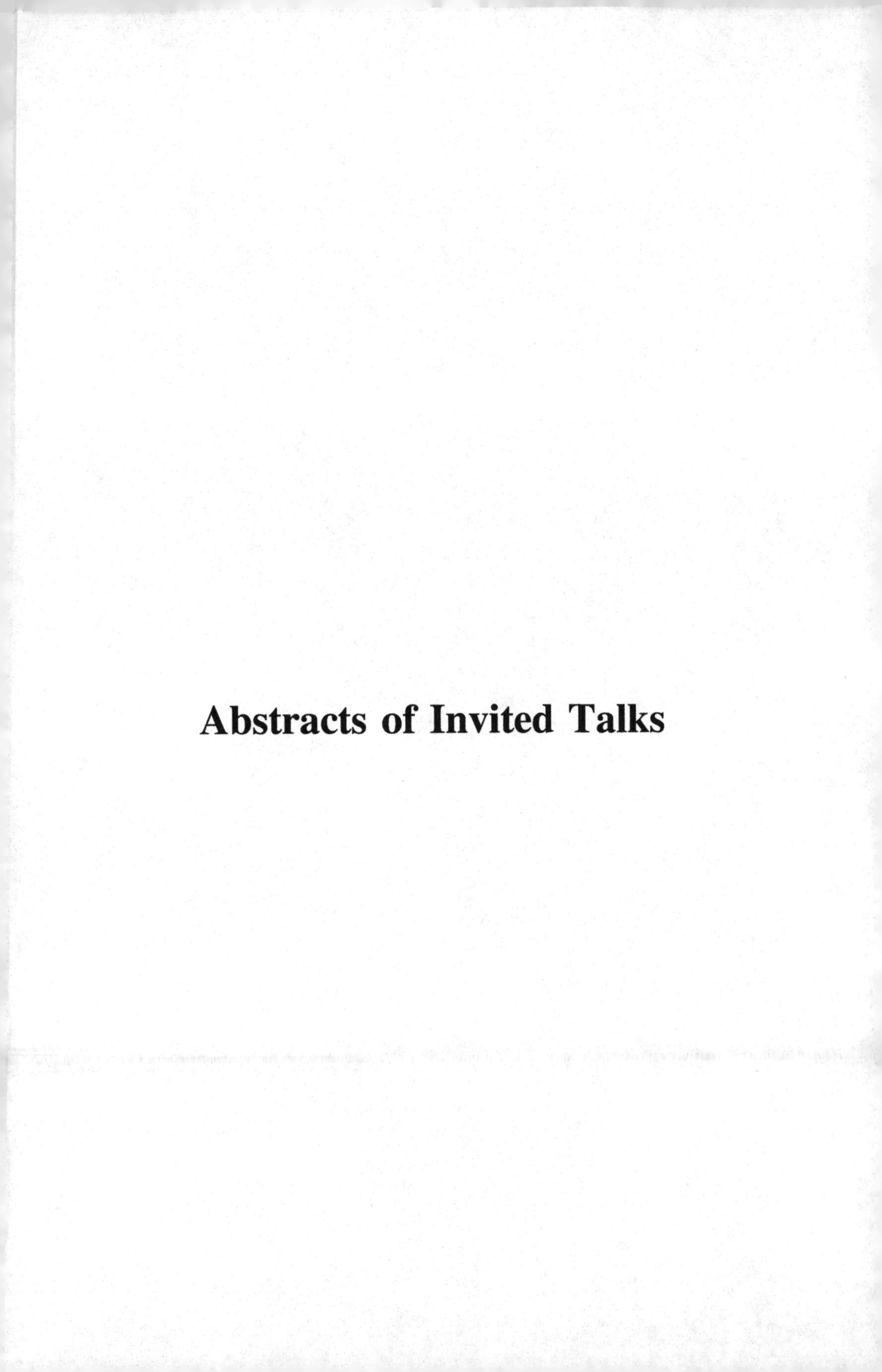

Abstracts of Invited Talks

Adaptive Influence Maximization: Adaptability via Non-adaptability

Ding-Zhu Du

University of Texas at Dallas, USA
dzdu@utdallas.edu

Adaptive influence maximization is an attractive research topic which obtained many researchers' attention. To enhance the role of adaptability, new information diffusion models, such as the dynamic independent cascade model, are proposed. In this talk, the speaker would like to present a recent discovery that in some models, the adaptive influence maximization can be transformed into a non-adaptive problem in another model. This reveals an interesting relationship between Adaptability and Non-adaptability.

Bounded Hanoi

Kazuo Iwama

National Ting Hua University, Taiwan
iwama@ie.nthu.edu.tw

The classic Towers of Hanoi puzzle involves moving a set of disks on three pegs. The number of moves required for a given number of disks is easy to determine, but when the number of pegs is increased to four or more, this becomes more challenging. After 75 years, the answer for four pegs was resolved only recently, and this *time complexity* question remains open for five or more pegs. In this article, the *space complexity*, i.e., how many disks need to be accommodated on the pegs involved in the transfer, is considered for the first time. Suppose m disks are to be transferred from some peg L to another peg R using k intermediate *work pegs* of heights $j_1, \ldots, j_k$, then how large can m be? We denote this value by $H(j_1, \ldots, j_k)$. We have the exact value for two work pegs, but so far only very partial results for three or more pegs. For example, $H(10!, 10!) = 26336386137601$ and $H(0!, 1!, 2!, \ldots, 10!) = 16304749471397$, but we still do not know the value for $H(1, 3, 3)$. Some new developments for three pegs are also mentioned. This is joint work with Mike Paterson.

Online Algorithms with Advice

Peter Rossmanith

Department of Computer Science, RWTH Aachen, Germany
rossmani@cs.rwth-aachen.de

Online algorithms have to make decisions without knowing the full input. A well-known example is caching: Which page should we discard from a cache when there is no more space left? The classical way to analyze the performance of online algorithms is the competitive ratio: How much worse is the result of an optimization problem relative to the optimal result. A different and relatively new viewpoint is the advice complexity: How much information about the future does an online algorithm need in order to become optimal? Or, similarly, how much can we improve the competitive ratio with a given amount of information about the future? In this talk we will look at the growing list of known results about online algorithms with advice and the interesting problems that occurred when the pioneers of the field came up with the definition of this model.

The Art of Big Data: Accomplishments and Research Needs

Weili Wu

University of Texas at Dallas, USA
weiliwu@utdallas.edu

Online social platforms have become more and more popular, and the dissemination of information on social networks has attracted wide attention in industry and academia. Aiming at selecting a small subset of nodes with maximum influence on networks, the Influence Maximization (IM) problem has been extensively studied. Since it is #P-hard to compute the influence spread given a seed set, the state-of-the-art methods, including heuristic and approximation algorithms, are with great difficulties such as theoretical guarantee, time efficiency, generalization, etc. This makes them unable to adapt to large-scale networks and more complex applications. With the latest achievements of Deep Reinforcement Learning (DRL) in artificial intelligence and other fields, a lot of work has focused on exploiting DRL to solve combinatorial optimization problems. Inspired by this, we propose a novel end-to-end DRL framework, ToupleGDD, to address the IM problem which incorporates three coupled graph neural networks for network embedding and double deep Q-networks for parameter learning. Previous efforts to solve the IM problem with DRL trained their models on a subgraph of the whole network, and then tested their performance on the whole graph, which makes the performance of their models unstable among different networks. However, our model is trained on several small randomly generated graphs and tested on completely different networks, and can obtain results that are very close to the state-of-the-art methods. In addition, our model is trained with a small budget, and it can perform well under various large budgets in the test, showing strong generalization ability. Finally, we conduct extensive experiments on synthetic and realistic datasets, and the experimental results prove the effectiveness and superiority of our model.

Contents

Multi-priority Graph Sparsification

Reyan Ahmed[1](✉), Keaton Hamm[2], Stephen Kobourov[1], Mohammad Javad Latifi Jebelli[1], Faryad Darabi Sahneh[1], and Richard Spence[1]

[1] University of Arizona, Tucson, AZ, USA
abureyanahmed@arizona.edu
[2] University of Texas at Arlington, Arlington, TX, USA

Abstract. A *sparsification* of a given graph G is a sparser graph (typically a subgraph) which aims to approximate or preserve some property of G. Examples of sparsifications include but are not limited to spanning trees, Steiner trees, spanners, emulators, and distance preservers. Each vertex has the same priority in all of these problems. However, real-world graphs typically assign different "priorities" or "levels" to different vertices, in which higher-priority vertices require higher-quality connectivity between them. Multi-priority variants of the Steiner tree problem have been studied previously, but have been much less studied for other types of sparsifiers. In this paper, we define a generalized multi-priority problem and present a rounding-up approach that can be used for a variety of graph sparsifications. Our analysis provides a systematic way to compute approximate solutions to multi-priority variants of a wide range of graph sparsification problems given access to a single-priority subroutine.

Keywords: graph spanners · sparsification · approximation algorithms

1 Introduction

A *sparsification* of a graph G is a graph H which preserves some property of G. Examples of sparsifications include spanning trees, Steiner trees, spanners, emulators, distance preservers, t–connected subgraphs, and spectral sparsifiers. Many sparsification problems are defined with respect to a given subset of vertices $T \subseteq V$ which we call *terminals*: e.g., a *Steiner tree* over (G, T) requires a tree in G which spans T. Most of their corresponding optimization problems (e.g., finding a minimum weight Steiner tree or spanner) are NP-hard to compute optimally, so one often seeks approximate solutions in practice.

In real-world networks, not all vertices or edges are created equal. For example, a road network may wish to not only connect its cities with roads, but also ensure that pairs of larger cities enjoy better connectivity (e.g., with major highways). In this paper, we are interested in generalizations of sparsification problems where each vertex possesses one of $k+1$ different *priorities* (between 0 and k, where k is the highest), in which the goal is to construct a graph H

Supported in part by NSF grants CCF-1740858, CCF-1712119, and CCF-2212130.

S.-Y. Hsieh et al. (Eds.): IWOCA 2023, LNCS 13889, pp. 1–12, 2023.
https://doi.org/10.1007/978-3-031-34347-6_1

such that (i) every edge in H has a *rate* between 1 and k inclusive, and (ii) for all $i \in \{1, \ldots, k\}$, the edges in H of rate $\geq i$ constitute a given type of sparsifier over the vertices whose priority is at least i. Throughout, we assume a vertex with priority 0 need not be included and all other vertices are *terminals.*

1.1 Problem Definition

A sparsification H is *valid* if it satisfies a set of constraints that depends on the type of sparsification. Given G and a set of terminals T, let $\mathcal{F}$ be the set of all valid sparsifications H of G over T. Throughout, we will assume that $\mathcal{F}$ satisfies the following *general constraints* that must hold for all types of sparsification we consider in this article: for all $H \in \mathcal{F}$: (a) H contains all terminals T in the same connected component, and (b) H is a subgraph of G. Besides these general constraints, there are additional constraints that depend on the specific type of sparsification as described below.

Tree Constraint of a Steiner Tree Sparsification: A Steiner tree over (G, T) is a subtree H that spans T. Here the specific constraint is that H must be a tree and we refer to it as the *tree constraint.*

Distance Constraints of Spanners and Preservers: A spanner is a subgraph H which approximately preserves distances in G (e.g., if $d_H(u,v) \leq \alpha d_G(u,v)$ for all $u, v \in V$ and $\alpha \geq 1$ then H is called a multiplicative α-spanner of G). A *subset spanner* needs only approximately preserve distances between a subset $T \subseteq V$ of vertices. A *distance preserver* is a special case of the spanner where $\alpha = 1$. The specific constraints are the distance constraints applied from the problem definition. For example, the inequality above is for so-called multiplicative α-spanners. We refer to these types of constraints as *distance constraints.*

The above problems are widely studied in literature; see surveys [5, 22]. In this paper, we study k-priority sparsification which is a generalization of the above problems. An example sparsification which we will not consider in the above framework is the *emulator*, which approximates distances but is not necessarily a subgraph. We now define a k-priority sparsification as follows, where $[k] := \{1, 2, \ldots, k\}$.

Definition 1 (k-priority sparsification). *Let $G(V, E)$ be a graph, where each vertex $v \in V$ has priority $\ell(v) \in [k] \cup \{0\}$. Let $T_i := \{v \in V \mid \ell(v) \geq i\}$. Let $w(e)$ be the edge weight of edge e. The weight of an edge with rate i is denoted by $w(e,i) = i\ w(e,1) = i\ w(e)$. For $i \in [k]$, let $\mathcal{F}_i$ denote the set of all valid sparsifications over T_i. A subgraph H with edge rates $R : E(H) \to [k]$ is a k-priority sparsification if for all $i \in [k]$, the subgraph of H induced by all edges of rate $\geq i$ belongs to $\mathcal{F}_i$. We assess the quality of a sparsification H by its weight, weight(H) $:= \sum_{e \in E(H)} w(e, R(e))$.*

Note that H induces a nested sequence of k subgraphs, and can also be interpreted as a *multi-level* graph sparsification [3]. A road map (Fig. 1(a)) serves as a good analogy of a multi-level sparsification, as zooming out filters out smaller roads. Figure 1(b) shows an example of 2-priority sparsification with distance

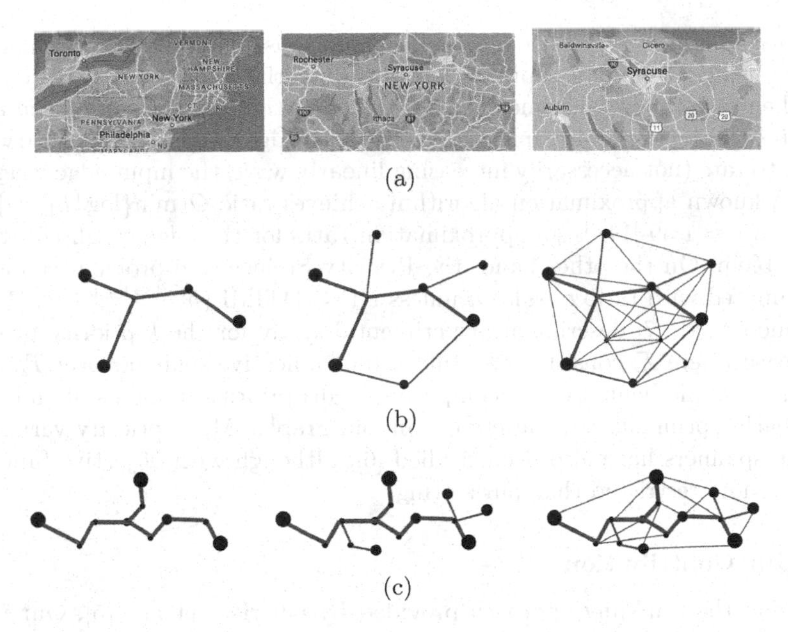

Fig. 1. Different levels of detail on a road map of New York (a), 2-priority sparsifications with distance constraints (b), and with a tree constraint (c). On the left side, we have the most important information (indicated using edge thickness). As we go from left to right, more detailed information appears on the screen.

constraints where $\mathcal{F}_i$ is the set of all subset +2 spanners over T_i; that is, the vertex pairs of T_i is connected by a path in H_i at most 2 edges longer than the corresponding shortest path in G. Similarly, Fig. 1(c) shows an example of 2-priority sparsification with a tree constraint.

Definition 1 is intentionally open-ended to encompass a wide variety of sparsification problems. The k-priority problem is a generalization of many NP-hard problems, for example, Steiner trees, spanners, distance preservers, etc. These classical problems can be considered different variants of the 1-priority problem. Hence, the k-priority problem cannot be simpler than the 1-priority problem. Let OPT be an optimal solution to the k-priority problem and the weight of OPT be weight(OPT). In this paper, we are mainly interested in the following problem.

Problem. *Given $\langle G, \ell, w \rangle$ consisting of a graph G with vertex priorities $\ell : V \to [k] \cup \{0\}$, can we compute a k-priority sparsification whose weight is small compared to weight(OPT)?*

1.2 Related Work

The case where $\mathcal{F}_i$ consists of all Steiner trees over T_i is known under different names including Priority Steiner Tree [18], Multi-level Network Design [11], Quality-of-Service Multicast Tree [15,23], and Multi-level Steiner Tree [3].

Charikar et al. [15] give two $O(1)$-approximations for the Priority Steiner Tree problem using a rounding approach which rounds the priorities of each terminal up to the nearest power of some fixed base (2 or e), then using a subroutine which computes an exact or approximate Steiner tree. If edge weights are arbitrary with respect to rate (not necessarily increasing linearly w.r.t. the input edge weights), the best known approximation algorithm achieves ratio $O(\min\{\log|T|, k\rho\}$ [15, 31] where $\rho \approx 1.39$ [14] is an approximation ratio for the edge-weighted Steiner tree problem. On the other hand, the Priority Steiner tree problem cannot be approximated with ratio $c \log\log n$ unless NP $\subseteq$ DTIME$(n^{O(\log\log\log n)})$ [18].

Ahmed et al. [7] describe an experimental study for the k-priority problem in the case where $\mathcal{F}_i$ consists of all subset multiplicative spanners over T_i. They show that simple heuristics for computing multi-priority spanners already perform nearly optimally on a variety of random graphs. Multi-priority variants of additive spanners have also been studied [6], although with objective functions that are more restricted than our setting.

1.3 Our Contribution

We extend the rounding approach provided by Charikar et al. [15]. Our result not only works for Steiner trees but also for graph spanners. We prove our result using proof by induction.

2 A General Approximation for *k*-Priority Sparsification

In this section, we generalize the rounding approach of [15]. The approach has two main steps: the first step rounds up the priority of all terminals to the nearest power of 2; the second step computes a solution independently for each rounded-up priority and merges all solutions from the highest priority to the lowest priority. Each of these steps can make the solution at most two times worse than the optimal solution. Hence, overall the algorithm is a 4-approximation. We provide the pseudocode of the algorithm below, here S_i in a partitioning is a set of terminals.

Algorithm 1. Algorithm k-priority Approximation($G = (V, E)$)

```
// Round up the priorities
for each terminal v ∈ V do
    Round up the priority of v to the nearest power of 2
// Independently compute the solutions
Compute a partitioning S_1, S_2, S_4, ··· , S_k from the rounded-up terminals
for each partition component S_i do
    Compute a 1-priority solution on partition component S_i
// Merge the independent solutions
for i ∈ {k, k − 1, ··· , 1} do
    Merge the solution of S_i to the solutions of lower priorities
```

We now propose a partitioning technique that will guarantee valid solutions[1].

Definition 2. *An inclusive partitioning of the terminal vertices of a k-priority instance assigns each terminal t_j to each partition component in $\{S_i : i \leq \ell(t_j), i = 2^k, k \geq 0\}$.*

We compute partitioning $S_1, S_2, \cdots, S_k$ from the rounded-up terminal sets and use them to compute the independent solutions. Here, we require one more assumption: given $1 \leq i < j \leq k$ and two partition components S_i, S_j, any two sparsifications of rate i and j can be "merged" to produce a third sparsification of rate i. Specifically, if $H_i \in \mathcal{F}_i$, and $H_j \in \mathcal{F}_j$, then there is a graph $H_{i,j} \in \mathcal{F}_i$ such that $H_j \subseteq H_{i,j}$. For the above sparsification problems (e.g., Steiner tree, spanners), we can often let $H_{i,j}$ be the union of the edges in H_i and H_j, though edges may need to be pruned to satisfy a tree constraint (by removing cycles).

Definition 3. *Let S_i and S_j be two partition components of a partitioning where $i < j$. Let H_i and H_j be the independently computed solution for S_i and S_j respectively. We say that the solution H_j is merged with solution H_i if we complete the following two steps:*

1. *If an edge e is not present in H_i but present in H_j, then we add e to H_i.*
2. *If there is a tree constraint, then prune some lower-rated edges to ensure there is no cycle.*

We need the second step of merging particularly for sparsifications with tree constraints. Although the merging operation treats these sparsifications differently, we will later show that the pruning step does not play a significant role in the approximation guarantee. Algorithm k-priority Approximation computes a partitioning from the rounded-up terminals. We now provide an approximation guarantee for Algorithm k-priority Approximation that is independent of the partitioning method.

Theorem 1. *Consider an instance $\varphi = \langle G, \ell, w \rangle$ of the k-priority problem. If we are given an oracle that can compute the minimum weight sparsification of G over a partition set S, then with at most $\log_2 k + 1$ queries to the oracle, Algorithm k-priority computes a k-priority sparsification with weight at most* 4 *weight*(OPT). *If instead of an oracle a ρ-approximation is given, then the weight of k-priority sparsification is at most 4ρ.*

Proof. Given φ, construct the rounded-up instance φ' which is obtained by rounding up the priority of each vertex to the nearest power of 2. Let OPT′ be an optimum solution to the rounded-up instance. We can obtain a feasible solution for the rounded-up instance from OPT by raising the rate of each edge to the nearest power of 2, and the rate of an edge will not increase more than two times the original rate. Hence weight(OPT′) $\leq$ 2 weight(OPT).

Then for each rounded-up priority $i \in \{1, 2, 4, 8, \ldots, k\}$, compute a sparsification independently over the partition component S_i, creating $\log_2 k + 1$ graphs.

[1] A detailed discussion can be found in the full version [8].

We denote these graphs by ALG_1, ALG_2, ALG_4, ..., ALG_k. Combine these sparsifications into a single subgraph ALG. This is done using the "merging" operation described earlier in this section: (i) add each edge of ALG_i to all sparsification of lower priorities $\text{ALG}_{i-1}, \text{ALG}_{i-2}, \cdots, \text{ALG}_1$ and (ii) prune some edges to make sure that there is exactly one path between each pair of terminals if we are computing priority Steiner tree.

It is not obvious why after this merging operation we have a k-priority sparsification with cost no more than 4 weight(OPT). The approximation algorithm computes solutions independently by querying the oracle, which means it is unaware of the terminal sets at the lower levels. Consider the topmost partition component S_k of the rounded-up instance. The approximation algorithm computes an optimal solution for that partition component. The optimal algorithm of the k-priority sparsification computes the solution while considering all the terminals and all priorities. Let OPT_i be the minimum weighted subgraph in an optimal k-priority solution OPT to generate a valid sparsification on partition component S_i. Then $\text{weight}(\text{ALG}_k) \leq \text{weight}(\text{OPT}_k)$, i.e., if we only consider the top partition component S_k, then the approximation algorithm is no worse than the optimal algorithm. Similarly, $\text{weight}(\text{ALG}_i) \leq \text{weight}(\text{OPT}_i)$ for each i.

However, the approximation algorithm may incur an additional cost when merging the edges of ALG_k in lower priorities. In the worst case, merged edges might not be needed to compute the solutions of the lower partition components (if the merged edges are not used in the lower partition components in their independent solutions, then we do not need to pay extra cost for the merging operation). This is because the approximation algorithm computes the solutions independently. On the other hand, in the worst case, it may happen that OPT_k includes all the edges to satisfy all the constraints of lower partition components. In this case, the cost of the optimal k-priority solution is $k{\cdot}\text{weight}(\text{OPT}_k)$. If $\text{weight}(\text{ALG}_k) \approx \text{weight}(\text{ALG}_{k-1}) \approx \cdots \approx \text{weight}(\text{ALG}_1)$ and the edges of the sparsification of a particular priority do not help in the lower priorities, then it seems like the approximation algorithm can perform around k times worse than the optimal k-priority solution. However, such an issue (the edges of the sparsification of a particular priority do not help in the lower priorities) does not arise as we are considering a rounded-up instance. In a rounded-up instance $S_k = S_{k-1} = \cdots = S_{\frac{k}{2}+1}$. Hence $\text{weight}(\text{ALG}_k) = \text{weight}(\text{ALG}_i)$ for $i = k-1, k-2, \cdots, \frac{k}{2}+1$.

Lemma 1. *If we compute independent solutions of a rounded-up k-priority instance and merge them, then the cost of the solution is no more than* 2 *weight*(OPT).

Proof. Let $k = 2^i$. Let the partitioning be $S_{2^i}, S_{2^{i-1}}, \cdots, S_1$. Suppose we have computed the independent solution and merged them in lower priorities. We actually prove a stronger claim, and use that to prove the lemma. Note that in the worst case the cost of approximation algorithm is $2^i\text{weight}(\text{ALG}_{2^i}) + 2^{i-1}\text{weight}(\text{ALG}_{2^{i-1}}) + \cdots + \text{weight}(\text{ALG}_1) = \sum_{p=0}^{i} 2^p\text{weight}(\text{ALG}_{2^p})$. And the cost of the optimal algorithm is $\text{weight}(\text{OPT}_{2^i}) + \text{weight}(\text{OPT}_{2^i-1}) + \cdots + \text{weight}(\text{OPT}_1) = \sum_{p=1}^{2^i} \text{weight}(\text{OPT}_p)$. We show that $\sum_{p=0}^{i} 2^p \text{weight}(\text{ALG}_{2^p}) \leq 2 \sum_{p=1}^{2^i} \text{weight}(\text{OPT}_p)$. We provide a proof by induction on i.

Base step: If $i = 0$, then we have just one partition component S_1. The approximation algorithm computes a sparsification for S_1 and there is nothing to merge. Since the approximation algorithm uses an optimal algorithm to compute independent solutions, $\text{weight}(\text{ALG}_1) \leq \text{weight}(\text{OPT}_1) \leq 2\,\text{weight}(\text{OPT}_1)$.

Inductive step: We assume that the claim is true for $i = j$ which is the induction hypothesis. Hence $\sum_{p=0}^{j} 2^p \text{weight}(\text{ALG}_{2^p}) \leq 2 \sum_{p=1}^{2^j} \text{weight}(\text{OPT}_p)$. We now show that the claim is also true for $i = j + 1$. In other words, we have to show that $\sum_{p=0}^{j+1} 2^p \text{weight}(\text{ALG}_{2^p}) \leq 2 \sum_{p=1}^{2^{j+1}} \text{weight}(\text{OPT}_p)$. We know,

$$
\begin{aligned}
\sum_{p=0}^{j+1} 2^p \text{weight}(\text{ALG}_{2^p}) &= 2^{j+1} \text{weight}(\text{ALG}_{2^{j+1}}) + \sum_{p=0}^{j} 2^p \text{weight}(\text{ALG}_{2^p}) \\
&\leq 2^{j+1} \text{weight}(\text{OPT}_{2^{j+1}}) + \sum_{p=0}^{j} 2^p \text{weight}(\text{ALG}_{2^p}) \\
&= 2 \times 2^j \text{weight}(\text{OPT}_{2^{j+1}}) + \sum_{p=0}^{j} 2^p \text{weight}(\text{ALG}_{2^p}) \\
&= 2 \sum_{p=2^j+1}^{2^{j+1}} \text{weight}(\text{OPT}_p) + \sum_{p=0}^{j} 2^p \text{weight}(\text{ALG}_{2^p}) \\
&\leq 2 \sum_{p=2^j+1}^{2^{j+1}} \text{weight}(\text{OPT}_p) + 2 \sum_{p=1}^{2^j} \text{weight}(\text{OPT}_p) \\
&= 2 \sum_{p=1}^{2^{j+1}} \text{weight}(\text{OPT}_p)
\end{aligned}
$$

Here, the second equality is just a simplification. The third inequality uses the fact that an independent optimal solution has a cost lower than or equal to any other solution. The fourth equality is a simplification, the fifth inequality uses the fact that the input is a rounded up instance. The sixth inequality uses the induction hypothesis. □

We have shown earlier that the solution of the rounded up instance has a cost of no more than 2 weight(OPT). Combining that claim and the previous claim, we can show that the solution of the approximation algorithm has cost no more than 4 weight(OPT). In most cases, computing the optimal sparsification is computationally difficult. If an oracle is instead replaced with a ρ-approximation, the rounding-up approach is a 4ρ-approximation, by following the same proof as above. □

3 Subset Spanners and Distance Preservers

Here we provide a bound on the size of subsetwise graph spanners, where lightness is expressed with respect to the weight of the corresponding Steiner tree. Given a (possibly edge-weighted) graph G and $\alpha \geq 1$, we say that H is a (multiplicative) α-spanner if $d_H(u,v) \leq \alpha \cdot d_G(u,v)$ for all $u, v \in V$, where α is the *stretch factor* of the spanner and $d_G(u,v)$ is the graph distance between u and v in G. A *subset spanner* over $T \subseteq V$ approximates distances between pairs of vertices in T (e.g., $d_H(u,v) \leq \alpha \cdot d_G(u,v)$ for all $u, v \in T$). For clarity, we refer to the case where $T = V$ as an *all-pairs spanner*. The *lightness* of an all-pairs spanner is defined as its total edge weight divided by $w(MST(G))$. A *distance preserver* is a spanner with $\alpha = 1$.

Althöfer et al. [10] give a simple greedy algorithm which constructs an all-pairs $(2k-1)$-spanner H of size $O(n^{1+1/k})$ and lightness $1 + \frac{n}{2k}$. The lightness has been subsequently improved; in particular Chechik and Wulff-Nilsen [17] give a $(2k-1)(1+\varepsilon)$ spanner with size $O(n^{1+1/k})$ and lightness $O_\varepsilon(n^{1/k})$. Up to ε dependence, these size and lightness bounds are conditionally tight assuming a girth conjecture by Erdős [21], which states that there exist graphs of girth $2k+1$ and $\Omega(n^{1+1/k})$ edges.

For subset spanners over $T \subseteq V$, the lightness is defined with respect to the minimum Steiner tree over T, since that is the minimum weight subgraph which connects T. We remark that in general graphs, the problem of finding a light multiplicative subset spanner can be reduced to that of finding a light spanner:

Lemma 2. *Let G be a weighted graph and let $T \subseteq V$. Then there is a poly-time constructible subset spanner with stretch $(2k-1)(1+\varepsilon)$ and lightness $O_\varepsilon(|T|^{1/k})$.*

Proof. Let $\tilde{G}$ be the metric closure over (G, T), namely the complete graph $K_{|T|}$ where each edge $uv \in E(\tilde{G})$ has weight $d_G(u,v)$. Let H' be a $(2k-1)(1+\varepsilon)$-spanner of $\tilde{G}$. By replacing each edge of H' with the corresponding shortest path in G, we obtain a subset spanner H of G with the same stretch and total weight. Using the spanner construction of [17], the total weight of H' is $O_\varepsilon(|T|^{1/k}) w(MST(\tilde{G}))$. Using the well-known fact that the MST of $\tilde{G}$ is a 2-approximation for the minimum Steiner tree over (G, T), it follows that the total weight of H' is also $O_\varepsilon(|T|^{1/k})$ times the minimum Steiner tree over (G, T). □

Thus, the problem of finding a subset spanner with multiplicative stretch becomes more interesting when the input graph is restricted (e.g., planar, or H-minor free). Klein [25] showed that every *planar* graph has a subset $(1+\varepsilon)$-spanner with lightness $O_\varepsilon(1)$. Le [28] gave a poly-time algorithm which computes a subset $(1+\varepsilon)$-spanner with lightness $O_\varepsilon(\log |T|)$, where G is restricted to be H-minor free.

On the other hand, subset spanners with additive $+\beta$ error are more interesting, as one cannot simply reduce this problem to the all-pairs spanner as in Lemma 2. It is known that every unweighted graph G has $+2$, $+4$, and $+6$

spanners with $O(n^{3/2})$ edges [9], $\widetilde{O}(n^{7/5})$ edges [16], and $O(n^{4/3})$ edges [12,26] respectively, and that the upper bound of $O(n^{4/3})$ edges cannot be improved even with $+n^{o(1)}$ additive error [2].

3.1 Subset Distance Preservers

Unlike spanners, general graphs do not contain sparse distance preservers that preserve all distances exactly; the unweighted complete graph has no nontrivial distance preserver and thus $\Theta(n^2)$ edges are needed. Similarly, subset distance preservers over a subset $T \subseteq V$ may require $\Theta(|T|^2)$ edges. It is an open question whether there exists $c > 0$ such that any undirected, unweighted graph and subset of size $|T| = O(n^{1-c})$ has a distance preserver on $O(|T|^2)$ edges [13]. Moreover, when $|T| = O(n^{2/3})$, there are graphs for which any subset distance preserver requires $\Omega(|T|n^{2/3})$ edges, which is $\omega(|T|^2)$ when $|T| = o(n^{2/3})$ [13].

Theorem 2. *If the above open question is true, then every unweighted graph with* $|T| = O(n^{1-c})$ *and terminal priorities in* $[k]$ *has a priority distance preserver of size* 4 *weight*(OPT).

4 Multi-priority Approximation Algorithms

In this section, we illustrate how the subset spanners mentioned in Sect. 3 can be used in Theorem 1, and show several corollaries of the kinds of guarantees one can obtain in this manner. In particular, we give the first weight bounds for multi-priority graph spanners. The case of Steiner trees was discussed [3].

4.1 Spanners

If the input graph is planar, then we can use the algorithm by Klein [25] to compute a subset spanner for the set of priorities we get from the rounding approach. The polynomial-time algorithm in [25] has constant approximation ratio, assuming constant stretch factor, yielding the following corollary.

Corollary 1. *Given a planar graph* G *and* $\varepsilon > 0$, *there exists a rounding approach based algorithm to compute a multi-priority multiplicative* $(1+\varepsilon)$*-spanner of* G *having* $O(\varepsilon^{-4})$ *approximation. The algorithm runs in* $O(\frac{|T|\log|T|}{\varepsilon})$ *time, where* T *is the set of terminals.*

The proof of this corollary follows from combining the guarantee of Klein [25] with the bound of Theorem 1. Using the approximation result for subset spanners provided in Lemma 2, we obtain the following corollary.

Corollary 2. *Given an undirected weighted graph* G, $t \in \mathbb{N}$, $\varepsilon > 0$, *there exists a rounding approach based algorithm to compute a multi-priority multiplicative* $(2t-1)(1+\varepsilon)$*-spanner of* G *having* $O(|T|^{\frac{1}{t}})$ *approximation, where* T *is the set of terminals. The algorithm runs in* $O(|T|^{2+\frac{1}{k}+\varepsilon})$ *time.*

For additive spanners, there are algorithms to compute subset spanners of size $O(n|T|^{\frac{2}{3}})$, $\tilde{O}(n|T|^{\frac{4}{7}})$ and $O(n|T|^{\frac{1}{2}})$ for additive stretch 2, 4 and 6, respectively [1, 24]. Similarly, there is an algorithm to compute a near-additive subset $(1+\varepsilon, 4)$–spanner of size $O(n\sqrt{\frac{|T|\log n}{\varepsilon}})$ [24]. If we use these algorithms as subroutines in Lemma 2 to compute subset spanners for different priorities, then we have the following corollaries.

Corollary 3. *Given an undirected weighted graph G, there exist polynomial-time algorithms to compute multi-priority graph spanners with additive stretch 2, 4 and 6, of size* $O(n|T|^{\frac{2}{3}})$, $\tilde{O}(n|T|^{\frac{4}{7}})$, *and* $O(n|T|^{\frac{1}{2}})$, *respectively.*

Corollary 4. *Given an undirected unweighted graph G, there exists a polynomial-time algorithm to compute multi-priority* $(1+\varepsilon, 4)$*–spanners of size* $O(n\sqrt{\frac{|T|\log n}{\varepsilon}})$.

Several of the above results involving additive spanners have been recently generalized to weighted graphs; more specifically, there are algorithms to compute subset spanners in weighted graphs of size $O(n|T|^{\frac{2}{3}})$, and $O(n|T|^{\frac{1}{2}})$ for additive stretch $2W(\cdot,\cdot)$, and $6W(\cdot,\cdot)$, respectively [4,19,20], where $W(u,v)$ denotes the maximum edge weight along the shortest u-v path in G. Hence, we have the following corollary.

Corollary 5. *Given an undirected weighted graph G, there exist polynomial-time algorithms to compute multi-priority graph spanners with additive stretch* $2W(\cdot,\cdot)$, *and* $6W(\cdot,\cdot)$, *of size* $O(n|T|^{\frac{2}{3}})$, *and* $O(n|T|^{\frac{1}{2}})$, *respectively.*

4.2 t–Connected Subgraphs

Another example which fits the framework of Sect. 1.1 is that of finding t–connected subgraphs [27,29,30], in which (similar to the Steiner tree problem) a set $T \subseteq V$ of terminals is given, and the goal is to find the minimum-cost subgraph H such that each pair of terminals is connected with at least t vertex-disjoint paths in H. If we use the algorithm of [27] (that computes a t–connected subgraph with approximation guarantee to $O(t\log t)$) in Theorem 1 to compute subsetwise t–connected subgraphs for different priorities, then we have the following corollary.

Corollary 6. *Given an undirected weighted graph G, using the algorithm of [27] as a subroutine in Theorem 1 yields a polynomial-time algorithm which computes a multi-priority* t*–connected subgraph over the terimals with approximation ratio* $O(t\log t)$ *provided* $|T| \geq t^2$.

5 Conclusions and Future Work

We study the k-priority sparsification problem that arises naturally in large network visualization since different vertices can have different priorities. Our problem relies on a subroutine for the single priority sparsification. A nice open problem is whether we can solve it directly without relying on a subroutine.

References

1. Abboud, A., Bodwin, G.: Lower bound amplification theorems for graph spanners. In: Proceedings of the 27th ACM-SIAM Symposium on Discrete Algorithms (SODA), pp. 841–856 (2016)
2. Abboud, A., Bodwin, G.: The 4/3 additive spanner exponent is tight. J. ACM (JACM) **64**(4), 28 (2017)
3. Ahmed, A.R., et al.: Multi-level Steiner trees. In: 17th International Symposium on Experimental Algorithms (SEA), pp. 15:1–15:14 (2018). https://doi.org/10.4230/LIPIcs.SEA.2018.15
4. Ahmed, R., Bodwin, G., Hamm, K., Kobourov, S., Spence, R.: On additive spanners in weighted graphs with local error. In: Kowalik, Ł, Pilipczuk, M., Rzążewski, P. (eds.) WG 2021. LNCS, vol. 12911, pp. 361–373. Springer, Cham (2021). https://doi.org/10.1007/978-3-030-86838-3_28
5. Ahmed, R., et al.: Graph spanners: a tutorial review. Comput. Sci. Rev. **37**, 100253 (2020)
6. Ahmed, R., Bodwin, G., Sahneh, F.D., Hamm, K., Kobourov, S., Spence, R.: Multilevel weighted additive spanners. In: Coudert, D., Natale, E. (eds.) 19th International Symposium on Experimental Algorithms (SEA 2021). Leibniz International Proceedings in Informatics (LIPIcs), Dagstuhl, Germany, vol. 190, pp. 16:1–16:23. Schloss Dagstuhl - Leibniz-Zentrum für Informatik (2021). https://doi.org/10.4230/LIPIcs.SEA.2021.16. https://drops.dagstuhl.de/opus/volltexte/2021/13788
7. Ahmed, R., Hamm, K., Latifi Jebelli, M.J., Kobourov, S., Sahneh, F.D., Spence, R.: Approximation algorithms and an integer program for multi-level graph spanners. In: Kotsireas, I., Pardalos, P., Parsopoulos, K.E., Souravlias, D., Tsokas, A. (eds.) SEA 2019. LNCS, vol. 11544, pp. 541–562. Springer, Cham (2019). https://doi.org/10.1007/978-3-030-34029-2_35
8. Ahmed, R., Hamm, K., Kobourov, S., Jebelli, M.J.L., Sahneh, F.D., Spence, R.: Multi-priority graph sparsification. arXiv preprint arXiv:2301.12563 (2023)
9. Aingworth, D., Chekuri, C., Indyk, P., Motwani, R.: Fast estimation of diameter and shortest paths (without matrix multiplication). SIAM J. Comput. **28**, 1167–1181 (1999). https://doi.org/10.1137/S0097539796303421
10. Althöfer, I., Das, G., Dobkin, D., Joseph, D., Soares, J.: On sparse spanners of weighted graphs. Discret. Comput. Geom. **9**(1), 81–100 (1993). https://doi.org/10.1007/BF02189308
11. Balakrishnan, A., Magnanti, T.L., Mirchandani, P.: Modeling and heuristic worst-case performance analysis of the two-level network design problem. Manag. Sci. **40**(7), 846–867 (1994). https://doi.org/10.1287/mnsc.40.7.846
12. Baswana, S., Kavitha, T., Mehlhorn, K., Pettie, S.: Additive spanners and (α, β)-spanners. ACM Trans. Algorithms (TALG) **7**(1), 5 (2010)
13. Bodwin, G.: New results on linear size distance preservers. SIAM J. Comput. **50**(2), 662–673 (2021). https://doi.org/10.1137/19M123662X
14. Byrka, J., Grandoni, F., Rothvoß, T., Sanità, L.: Steiner tree approximation via iterative randomized rounding. J. ACM **60**(1), 6:1–6:33 (2013). https://doi.org/10.1145/2432622.2432628
15. Charikar, M., Naor, J., Schieber, B.: Resource optimization in QoS multicast routing of real-time multimedia. IEEE/ACM Trans. Netw. **12**(2), 340–348 (2004). https://doi.org/10.1109/TNET.2004.826288
16. Chechik, S.: New additive spanners. In: Proceedings of the Twenty-Fourth Annual ACM-SIAM Symposium on Discrete Algorithms, pp. 498–512. Society for Industrial and Applied Mathematics (2013)

17. Chechik, S., Wulff-Nilsen, C.: Near-optimal light spanners. ACM Trans. Algorithms (TALG) **14**(3), 33 (2018)
18. Chuzhoy, J., Gupta, A., Naor, J.S., Sinha, A.: On the approximability of some network design problems. ACM Trans. Algorithms **4**(2), 23:1–23:17 (2008). https://doi.org/10.1145/1361192.1361200
19. Elkin, M., Gitlitz, Y., Neiman, O.: Almost shortest paths and PRAM distance oracles in weighted graphs. arXiv preprint arXiv:1907.11422 (2019)
20. Elkin, M., Gitlitz, Y., Neiman, O.: Improved weighted additive spanners. arXiv preprint arXiv:2008.09877 (2020)
21. Erdős, P.: Extremal problems in graph theory. In: Proceedings of the Symposium on Theory of Graphs and its Applications, p. 2936 (1963)
22. Hauptmann, M., Karpiński, M.: A compendium on Steiner tree problems. Inst. für Informatik (2013)
23. Karpinski, M., Mandoiu, I.I., Olshevsky, A., Zelikovsky, A.: Improved approximation algorithms for the quality of service multicast tree problem. Algorithmica **42**(2), 109–120 (2005). https://doi.org/10.1007/s00453-004-1133-y
24. Kavitha, T.: New pairwise spanners. Theory Comput. Syst. **61**(4), 1011–1036 (2016). https://doi.org/10.1007/s00224-016-9736-7
25. Klein, P.N.: A subset spanner for planar graphs, with application to subset TSP. In: Proceedings of the Thirty-Eighth Annual ACM Symposium on Theory of Computing, STOC 2006, pp. 749–756. ACM, New York (2006). https://doi.org/10.1145/1132516.1132620. http://doi.acm.org/10.1145/1132516.1132620
26. Knudsen, M.B.T.: Additive spanners: a simple construction. In: Ravi, R., Gørtz, I.L. (eds.) SWAT 2014. LNCS, vol. 8503, pp. 277–281. Springer, Cham (2014). https://doi.org/10.1007/978-3-319-08404-6_24
27. Laekhanukit, B.: An improved approximation algorithm for minimum-cost subset k-connectivity. In: Aceto, L., Henzinger, M., Sgall, J. (eds.) ICALP 2011. LNCS, vol. 6755, pp. 13–24. Springer, Heidelberg (2011). https://doi.org/10.1007/978-3-642-22006-7_2
28. Le, H.: A PTAS for subset TSP in minor-free graphs. In: Proceedings of the Thirty-First Annual. Society for Industrial and Applied Mathematics, USA (2020)
29. Nutov, Z.: Approximating minimum cost connectivity problems via uncrossable bifamilies and spider-cover decompositions. In: IEEE 50th Annual Symposium on Foundations of Computer Science (FOCS 2009), Los Alamitos, CA, USA. IEEE Computer Society (2009). https://doi.org/10.1109/FOCS.2009.9. https://doi.ieeecomputersociety.org/10.1109/FOCS.2009.9
30. Nutov, Z.: Approximating subset k-connectivity problems. J. Discret. Algorithms **17**, 51–59 (2012)
31. Sahneh, F.D., Kobourov, S., Spence, R.: Approximation algorithms for the priority Steiner tree problem. In: 27th International Computing and Combinatorics Conference (COCOON) (2021). http://arxiv.org/abs/1811.11700

Point Enclosure Problem for Homothetic Polygons

Waseem Akram(✉) and Sanjeev Saxena

Department of Computer Science and Engineering, Indian Institute of Technology, Kanpur, Kanpur 208 016, India
{akram,ssax}@iitk.ac.in

Abstract. In this paper, we investigate the following problem: "given a set $\mathcal{S}$ of n homothetic polygons, preprocess $\mathcal{S}$ to efficiently report all the polygons of $\mathcal{S}$ containing a query point." A set of polygons is said to be homothetic if each polygon in the set can be obtained from any other polygon of the set using scaling and translating operations. The problem is the counterpart of the homothetic range search problem discussed by Chazelle and Edelsbrunner (Chazelle, B., and Edelsbrunner, H., Linear space data structures for two types of range search. Discrete & Computational Geometry 2, 2 (1987), 113–126). We show that after preprocessing a set of homothetic polygons with constant number of vertices, the queries can be answered in $O(\log n + k)$ optimal time, where k is the output size. The preprocessing takes $O(n \log n)$ space and time. We also study the problem in dynamic setting where insertion and deletion operations are also allowed.

Keywords: Geometric Intersection · Algorithms · Data Structures · Dynamic Algorithms

1 Introduction

The point enclosure problem is one of the fundamental problems in computational geometry [3,6,20,24]. Typically, a point enclosure problem is formulated as follows:

> Preprocess a given set of geometrical objects so that for an arbitrary query point, all objects of the set containing the point can be reported efficiently.

In the counting version of the problem, we need only to compute the number of such objects. The point enclosure problems with orthogonal input objects (e.g. intervals, axes-parallel rectangles) have been well studied [1,3,5,6,24], but less explored for non-orthogonal objects.

We consider the point enclosure problem for homothetic polygons. A family of polygons is said to be homothetic if each polygon can be obtained from any other polygon in the family using scaling and translating operations. More precisely, a polygon P' is said to be homothetic to another polygon P [8], if there

S.-Y. Hsieh et al. (Eds.): IWOCA 2023, LNCS 13889, pp. 13–24, 2023.
https://doi.org/10.1007/978-3-031-34347-6_2

exists a point q and a real value c such that $P' = \{p \in \mathbb{R}^2 |$ there is a point $v \in P$ such that $p_x = q_x + cv_x$ and $p_y = q_y + cv_y\}$. Note that polygons homothetic to P are also homothetic to each other. Finding a solution to the point enclosure problem for homothetic triangles is sufficient. We can triangulate homothetic polygons into several sets of homothetic triangles and process each such set separately. Thus, our primary goal is to find an efficient solution for the triangle version.

The problem has applications in chip design. In VLSI, one deals with orthogonal and c-oriented objects (objects with sides parallel to previously defined c-directions) [4]. Earlier, Chazelle and Edelsbrunner [8] studied a range search problem in which query ranges are homothetic triangles (closed under translating and scaling). The problem we are considering is its "dual" in that the roles of input and query objects have swapped.

We study the problem in the static and dynamic settings. In the static version, the set of input polygons can not be changed while in the dynamic setting, new polygons may be inserted and existing ones may be deleted. For the static version, we propose a solution that can support point enclosure queries in optimal $O(\log n + k)$ time, where k is the output size. Preprocessing takes $O(n \log n)$ space and time. In the dynamic setting, we present a data structure that can answer a point enclosure query in $O(\log^2 n + k)$ time, where k is the output size and n is the current size of the dynamic set. An insertion or deletion operation takes $O(\log^2 n)$ amortized time. The total space used by the structure is $O(n \log n)$.

Remark 1. The point enclosure problem for homothetic triangles can also be solved by transforming it into an instance of the 3-d dominance query problem [2]. Here, we present a direct approach to solve the problem (without using the machinery of the 3-d dominance problem [16,21,23]).

The point enclosure problem for general triangles has been studied see, e.g. [10,19]. Overmars et al. [19] designed a data structure using a segment partition tree. Its preprocessing takes $O(n \log^2 n)$ space and $O(n \log^3 n)$ time. The triangles containing a query point can be reported in $O(n^\lambda + k)$ time and counted in $O(n^\lambda)$ time, where k is the output size and $\lambda \approx 0.695$. Cheng and Janardan [10] improved the query and space bounds to $O(\min\{\sqrt{n} \log n + k \log n \, , \sqrt{n} \log^2 n + k\})$ and $O(n \log^2 n)$ respectively, again k is the output size. However, the preprocessing time increases to $O(n\sqrt{n} \log^2 n)$.

The point enclosure problems for triangles with other constraints have also been studied. Katz [14] gave a solution for the point enclosure problem for convex simply-shaped fat objects. The query time is $O(\log^3 n + k \log^2 n)$, where k is the output size. Gupta et al. [12] provided solutions for the problem with fat input triangles. Sharir [22] gave a randomised algorithm to preprocess a set of discs in the plane such that all discs containing a query point are reported in $O((k + 1) \log n)$ (the worst-case) time.

Katz and Nielsen [15] considered a related decision problem of determining whether a given set of geometric objects is k-pierceable or not. A set of objects S is k-pierceable if there exists a set P of k points in the plane such that each object in S is pierced by (contains) at least one point of P. They solved the 3-piercing problem for a set of n homothetic triangles in $O(n \log n)$ time. A similar problem has been studied by Nielsen in [18].

Güting [13] considered the point enclosure problem for c-oriented polygons: polygons whose edges are oriented in only a constant number of previously defined directions. He gave an optimal $O(\log n)$ query time solution for its counting version. Chazelle and Edelsbrunner [8] gave a linear space and $O(\log n + \#output)$ query time solution for the problem of reporting points lying in a query homothetic triangle.

First, we consider a more specific problem where the input set $\mathcal{S}$ contains homothetic isosceles right-angled triangles. We can assume, without loss of generality, that the right-angled isosceles triangles in $\mathcal{S}$ are axes-parallel.

In our static optimal solution, we build a segment tree $\mathcal{T}$ over the x-projections of the triangles in $\mathcal{S}$. We represent the canonical set $\mathcal{S}(v)$ of triangles, for each node v, by a linear list. For each triangle $T \in \mathcal{S}(v)$, $v \in \mathcal{T}$, we define two geometric objects (right-angled triangle and rectangle) such that a point in the plane lies in one of these objects iff it lies in T. We call them trimmed triangles and trimmed rectangles for (T, v). We store them in different structures separately that can quickly answer the point enclosure queries. Further, we employ the fractional cascading technique to achieve $O(\log n + k)$ query time, where k is the output size. $O(n \log n)$ space and time are needed to build the data structure.

For the dynamic setting, we maintain the triangles in $\mathcal{S}$ in an augmented dynamic segment tree [17]. At each node v in the tree, we maintain an augmented list $A(v)$ of trimmed triangles sorted by an order $\preceq$ (to be defined later) and a dynamic interval tree [11] for storing y-projections of trimmed rectangles. For a given query point, we obtain all the trimmed triangles (the trimmed rectangles) containing the point using the augmented lists (the interval trees) associated with the nodes on the search path for the query point. The query time is $O(\log^2 n + k)$. We can insert (delete) a triangle into (from) the structure in $O(\log^2 n)$ time. The space used by the data structure is $O(n \log n)$.

We show that, without increasing any bound, a solution for the special case can be extended for general homothetic triangles. For homothetic polygons, by similarly triangulating all polygons, we get several instances of the problem for homothetic triangles.

The paper is organised as follows. In Sect. 2, we consider a special case where the input objects are homothetic isosceles right-angled triangles and present solutions for both the static setting and the dynamic setting. Section 3 describes a transformation to obtain a solution for the general problem. In Sect. 4, we study the problem for homothetic polygons. Finally, some conclusions are in Sect. 5.

1.1 Preliminaries

One-Dimensional Point Enclosure Problem: The problem asks to preprocess a set of intervals on the line so that all the intervals containing a query point (real value) can be found efficiently. Chazelle [6] describes a linear space structure, namely the window-list, that can find all the intervals containing a query point in $O(\log n + k)$ time, where n is the number of input intervals, and k is the output size. There is exactly one window in the window-list containing a particular point. Knowing the query-point window allows us to find the required intervals in $O(k)$ time. The preprocessing time is $O(n \log n)$. The window-list structure can be built in linear time if the interval endpoints are already sorted.

Interval Tree: The interval tree [3] is a classical data structure that can be used to store a set of n intervals on the line such that, given a query point (a real value), one can find all intervals containing the point in optimal $O(\log n + \#output)$ time. The dynamic interval tree [11] also supports the insertions of new intervals and deletions of the existing ones, each one in $O(\log n)$ amortized time. The space used in both settings is $O(n)$.

Segment Tree: The segment tree data structure [3] for a set B of n segments on the real line supports the following operations.

- report all the segments in B that contain a query point x in $O(\log n + \#output)$ time.
- count all the segments in B that contain a query point x in $O(\log n)$ time.

It can be built in $O(n \log n)$ time and space. Each node v in the segment tree corresponds to a vertical slab $H(v) = [x, x') \times \mathbb{R}^2$, where $[x, x')$ is an interval on the x-axis. The union of vertical slabs corresponding to all nodes at any level in the segment tree is the entire plane. We say that a segment $s_i \in B$ spans node v if s_i intersects $H(v)$ and none of its endpoints lies in the interior of $H(v)$. The canonical set $B(v)$, at node v, contains those segments of B that span v but do not span u, where u is the parent of v in the segment tree. A segment $s_i \in B$ can belong to canonical set $B(v)$ of at most two nodes at any tree level. As there are $O(\log n)$ levels in the segment tree, segment s_i may belong to $O(\log n)$ nodes. So, the total space required to store all the segments in the segment tree will be $O(n \log n)$.

Fractional Cascading Technique: Chazelle and Guibas [9] introduced the *fractional cascading technique*. Suppose there is an ordered list $C(v)$ in each node v of a binary tree of height $O(\log n)$. Using the fractional cascading method, we can search for an element in all lists $C(v)$ along a root-to-leaf path in the tree in $O(\log n)$ time, where $\Sigma_v(|C(v)|) = O(n)$.

Mehlhorn and Näher [17] showed that fractional cascading also supports insertions into and deletion from the lists efficiently. Specifically, they showed that a search for a key in t lists takes $O(\log n + t \log \log n)$ time and an insertion or deletion takes $O(\log \log n)$ amortized time, where n is the total size of all lists. As an application of the dynamic fractional cascading, they gave the following theorem.

Theorem 1. *[17] Let S be a set of n horizontal segments with endpoints in $\mathbb{R} \times \mathbb{R}$. An augmented dynamic segment tree for S can be built in $O(n \log n \log \log n)$*

time and $O(n \log n)$ *space. It supports an insertion or deletion operation in* $O(\log n \log\log n)$ *amortized time. An orthogonal segment intersection search query can be answered in* $O(\log n \log\log n + \#output)$*. If only insertions or only deletions operations are to be supported, then* $\log\log n$ *factors can be replaced by* $O(1)$ *from all the bounds mentioned earlier.*

We use the following definitions and notations. A right-angled triangle with two sides parallel to the axes will be called an *axes-parallel right-angled triangle*. We will characterise a point p in the plane by coordinates (p_x, p_y). By x-projection (resp. y-projection) of an object O, we mean the projection of a geometric object O on the x-axis (resp. y-axis). For any pair of points p and q in the plane, we say point q *dominates* point p if $q_x \geq p_x$ and $q_y \geq p_y$, and at least one of the inequality is strict.

2 Isosceles Right-Angled Triangles

This section considers a particular case of the point enclosure problem where the input triangles are isosceles right-angled. Let $\mathcal{S}$ be a set of n homothetic isosceles right-angled triangles in the plane. Without loss of generality, we can assume that the triangles in set $\mathcal{S}$ are axes-parallel triangles, and have their right-angled vertices at bottom-left (with minimum x and minimum y coordinates). As the hypotenuses of the triangles in $\mathcal{S}$ are parallel, we can define an ordering among the triangles. Let h_i and h_j be the hypotenuses of triangles T_i and $T_j \in \mathcal{S}$, respectively. We say that $T_i \preceq T_j$ if the hypotenuse of h_i lies below or on the line through h_j (i.e. if h_j is extended in both directions). We use the notations $T_i \preceq T_j$ and $h_i \preceq h_j$ interchangeably in this and the next section. By the position of a point q in a list of triangles (sorted by $\preceq$), we mean the position of the line through q and parallel to the hypotenuse.

2.1 (Static) Optimal Algorithm

Let us assume that we have a segment tree $\mathcal{T}$ built on the x-projections of the triangles in $\mathcal{S}$. Instead of x-projections, let the canonical sets $\mathcal{S}(.)$ contain the corresponding triangles. Consider a node v on the search path for an arbitrary (but fixed) q in the plane. A triangle $T \in \mathcal{S}(v)$ will contain q if the point q lies in $T \cap H(v)$, where $H(v)$ is the vertical slab corresponding to node v in the segment tree $\mathcal{T}$. Observe that the region $T \cap H(v)$ is a trapezoid. We can partition it into an axes-parallel right-angled triangle and a (possibly empty) axes-parallel rectangle. We next define a *trimmed triangle* for (T, v), where triangle $T \in \mathcal{S}(v)$. Let A' and C' be the points at which the hypotenuse of the triangle T intersects the boundary of the slab $H(v)$ (see Fig. 1). The trimmed triangle for (T, v) is a right-angled triangle in $T \cap H(v)$ such that A' and C' are the endpoints of the hypotenuse. The remaining portion of $T \cap H(v)$ is a (possibly empty) rectangle lying below the trimmed triangle. We call it the *trimmed rectangle* for (T, v). Thus, a triangle $T \in \mathcal{S}(v)$ will contain point q in $H(v)$ iff q lies in either the trimmed triangle or the trimmed rectangle for (T, v).

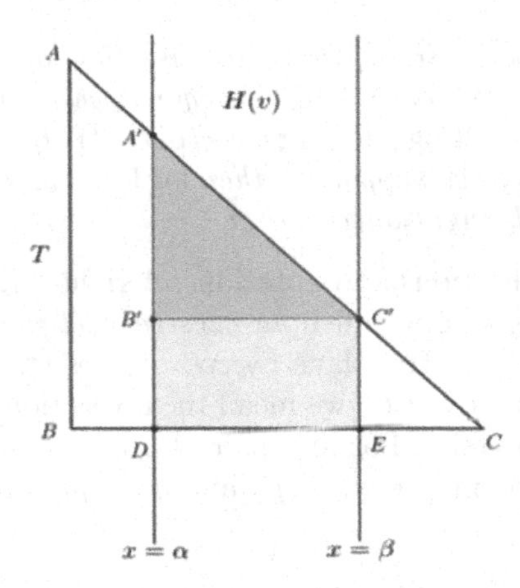

Fig. 1. Triangle $A'B'C'$ is the trimmed triangle for (T, v) and rectangle $B'C'ED$ is the trimmed rectangle for (T, v).

Lemma 1. *Let $L(v)$ be the sorted list (by order $\preceq$) of trimmed triangles at node v. The trimmed triangles at node v containing a point $q \in H(v)$ will be contiguous in the list $L(v)$.*

Proof. Let $[\alpha, \beta] \times \mathbb{R}^2$ be the slab of node v and $T'_1, T'_2, ..T'_r$ be the sorted list of trimmed triangles stored at v such that $T'_i \preceq T'_{i+1}$, $i \in [r-1]$. For the sake of contradiction, let us assume there exist three integers $i < k < j$ such that trimmed triangles T'_i and T'_j contain q but the trimmed triangle T'_k does not.

By definition, all trimmed triangles at node v are right-angled triangles with two sides parallel to the axes. As triangles in $\mathcal{S}$ are isosceles right-angled triangles, the horizontal and vertical sides of each trimmed triangle at v will be equal to $\beta - \alpha$; hence the trimmed triangles at node v are congruent with the same orientation. So for any triplet $T'_i \preceq T'_k \preceq T'_j$, their horizontal sides will hold the same order.

A trimmed triangle will contain a point $q \in H(v)$ if q lies above its horizontal side and below the hypotenuse. The horizontal side (and hypotenuse) of T'_k lies between the horizontal sides (hypotenuses) of T'_i and T'_j. Since T'_k does not contain q, its horizontal side lies above q or its hypotenuse lies below the point q. The former one can not be true as in that case, point q would lie outside of T'_j; a contradiction. If q lies above the hypotenuse of T'_k, q would also lie outside T'_i. Therefore, no such triplet can exist. Thus, all trimmed triangles at v containing the point $q \in H(v)$ will be contiguous in the list $L(v)$. □

Lemma 2. *At any node $v \in \mathcal{T}$, the problem of computing trimmed rectangles containing a point $q \in H(v)$ can be transformed to an instance of the 1-d point enclosure problem.*

Proof. By definition, all trimmed rectangles at node v are axes-parallel rectangles with $x_1 = \alpha$ and $x_2 = \beta$. Here $[\alpha, \beta] \times \mathbb{R}^2$ is the slab corresponding to node v.

Thus, the two y-coordinates of a trimmed rectangle can be used to know whether point q lies in the rectangle. As point q is inside the slab $[\alpha, \beta] \times \mathbb{R}^2$, we have $\alpha \leq q_x \leq \beta$. Point q will be inside a trimmed rectangle $[\alpha, \beta] \times [y_1, y_2]$ if and only if $y_1 \leq q_y \leq y_2$. Hence, the problem of computing trimmed rectangles at node v containing q transforms to an instance of the 1-d point enclosure problem. □

First, we sort the given set $\mathcal{S}$ by order $\preceq$. Next, we build a segment tree $\mathcal{T}$ over the x-projections of triangles in $\mathcal{S}$. We pick each triangle from $\mathcal{S}$ in order and store it in the canonical sets of corresponding $O(\log n)$ nodes. As a result, triangles in canonical set $\mathcal{S}(v)$, for each node v, will also be sorted by order $\preceq$. We realise the canonical set of each node by a linear list. For each node $v \in \mathcal{T}$, we store the trimmed triangles in a linear list $L(v)$ sorted by order $\preceq$ and the y-projections of the trimmed rectangles in a window-list structure $I(v)$. The preprocessing takes $O(n \log n)$ time and space.

For a query point $q = (q_x, q_y)$, we find the search path Π for q_x in the segment tree $\mathcal{T}$. Path Π is a root-to-leaf path in the tree $\mathcal{T}$. We traverse the path Π in the root-to-leaf fashion and compute trimmed triangles and trimmed rectangles of each traversed node. Let v be the current node being traversed. In order to compute the required trimmed triangles at v, we locate the position of point q among the hypotenuses of the triangles in the sorted list $L(v)$ using binary search. We next move upwards in the list $L(v)$ and keep reporting the triangles as long as the horizontal side is below point q and stop as soon as we encounter a triangle whose horizontal side lies above point q. We query the window-list structure $I(v)$ with query value q_y, and report the rectangles corresponding to returned y-projections as the trimmed rectangles containing q.

Lemma 3. *The query procedure computes all the triangles in $\mathcal{S}$ containing q in $O(\log^2 n + k)$ time, where k is the output size.*

Proof. The correctness proof is immediate from Lemmas 1 and 2. At each node on the search path, $O(\log n)$ time is needed to find the location q in $L(v)$ and $O(1+k_t)$ time is needed to report the k_t triangles containing q. The query to the window-list $I(v)$ takes $O(\log n)$ time to find the target window and $O(k_r)$ time to report k_r y-projections containing q_y. Thus at each node on the search path for q, we are spending $O(\log n + k_v)$ time to find triangles in $\mathcal{S}(v)$ containing q, where $k_v = k_t + k_r$. Since there are $O(\log n)$ nodes on the search path, the query procedure would take $O(\log^2 n + k)$ time in total, where k is the output size. □

So we have the following result.

Theorem 2. *We can process a set of n homothetic isosceles right-angled triangles so that, for a given query point, we can find all the triangles containing the query point in $O(\log^2 n + k)$, where k is the output size. The structure can be built in $O(n \log n)$ space and time.*

We use the fractional cascading technique [9] to reduce the query time to $O(\log n + k)$. Recall that if we have an ordered list $C(v)$ in each node v of a binary tree, then using the fractional cascading technique, we can search for an element in all lists $C(v)$ along a root-to-leaf path in the tree in $O(\log n)$ time, where $\Sigma_v(|C(v)|) = O(n)$.

As the list $L(v)$, for every node v, is sorted by order $\preceq$, we can use the fractional cascading technique. We search for point q in the list $L(root)$ in $O(\log n)$ time. Then, using the fractional cascading pointers, we can get the position of point q in the list of any child in $O(1)$ additional time. Thus, for a given point, we can search the point in all lists $L(v)$ along a root-to-leaf path in $O(\log n)$ time in total.

The window-list structure $I(v)$ is a sorted list of contiguous intervals (or windows) on the y-axis, and y-projections of the trimmed rectangles are stored in these windows. As the structure $I(v)$, for every node v, is also a sorted list, we can use the fractional cascading technique for structures $I(v)$ as well. After locating point q in $I(root)$ in $O(\log n)$ time, we can identify the correct window for each child in $O(1)$ additional time. Thus, for a given point q_y, we can identify the window containing q_y in all $I(v)$ structures along a root-to-leaf path $O(\log n)$ time in total. Therefore, we have the following.

Theorem 3. *We can preprocess a given set of n of homothetic isosceles right-angled triangles, in $O(n \log n)$ time and space to report all triangles that contain a query point in $O(\log n + k)$ time, where k is the the output size.*

2.2 Dynamic Data Structure

We next describe a dynamic data structure for homothetic isosceles right-angled triangles. In Sect. 3, we discuss general homothetic triangles.

We build an augmented dynamic segment tree for $\mathcal{S}$ using the method of Mehlhorn and Näher (see Theorem 6 in [17]) with the ordering $\preceq$. Recall that we say $T_i \preceq T_j$ if the hypotenuse of T_i lies below or on the line through the hypotenuse of T_j. Each node v in the segment tree maintains an augmented list $A(v)$ of (trimmed) triangles. Moreover, for each node v in the segment tree, we store the y-projections of the trimmed rectangles in a dynamic interval tree [11]. We denote this structure by $\mathcal{T_D}$.

The segment tree with augmented lists $A(.)$ uses $O(n \log n)$ space [17]. As a dynamic interval tree uses linear space, so the total space used in maintaining the interval trees associated with all nodes will be $O(n \log n)$.

Lemma 4. *The total space used by $\mathcal{T_D}$ is $O(n \log n)$, where n is the number of triangles currently stored.*

For a given query point q in the plane, let $v_0, v_1, v_2, ..., v_l$ be the nodes on the search path for q_x in the tree $\mathcal{T_D}$. Note that $l = O(\log n)$. Using a binary search, we find the position of q in the augmented list $A(v_0)$ in $O(\log n)$ time. For each $i \in \{1, 2, ..l\}$, we locate the position of q in $A(v_i)$ from its known position in $A(v_{i-1})$, in $O(\log \log n)$ time, using the dynamic fractional cascading technique [17]. Having the position of q in an augmented list $A(v_i)$, we can find the trimmed triangles containing q, in a similar fashion as we have done in the static case, in time proportional to the number of reported triangles. Thus, we can report all trimmed triangles that contain q in $O(\log n \log\log n + k_t)$ time, where k_t is the number of reported trimmed triangles.

For each node v_i, we query the associated dynamic interval tree with q_y. We report the triangles corresponding to y-projections (of trimmed rectangles) returned as the query response. A query takes $O(\log |\mathcal{S}(v)|) = O(\log n)$ time at each node; the total time spent in finding all trimmed rectangles (hence homothetic triangles) containing the query point q will be $O(\log^2 n + k_r)$. Here, k_r is the number of trimmed rectangles containing q. Thus, the query time is dominated by the time needed to compute trimmed rectangles.

Let T be the homothetic triangle we want to insert. We find $O(\log n)$ nodes in tree $\mathcal{T_D}$ corresponding to the new triangle T. We simply insert T into the augmented lists of these $O(\log n)$ nodes in $O(\log n \log\log n)$ amortized time. Moreover, for each of the $O(\log n)$ nodes, we insert the y-projection of the trimmed rectangle at the node in the associated dynamic interval tree in $O(\log n)$ time; hence $O(\log^2 n)$ amortized time is needed in total. Thus, the total time spent during an insertion operation is $O(\log^2 n)$.

Let T' be the triangle to be deleted. Again, we find the $O(\log n)$ nodes of $\mathcal{T_D}$ storing T'. Deleting T' from the augmented lists $A(.)$ of these nodes will take $O(\log n \log\log n)$ time in total, while deletion from the interval trees associated with these nodes will take $O(\log^2 n)$ time in total. Thus, a deletion operation will take $O(\log^2 n)$ amortized time. Thus, we have the following theorem.

Theorem 4. *We can build a dynamic data structure for a set of homothetic triangles that supports a point enclosure query in $O(\log^2 n + k)$ worst-case time and an update operation in $O(\log^2 n)$ amortized time. Here, n is the number of triangles currently in the set, and k is the number of triangles containing the query point.*

3 General Homothetic Triangles

First, we consider the case where homothetic triangles are arbitrary right-angled triangles, not necessary isosceles. As triangles are homothetic, all hypotenuses will have the same slope. We can make each triangle isosceles by scaling x and y coordinates. Let the hypotenuses of all input triangles be parallel to the line $x/a + y/b = 1$. By transformation $x = ax'$ and $y = by'$, the hypotenuses become parallel to the line $x + y = 1$. The transformed triangles are processed as before. For a query point $q = (q_x, q_y)$, we find the transformed triangles containing the point $(\frac{q_x}{a}, \frac{q_y}{b})$. The preprocessing bounds and query time-bound remain the same.

Let us now consider the general case of homothetic triangles. We only describe the case where homothetic triangles are acute-angled triangles; the obtuse-angled case can be handled analogously. Without loss of generality, let us assume that the triangles of the input set $\mathcal{S}$ are in the first quadrant and have one side parallel to the x-axis. Let m be the slope of the sides of the triangles that make a positive angle with the x-axis. We make these triangles right-angled triangles by

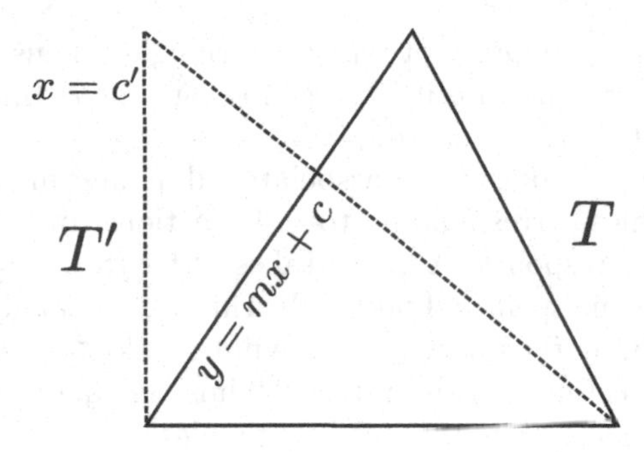

Fig. 2. Dashed triangle T' is the transformed triangle of triangle $T \in \mathcal{S}$.

a linear shear transformation: $x' = x - y/m$ and $y' = y$. See Fig. 2. A triangle $T \in \mathcal{S}$ will contain a point q if and only if its transformed triangle T' contains the transformed query point q' (see Appendix for proof). The transformed homothetic right-angled triangles are handled as described earlier. For a query point (q_x, q_y), we find the transformed triangles containing the point $(q'_x + \frac{q'_y}{m}, q'_y)$. Again, the bounds remain the same. Hence,

Theorem 5. *We can process a given set of n homothetic triangles so that for a given query point, all the triangles containing the query point can be found in $O(\log n + k)$ time, where k is the output size. The structure can be built in $O(n \log n)$ space and time.*

4 Homothetic Polygons

Assume that we are given a set of homothetic polygons, each having a constant number of vertices. We partition each homothetic polygon into homothetic triangles by adding "similar" diagonals, see Fig. 3. Let $v_1, v_2, ...v_m$ be the vertices of a particular polygon in clockwise order starting from the vertex with the smallest x-coordinate. If there is a diagonal between v_i and v_j in the triangulation, all other polygons will have a diagonal joining the corresponding vertices. As a result, we get several sets of homothetic triangles. The problem breaks into several instances of the point enclosure problem for homothetic triangles. We process each instance as described earlier. For a query point q, we find the triangles (and report the corresponding polygons) containing the point q using the computed solutions of the instances. If the polygons are m-sided, a triangulation will result in $m-2$ triangles. Thus, there will be $O(m)$ instances of the problem for homothetic triangles, each with n triangles. The triangulation of a polygon takes linear time [7], and adding similar diagonals to all the polygons requires $O(mn)$ time. Preprocessing each instance takes $O(n \log n)$ time. As the number of vertices (m) in each polygon is constant, thus total preprocessing time will be $O(n \log n)$.

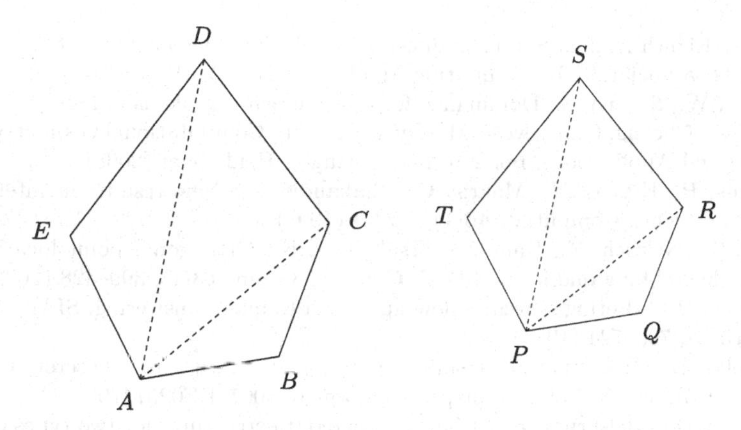

Fig. 3. Two homothetic polygons $ABCDE$ and $PQRST$ triangulated with similar diagonals (AC, PR) and (AD, PS).

Remark 2. If number of vertices, m, is a parameter, then preprocessing all $m-2$ instances will take $O(mn \log n)$ space and time. Thus, the total preprocessing time and space will be $O(mn \log n)$. The query time will be $O(m \log n + k)$, where k is the number of polygons containing a query point. Hence, the bounds from Theorem 3 also hold for homothetic polygons with $m = O(1)$-size description complexity.

5 Conclusions

In this work, we studied the problem of finding all homothetic triangles containing a query point and gave solutions in the static and the dynamic settings. For the static case, we have given a near-linear space solution with optimal query time. We believe that the bounds presented for the dynamic case can be improved. We extended the solutions for a more general problem where input objects are homothetic polygons.

As points in an arbitrary homothetic polygon can have arbitrary coordinates, at first sight it appears difficult to work in rank space (and the word-RAM model). We leave the problem of improving the results in the word-RAM model as an open problem.

Acknowledgements. We wish to thank anonymous referees for careful reading of the manuscript, their comments and suggestions. We believe the suggestions have helped in improving the manuscript.

References

1. Afshani, P., Arge, L., Larsen, K.G.: Higher-dimensional orthogonal range reporting and rectangle stabbing in the pointer machine model. In: Proceedings of the

Twenty-Eighth Annual Symposium on Computational Geometry, SoCG 2012, pp. 323–332. Association for Computing Machinery, New York (2012)
2. Akram, W., Saxena, S.: Dominance for containment problems (2022)
3. Berg, M., Cheong, O., Kreveld, M., Overmars, M.: Computational Geometry: Algorithms and Applications, pp. 219–241. Springer, Heidelberg (2008)
4. Bozanis, P., Kitsios, N., Makris, C., Tsakalidis, A.: New results on intersection query problems. Comput. J. **40**(1), 22–29 (1997)
5. Chan, T., Nekrich, Y., Rahul, S., Tsakalidis, K.: Orthogonal point location and rectangle stabbing queries in 3-D. J. Comput. Geom. **13**(1), 399–428 (2022)
6. Chazelle, B.: Filtering search: a new approach to query-answering. SIAM J. Comput. **15**(3), 703–724 (1986)
7. Chazelle, B.: Triangulating a simple polygon in linear time. Discret. Comput. Geom. **6**(3), 485–524 (1991). https://doi.org/10.1007/BF02574703
8. Chazelle, B., Edelsbrunner, H.: Linear space data structures for two types of range search. Discret. Comput. Geom. **2**(2), 113–126 (1987). https://doi.org/10.1007/BF02187875
9. Chazelle, B., Guibas, L.: Fractional cascading: I. A data structuring technique. Algorithmica **1**, 133–162 (1986)
10. Cheng, S.W., Janardan, R.: Algorithms for ray-shooting and intersection searching. J. Algorithms **13**(4), 670–692 (1992)
11. Chiang, Y.-J., Tamassia, R.: Dynamic algorithms in computational geometry. Proc. IEEE **80**(9), 1412–1434 (1992)
12. Gupta, P., Janardan, R., Smid, M.: Further results on generalized intersection searching problems: counting, reporting, and dynamization. J. Algorithms **19**(2), 282–317 (1995)
13. Güting, R.H.: Stabbing C-oriented polygons. Inf. Process. Lett. **16**(1), 35–40 (1983)
14. Katz, M.J.: 3-D vertical ray shooting and 2-D point enclosure, range searching, and arc shooting amidst convex fat objects. Comput. Geom. **8**(6), 299–316 (1997)
15. Katz, M.J., Nielsen, F.: On piercing sets of objects. In: Proceedings of the Twelfth Annual Symposium on Computational Geometry, pp. 113–121 (1996)
16. Makris, C., Tsakalidis, A.: Algorithms for three-dimensional dominance searching in linear space. Inf. Process. Lett. **66**(6), 277–283 (1998)
17. Mehlhorn, K., Näher, S.: Dynamic fractional cascading. Algorithmica **5**(1–4), 215–241 (1990)
18. Nielsen, F.: On point covers of C-oriented polygons. Theor. Comput. Sci. **263**(1–2), 17–29 (2001)
19. Overmars, M.H., Schipper, H., Sharir, M.: Storing line segments in partition trees. BIT Numer. Math. **30**(3), 385–403 (1990)
20. Preparata, F.P., Shamos, M.I.: Computational Geometry: An Introduction, pp. 323–373. Springer, New York (1985)
21. Saxena, S.: Dominance made simple. Inf. Process. Lett. **109**(9), 419–421 (2009)
22. Sharir, M.: On k-sets in arrangements of curves and surfaces. Discret. Comput. Geom. **6**(4), 593–613 (1991)
23. Shi, Q., JaJa, J.: Fast algorithms for 3-D dominance reporting and counting. Int. J. Found. Comput. Sci. **15**(04), 673–684 (2004)
24. Vaishnavi, V.K.: Computing point enclosures. IEEE Trans. Comput. **31**(01), 22–29 (1982)

Hardness of Balanced Mobiles

Virginia Ardévol Martínez[1](✉), Romeo Rizzi[2], and Florian Sikora[1]

[1] Université Paris-Dauphine, PSL University, CNRS, LAMSADE, 75016 Paris, France
{virginia.ardevol-martinez,florian.sikora}@dauphine.fr

[2] Department of Computer Science, University of Verona, Verona, Italy
romeo.rizzi@univr.it

Abstract. Measuring tree dissimilarity and studying the shape of trees are important tasks in phylogenetics. One of the most studied shape properties is the notion of tree imbalance, which can be quantified by different indicators, such as the Colless index. Here, we study the generalization of the Colless index to mobiles, i.e., full binary trees in which each leaf has been assigned a positive integer weight. In particular, we focus on the problem Balanced Mobiles, which given as input n weights and a full binary tree on n leaves, asks to find an assignment of the weights to the leaves that minimizes the Colless index, i.e., the sum of the imbalances of the internal nodes (computed as the difference between the total weight of the left and right subtrees of the node considered). We prove that this problem is strongly NP-hard, answering an open question given at IWOCA 2016.

Keywords: Phylogenetic trees · Colless Index · Balanced Mobiles · Strong NP-hardness

1 Introduction

Phylogenetics is the study of evolutionary relationship among biological entities (taxa). Its main task is to infer trees whose leaves are bijectively labeled by a set of taxa and whose patterns of branching reflect how the species evolved from their common ancestors (*phylogenetic trees*). The inferred trees are often studied by comparing them to other phylogenetic trees or to existing models. Thus, it is important to be able to formally quantify how different trees differ from each other and to have measures that give information about the shape of the trees. With respect to the latter, one of the most studied shape properties of phylogenetic trees is that of *tree balance*, measured by metrics such as the Sackin index [12] or the Colless index [3] (see also the survey of Fischer et al. [6]). The Colless index is defined for binary trees as the sum, over all internal nodes v of the tree, of the absolute value of the difference of the number of leaves in the two children of v. It is one of the most popular and used metrics, see for example [1,2,5,11,13].

S.-Y. Hsieh et al. (Eds.): IWOCA 2023, LNCS 13889, pp. 25–35, 2023.
https://doi.org/10.1007/978-3-031-34347-6_3

The natural generalization with *weights* on the leaves has later been studied within *mobiles*[1], defined as full binary trees with positive weights on their leaves. In particular, given a set of n integer weights $\{w_1, \ldots, w_n\}$, the problem BALANCED MOBILES asks to find a mobile whose n leaves have weights $w_1, \ldots, w_n$, and which minimizes the total Colless index (i.e., the sum of the imbalances $|x - y|$ of every internal node, where x and y represent the total weight of the leaves on the left and right subtrees of the node considered) [9]. Despite being a natural generalization, the complexity of this problem is still not yet known. In fact, it was proposed as an open problem by Hamoudi, Laplante and Mantaci in IWOCA 2016 [8].

Still, some results are known for some specific cases. For example, if all the leaves have unit weight, it is known that building a partition tree or a left complete tree are both optimal solutions, and their imbalance can be computed in polynomial time using a recursive formula. On the other hand, if all the weights are powers of two or if a perfectly balanced mobile can be constructed, the well known Huffman's algorithm [10] is optimal. This algorithm recursively builds a mobile by grouping the two smallest weights together (where the weight of the constructed subtree is added to the list of weights in each step).

With respect to the complexity, it is only known that the problem is in the parameterized class XP, parameterized by the optimal imbalance [9] (i.e. it is polynomial for constant values of the parameter). This result was obtained by using a relaxation of Huffman's algorithm, which gives an algorithm of complexity $\mathcal{O}(\log(n)n^{C^*})$, where C^* is the optimal imbalance. An ILP is also given to solve the problem [9]. However, no polynomial time approximation algorithm has been proposed for this problem, although it is known that Huffman's algorithm does not construct an approximate solution in the general case, being arbitrarily far away from the optimum for some instances [9].

In this paper, we shed some light into the complexity of the problem by showing that BALANCED MOBILES is strongly NP-hard when both the full binary tree and the weights are given as input.

2 Preliminaries

We first give the necessary definitions to present the problem.

Definition 1. *A* full binary tree *is a rooted tree where every node that has at least one child has precisely two children. A full binary tree is said to be* perfect *when all its leaves are at the same depth. The depth $d(v)$ of a node v is defined by*

$$d(v) := \begin{cases} 0 \text{ if } v = r, \text{ the root,} \\ 1 + d(F(v)) \text{ otherwise,} \end{cases}$$

where $F(v)$ denotes the father of node v. Also, for every non-leaf node v, $L(v)$ (resp., $R(v)$) denotes the left (resp., right) child of node v.

[1] The term "mobile" comes from what can be found in modern art (e.g. the ones of Calder, well known in TCS being the illustration of the cover of the famous book CLRS' Introduction to Algorithms [4]) or the toy above toddler beds [9].

Definition 2. *A binary tree is said to be* leaf-weighted *when a natural number $w(v)$ is assigned to each one of its leaf nodes v. Then, the recurrence $w(v) := w(L(v)) + w(R(v))$ extends w defining it also on every internal node v as the total weight over the leaves of the subtree rooted at v. A leaf-weighted full binary tree is also called a* mobile.

In this paper, we focus only on the Colless index to measure the balance of mobiles. Thus, we will just refer to the cost at each node as *imbalance*, and to the total Colless index of the tree as the *total cost.*

Definition 3. *The* imbalance *of an internal node v is defined as $imb(v) := |w(L(v)) - w(R(v))|$. The* total cost *of a leaf-weighted full binary tree (mobile) is the sum of the imbalances over the internal nodes. If the total cost is equal to 0, the mobile is said to be* perfectly balanced.

We can now define the problem BALANCED MOBILES studied in this paper.

> BALANCED MOBILES
> **Input:** n natural numbers and a full binary tree $\mathcal{T}$ with n leaves.
> **Task:** Assign each number to a different leaf of the given full binary tree in such a way that the sum of the imbalance over the internal nodes of the resulting leaf-weighted binary tree is minimum.

3 BALANCED MOBILES is NP-Hard in the Strong Sense

We prove that BALANCED MOBILES as formulated above is NP-hard in the strong sense.

To do so, we will reduce from ABC-PARTITION, a variant of the problem 3-PARTITION which we define below.

> ABC-PARTITION
> **Input:** A target integer T, three sets A, B, C containing n integers each such that the total sum of the $3n$ numbers is nT.
> **Task:** Construct n triplets, each of which contains one element from A, one from B and one from C, and such that the sum of the three elements of each triplet is precisely the target value T.

The ABC-PARTITION problem is known to be strongly NP-hard, that is, it is NP-hard even when restricted to any class of instances in which all numbers have magnitude $\mathcal{O}(poly(n))$. This fact is reported in [7], where the problem, labeled as [SP16], is also called NUMERICAL 3-D MATCHING, as it can also be reduced to the 3-DIMENSIONAL MATCHING problem.

3.1 Preliminary Steps on the ABC-PARTITION Problem

As a first step in the reduction, given an instance of ABC-PARTITION, we will reduce it to an equivalent instance of the same problem with some specific properties that will be useful for the final reduction.

A class of instances is called *shallow* when it comprises only instances all of whose numbers have magnitude $\mathcal{O}(poly(n))$. Since we aim at proving strong NP-hardness of the target problem, we need to make sure that, starting from any shallow class of instances, the classes of instances produced at every step remain shallow.

We start with some easy observations.

Observation 1. *For any natural constant k, we can assume that all numbers are divisible by 2^k, simply by multiplying all of them by 2^k.*

Note that, since k is a constant, after this first reduction we are still dealing with a shallow class of instances.

For the next step, we assume all numbers are greater than 1, which follows from the above with $k = 1$.

Observation 2. *We can then assume that n is a power of two, otherwise, let h be the smallest natural such that $2^h > n$, we can just add $2^h - n$ copies of the number $T - 2$ to the set A, and $2^h - n$ copies of the number 1 to both sets B and C.*

Note that we are still dealing with a shallow class of instances.

The next step requires to be more formal. Assume the three given sets of natural numbers to be $A = \{a_1^0, a_2^0, \ldots, a_n^0\}$, $B = \{b_1^0, b_2^0, \ldots, b_n^0\}$ and $C = \{c_1^0, c_2^0, \ldots, c_n^0\}$, the target value to be T^0 and let M^0 be the maximum number in $A \cup B \cup C$. Here, $M^0 = \mathcal{O}(poly(n))$ since this generic instance is taken from a shallow class. Consider the instance of the problem where the n input numbers and the target value T^0 are transformed as follows:

$$
\begin{aligned}
a_i^1 &:= a_i^0 + 8n^2M^0 && \text{for every } i = 1, 2, ..., n \\
b_i^1 &:= b_i^0 + 4n^2M^0 && \text{for every } i = 1, 2, ..., n \\
c_i^1 &:= c_i^0 + 2n^2M^0 && \text{for every } i = 1, 2, ..., n \\
T^1 &:= T^0 + 14n^2M^0 && \\
M^1 &:= T^1 &&
\end{aligned}
$$

Notice that the new M^1 does not represent any longer the maximum value of the numbers of the input instance because it is equal to $T^0 + 14n^2M^0$, while the value of ever number is bounded above by $a_i^0 + 8n^2M^0$. The role that parameter M^1 plays in our reduction will be seen only later.

Clearly, this new instance is equivalent to the previous one in the sense that either both or none of them are yes-instances of ABC-PARTITION. Moreover, this reduction yields a shallow class of instances $\mathcal{C}^1$ when applied to a shallow class of instances $\mathcal{C}^0$. Therefore, thanks to Observation 2 and this transformation, we can assume that we are dealing with a shallow class of instances each of which satisfies the following two properties:

$$n \text{ is a power of two} \tag{1}$$
$$b > c \text{ and } a > b + c \text{ for every } a \in A,\ b \in B \text{ and } c \in C. \tag{2}$$

3.2 ABCDE-Partition Problem

Once Properties (1) and (2) are in place, we create a next equivalent instance through one further reduction, this time yielding an instance of a slightly different version of the multi-dimensional partition problem, the ABCDE-PARTITION problem, which we define below.

ABCDE-PARTITION
Input: A target integer T, five sets A, B, C, D, E, with n integers in each, such that the sum of the numbers of all sets is nT.
Task: Construct n 5-tuples, each of which contains one element of each set, with the sum of these five elements being precisely T.

If not known, the next transformation in our reduction proves that this variant is also strongly NP-hard. In fact, where $a_i^1, b_i^1, c_i^1, M^1, T^1$ comprise the modified input of the ABC-PARTITION problem after the last transformation detailed just above, consider the equivalent instance of the ABCDE-PARTITION problem where the input numbers and the target value are defined as follows:

$$\begin{aligned}
a_i &:= a_i^1 + 8n^2M^1 && \textit{for every } i = 1, 2, ..., n\\
b_i &:= b_i^1 + 4n^2M^1 && \textit{for every } i = 1, 2, ..., n\\
c_i &:= c_i^1 + 2n^2M^1 && \textit{for every } i = 1, 2, ..., n\\
d_i &:= n^2M^1 && \textit{for every } i = 1, 2, ..., n\\
e_i &:= n^2M^1 && \textit{for every } i = 1, 2, ..., n\\
T &:= T^1 + 16n^2M^1 &&\\
M &:= M^1 &&
\end{aligned}$$

Notice that, once again, the new M parameter does not represent the maximum value of the numbers comprising the new input instance. In fact, it is significantly smaller.

Thanks to this last transformation, we see that the ABCDE-PARTITION problem is NP-hard even when restricted to a shallow class of instances each of which satisfies the following three properties:

$$n \text{ is a power of two (say } n = 2^h) \tag{3}$$
$$\text{the numbers in } D \cup E \text{ are all equal} \tag{4}$$
$$c > d + e, b > c + d + e \text{ and } a > b + c + d + e,$$
$$\text{for every } (a, b, c, d, e) \in A \times B \times C \times D \times E \tag{5}$$

This instance also possesses other useful properties that we will exploit in the reduction to the BALANCED MOBILES problem we are going to describe next.

3.3 Reduction to BALANCED MOBILES

To the above instance (T, A, B, C, D, E) of ABCDE-PARTITION, we associate the following instance $(\mathcal{T}, W)$ of BALANCED MOBILES:

Weights (W). Besides the weights a_i, b_i, c_i, d_i, e_i defined above for every $i = 1, 2, ..., n$, we also introduce $n = 2^h$ copies of the number T. Notice that all these numbers have magnitude $\mathcal{O}(poly(n))$ since it was assumed that $M = \mathcal{O}(poly(n))$.

Full Binary Tree ($\mathcal{T}$). Before describing how to construct the tree $\mathcal{T}$, which completes the description of the instance and the reduction, we still have one proviso.

While describing how to obtain the instance of the target problem from the instance of the source problem, it often helps to describe simultaneously how to obtain a yes-certificate for the target instance from a hypothetical yes-certificate for the source instance. Hence, let σ_B and σ_C be two permutations in S_n meant to encode a generic possible solution to the generic ABCDE-PARTITION problem instance (since all the elements in D and E are equal, it is enough to consider these two permutations). The pair (σ_B, σ_C) is a truly valid solution, i.e., a yes-certificate, iff $a_i + b_{\sigma_B(i)} + c_{\sigma_C(i)} + d_i + e_i = T$ for every $i = 1, 2, ..., n$. We are now going to describe not only how to construct an instance of the target problem but also a solution $S = S(\sigma_B, \sigma_C)$ for it, which depends solely on the hypothetical yes-certificate (σ_B, σ_C).

The tree $\mathcal{T}$ and the solution $S = S(\sigma_B, \sigma_C)$ are constructed as follows:

1. Start with a perfect binary tree of height $h+1$, with $2n = 2^{(h+1)}$ leaves. Its n internal nodes at depth h are called test nodes, denoted $t_i, i \in [n]$ (also called r_i^0). This tree is a full binary tree thanks to Property 3. Moreover, each test node t_i will next become the root of a subtree of depth 5, all these subtrees having the very same topology, described in the following and illustrated in Fig. 1.
2. For $i = 1, ..., n$, the left child of the test node t_i is a leaf of the subtree rooted at t_i (and hence also of the global tree under construction). The certificate assigns one different copy of T to each left child of a test node. These n nodes will be the only leaves at depth $2n = 2^{(h+1)}$ in the final tree under construction. However, the right child of t_i, called r_i^1, has two children described next.

3. The left child of r_i^1 is a leaf, and the certificate assigns to this leaf the number a_i. On the other hand, the right child of r_i^1, which we denote r_i^2, will not be a leaf, which means that it has two children, described next.
4. In the next step, we also let the left child of r_i^2 be a leaf. The certificate assigns the number $b_{\sigma_B(i)}$ to the left child. On the other hand, the right child of r_i^2, called r_i^3, will have two children, described next.
5. As before, the left child of r_i^3 is also a leaf, and the certificate assigns the number $c_{\sigma_C(i)}$ to it. The right child of r_i^3, called r_i^4, will also have two children, but, finally, both of them are leaves: to the left (resp., right) child leaf, the certificate associates the number d_i (resp., e_i).

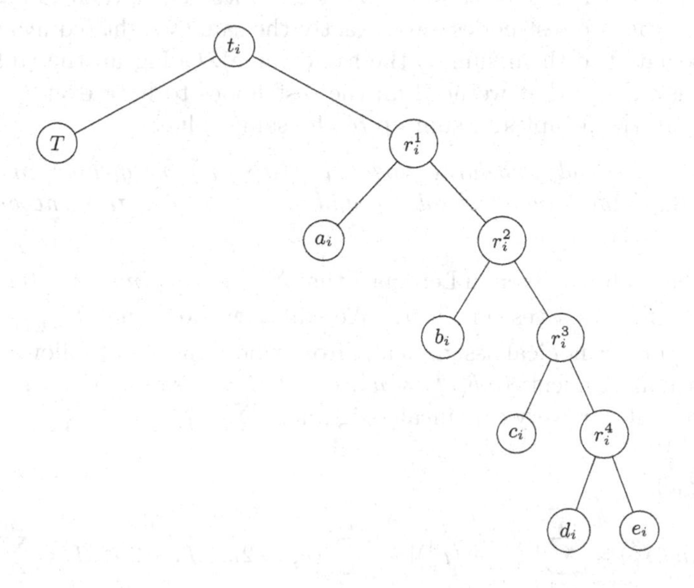

Fig. 1. Subtree rooted at the test node t_i with a canonical weight assignment. Recall that in the full tree $\mathcal{T}$, a full binary tree connects all the test nodes t_i.

The set I of the internal nodes of $\mathcal{T}$ partitions into $I_<$ and $I_\geq$, those of depth less than h and those of depth at least h, respectively. In other words, all the strict ancestors of test nodes t_i versus $I_\geq = \bigcup_{i=1,\dots,n}\{t_i, r_i^1, r_i^2, r_i^3, r_i^4\}$. We also define $I_>$ as the set of internal nodes at depth strictly greater than h.

Definition 4. *A weight assignment is called* canonical *if it is of the form $S(\sigma_B, \sigma_C)$ for some pair (σ_B, σ_C). Equivalently, the canonical assignments are those where all $2n$ leaf nodes at depth $h+5$ have been assigned one different copy of weight Mn^2, and all n leaf nodes at depth $h+1$ (resp., $h+2$, $h+3$, or $h+4$) have weight precisely T (resp., falling in the interval $(8n^2M, 8n^2M+M)$, $(4n^2M, 4n^2M+M)$, or $(2n^2M, 2n^2M+M)$).*

The NP-hardness result now follows from the following discussion.

Lemma 5. *The total imbalance cost of $S(\sigma_B, \sigma_C)$ in the nodes $I_< \cup \{t_i \mid i \in [n]\}$ is equal to 0 if and only if $S(\sigma_B, \sigma_C)$ encodes a yes-certificate.*

Proof. First of all, the imbalance at the internal node t_i is equal to 0 if and only if

$$T = a_i + b_i + c_i + d_i + e_i$$

or equivalently,

$$T^1 = a_i^1 + b_i^1 + c_i^1$$

for every $i \in [n]$. That is, every 5-tuple (resp., every triplet) needs to sum up to T (resp., T^1), the target value. To complete the proof, we just need to observe that nodes at depth $h-1$ have as children two test nodes. Thus, their imbalance is 0 if and only if the two test nodes have exactly the same weight (equivalently, the triplets associated to them sum to the same value). Going up the (full binary) tree, it is easy to see that we need all the test nodes to have exactly the same weight, i.e., all the 5-tuples to sum up to the same value. □

Lemma 6. *The total imbalance cost of $S(\sigma_B, \sigma_C)$ is greater or equal to $\sum_{i=1}^n (a_i^1 - c_i^1)$ and equality holds if and only if $S(\sigma_B, \sigma_C)$ encodes a yes-certificate.*

Proof. We have already seen in Lemma 5 that $\sum_{v \in I_< \cup \{t_i\}} imb(v) = 0$ if and only if $S(\sigma_B, \sigma_C)$ encodes a yes-certificate. We will now prove that $\sum_{v \in I_>} imb(v) = \sum_{i=1}^n (a_i^1 - c_i^1)$ for canonical assignments, from where the result follows. First, for any canonical assignment, $imb(r_i^4) = n^2M - n^2M = 0$ for every $i = 1, \ldots, n$. We will next see that, for every canonical assignment, $\sum_{v \in \{r_i^1, r_i^2, r_i^3 : i=1,\ldots,n\}} imb(v) = \sum_{i=1}^n (a_i^1 - c_i^1)$.

First of all,

$$\sum_{i=1}^n imb(r_i^3) = \sum_{i=1}^n |c_i - w(r_i^4)| = \sum_{i=1}^n |(c_i^1 + 2n^2M) - 2n^2M| = \sum_{i=1}^n c_i^1$$

Similarly,

$$\sum_{i=1}^n imb(r_i^2) = \sum_{i=1}^n |b_i - w(r_i^3)| = \sum_{i=1}^n |b_i^1 - c_i^1| = \sum_{i=1}^n (b_i^1 - c_i^1)$$

where the last equality follows from the property that $b_i > c_i$. Finally,

$$\sum_{i=1}^n imb(r_i^1) = \sum_{i=1}^n |a_i - w(r_i^2)| = \sum_{i=1}^n |a_i^1 - (b_i^1 + c_i^1)| = \sum_{i=1}^n (a_i^1 - b_i^1 - c_i^1)$$

where again we use the property that $a_i^1 > b_i^1 + c_i^1$. Thus, summing all the costs below every test node, we get that the total cost is

$$\sum_{i=1}^n ((a_i^1 - b_i^1 - c_i^1) + (b_i^1 - c_i^1) + c_i^1) = \sum_{i=1}^n (a_i^1 - c_i^1)$$

□

Before continuing, note that $\sum_{i=1}^{n}(a_i^1 - c_i^1) \ll n^2M$. Indeed,

$$\sum_{i=1}^{n}(a_i^1 - c_i^1) \leq \sum_{i=1}^{n}(M^0 + 8n^2M^0 - 2n^2M^0) \leq nM$$

The last inequality follows since $M = T^0 + 14n^2M^0$, which is clearly greater than the term in the sum.

Lemma 7. *Any assignment f of cost less than n^2M is canonical.*

Proof. Let f be any assignment of cost less than n^2M. Notice that the constructed tree has precisely n internal nodes whose two children are both leaves (those labeled r_i^4). At the same time, we have $2n$ weights of value n^2M, whereas all other weights exceed $2n^2M$. Therefore, all copies of weight n^2M should be assigned to the leaves that are children of some r_i^4, that is, to the $2n$ nodes of largest depth $h+5$. Indeed, if at least one of the copies of weight n^2M were assigned to a node which is not at largest depth, then the imbalance at its parent node would be $|n^2M - w_r|$, with w_r being the weight of the right child, which is always greater or equal than $2n^2M$. Thus, the imbalance would be already at least n^2M.

After this, we can go up the tree. The n nodes of weight in the interval $[2n^2M, 2n^2M + M]$ are mapped to nodes that are left children of r_i^3 nodes (all leaf nodes at depth $h+4$). Otherwise, the weight of that node would be at least $4n^2M$, yielding an imbalance of at least $2n^2M$. Similarly, the n nodes of weight in the interval $[4n^2M, 4n^2M + M]$ are mapped to nodes that are left children of r_i^2 nodes (all leaf nodes at depth $h+3$), and the n nodes of weight in the interval $[8n^2M, 8n^2M + M]$ are mapped to nodes that are left children of r_i^2 nodes (all leaf nodes at depth $h+2$). Finally, the n nodes of weight T are mapped to nodes that are left children of test nodes t_i (all leaf nodes at depth $h+1$).

This shows that if we want an assignment of cost less than n^2M, then every weight, while it can be assigned to the leaves of a subtree rooted at any of the test nodes t_i, it has to be assigned to a leaf node of the right depth/category. But then f is a canonical assignment. □

Theorem 8. BALANCED MOBILES *is strongly* NP*-hard.*

Proof. We have described a log-space reduction that, given a generic instance I of $ABCDE$-PARTITION yields an instance $(\mathcal{T}, W)$ of BALANCED MOBILES and a lower bound $L := \sum_{i=1}^{n}(a_i^1 - c_i^1)$ such that:

1. Every possible solution to $(\mathcal{T}, W)$ has total imbalance cost at least L.
2. $(\mathcal{T}, W)$ admits a solution of cost L iff I is a yes-instance of the $ABCDE$-PARTITION problem.

This already shows that the BALANCED MOBILES optimization problem is NP-hard. The BALANCED MOBILES problem is strongly NP-hard because the $ABCDE$-PARTITION problem is strongly NP-complete, and when the reduction is applied on top of any shallow class of instances of $ABCDE$-PARTITION, it yields a shallow class of instances of BALANCED MOBILES. □

Note that this implies that the decision version of the problem is (strongly) NP-complete. Indeed, one can check that the problem is in NP because given a potential solution, it can be verified in polynomial time whether it is valid or not.

4 Conclusion

We have shown that Balanced Mobiles is strongly NP-hard when the full binary tree is given as input. However, note that the complexity when the tree is not given remains open. Indeed, our reduction cannot be directly extended to this case since then, there is no structure to ensure that weights of set A are grouped with weights of sets B and C. On the other hand, the complexity when the weights are constant is also unknown, as in our proof, the constructed weights depend on n. Finally, with respect to the parameterized complexity, as we mentioned before, it is only known that the problem is in the parameterized class XP, parameterized by the optimal imbalance [9], so other future work includes to study whether there exists a fixed parameter algorithm or not.

Acknowledgements. Part of this work was conducted when RR was an invited professor at Université Paris-Dauphine. This work was partially supported by the ANR project ANR-21-CE48-0022 ("S-EX-AP-PE-AL").

References

1. Bartoszek, K., Coronado, T.M., Mir, A., Rosselló, F.: Squaring within the Colless index yields a better balance index. Math. Biosci. **331**, 108503 (2021)
2. Blum, M.G., François, O., Janson, S.: The mean, variance and limiting distribution of two statistics sensitive to phylogenetic tree balance. Ann. Appl. Probab. **16**(4), 2195–2214 (2006)
3. Colless, D.H.: Phylogenetics: The Theory and Practice of Phylogenetic Systematics (1982)
4. Cormen, T.H., Leiserson, C.E., Rivest, R.L., Stein, C.: Introduction to Algorithms, 3rd edn. MIT Press, Cambridge (2009)
5. Coronado, T.M., Fischer, M., Herbst, L., Rosselló, F., Wicke, K.: On the minimum value of the Colless index and the bifurcating trees that achieve it. J. Math. Biol. **80**(7), 1993–2054 (2020)
6. Fischer, M., Herbst, L., Kersting, S., Khn, L., Wicke, K.: Tree balance indices: a comprehensive survey (2021)
7. Garey, M.R., Johnson, D.S.: Computers and Intractability: A Guide to the Theory of NP-Completeness. W.H. Freeman (1979)
8. Hamoudi, Y., Laplante, S., Mantaci, R.: Balanced mobiles (2016). https://www.iwoca.org. Problems Section
9. Hamoudi, Y., Laplante, S., Mantaci, R.: Balanced mobiles with applications to phylogenetic trees and Huffman-like problems. Technical report (2017). https://hal.science/hal-04047256
10. Huffman, D.A.: A method for the construction of minimum-redundancy codes. Proc. IRE **40**(9), 1098–1101 (1952)

11. Mir, A., Rotger, L., Rosselló, F.: Sound Colless-like balance indices for multifurcating trees. PLoS ONE **13**(9), e0203401 (2018)
12. Sackin, M.J.: "Good" and "bad" phenograms. Syst. Biol. **21**(2), 225–226 (1972)
13. Shao, K.-T., Sokal, R.R.: Tree balance. Syst. Biol. **39**(3), 266–276 (1990)

Burn and Win

Pradeesha Ashok[1], Sayani Das[2], Lawqueen Kanesh[3], Saket Saurabh[2,4], Avi Tomar[1](✉), and Shaily Verma[2]

[1] International Institute of Information Technology Bangalore, Bengaluru, India
{pradeesha,Avi.Tomar}@iiitb.ac.in

[2] The Institute of Mathematical Sciences, Chennai, India
{sayanidas,saket,shailyverma}@imsc.res.in

[3] Indian Institute of Technology Jodhpur, Jodhpur, India
lawqueen@iitj.ac.in

[4] University of Bergen, Bergen, Norway

Abstract. Given a graph G and an integer k, the Graph Burning problem asks whether the graph G can be burned in at most k rounds. Graph burning is a model for information spreading in a network, where we study how fast the information spreads in the network through its vertices. In each round, the fire is started at an unburned vertex, and fire spreads from every burned vertex to all its neighbors in the subsequent round burning all of them and so on. The minimum number of rounds required to burn the whole graph G is called the *burning number* of G. Graph Burning is NP-hard even for the union of disjoint paths. Moreover, Graph Burning is known to be W[1]-hard when parameterized by the burning number and para-NP-hard when parameterized by treewidth. In this paper, we prove the following results:

- In this paper, we give an explicit algorithm for the problem parameterized by treewidth, τ and k, that runs in time $k^{2\tau}4^k5^\tau n^{O(1)}$. This also gives an FPT algorithm for Graph Burning parameterized by burning number for apex-minor-free graphs.
- Y. Kobayashi and Y. Otachi [Algorithmica 2022] proved that the problem is FPT parameterized by distance to cographs and gave a double exponential time FPT algorithm parameterized by distance to split graphs. We improve these results partially and give an FPT algorithm for the problem parameterized by distance to cographs $\cap$ split graphs (threshold graphs) that runs in $2^{\mathcal{O}(t \ln t)}$ time.
- We design a kernel of exponential size for Graph Burning in trees.
- Furthermore, we give an exact algorithm to find the burning number of a graph that runs in time $2^n n^{\mathcal{O}(1)}$, where n is the number of vertices in the input graph.

Keywords: Burning number · fixed-parameter tractability · treewidth · threshold graphs · exact algorithm

1 Introduction

Given a simple undirected graph $G = (V, E)$, the graph burning problem is defined as follows. Initially, at round $t = 0$, all the nodes are unburned. At each round $t \geq 1$,

S.-Y. Hsieh et al. (Eds.): IWOCA 2023, LNCS 13889, pp. 36–48, 2023.
https://doi.org/10.1007/978-3-031-34347-6_4

one new unburned vertex is chosen to burn, if such a node exists, and is called a fire source. When a node is burned, it remains burned until the end of the process. Once a node is burned in round t, its unburned neighbors become burned in round $t+1$. The process ends when there are no unburned vertices in the graph. The *burning number* of a graph G is the minimum number of rounds needed to burn the whole graph G, denoted by $b(G)$. The sources chosen in each round form a sequence of vertices called a *burning sequence* of the graph. Let $\{b_1, b_2, \cdots, b_k\}$ be a burning sequence of graph G. For $v \in V$, $N_k[v]$ denotes the set of all vertices within distance k from v, including v. Then, $\bigcup_{1 \leq i \leq k} N_{k-i}[b_i] = V$.

Given a graph G and an integer k, the GRAPH BURNING problem asks if $b(G) \leq k$? This problem was first introduced by Bonato, Janssen, and Roshanbin [3,4,13]. For any graph G with radius r and diameter d, $\lceil (d+1)^{1/2} \rceil \leq b(G) \leq r+1$. Both bounds are tight, and paths achieve the lower bound.

The GRAPH BURNING is not only NP-complete on general graphs but for many restricted graph classes. It has been shown that GRAPH BURNING is NP-complete when restricted to trees of maximum degree 3, spider and path-forests [1]. It was also shown that this problem is NP-complete for caterpillars of maximum degree 3 [8,12]. In [7], authors have shown that GRAPH BURNING is NP-complete when restricted to interval graphs, permutation graphs, or disk graphs. Moreover, the GRAPH BURNING problem is known to be polynomial time solvable on cographs and split graphs [10].

The burning number has also been studied in directed graphs. Computing the burning number of a directed tree is NP-hard. Furthermore, the GRAPH BURNING problem is W[2]-complete for directed acyclic graphs [9]. For further information about GRAPH BURNING, the survey by Bonato [2] can be referred to.

The parameterized complexity of GRAPH BURNING was first studied by Kare and Reddy [10]. They showed that GRAPH BURNING on connected graphs is fixed-parameter tractable parameterized by distance to cluster graphs and by neighborhood diversity. In [11], the authors showed that GRAPH BURNING is fixed-parameter tractable when parameterized by the clique-width and the maximum diameter among all connected components, which also implies that GRAPH BURNING is fixed-parameter tractable parameterized by modular-width, by tree-depth, and by distance to cographs. They also showed that this problem is fixed-parameter tractable parameterized by distance to split graphs. It has also been shown that GRAPH BURNING parameterized by solution size, k, is W[2]-complete. The authors also showed that GRAPH BURNING parameterized by vertex cover number does not admit a polynomial kernel unless NP $\subseteq$ coNP/poly.

Our Results: In Sect. 2 we add all the necessary definitions. In Sect. 3, we use nice tree decomposition of G to give an FPT algorithm for GRAPH BURNING parameterized by treewidth and solution size. This result also implies that GRAPH BURNING parameterized by burning number is FPT on apex-minor-free graphs. In Sect. 4, we show that GRAPH BURNING is fixed-parameter tractable when parameterized by distance to cographs $\cap$ split graphs, also known as *threshold graphs*, which partially improve the results given in [11]. In Sect. 5, we design an exponential kernel for Graph Burning in trees. In Sect. 6, we give a non trivial exact algorithm for finding the burning number in general graphs.

2 Preliminaries

We consider the graph, $G = (V, E)$, to be simple, finite, and undirected throughout this paper. $G[V \backslash X]$ represents the subgraph of G induced by $V \backslash X$. $N_G(v)$ represents the set of neighbors of the vertex v in graph G. We simply use $N(v)$ if there is no ambiguity about the corresponding graph. $N_G[S] = \{u : u \in N_G(v), \forall v \in S\}$. For $v \in V$, $N_k[v]$ denotes the set of all vertices within distance k from v, including v itself. $N_1[v] = N[v]$, the closed neighborhood of v. For any pair of vertices $u, v \in V$, $dist_G(v, u)$ represents the length of the shortest path between vertices u and v in G. The set $\{1, 2, \cdots, n\}$ is denoted by $[n]$. For definitions related to parameterized complexity, refer the book by Cygan et al. [5].

A graph G is an *apex graph* if G can be made planar by removing a vertex. For a fixed apex graph H, a class of graphs $\mathcal{S}$ is *apex-minor-free* if every graph in $\mathcal{S}$ does not contain H as a minor. A *threshold graph* can be built from a single vertex by repeatedly performing the following operations.

(i) Add an isolated vertex.
(ii) Add a dominating vertex, i.e., add a vertex that is adjacent to every other vertex.

Thus for a threshold graph G, there exists an ordering of $V(G)$ such that any vertex is either adjacent to every vertex that appears before that in the ordering or is adjacent to none of them.

3 Parameterized by Treewidth and Burning Number

We prove the following result in this section.

Theorem 1. Graph Burning *is FPT for apex-minor-free graphs when parameterized by burning number.*

To prove this result, we first give an FPT algorithm for Graph Burning parameterized by treewidth+burning number. For definitions and notations related to treewidth, refer the book by Cygan et al. [5].

Theorem 2. Graph Burning *admits an FPT algorithm that runs in* $k^{2\tau}4^k5^\tau n^{\mathcal{O}(1)}$ *time, when parameterized by the combined parameter, treewidth* τ *and burning number* k.

Proof. We use dynamic programming over a nice tree decomposition T of G. We shall assume a nice tree decomposition of G of width τ is given.

We use the following notation as in [5]. For every node t in the nice tree decomposition, we define G_t to be a subgraph of G where $G_t = (V_t, E_t = \{e : e \text{ is introduced in the subtree of } t\})$. V_t is the union of all vertices introduced in the bags of the subtree rooted at t. For a function $f : X \to Y$ and $\alpha \in Y$, we define a new function $f_{v\to\alpha} : X \cup \{v\} \to Y$ as follows:

$$f_{v\to\alpha}(x) = \begin{cases} f(x), & \text{when } x \neq v \\ \alpha, & \text{when } x = v \end{cases}$$

We define subproblems on every node $t \in V(T)$. We consider the partitioning of the bag X_t by a mapping $\Psi : X_t \to \{B, R, W\}$, where B, R, and W respectively represent assigning black, grey, and white colors to vertices. Intuitively, a black vertex represents a fire source, a grey vertex is not yet burned, and a white vertex is burned by another fire source through a path that is contained in G_t. Each vertex is further assigned two integer values by two functions $FS : X_t \to [k]$ and $D : X_t \to [k-1] \cup \{0\}$. For a vertex $v \in X_t$, $FS(v)$ intuitively stores the index of the fire source that will burn v, and $D(v)$ stores the distance between the fire source and v.

Let S be the set $\{*, \uparrow, \downarrow\}$. For every bag, we consider an array $\gamma \in S^k$. Here the entries in γ represent the location of each fire source with respect to the bag X_t. More precisely, for $1 \leq i \leq k$, $\gamma[i]$ is $*$ when the i-th fire source is in X_t, is $\downarrow$ when the i-th fire source is in $V_t \setminus X_t$ and $\uparrow$ otherwise.

For a tuple $f[t, \gamma, FS, D, \Psi]$, we define that a burning sequence of G *realizes* the tuple if the fire sources in the burning sequence match γ i.e., for $1 \leq i \leq k$, the i-th fire source is in X_t, $V_t \setminus X_t$ and $V \setminus V_t$ if $\gamma[i]$ is $*, \downarrow$ and $\uparrow$ respectively and the following conditions are met.

1. A black vertex $v \in X_t$ is part of the burning sequence at index $FS(v)$.
2. A white vertex $v \in X_t$ is burned by a fire source with index $FS(v)$ by a path of length $D(v)$ that lies entirely in G_t.
3. A grey vertex $v \in X_t$ is not burned in G_t.
4. For a vertex v in $V_t \setminus X_t$, v is either burned or there exists a path from v to a grey vertex $u \in X_t$ such that $FS(u) + D(u) + dist_G(u, v) \leq k$.

We now define a sub-problem $f[t, \gamma, FS, D, \Psi]$ that returns *True* if and only if there exists a burning sequence that realizes $[t, \gamma, FS, D, \Psi]$. From the above definition, it is easy to see that G admits a burning sequence of length k if and only if $f[t, \gamma, FS, D, \Psi]$, where t is the root node (and therefore empty), γ contains all entries as $\downarrow$ and FS, D, Ψ are null, returns *True*.

A tuple (t, FS, D, γ, Ψ) is *valid* if the following conditions hold for every vertex $v \in X_t$

- $\Psi(v) = B$ if and only if $D(v) = 0$ and $\gamma[i] = *$, where $i = FS(v)$.
- $\Psi(v) = W$, if and only if $D(v) > 0$ and $\gamma[i] = *$ or $\gamma[i] = \downarrow$, where $i = FS(v)$.
- For all vertices v, $D(v) \leq k - FS(v)$.
- For all $1 \leq i \leq k$ such that $\gamma[i] = *$, there exists exactly one vertex v in X_t such that $FS(v) = i$ and $D(v) = 0$.

For an invalid tuple, $f[.]$, by default, returns *False*. We now define the values of $f[.]$ for different types of nodes in T.

Leaf Node: In this case, $X_t = \emptyset$. So $f[t, \gamma, FS, D, \Psi]$ returns $True$ if $\gamma[i] = \uparrow$, for all $1 \leq i \leq k$ and FS, D, Ψ are null functions. Otherwise, this returns $False$.

Introduce Vertex Node: Let v be the vertex being introduced and t' be the only child node of t such that $v \notin X_{t'}$ and $X_t = X_{t'} \cup \{v\}$.

$$f[t, \gamma, FS, D, \Psi] = \begin{cases} f[t', \gamma', FS_{|X_{t'}}, D_{|X_{t'}}, \Psi_{|X_{t'}}], & \text{if } \Psi(v) = B \\ f[t', \gamma, FS_{|X_{t'}}, D_{|X_{t'}}, \Psi_{|X_{t'}}], & \text{if } \Psi(v) = R \\ False, & \text{if } \Psi(v) = W \end{cases}$$

In the recurrence, γ' is the same as γ except that $\gamma'[FS(v)] = \uparrow$. The correctness of the recurrence follows from the fact that v is an isolated vertex in G_t and v is not present in $G_{t'}$.

Forget Vertex Node: Let t' be the only child node of t such that $\exists v \notin X_t$ and $X_{t'} = X_t \cup \{v\}$.

We now give the recurrence as follows.

$$f[t, \gamma, FS, D, \Psi] = \begin{cases} \bigvee_{i:\ \gamma[i]=\downarrow} f[t', \gamma_{\gamma[i]\to *}, FS_{v\to i}, D_{v\to 0}, \Psi_{v\to B}] \\ \bigvee_{\substack{i:\ \gamma[i]\neq\uparrow, \\ 1\leq j\leq k-i}} f[t', \gamma, FS_{v\to i}, D_{v\to j}, \Psi_{v\to W}] \\ \bigvee_{\substack{1\leq j\leq k \\ \exists w\in X_t \text{ such that} \\ j\leq k-FS(w), D(w)=j-dist_G(v,w) \\ \text{and } \Psi(w)=R}} f[t', \gamma, FS_{v\to FS(w)}, D_{v\to j}, \Psi_{v\to R}] \end{cases}$$

In the last case, we consider the case where v is burned by a path P that lies outside G_t, at least partially. The feasibility of P is tracked by a vertex $w \in X_t$ that is closer to the fire source in P[1].

Lemma 1. *[*] The recurrence for Forget Vertex Node is correct.*

Introduce Edge Node: Let t be an introduce edge node with child node t' and let (u, v) be the edge introduced at t. We compute the value of f based on the following cases.

1. If $\Psi(u) = \Psi(v) = R$, set $f[t, \gamma, FS, D, \Psi] = f[t', \gamma, FS, D, \Psi]$
2. If $FS(u) + D(u) = FS(v) + D(v)$, set $f[t, \gamma, FS, D, \Psi] = f[t', \gamma, FS, D, \Psi]$
3. If $FS(u) + D(u) + 1 = FS(v) + D(v)$ and $FS(u) \neq FS(v)$, set $f[t, \gamma, FS, D, \Psi] = f[t', \gamma, FS, D, \Psi]$
4. If $\Psi(v) \in \{B, W\}, \Psi(u) = W, FS(u) = FS(v)$ and $D(u) = D(v) + 1$, then set $f[t, \gamma, FS, D, \Psi] = f[t', \gamma, FS, D, \Psi_{u\to R}] \bigvee f[t', \gamma, FS, D, \Psi]$
5. For all other cases, set $f[t, \gamma, FS, D, \Psi] = False$

Lemma 2. *[*] The recurrence for Introduce Edge Node is correct.*

[1] Proofs of results that are marked with [*] are omitted due to the space constraint.

Join Node: Let t be a join node and t_1, t_2 be the child nodes of t such that $X_t = X_{t_1} = X_{t_2}$. We call tuples $[t_1, \gamma_1, FS, D, \Psi_1]$ and $[t_2, \gamma_2, FS, D, \Psi_2]$ as $[t, \gamma, FS, D, \Psi]$-consistent if the following conditions hold.
For all values of $i, 1 \leq i \leq k$,

if $\gamma[i] = *$ then $\gamma_1[i] = \gamma_2[i] = *$
if $\gamma[i] = \uparrow$ then $\gamma_1[i] = \gamma_2[i] = \uparrow$
if $\gamma[i] = \downarrow$ then either $\gamma_1[i] = \downarrow$, $\gamma_2[i] = \uparrow$ or $\gamma_1[i] = \uparrow$, $\gamma_2[i] = \downarrow$

For all $v \in X_t$,

if $\Psi(v) = B$ then $\Psi_1(v) = \Psi_2(v) = B$
if $\Psi(v) = R$ then $\Psi_1(v) = \Psi_2(v) = R$
if $\Psi(v) = W$ then $(\Psi_1(v), \Psi_2(v)) \in \{(W, W), (W, R), (R, W)\}$

We give a short intuition for the above. The cases where $\gamma[i] = *$ and $\gamma[i] = \uparrow$ are easy to see. When $\gamma[i] = \downarrow$, the i-th fire source is below the bag. By the property of the tree decomposition, $V_{t_1} \setminus X_t$ and $V_{t_2} \setminus X_t$ are disjoint. Therefore, exactly one of $\gamma_1[i]$ and $\gamma_2[i]$ is set to $\downarrow$. Similarly, $\Psi(v) = B$ and $\Psi(v) = R$ are easy to see. When $\Psi(v) = W$, the vertex is already burned below. Here again, there are two possibilities: v is burned in exactly one of G_{t_1} and G_{t_2} and v is burned in both of them (possibly by different paths). Therefore, $(\Psi_1(v), \Psi_2(v)) \in \{(W, W), (W, R), (R, W)\}$.

Then, the recurrence is as follows, where the OR operations are done over all pairs of tuples, which are $[t, \gamma, FS, D, \Psi]$-consistent.

$$f[t, \gamma, FS, D, \Psi] = \bigvee (f[t_1, \gamma_1, FS, D, \Psi_1] \wedge f[t_2, \gamma_2, FS, D, \Psi_2])$$

Lemma 3. *[*] The recurrence for Join node is correct.*

Running Time: Note that we can compute each entry for $f[\cdot]$ in time $k^{2\tau} 3^k 3^\tau n^{\mathcal{O}(1)}$, except for join nodes. For join nodes, we require extra time as we are computing over all possible consistent tuples. Let $(\gamma, \gamma_1, \gamma_2)$ and (Ψ, Ψ_1, Ψ_2) be such that $(t_1, \gamma_1, FS, D, \Psi_1)$ and $(t_2, \gamma_2, FS, D, \Psi_2)$ are (t, γ, FS, D, Ψ) consistent then, $\forall i \in [k]$, $(\gamma[i], \gamma_1[i], \gamma_2[i]) \in \{(*, *, *), (\uparrow, \uparrow, \uparrow), (\downarrow, \downarrow, \uparrow), (\downarrow, \uparrow, \downarrow)\}$ and $\forall v \in X_t$, $(\Psi[v], \Psi_1[v], \Psi_2[v]) \in \{(B, B, B), (R, R, R), (W, W, W), (W, W, R), (W, R, W)\}$. Therefore, the total number of consistent tuples over all join nodes is upper bounded by $k^{2\tau} 4^k 5^\tau n^{\mathcal{O}(1)}$. Hence the running time of the algorithm can be bounded by $k^{2\tau} 4^k 5^\tau n^{\mathcal{O}(1)}$. □

For apex-minor-free graphs, the treewidth is bounded by the diameter of the graph, as shown in [6]. It has been established that the diameter of a graph is bounded by a function of the burning number of the graph [4]. As a result, the treewidth of apex-minor-free graphs is bounded by a function of the burning number. This observation, along with Theorem 2, proves Theorem 1.

4 Parameterized by Distance to Threshold Graphs

In this section, we give an FPT algorithm for GRAPH BURNING parameterized by the distance to threshold graphs. Recall that the problem is known to be in FPT; the paper [11] shows that GRAPH BURNING is FPT parameterized by distance to cographs and gives a double-exponential time FPT algorithm when parameterized by distance to split graphs. Since both these parameters are smaller than the distance to threshold graphs that are precisely the graphs in graph-class cographs $\cap$ split, these results imply fixed-parameter tractability when the distance to threshold graphs is a parameter. Here, we give an FPT algorithm that runs in single-exponential time, which improves the previously known algorithms. We will consider a connected graph $G = (V, E)$ and a subset $X \subseteq V$ with $|X| = t$, such that the induced subgraph $G[V \setminus X]$ is a threshold graph. It is assumed that the set X is given as part of the input.

Theorem 3. GRAPH BURNING *on G can be solved in time $t^{2t}n^{\mathcal{O}(1)}$.*

Proof. Since $G[V \setminus X]$ is a threshold graph, there exists an ordering Π of vertices such that every vertex is either a dominating vertex or an isolated vertex for the vertices preceding it in Π. Let $v_d \in V \setminus X$ be the last dominating vertex in Π and (D, I) be the partition of $V \setminus X$ such that D is a set that contains the vertices in Π till the vertex v_d and I is the set containing all remaining vertices. Thus, in $G[V \setminus X]$, I is a maximal set of isolated vertices.

We observe that G can be burned in at most $t + 3$ steps. In at most t steps, we set each vertex in X as a fire source. Since every vertex in I has at least one neighbor in X, all vertices in I are burned in at most $t + 1$ steps. Similarly, since at least one vertex from D has a neighbor in X and D induces a graph of diameter 2, every vertex in D is burned in at most $t + 3$ steps. Therefore, we assume $k < t + 3$ for the rest of the proof.

For a valid burning sequence of length k of the graph G, for every vertex $v \in V$, let $(fs(v), d(v))$ be a pair where $1 \leq fs(v) \leq k$ and $0 \leq d(v) \leq k - fs(v)$, such that $fs(v)$ is the index of the fire source that burns the vertex v and $d(v)$ is the distance between that fire source and v. It also implies that v is going to burn at the $(fs(v) + d(v))$-th round. When two fire sources can simultaneously burn v in the same round, $fs(v)$ is assigned the index of the earlier fire source. The basic idea of the algorithm is to guess the pair $(fs(v), d(v))$ for every $v \in X$ and then extend this guess into a valid burning sequence for G in polynomial time.

Consider two families of functions $\mathcal{F} = \{fs : X \to [k]\}$ and $\mathcal{D} = \{d : X \to \{0\} \cup [k-1]\}$ on X. A pair $(fs, d) \in \mathcal{F} \times \mathcal{D}$ corresponds to a *guess* that decides how a vertex in X is burnt. We further extend this to an *augmented guess* by guessing the fire sources in D. We make the following observation based on the fact that every pair of vertices in D is at a distance at most two since v_d is adjacent to every vertex in D.

Observation 1. *There can be at most two fire sources in D. Moreover, if there are two fire sources in D, they are consecutive in the burning sequence.*

Thus, an augmented guess can be considered as extending the domain of the functions fs and d to $X \cup D'$ where $D' \subseteq D$ and $0 \leq |D| \leq 2$. Note that, for all $v \in D'$, $d(v) = 0$ since we are only guessing the fire sources in D.
An augmented guess is considered *valid* if the following conditions are true.

1. For all v in the domain of the functions, $0 \leq d(v) \leq k - fs(v)$.
2. For all $1 \leq i \leq k$, there exists at most one vertex v such that $fs(v) = i$ and $d(v) = 0$.
3. For all u, v in the domain of the functions, $|(fs(u)+d(u)) - (fs(v)+d(v))| \leq dist_G(u, v)$.

Algorithm 1 gives the procedure to extend a valid augmented guess to a valid burning sequence by identifying the firesources in I. Specifically, Algorithm 1 takes a valid augmented guess as input and returns YES if it can be extended to a burning sequence of length k.

Algorithm 1.

1: **for** $1 \leq i < k$ such that $\nexists v$ such that $fs(v) = i$ and $d(v) = 0$ **do**
2: $X_i = \{v \in X : fs(v) = i, d(v) = 1\}$
3: **if** X_i is not empty **then**
4: $I_i = \{u \in I : X_i \subseteq N(u)\}$; $F_i = I_i$
5: **for** $u \in I_i$ **do**
6: **if** there exists $w \in N(u) \setminus X_i$ such that $(fs(w)+d(w) = i+2) \vee (fs(w) + d(w) = i-1) \vee (fs(w) = i+1 \wedge d(w) = 0)$ **then**
7: Delete u from F_i.
8: **if** $(i \neq k-1)$ **then**
9: **if** F_i is not empty **then**
10: Let v be an arbitrary vertex in F_i. Set v as the i-th fire source.
11: **else** return NO.
12: **else**
13: $F' = \{u \in F_i : \forall w \in N(u) \setminus X_i, fs(w) + d(w) = k\}$
14: **if** $|F'| > 2$ **then** return NO.
15: **else if** $F' \neq \emptyset$ **then**
16: set an arbitrary vertex in F' as the i-th fire source.
17: **else**
18: set an arbitrary vertex in F_i as the i-th fire source.
19: **if** CheckValidity() = True **then**
20: Return YES.
21: **else** return NO.

Lemma 4. *Algorithm 1 is correct.*

Proof. It is enough to prove that the fire sources in I are correctly identified.

For every $1 \leq i < k$ such that the i-th fire source is not "discovered" yet, we consider the vertices in X_i. $I_i \subseteq I$ is the set of vertices adjacent to every vertex in X_i. The i-th fire source, if exists, should be one of the vertices from I_i. The algorithm further considers a set F_i that is obtained from I_i by filtering out vertices that cannot be the i-th fire source. Specifically, the set F_i contains vertices v such that, $X_i \subseteq N(v)$ and for all $w \in N(v) \setminus X_i$, $i \leq fs(w) + d(w) \leq i+1$. We shall show that a vertex outside this set cannot be the i-th fire source.

Let $v \in I_i$ and $w \in N(v) \setminus X_i$. For all $u \in X_i$, $fs(u) + d(u) = i+1$ and $dist_G(u, w) \leq 2$, since they have a common neighbor v. By the constraint given in the definition of a valid augmented guess, $i - 1 \leq fs(w) + d(w) \leq i + 3$. If $fs(w) + d(w) \geq i+2$, then $fs(v) + d(v) \geq i+1$ and v is not the i-th fire source. Also, if $fs(w) + d(w) = i - 1$, then $fs(w) < i$ and v is not the i-th fire source since an earlier fire source can burn v in the i-th round. Further, if v is the i-th fire source, then the case where $fs(w) = i + 1$ and $d(w) = 0$ is not possible since w will be burned by v in the $(i+1)$-th round. Note that, by definition, if a vertex v can be burned simultaneously by two different fire sources, then $fs(v)$ is assigned the index of the earlier fire source. Thus, the i-th fire source, if exists, should belong to the set F_i.

Assume $i < k-1$. Let v_1 and v_2 be arbitrary vertices in the set F_i and let v_1 be the i-th fire source in a burning sequence γ of G. Now, we will prove that a sequence γ' obtained by replacing v_1 with v_2 as the i-th fire source in γ is also a valid burning sequence. Note that, in γ, X_i is exactly the set of vertices that are burned by v_1 in the $(i+1)$-th round since any other neighbor of v_1 is burned in the i-th or $(i+1)$-th round by a different fire source. Now, since $X_i \subseteq N(v_2)$, X_i is burned in the $(i+1)$-th round by γ' also. Also, any other neighbor of v_1 or v_2 is burned in the i-th or $(i+1)$-th round by a fire source that is not the i-th fire source, which also ensures v_1 gets burned before the k-th round. Hence, if there exists a fire source in I_i, then any arbitrary vertex in I_i can be the fire source.

Assume $i = k - 1$. Now we consider the subset $F' = \{u \in F_i : \forall w \in N(u) \setminus X_i, fs(w) + d(w) = k\}$ of F_i. A vertex in F' can be burned only if it is the k-th or $(k-1)$-th fire source. Hence, we return No if $|F'| > 2$. Otherwise an arbitrary vertex is set as the $(k-1)$-th fire source.

Finally, once the fire sources are set, we can check the validity of the burning sequence in polynomial time. □

Extending a valid augmented guess can be done in polynomial time. Thus the running time of the algorithm is determined by the number of valid augmented guesses which is bounded by $t^{2t}n^{O(1)}$. □

5 A Kernel for Trees

In this section, we design a kernel for the Graph Burning problem on trees. Let (T, k) be the input instance of Graph Burning where T is a tree. First, we arbitrarily choose a vertex $r \in V(T)$ to be the *root* and make a rooted tree. Let $L_0, L_1, L_2, \ldots, L_p$ be the levels of tree T rooted at r, where $L_0 = \{r\}$. To give a kernel, we give a marking procedure to mark some vertices in each level and show that we can remove the unmarked vertices. In bottom-up fashion, starting from the last level, in each iteration, we mark at most $k+1$ children for each vertex in the level, and we remove the subtree rooted at any unmarked children. Observe that while doing that, we maintain the connectedness of the tree. We show that the removal of subtrees rooted on unmarked vertices does not affect the burning number of the tree. Let T_z be the subtree rooted at a vertex z and M_i be the set of marked vertices at level L_i.

Marking Procedure: For all $i \in [p]$, we initialise $M_i = \emptyset$. For a fixed i, $i \in [p]$, do as follows: For each vertex $x \in L_{p-i}$, mark at most $k+1$ children of x such that the depth of the subtrees rooted on marked children is highest and add them into M_{p-i+1}.

Reduction Rule 1. *If $z \in L_{p-i}$ such that $z \notin M_{p-i}$, then remove the subtree T_z.*

Lemma 5. *The Reduction Rule 1 is safe.*

Proof. To show the safeness of Reduction Rule 1, we show that (T, k) is a Yes-instance of Graph Burning if and only if $(T - T_z, k)$ is a Yes-instance.

For the forward direction, assume that (T, k) is a Yes-instance. Note that $T - T_z$ is a tree. Let $(b_1, b_2, \ldots, b_k)$ be a burning sequence for T. If any of the b_i belongs to T_z, then we replace b_i by placing a fire source on the first unburned ancestor of z in $T - T_z$. Therefore, we can have a burning sequence of size k. Hence, $(T - T_z, k)$ is a Yes-instance.

For the backward direction, assume that $(T - T_z, k)$ is a Yes-instance, and we need to show that (T, k) is a Yes-instance. Let $P(z)$ be the parent of vertex z in T. Suppose that $x_1, x_2, \ldots, x_{k+1}$ be the set of marked children (neighbors) of $P(z)$. We have to show that any vertex u in the subtree T_z can also be burned by the same burning sequence.

Since the burning number of $T - T_z$ is k, there is at least one marked child x_j of $P(z)$ such that there is no fire source placed in the subtree T_{x_j}. Observe that there exists a vertex u' in the subtree T_{x_j} such that the distances $d(u, P(z))$ and $d(u', P(z))$ are the same since the height of T_{x_j} is at least the height of subtree T_z by the marking procedure. Let the vertex u' get burned by a fire source s. Note that the fire source s is either placed on some ancestor of x_j or some subtree rooted at a sibling of x_j. In both cases, the s-u' path contains the vertex $P(z)$. Since $d(u, P(z)) = d(u', P(z))$, the vertex u also gets burned by the same fire source in the same round. Thus, every vertex in T_z can be burned by some fire source from the same burning sequence as $T \setminus T_z$. Hence, (T, k) is also a Yes-instance. □

Iteratively, for each fixed value of i, $i \in [p]$ (starting from $i = 1$), We apply the marking procedure once, and the Reduction Rule 1 exhaustively for each unmarked vertex. After the last iteration ($i = p$), we get a tree T'. Observe that we can complete the marking procedure in polynomial time, and the Reduction Rule 1 will be applied at most n times. Therefore, we can obtain the kernel T' in polynomial time.

Kernel Size: Note that the obtained tree T' is a $(k+1)$-ary tree. Let b_1 be the first fire source in T'; then we know that b_1 will burn vertices up to $(k-1)$ distance. Therefore, we count the maximum number of vertices b_1 can burn.

First, note that b_1 will burn the vertices in the subtree rooted at b_1 up to height $k-1$. Let n_0 be the number of vertices in the subtree rooted at b_1. It follows that $n_0 \leq \frac{(k+1)^{k-1}-1}{k}$, that is, $n_0 \leq (k+1)^{k-1}$. Note that b_1 also burns the vertices on the path between b_1 to root r up to distance $k-1$ and the vertices rooted on these vertices. Let $P = b_1 v_1 v_2 \ldots v_{k-1} \ldots r$ be b_1-r path in T'. Then b_1 also burns the vertices in the subtree rooted at v_i, say T_{v_i}, upto height $k-1-i$, where $i \in [k-1]$. Let $n_i = |V(T_{v_i})|$. Therefore, for any $i \in [k-1]$, $n_i \leq (k+1)^{k-1}$ as $n_i < n_0$. Thus, the total number of vertices b_1 can burn is at most $(k+1)^k$. Since each fire source b_i can burn only fewer vertices than the maximum number of vertices that can be burned by source b_1, the total number of vertices any burning sequence of size k can burn is at most $(k+1)^{k+1}$.

Therefore, if there are more than $(k+1)^{k+1}$ vertices in T', then we can conclude that (T, k) is a No-instance of Graph Burning problem. This gives us the following result.

Theorem 4. *In trees,* Graph Burning *admits a kernel of size* $(k+1)^{k+1}$.

6 Exact Algorithm

In this section, we design an exact algorithm for the Graph Burning problem. Here, we reduce the Graph Burning problem to the shortest path problem in a *configuration graph.*

Construction of a Configuration Graph: Given a graph $G = (V, E)$, we construct a directed graph $G' = (V', E')$ as follows:

(i) For each set $S \subseteq V(G)$, add a vertex $x_S \in V'$.
(ii) For each pair of vertices $x_S, x_{S'} \in V'$ such that there exists a vertex $w \notin S$ and $N_G[S] \cup \{w\} = S'$, add an arc from x_S to $x_{S'}$.

We call the graph G' as the configuration graph of G.

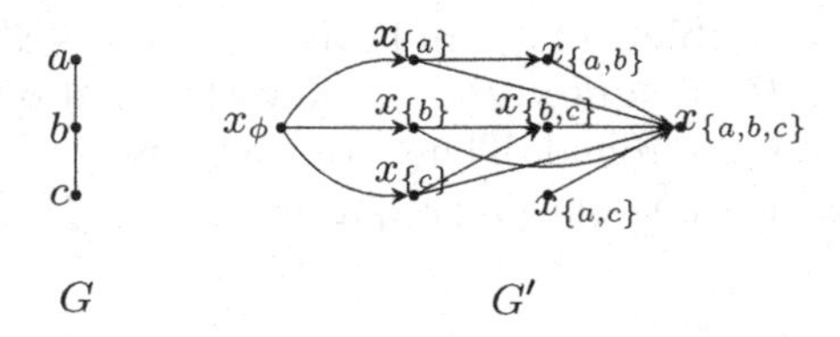

Fig. 1. An illustration of G', the configuration graph of G.

Figure 1 shows an example of a configuration graph.

We have constructed G' in such a way that a shortest path between the vertices x_S and x_T, where $S = \emptyset$ and $T = V(G)$, gives a burning sequence for the original graph G. The following result proves this fact.

Lemma 6. *[*] Let $G' = (V', E')$ be the configuration graph of a given graph G and $S = \emptyset$ and $T = V(G)$. There exists a path of length k between the vertices x_S and x_T in G' if and only if there is a burning sequence for G of size k.*

Lemma 6 shows that a shortest path between x_S and x_T in G' gives the burning sequence for graph G with minimum length. Thus, we can find a minimum size burning sequence in two steps:

(i) We construct a configuration graph G' from G.
(ii) Find a shortest path between the vertices x_S and x_T in G', where $S = \emptyset$ and $T = V(G)$.

Observe that we can construct the graph G' in $(|V(G')| + |E(G')|)$-time and find a shortest path in G' in $\mathcal{O}(|V(G')| + |E(G')|)$-time. We know that $|V(G')| = 2^n$ and note that the total degree (in-degree+out-degree) of each vertex in G' is at most n. Therefore, $|E(G')| \leq n \cdot 2^n$. Therefore, the total running time of the algorithm is $2^n n^{\mathcal{O}(1)}$. Thus, we have proved the next theorem.

Theorem 5. *Given a graph G, the burning number of G can be computed in $2^n n^{\mathcal{O}(1)}$ time, where n is the number of vertices in G.*

References

1. Bessy, S., Bonato, A., Janssen, J., Rautenbach, D., Roshanbin, E.: Burning a graph is hard. Discret. Appl. Math. **232**, 73–87 (2017)
2. Bonato, A.: A survey of graph burning. arXiv preprint arXiv:2009.10642 (2020)
3. Bonato, A., Janssen, J., Roshanbin, E.: Burning a graph as a model of social contagion. In: Bonato, A., Graham, F.C., Prałat, P. (eds.) WAW 2014. LNCS, vol. 8882, pp. 13–22. Springer, Cham (2014). https://doi.org/10.1007/978-3-319-13123-8_2
4. Bonato, A., Janssen, J., Roshanbin, E.: How to burn a graph. Internet Math. **12**(1–2), 85–100 (2016)
5. Cygan, M., et al.: Parameterized Algorithms, vol. 5. Springer, Heidelberg (2015)
6. Eppstein, D.: Diameter and treewidth in minor-closed graph families. Algorithmica **27**, 275–291 (2000)
7. Gupta, A.T., Lokhande, S.A., Mondal, K.: NP-completeness results for graph burning on geometric graphs. arXiv preprint arXiv:2003.07746 (2020)
8. Hiller, M., Koster, A.M., Triesch, E.: On the burning number of p-caterpillars. In: Gentile, C., Stecca, G., Ventura, P. (eds.) Graphs and Combinatorial Optimization: from Theory to Applications, pp. 145–156. Springer, Cham (2021). https://doi.org/10.1007/978-3-030-63072-0_12
9. Janssen, R.: The burning number of directed graphs: bounds and computational complexity. arXiv preprint arXiv:2001.03381 (2020)

10. Kare, A.S., Vinod Reddy, I.: Parameterized algorithms for graph burning problem. In: Colbourn, C.J., Grossi, R., Pisanti, N. (eds.) IWOCA 2019. LNCS, vol. 11638, pp. 304–314. Springer, Cham (2019). https://doi.org/10.1007/978-3-030-25005-8_25
11. Kobayashi, Y., Otachi, Y.: Parameterized complexity of graph burning. Algorithmica 1–15 (2022)
12. Land, M.R., Lu, L.: An upper bound on the burning number of graphs. In: Bonato, A., Graham, F.C., Prałat, P. (eds.) WAW 2016. LNCS, vol. 10088, pp. 1–8. Springer, Cham (2016). https://doi.org/10.1007/978-3-319-49787-7_1
13. Roshanbin, E.: Burning a graph as a model for the spread of social contagion. Ph.D. thesis, Dalhousie University (2016)

Min-Max Relative Regret for Scheduling to Minimize Maximum Lateness

Imad Assayakh(✉), Imed Kacem, and Giorgio Lucarelli

LCOMS, University of Lorraine, Metz, France
{imad.assayakh,imed.kacem,giorgio.lucarelli}@univ-lorraine.fr

Abstract. We study the single machine scheduling problem under uncertain parameters, with the aim of minimizing the maximum lateness. More precisely, the processing times, the release dates and the delivery times of the jobs are uncertain, but an upper and a lower bound of these parameters are known in advance. Our objective is to find a robust solution, which minimizes the maximum relative regret. In other words, we search for a solution which, among all possible realizations of the parameters, minimizes the worst-case ratio of the deviation between its objective and the objective of an optimal solution over the latter one. Two variants of this problem are considered. In the first variant, the release date of each job is equal to 0. In the second one, all jobs are of unit processing time. In all cases, we are interested in the sub-problem of maximizing the (relative) regret of a given scheduling sequence. The studied problems are shown to be polynomially solvable.

Keywords: Scheduling · Maximum lateness · Min-max relative regret · Interval uncertainty

1 Introduction

Uncertainty is a crucial factor to consider when dealing with combinatorial optimization problems and especially scheduling problems. Thus, it is not sufficient to limit the resolution of a given problem to its deterministic version for a single realisation of the uncertain parameters, i.e., a scenario. In our study, we investigate a widely used method of handling uncertainty that relies on a set of known possible values of the uncertain parameters without any need of probabilistic description, namely the *robustness approach* or *worst-case approach* [12]. The aim of this approach is to generate solutions that will have a good performance under any possible scenario and particularly in the most unfavorable one.

The use of the robustness approach involves specifying two key components. The first component is the choice of the type of uncertainty set. Literature has proposed various techniques for describing the uncertainty set [6], with *the discrete uncertainty* and *the interval uncertainty* being the most well-examined. Indeed, the most suitable representation of uncertainty in scheduling problems is the interval uncertainty, where the value of each parameter is restricted within a

S.-Y. Hsieh et al. (Eds.): IWOCA 2023, LNCS 13889, pp. 49–61, 2023.
https://doi.org/10.1007/978-3-031-34347-6_5

specific closed interval defined by a lower and an upper bound. These bounds can be estimated through a data analysis on traces of previous problem executions.

The second component is the choice of the appropriate robustness criterion [1, 16]. The *absolute robustness* or *min-max* criterion seeks to generate solutions that provide the optimal performance in the worst case scenario. This criterion can be seen as overly pessimistic in situations where the worst-case scenario is unlikely, causing decision makers to regret not embracing a moderate level of risk. The *robust deviation* or *min-max regret* criterion aims at minimizing the maximum *absolute regret*, which is the most unfavorable deviation from the optimal performance among all scenarios. The *relative robust deviation* or *min-max relative regret* criterion seeks to minimize the maximum *relative regret*, which is the worst percentage deviation from the optimal performance among all possible scenarios. Averbakh [4] remarks that the relative regret objective is more appropriate compared to the absolute regret objective in situations where the statement "10% more expensive" is more relevant than "costs $30 more". However, the min-max relative regret criterion has a complicated structure and this may explain why limited knowledge exists about it.

The focus of this paper is to investigate the min-max relative regret criterion for the fundamental single machine scheduling problem with the maximum lateness objective. The interval uncertainty can involve the processing times, the release dates or the delivery times of jobs. In Sect. 2, we formally define our problem and the used criteria. In Sect. 3, we give a short review of the existing results for scheduling problems with and without uncertainty consideration. We next consider two variants of this problem.

In Sect. 4, we study the variant where all jobs are available at time 0 and the interval uncertainty is related to processing and delivery times. Kasperski [11] has applied the min-max regret criterion to this problem and developed a polynomial time algorithm to solve it by characterizing the worst-case scenario based on a single guessed parameter through some dominance rules. We prove that this problem is also polynomial for the min-max relative regret criterion. An iterative procedure is used to prove some dominance rules based on three guessed parameters in order to construct a partial worst-case scenario. To complete this scenario, we formulate a linear fractional program and we explain how to solve it in polynomial time.

In Sect. 5, we study the maximum relative regret criterion for the variant of the maximum lateness problem where the processing times of all jobs are equal to 1 and interval uncertainty is related to release dates and delivery times. For a fixed scenario, Horn [8] proposed an optimal algorithm for this problem. For the uncertainty version, we simulate the execution of Horn's algorithm using a guess of five parameters, in order to create a worst-case scenario along with its optimal schedule. Note that, we also give a much simpler analysis for the maximum regret criterion of this variant of our scheduling problem.

We conclude in Sect. 6.

2 Problem Definition and Notations

In this paper, we consider the problem of scheduling a set $\mathcal{J}$ of n non-preemptive jobs on a single machine. In the standard version of the problem, each job is characterized by a *processing time*, a *release date* and a *due date*. However, in this work we use a known equivalent definition of this problem in which the due dates are replaced by *delivery times*. In general, the values of the input parameters are not known in advance. However, an estimation interval for each value is known. Specifically, given a job $j \in \mathcal{J}$, let $[p_j^{\min}, p_j^{\max}]$, $[r_j^{\min}, r_j^{\max}]$ and $[q_j^{\min}, q_j^{\max}]$ be the uncertainty intervals for its characteristics.

A *scenario* $s = (p_1^s, ..., p_n^s, r_1^s, ..., r_n^s, q_1^s, ..., q_n^s)$ is a possible realisation of all values of the instance, such that $p_j^s \in [p_j^{\min}, p_j^{\max}]$, $r_j^s \in [r_j^{\min}, r_j^{\max}]$ and $q_j^s \in [q_j^{\min}, q_j^{\max}]$, for every $j \in \mathcal{J}$. The set of all scenarios is denoted by $\mathcal{S}$. A solution is represented by a *sequence* of jobs, $\pi = (\pi(1), ..., \pi(n))$ where $\pi(j)$ is the jth job in the sequence π. The set of all sequences is denoted by Π.

Consider a schedule represented by its sequence $\pi \in \Pi$ and a scenario $s \in \mathcal{S}$. The *lateness* of a job $j \in \mathcal{J}$ is defined as $L_j^s(\pi) = C_j^s(\pi) + q_j^s$, where $C_j^s(\pi)$ denotes the *completion time* of j in the schedule represented by π under the scenario s. The *maximum lateness* of the schedule is defined as $L(s, \pi) = \max_{j \in J} L_j^s(\pi)$. The job $c \in \mathcal{J}$ of maximum lateness in π under s is called *critical*, i.e., $L_c^s(\pi) = L(s, \pi)$. The set of all critical jobs in π under s is denoted by $Crit(s, \pi)$. We call *first critical job*, the critical job which is processed before all the other critical jobs. By considering a given scenario s, the optimal sequence is the one leading to a schedule that minimizes the maximum lateness, i.e., $L^*(s) = \min_{\pi \in \Pi} L(s, \pi)$. This is a classical scheduling problem, denoted by $1|r_j|L_{max}$ using the standard three-field notation, and it is known to be NP-hard [15].

In this paper, we are interested in the min-max regret and the min-max relative regret criteria whose definitions can be illustrated by a game between two agents, Alice and Bob. Alice selects a sequence π of jobs. The problem of Bob has as input a sequence π chosen by Alice, and it consists in selecting a scenario s such that the regret R of Alice $R(s, \pi) = L(s, \pi) - L^*(s)$ or respectively the relative regret RR of Alice $RR(s, \pi) = \frac{L(s,\pi) - L^*(s)}{L^*(s)} = \frac{L(s,\pi)}{L^*(s)} - 1$ is maximized. The value of $Z(\pi) = \max_{s \in S} R(s, \pi)$ (resp. $ZR(\pi) = \max_{s \in S} RR(s, \pi)$) is called *maximum regret* (resp. *maximum relative regret*) for the sequence π. In what follows, we call the problem of maximizing the (relative) regret, given a sequence π, as the *Bob's problem*. Henceforth, by slightly abusing the definition of the relative regret, we omit the constant -1 in $RR(s, \pi)$, since a scenario maximizing the fraction $\frac{L(s,\pi)}{L^*(s)}$ maximizes also the value of $\frac{L(s,\pi)}{L^*(s)} - 1$. Then, Alice has to find a sequence π which minimizes her maximum regret (resp. maximum relative regret), i.e., $\min_{\pi \in \Pi} Z(\pi)$ (resp. $\min_{\pi \in \Pi} ZR(\pi)$). This problem is known as the *min-max (relative) regret* problem and we call it as *Alice's problem*.

Given a sequence π, the scenario that maximises the (relative) regret over all possible scenarios is called *the worst-case scenario* for π. A *partial (worst-case) scenario* is a scenario defined by a fixed subset of parameters and can be *extended* to a (worst-case) scenario by setting the remaining unknown parameters. For a fixed scenario s, any schedule may consist of several *blocks*, i.e., a maximal set of jobs, which are processed without any *idle time* between them. A job u_j is said to be *first-block* for the job j if it is the first job processed in the block containing j in a given schedule.

3 Related Work

In the deterministic version, the problem $1|r_j|L_{\max}$ has been proved to be strongly NP-hard [15]. For the first variant where all release dates are equal, the problem can be solved in polynomial time by applying *Jackson's rule* [9], i.e., sequencing the jobs in the order of non-increasing delivery times. For the second variant with unit processing time jobs, the rule of scheduling, at any time, an available job with the biggest delivery time is shown to be optimal by Horn [8].

For the discrete uncertainty case, the min-max criterion has been studied for several scheduling problems with different objectives. Kouvelis and Yu [12] proved that the min-max resource allocation problem is NP-hard and admits a pseudo-polynomial algorithm. Aloulou and Della Croce [2] showed that the min-max $1||\sum U_j$ problem of minimizing the number of late jobs is NP-hard, while the min-max problem of the single machine scheduling is polynomially solvable for many objectives like makespan, maximum lateness and maximum tardiness even in the presence of precedence constraints. The only scheduling problem studied under discrete uncertainty for min-max (relative) regret is the $1||\sum C_j$ for which Yang and Yu [17] have proved that it is NP-hard for all the three robustness criteria.

For the interval uncertainty case, the min-max criterion has the same complexity as the deterministic problem since it is equivalent to solve it for an extreme well-known scenario. Considerable research has been dedicated to the min-max regret criterion for different scheduling problems. Many of these problems have been proved to be polynomially solvable. For instance, Averbakh [3] considered the min-max regret $1||\max w_jT_j$ problem to minimize the maximum weighted tardiness, where weights are uncertain and proposed a $O(n^3)$ algorithm. He also presented a $O(m)$ algorithm for the makespan minimization for a permutation flow-shop problem with 2 jobs and m machines with interval uncertainty related to processing times [5]. The min-max regret version of the first variant of our problem has been considered by Kasperski [11] under uncertain processing times and due dates. An $O(n^4)$ algorithm has been developed which works even in the presence of precedence constraints. On the other hand, Lebedev and Averbakh. [14] showed that the min-max regret $1||\sum C_j$ problem is NP-hard. Kacem and Kellerer [10] considered the single machine problem of scheduling jobs with a common due date with the objective of maximizing the number of early jobs and they proved that the problem is NP-hard.

Finally, for the min-max relative regret criterion for scheduling problems with interval uncertainty, the only known result is provided by Averbakh [4] who considered the problem $1||\sum C_j$ with uncertain processing times and he proved that it is NP-hard.

4 Min-Max Relative Regret for $1 \mid\mid L_{max}$

In this section, we consider the min-max relative regret criterion for the maximum lateness minimization problem, under the assumption that each job is available at time 0, i.e., $r_j^s = 0$ for all jobs $j \in \mathcal{J}$ and all possible scenarios $s \in \mathcal{S}$. For a fixed scenario, this problem can be solved by applying the Jackson's rule, i.e., sequencing the jobs in order of non-increasing delivery times.

4.1 The Bob's Problem

We denote by $B(\pi, j)$ the set of all the jobs processed before job $j \in \mathcal{J}$, including j, in the sequence π and by $A(\pi, j)$ the set of all the jobs processed after job j in π. The following lemma presents some properties of a worst-case scenario for a given sequence of jobs.

Lemma 1. *Let π be a sequence of jobs. There exists (1) a worst case scenario s for π, (2) a critical job $c_\pi \in Crit(s, \pi)$ in π under s, and (3) a critical job $c_\sigma \in Crit(s, \sigma)$ in σ under s, where σ is the optimal sequence for s, such that:*

i for each job $j \in A(\pi, c_\pi)$, it holds that $p_j^s = p_j^{\min}$,
ii for each job $j \in \mathcal{J} \backslash \{c_\pi\}$, it holds that $q_j^s = q_j^{\min}$,
iii for each job $j \in B(\pi, c_\pi) \cap B(\sigma, c_\sigma)$, it holds that $p_j^s = p_j^{\min}$, and
iv c_σ is the first critical job in σ under s.

Consider the sequence π chosen by Alice. Bob can guess the critical job c_π in π and the first critical job c_σ in σ. Then, by Lemma 1 (i)–(ii), he can give the minimum processing times to all jobs in $A(\pi, c_\pi)$, and the minimum delivery times to all jobs except for c_π. Since the delivery times of all jobs except c_π are determined and the optimal sequence σ depends only on the delivery times according to the Jackson's rule, Bob can obtain σ by guessing the position $k \in [\![1, n]\!]$ of c_π in σ. Then, by Lemma 1 (iii), he can give the minimum processing times to all jobs in $B(\pi, c_\pi) \cap B(\sigma, c_\sigma)$. We denote by the triplet (c_π, c_σ, k) the guess made by Bob. Based on the previous assignments, Bob gets a partial scenario $\bar{s}^\pi_{c_\pi, c_\sigma, k}$. It remains to determine the exact value of q_{c_π} and the processing times of jobs in $B(\pi, c_\pi) \cap A(\sigma, c_\sigma)$ in order to extend $\bar{s}^\pi_{c_\pi, c_\sigma, k}$ to a scenario $s^\pi_{c_\pi, c_\sigma, k}$. At the end, Bob will choose, among all the scenarios $s^\pi_{c_\pi, c_\sigma, k}$ created, the worst case scenario s_π for the sequence π, i.e., $s_\pi = \arg\max_{i,j,k} \{\frac{L(s^\pi_{i,j,k}, \pi)}{L^*(s^\pi_{i,j,k})}\}$.

In what follows, we propose a *linear fractional program* (P) in order to find a scenario $s^\pi_{c_\pi, c_\sigma, k}$ which extends $\bar{s}^\pi_{c_\pi, c_\sigma, k}$ and maximizes the relative regret for the given sequence π. Let p_j, the processing time of each job $j \in B(\pi, c_\pi) \cap A(\sigma, c_\sigma)$, and q_{c_π}, the delivery time of job c_π, be the continuous decision variables in (P).

All other processing and delivery times are constants and their values are defined by $\bar{s}^{\pi}_{c_\pi,c_\sigma,k}$. Recall that $\sigma(j)$ denotes the j-th job in the sequence σ. To simplify our program, we consider two fictive values $q_{\sigma(n+1)} = q^{\min}_{c_\pi}$ and $q_{\sigma(0)} = q^{\max}_{c_\pi}$.

$$\text{maximize} \ \frac{\sum_{i \in B(\pi,c_\pi)} p_i + q_{c_\pi}}{\sum_{i \in B(\sigma,c_\sigma)} p_i + q_{c_\sigma}} \qquad (P)$$

$$\text{subject to} \quad \sum_{i \in B(\pi,j)} p_i + q_j \leq \sum_{i \in B(\pi,c_\pi)} p_i + q_{c_\pi} \qquad \forall j \in \mathcal{J} \quad (1)$$

$$\sum_{i \in B(\sigma,j)} p_i + q_j \leq \sum_{i \in B(\sigma,c_\sigma)} p_i + q_{c_\sigma} \qquad \forall j \in \mathcal{J} \quad (2)$$

$$p_j \in [p_j^{\min}, p_j^{\max}] \quad \forall j \in B(\pi, c_\pi) \cap A(\sigma, c_\sigma) \quad (3)$$

$$q_{c_\pi} \in [\max\{q^{\min}_{c_\pi}, q_{\sigma(k+1)}\}, \min\{q^{\max}_{c_\pi}, q_{\sigma(k-1)}\}] \quad (4)$$

The objective of (P) maximizes the relative regret for the sequence π under the scenario $s^{\pi}_{c_\pi,c_\sigma,k}$ with respect to the hypothesis that c_π and c_σ are critical in π and σ, respectively, i.e.,

$$ZR(\pi) = \frac{L(s^{\pi}_{c_\pi,c_\sigma,k}, \pi)}{L^*(s^{\pi}_{c_\pi,c_\sigma,k})} = \frac{L_{c_\pi}^{s^{\pi}_{c_\pi,c_\sigma,k}}(\pi)}{L_{c_\sigma}^{s^{\pi}_{c_\pi,c_\sigma,k}}(\sigma)}$$

Constraints (1) and (2) ensure this hypothesis. Constraints (3) and (4) define the domain of the continuous real variables p_j, $j \in B(\pi, c_\pi) \cap A(\sigma, c_\sigma)$, and q_{c_π}. Note that, the latter one is based also on the guess of the position of c_π in σ. The program (P) can be infeasible due to the Constraints (1) and (2) that impose jobs c_π and c_σ to be critical. In this case, Bob ignores the current guess of (c_π, c_σ, k) in the final decision about the worst case scenario s_π that maximizes the relative regret.

Note that the Constraint (1) can be safely removed when considering the whole procedure of Bob for choosing the worst case scenario s_π. Indeed, consider a guess (i, j, k) which is infeasible because the job i is not critical in π due to the Constraint (1). Let s be the scenario extended from the partial scenario $\bar{s}^{\pi}_{i,j,k}$ by solving (P) without using the Constraint (1). Let c_π be the critical job under the scenario s. Thus, $L^s_{c_\pi}(\pi) > L^s_i(\pi)$. Consider now the scenario s' of maximum relative regret in which c_π is critical. Since s_π is the worst case scenario chosen by Bob for the sequence π and by the definition of s' we have

$$\frac{L(s_\pi, \pi)}{L^*(s_\pi)} \geq \frac{L(s', \pi)}{L^*(s')} \geq \frac{L(s, \pi)}{L^*(s)} = \frac{L^s_{c_\pi}(\pi)}{L^*(s)} > \frac{L^s_i(\pi)}{L^*(s)}$$

In other words, if we remove the Constraint (1), (P) becomes feasible while its objective value cannot be greater than the objective value of the worst case scenario s_π and then the decision of Bob with respect to the sequence π is not affected. This observation is very useful in Alice's algorithm. However, a similar observation cannot hold for Constraint (2) which imposes c_σ to be critical in σ.

As mentioned before, the program (P) is a linear fractional program, in which all constraints are linear, while the objective function corresponds to a fraction of linear expressions of the variables. Moreover, the denominator of the objective function has always a positive value. Charnes and Cooper [7] proposed a polynomial transformation of such a linear fractional program to a linear program. Hence, (P) can be solved in polynomial time.

Note also that, in the case where $c_\pi \neq c_\sigma$, the value of the maximum lateness in the optimal sequence $(\sum_{j\in B(\sigma,c_\sigma)} p_j^s + q_{c_\sigma}^s)$ is fixed since the processing times of jobs processed before the job c_σ in σ, as well as, the delivery time $q_{c_\sigma}^s$ of the job c_σ are already determined in the partial scenario $\bar{s}^{\pi}_{c_\pi,c_\sigma,k}$. Therefore, if $c_\pi \neq c_\sigma$ then (P) is a linear program. Consequently, the Charnes-Cooper transformation is used only in the case where $c_\pi = c_\sigma$.

Theorem 1. *Given a sequence π, there is a polynomial time algorithm that returns a worst case scenario s_π of maximum relative regret $ZR(\pi)$ for the problem $1 \mid\mid L_{\max}$.*

4.2 The Alice's Problem

In this section, we show how Alice constructs an optimal sequence π minimizing the maximum relative regret, i.e., $\pi = \text{argmin}_{\sigma\in\Pi} ZR(\sigma)$. Intuitively, by starting from the last position and going backwards, Alice searches for an unassigned job that, if placed at the current position and it happens to be critical in the final sequence π, will lead to the minimization of the maximum relative regret for π.

In order to formalize this procedure we need some additional definitions. Assume that Alice has already assigned a job in each position $n, n-1, \ldots, r+1$ of π. Let B_r be the set of unassigned jobs and consider any job $i \in B_r$. If i is assigned to the position r in π, then the sets $B(\pi,i)$ and $A(\pi,i)$ coincide with B_r and $\mathcal{J}\setminus B_r$, respectively, and are already well defined, even though the sequence π is not yet completed (recall that $B(\pi,i)$ includes i). Indeed, $B(\pi,i)$ and $A(\pi,i)$ depend only on the position r, i.e., $B(\pi,i) = B(\pi,j) = B_r$ and $A(\pi,i) = A(\pi,j)$ for each couple of jobs $i,j \in B_r$. Hence, Alice can simulate the construction in Bob's procedure in order to decide which job to assign at position r. Specifically, for a given job $i \in B_r$, Alice considers all scenarios $s_{i,j,k}^{B_r}$, where $j \in \mathcal{J}$ is the first critical job in the optimal sequence σ for this scenario and $k \in [\![1,n]\!]$ is the position of i in σ, constructed as described in Bob's algorithm. Note that, we slightly modified the notation of the scenario constructed by Bob for a guess (i,j,k) to $s_{i,j,k}^{B_r}$ instead of $s_{i,j,k}^{\pi}$, since a partial knowledge ($B_r = B(\pi,i)$) of π is sufficient for his procedure. Moreover, the reason of omitting Constraint (1) in the program (P) is clarified here, since the job i is not imposed to be necessarily critical in π. For a job $i \in B_r$, let

$$f_i(\pi) = \max_{j\in\mathcal{J},k\in[\![1,n]\!]} \left\{ \frac{L(s_{i,j,k}^{B(\pi,i)})}{L^*(s_{i,j,k}^{B(\pi,i)})} \right\}$$

Then, Alice assigns to position r the job i which minimizes $f_i(\pi)$, and the following theorem holds.

Theorem 2. *There is a polynomial time algorithm which constructs a sequence* π *that minimizes the maximum relative regret for the problem* $1 \mid\mid L_{\max}$.

5 Min-Max Relative Regret for $1 \mid r_j, p_j = 1 \mid L_{\max}$

In this section, we consider the case of unit processing time jobs, i.e., $p_j^s = 1$ for all jobs $j \in \mathcal{J}$ and all possible scenarios $s \in \mathcal{S}$. In contrast to the previous section, the jobs are released on different dates whose values are also imposed to uncertainties. For a fixed scenario, Horn [8] proposes an extension of the Jackson's rule leading to an optimal schedule for this problem: at any time t, schedule the available job, if any, of the biggest delivery time, where a job j is called available at time t if $r_j \leq t$ and j is not yet executed before t.

5.1 The Bob's Problem

Since all jobs are of unit the processing times, a scenario s is described by the values of the release dates and the delivery times of the jobs, i.e., by $r_j^s \in [r_j^{\min}, r_j^{\max}]$ and $q_j^s \in [q_j^{\min}, q_j^{\max}]$, for each $j \in \mathcal{J}$. Recall that, in the presence of different release dates, the execution of the jobs is partitioned into blocks without any idle time, while, given a sequence π and a scenario s, the first job in the block of a job $j \in \mathcal{J}$ is called first-block job for j in π under s. The following lemma characterizes a worst case scenario for a given sequence of jobs π.

Lemma 2. *Let* π *be a sequence of jobs. There exists a worst case scenario* s*, a critical job* $c \in Crit(s, \pi)$ *and its first-block job* u_c *in* π *under* s *such that:*

i for each job $j \in \mathcal{J}\backslash\{c\}$*, it holds that* $q_j^s = q_j^{\min}$*,*
ii for each job $j \in \mathcal{J}\backslash\{u_c\}$*, it holds that* $r_j^s = r_j^{\min}$*.*

Consider the sequence π chosen by Alice. Bob can guess the critical job c in π and its first-block job u_c. Using Lemma 2, we get a partial scenario $\bar{s}$ by fixing the delivery times of all jobs except for c as well as the release dates of all jobs except for u_c to their minimum values. It remains to determine the values of q_c and r_{u_c} in order to extend the partial scenario $\bar{s}$ to a scenario s. At the end, Bob will choose, among all scenarios created, the worst case one for the sequence π, i.e., the scenario with the maximum value of relative regret.

In what follows, we explain how to construct a sequence σ which will correspond to an optimal schedule for the scenario s when the values of q_c and r_{u_c} will be fixed. The main idea of the proposed algorithm is that, once a couple of σ and s is determined, then σ corresponds to the sequence produced by applying Horn's algorithm with the scenario s as an input. The sequence σ is constructed from left to right along with an associated schedule which determines the starting time B_j and the completion time $C_j = B_j + 1$ of each job $j \in \mathcal{J}$. The assignment of a job j to a position of this schedule (time B_j) introduces additional constraints in order to respect the sequence produced by Horn's algorithm:

(C1) there is no idle time in $[r_j, B_j)$,

(C2) at time B_j, the job j has the biggest delivery time among all available jobs at this time, and

(C3) the delivery times of all jobs scheduled in $[r_j, B_j)$ should be bigger than q_j^s.

These constraints are mainly translated to a refinement of the limits of q_c or of r_{u_c}, i.e., updates on $q_c^{\min}$, $q_c^{\max}$, $r_{u_c}^{\min}$ and $r_{u_c}^{\max}$. If at any point of our algorithm the above constraints are not satisfied, then we say that the assumptions/guesses made become infeasible, since they cannot lead to a couple (σ, s) respecting Horn's algorithm. Whenever we detect an infeasible assumption/guess, we throw it and we continue with the next one.

Let $\ell[x]$ be the x-th job which is released after the time $r_{u_c}^{\min}$, that is, $r_{u_c}^{\min} \leq r_{\ell[1]}^{s} \leq r_{\ell[2]}^{s} \leq \ldots \leq r_{\ell[y]}^{\bar{s}}$. By convention, let $r_{\ell[0]}^{\bar{s}} = r_{u_c}^{\min}$ and $r_{\ell[y+1]}^{\bar{s}} = +\infty$. To begin our construction, we guess the positions k_c and k_{u_c} of the jobs c and u_c, respectively, in σ as well as the interval $[r_{\ell[x]}^{\bar{s}}, r_{\ell[x+1]}^{\bar{s}})$, $0 \leq x \leq y$, of B_{u_c} in the optimal schedule s for σ. Let $k_{\min} = \min\{k_c, k_{u_c}\}$. We start constructing σ and its corresponding schedule by applying Horn's algorithm with input the set of jobs $\mathcal{J}\backslash\{c, u_c\}$ for which all data are already determined by the partial scenario $\bar{s}$, until $k_{\min} - 1$ jobs are scheduled. Then, we set $\sigma(k_{\min}) = \arg\min\{k_c, k_{u_c}\}$. We now need to define the starting time of $\sigma(k_{\min})$ and we consider two cases:

Case 1: $k_c < k_{u_c}$. We set $B_c = \max\{C_{\sigma(k_{\min}-1)}, r_c^{\bar{s}}\}$. If $B_c = r_c^{\bar{s}}$ and there is an idle time and an available job $j \in \mathcal{J}\backslash\{c, u_c\}$ in $[C_{\sigma(k_{\min}-1)}, B_c)$, then we throw the guess k_c, k_{u_c}, $[r_{\ell[x]}^{\bar{s}}, r_{\ell[x+1]}^{\bar{s}})$ since we cannot satisfy constraint (C1) for j, and hence our schedule cannot correspond to the one produced by Horn's algorithm.

Let $q_a^{\bar{s}} = \max\{q_j^{\bar{s}} : j \in \mathcal{J}\backslash\{c, u_c\}$ is available at $B_c\}$. Then, in order to satisfy constraint (C2) we update $q_c^{\min} = \max\{q_c^{\min}, q_a^{\bar{s}}\}$. Let $q_b^{\bar{s}} = \min\{q_j : j \in \mathcal{J}\backslash\{c, u_c\}$ is executed in $[r_c, B_c)\}$. Then, in order to satisfy constraint (C3) we update $q_c^{\max} = \min\{q_c^{\max}, q_b^{\bar{s}}\}$. If $q_c^{\max} < q_c^{\min}$, then we throw the guess k_c, k_{u_c}, $[r_{\ell[x]}^{\bar{s}}, r_{\ell[x+1]}^{\bar{s}})$ since we cannot get a feasible value for q_c^s.

It remains to check if there is any interaction between c and u_c. Since $k_c < k_{u_c}$, u_c is not executed in $[r_c, B_c)$. However, u_c may be available at B_c, but we cannot be sure for this because the value of $r_{u_c}^s$ is not yet completely determined. For this reason, we consider two opposite assumptions. Note that B_c is already fixed by the partial scenario $\bar{s}$ in the following assumptions, while $r_{u_c}^s$ is the hypothetical release date of u_c in the scenario s.

Assumption 1.1: $r_{u_c}^s \leq B_c$. In order to impose this assumption, we update $r_{u_c}^{\max} = \min\{r_{u_c}^{\max}, B_c\}$.

Assumption 1.2: $r_{u_c}^s > B_c$. In order to impose this assumption, we update $r_{u_c}^{\min} = \max\{r_{u_c}^{\min}, B_c + 1\}$.

If in any of these cases we have that $r_{u_c}^{\max} < r_{u_c}^{\min}$, then we throw the corresponding assumption, since there is no feasible value for $r_{u_c}^s$. For each non-thrown assumption, we continue our algorithm separately, and we eventually get two different couples of sequence/scenario if both assumptions are maintained. More specifically, for each assumption, we continue applying Horn's algorithm with input the set of jobs $\mathcal{J}\backslash\{\sigma(1), \sigma(2), \ldots, \sigma(k_{\min} - 1), c, u_c\}$ starting from time

$C_c = B_c + 1$, until $k_{u_c} - k_c - 1$ additional jobs are scheduled. Then, we set $\sigma(k_{u_c}) = u_c$ and $B_{u_c} = \max\{C_{\sigma(k_{u_c}-1)}, r_{u_c}^{\min}, r_{\ell[x]}^{\bar{s}}\}$. Note that B_{u_c} depends for the moment on the (updated) $r_{u_c}^{\min}$ and not on the final value of $r_{u_c}^{s}$ which has not been determined at this point of the algorithm. If $B_{u_c} \geq r_{\ell[x+1]}^{\bar{s}}$, then we throw the guess on $[r_{\ell[x]}^{\bar{s}}, r_{\ell[x+1]}^{\bar{s}})$. We next check if the constraints (C1)-(C3) are satisfied for all jobs in $\mathcal{J}\backslash\{u_c\}$ with respect to the assignment of the job u_c at the position k_{u_c} of σ with starting time B_{u_c}. If not, we throw the current assumption. Otherwise, Horn's algorithm with input the jobs in $\mathcal{J}\backslash\{\sigma(1), \sigma(2), \ldots, \sigma(k_{u_c})\}$ and starting from time $B_{u_c} + 1$ is applied to complete σ.

Case 2: $k_c > k_{u_c}$. We set $B_{u_c} = \max\{C_{\sigma(k_{\min}-1)}, r_{u_c}^{\min}, r_{\ell[x]}^{\bar{s}}\}$. As before, B_{u_c} depends on $r_{u_c}^{\min}$ and not on the final value of $r_{u_c}^{s}$. If $B_{u_c} \geq r_{\ell[x+1]}^{\bar{s}}$ then we throw the current guess on $[r_{\ell[x]}^{\bar{s}}, r_{\ell[x+1]}^{\bar{s}})$. We need also to check if the constraints (C1)-(C3) are satisfied for all jobs in $\mathcal{J}\backslash\{u_c\}$ with respect to the assignment of the job u_c at the position k_{u_c} of σ with starting time B_{u_c}. If not, we throw the current guess k_c, k_{u_c}, $[r_{\ell(q)}^{\bar{s}}, r_{\ell(q)+1}^{\bar{s}})$. Note that the last check is also applied for c and eventually leads to update $q_c^{\max} = \min\{q_c^{\max}, q_{u_c}^{\bar{s}}\}$ if c is available at B_{u_c}. This can be easily verified because of the guess of the interval of B_{u_c}.

Next, we continue applying Horn's algorithm with input the set of jobs $\mathcal{J}\backslash\{\sigma(1), \sigma(2), \ldots, \sigma(k_{\min}-1), c, u_c\}$ starting from time $B_{u_c}+1$, until $k_c - k_{u_c} - 1$ additional jobs are scheduled. Then, we set $\sigma(k_c) = c$, $B_c = \max\{C_{\sigma(k_c-1)}, r_c^{\bar{s}}\}$, and we check if the constraints (C1)-(C3) are satisfied for all jobs in $\mathcal{J}\backslash\{c\}$ with respect to the assignment of the job c at the position k_c of σ with starting time B_c. If not, we throw the current guess k_c, k_{u_c}, $[r_{\ell[x]}^{\bar{s}}, r_{\ell[x+1]}^{\bar{s}})$. Moreover, an update on $q_c^{\min}$ and $q_c^{\max}$ is possible here, like the one in the begin of case 1. Finally, Horn's algorithm with input the jobs in $\mathcal{J}\backslash\{\sigma(1), \sigma(2), \ldots, \sigma(k_c)\}$ and starting from time $C_c = B_c + 1$ is applied to complete σ.

Note that after the execution of the above procedure for a given guess c, u_c, k_c, k_{u_c}, $[r_{\ell[x]}^{\bar{s}}, r_{\ell[x+1]}^{\bar{s}})$, and eventually an assumption 1.1 or 1.2, we get a sequence σ and its corresponding schedule, while the values of q_c^{s} and $r_{u_c}^{s}$ are still not defined but their bounds are probably limited to fit with this guess. Then, we apply the following three steps in order to get the scenario s:

1. Extend the partial scenario $\bar{s}$ to a scenario $s_{\min}$ by setting $q_c^{s_{\min}} = q_c^{\min}$ and $r_{u_c}^{s_{\min}} = r_{u_c}^{\min}$.
2. Extend the scenario $s_{\min}$ to the scenario s_1 by increasing the delivery time of c to its maximum without increasing the maximum lateness and without exceeding $q_c^{\max}$, i.e., $q_c^{s_1} = q_c^{s_{\min}} + \min\{q_c^{\max} - q_c^{s_{\min}}, L^*(s_{\min}) - L_c^{s_{\min}}(\sigma)\}$.
3. Extend the scenario s_1 to the scenario s by increasing the release date of u_c to its maximum without increasing the maximum lateness, without exceeding $r_{u_c}^{\max}$ and without violating the constraint (C1) and the current guess.

The following theorem holds since in an iteration of the above algorithm, the guess corresponding to an optimal sequence σ for the worst case scenario s will be considered, while Horn's algorithm guarantees the optimality of σ.

Theorem 3. *There is a polynomial time algorithm which, given a sequence* π, *constructs a worst case scenario* s_π *of maximum relative regret for the problem* $1|r_j, p_j = 1|L_{\max}$.

5.2 The Alice's Problem

In this section, we describe Alice's algorithm in order to construct an optimal sequence minimizing the maximum relative regret for $1|r_j, p_j = 1|L_{\max}$. Since Alice knows how Bob proceeds, she can do a guess g of the five parameters c, u_c, k_c, k_{u_c}, $[r_{\ell[x]}, r_{\ell[x+1]})$ in order to construct an optimal sequence σ_g for a scenario s_g corresponding to this guess. Then, she assumes that σ_g is provided as input to Bob. Bob would try to maximize its relative regret with respect to σ_g by eventually doing a different guess $\hat{g}$, obtaining a scenario $s_{\hat{g}}$, i.e.,

$$RR(s_{\hat{g}}, \sigma_g) = \max_{g'} \frac{L(s_{g'}, \sigma_g)}{L^*(s_{g'})}$$

Note that, if $g = \hat{g}$, then $RR(s_{\hat{g}}, \sigma_g) = 1$ since by definition σ_g is the optimal sequence for the scenario $s_g = s_{\hat{g}}$. Therefore, Alice can try all possible guesses in order to find the one that minimizes her maximum relative regret by applying Bob's algorithm to the sequence obtained by each guess, and hence the following theorem holds.

Theorem 4. *There is a polynomial time algorithm which constructs a sequence* π *minimizing the maximum relative regret for the problem* $1|r_j, p_j = 1|L_{\max}$.

Note that, Bob's guess for this problem defines almost all parameters of a worst case scenario, without really using the input sequence provided by Alice. This is not the case in Sect. 4 where, according to Lemma 1, the jobs that succeed the critical job in Alice's sequence should be known. For this reason Alice's algorithm is simpler here compared to the one in Sect. 4.2.

6 Conclusions

We studied the min-max relative regret criterion for dealing with interval uncertain data for the single machine scheduling problem of minimizing the maximum lateness. We considered two variants and we proved that they can be solved optimally in polynomial time. Our main technical contribution concerns the sub-problem of maximizing the relative regret for these variants. The complexity of our results justifies in a sense the common feeling that the min-max relative criterion is more difficult than the min-max regret criterion.

Note that our result for the variant without release dates can be extended even in the case where the jobs are subject to precedence constraints. Indeed, Lawler [13] proposed an extension of Jackson's rule for the deterministic version of this problem, while the monotonicity property still holds. Thus, the corresponding lemma describing a worst case scenario holds, and the determination of the optimal sequence depends only on the guess of the position of the critical job in this sequence which should be imposed to respect the precedence constraints.

In the future, it is interesting to clarify the complexity of the general maximum lateness problem with respect to min-max relative regret when all parameters are subject to uncertainty. We believe that this problem is NP-hard. If this is confirmed, the analysis of an approximation algorithm is a promising research direction.

Acknowledgement. This research has been partially supported by the ANR Lorraine Artificial Intelligence project (ANR-LOR-AI).

References

1. Aissi, H., Bazgan, C., Vanderpooten, D.: Min-max and min-max regret versions of combinatorial optimization problems: a survey. Eur. J. Oper. Res. **197**(2), 427–438 (2009)
2. Aloulou, M., Della Croce, F.: Complexity of single machine scheduling problems under scenario-based uncertainty. Oper. Res. Lett. **36**, 338–342 (2008)
3. Averbakh, I.: Minmax regret solutions for minimax optimization problems with uncertainty. Oper. Res. Lett. **27**(2), 57–65 (2000)
4. Averbakh, I.: Computing and minimizing the relative regret in combinatorial optimization with interval data. Discret. Optim. **2**(4), 273–287 (2005)
5. Averbakh, I.: The minmax regret permutation flow-shop problem with two jobs. Eur. J. Oper. Res. **169**(3), 761–766 (2006)
6. Buchheim, C., Kurtz, J.: Robust combinatorial optimization under convex and discrete cost uncertainty. EURO J. Comput. Optim. **6**(3), 211–238 (2018). https://doi.org/10.1007/s13675-018-0103-0
7. Charnes, A., Cooper, W.W.: Programming with linear fractional functionals. Nav. Res. Logistics Q. **9**(3–4), 181–186 (1962)
8. Horn, W.: Some simple scheduling algorithms. Nav. Res. Logistics Q. **21**(1), 177–185 (1974)
9. Jackson, J.: Scheduling a production line to minimize maximum tardiness. Research report, Office of Technical Services (1955)
10. Kacem, I., Kellerer, H.: Complexity results for common due date scheduling problems with interval data and minmax regret criterion. Discret. Appl. Math. **264**, 76–89 (2019)
11. Kasperski, A.: Minimizing maximal regret in the single machine sequencing problem with maximum lateness criterion. Oper. Res. Lett. **33**(4), 431–436 (2005)
12. Kouvelis, P., Yu, G.: Robust Discrete Optimization and Its Applications. Kluwer Academic Publishers, Amsterdam (1997)
13. Lawler, E.L.: Optimal sequencing of a single machine subject to precedence constraints. Manage. Sci. **19**(5), 544–546 (1973)
14. Lebedev, V., Averbakh, I.: Complexity of minimizing the total flow time with interval data and minmax regret criterion. Discret. Appl. Math. **154**, 2167–2177 (2006)
15. Lenstra, J.K., Rinnooy Kan, A.H.G., Brucker, P.: Complexity of machine scheduling problems. In: Studies in Integer Programming, Annals of Discrete Mathematics, vol. 1, pp. 343–362. Elsevier (1977)

16. Tadayon, B., Smith, J.C.: Robust Offline Single-Machine Scheduling Problems, pp. 1–15. Wiley, Hoboken (2015)
17. Yang, J., Yu, G.: On the robust single machine scheduling problem. J. Comb. Optim. **6**, 17–33 (2002)

Advice Complexity Bounds for Online Delayed $\mathcal{F}$-Node-, H-Node- and H-Edge-Deletion Problems

Niklas Berndt and Henri Lotze[(✉)]

RWTH Aachen University, Aachen, Germany
niklas.berndt@rwth-aachen.de, lotze@cs.rwth-aachen.de

Abstract. Let $\mathcal{F}$ be a fixed finite obstruction set of graphs and G be a graph revealed in an online fashion, node by node. The online Delayed $\mathcal{F}$-Node-Deletion Problem ($\mathcal{F}$-Edge-Deletion Problem) is to keep G free of every $H \in \mathcal{F}$ by deleting nodes (edges) until no induced subgraph isomorphic to any graph in $\mathcal{F}$ can be found in G. The task is to keep the number of deletions minimal.

Advice complexity is a model in which an online algorithm has access to a binary tape of infinite length, on which an oracle can encode information to increase the performance of the algorithm. We are interested in the minimum number of advice bits that are necessary and sufficient to solve a deletion problem optimally.

In this work, we first give essentially tight bounds on the advice complexity of the Delayed $\mathcal{F}$-Node-Deletion Problem and $\mathcal{F}$-Edge-Deletion Problem where $\mathcal{F}$ consists of a single, arbitrary graph H. We then show that the gadget used to prove these results can be utilized to give tight bounds in the case of node deletions if $\mathcal{F}$ consists of either only disconnected graphs or only connected graphs. Finally, we show that the number of advice bits that is necessary and sufficient to solve the general Delayed $\mathcal{F}$-Node-Deletion Problem is heavily dependent on the obstruction set $\mathcal{F}$. To this end, we provide sets for which this number is either constant, logarithmic or linear in the optimal number of deletions.

Keywords: Online Algorithms · Advice Complexity · Late Accept Model · Node-Deletion · Edge-Deletion · Graph Modification

1 Introduction

The analysis of online problems is concerned with studying the worst case performance of algorithms where the instance is revealed element by element and decisions have to be made immediately and irrevocably. To measure the performance of such algorithms, their solution is compared to the optimal solution of the same instance. The largest ratio between the size of an online algorithms solution and the optimal solution size over all instances is then called the (strict) *competitive ratio* of an algorithm. Finding an algorithm with the smallest possible competitive ratio is the common aim of online analysis. The study of online algorithms

S.-Y. Hsieh et al. (Eds.): IWOCA 2023, LNCS 13889, pp. 62–73, 2023.
https://doi.org/10.1007/978-3-031-34347-6_6

was started by Sleator and Tarjan [16] and has been active ever since. For a more thorough introduction on competitive analysis, we refer the reader to the standard book by Borodin and El-Yaniv [5].

The online problems studied in this work are each defined over a fixed family $\mathcal{F}$ of graphs. An induced online graph G is revealed iteratively by revealing its nodes. The solution set, i.e. a set of nodes (edges), of an algorithm is called S and we define $G - S$ as $V(G) \backslash V(S)$ or as $E(G) \backslash E(S)$ respectively, depending on whether S is a set of nodes or edges. When in some step i an induced subgraph of $G[\{v_1, \ldots, v_i\}] - S$ is isomorphic to a graph $H \in \mathcal{F}$, an algorithm is forced to delete nodes (edges) T by adding them to S until no induced graph isomorphic to some $H \in \mathcal{F}$ can be found in $G[\{v_1, \ldots, v_i\}] - \{S \cup T\}$. The competitive ratio of an algorithm is then measured by taking the ratio of its solution set size to the solution set size of an optimal offline algorithm.

Note that this problem definition is not compatible with the classical online model, as nodes (edges) do not immediately have to be added to S or ultimately not be added to S. Specifically, elements that are not yet part of S may be added to S at a later point, but no elements may be removed from S at any point. Furthermore, an algorithm is only forced to add elements to S whenever an $H \in \mathcal{F}$ is isomorphic to some induced subgraph of the current online graph. Chen et al. [9] showed that no algorithm for this problem can admit a constantly bounded competitive ratio in the classical online setting and that there are families $\mathcal{F}$ for which the competitive ratio is strict in the size of the largest forbidden graph $H \in \mathcal{F}$. This model, where only an incremental valid partial solution is to be upheld, was first studied by Boyar et al. [6] and coined "Late Accept" by Boyar et al. [7] in the following year. As we study the same problems as Chen et al., we use the term *delayed* for consistency.

When studying the competitive ratio of online algorithms, Dobrev et al. [10] asked the question which, and crucially *how much* information an online algorithm is missing in order to improve its competitive ratio. This model was revised by Hromkovič et al. [11], further refined by Böckenhauer et al. [4] and is known as the study of *advice complexity* of an online problem. In this setting, an online algorithm is given access to a binary advice tape that is infinite in one direction and initialized with random bits. An oracle may then overwrite a number of these random bits, starting from the initial position of the tape in order to encode information about the upcoming instance or to simply give instructions to an algorithm. An algorithm may then read from this tape and act on this information during its run. The maximum number of bits an algorithm reads from the tape over all instances to obtain a target competitive ratio of c is then called the *advice complexity* of an algorithm. For a more thorough introduction to the analysis of advice complexity and problems studied under this model, we refer the reader to the book by Komm [12] and the survey paper by Boyar et al. [8]. In this work, we are interested in the minimum needed and maximum necessary number of bits of information an online algorithm needs in order to solve the discussed problems optimally.

Table 1. Advice complexity of node-deletion problems.

Node-Deletion	Single graph H forbidden	Family $\mathcal{F}$ of graphs forbidden
All graphs connected	Chen et al. [9]	Chen et al. [9]
Arbitrary graphs	Essentially tight bound	Lower & Upper bounds
	Theorem 1, Corollary 1	Theorem 1, Theorem 6, Theorem 7, Theorem 8

Table 2. Advice complexity of edge-deletion problems.

Edge-Deletion	Single graph H forbidden	Family $\mathcal{F}$ of graphs forbidden
All graphs connected	Chen et al. [9]	Chen et al. [9]
Arbitrary graphs	Essentially tight bound	Open
	Theorem 2, Corollary 2, Theorem 5	

The analysis of advice complexity assumes the existence of an almighty oracle that can give perfect advice, which is not realistic. However, recent publications utilize such bounds, especially the information theoretic lower bounds. One field is that of machine-learned advice [1,14,15]. A related field is that of uncertain or untrusted advice [13], which analyses the performance of algorithms depending on how accurate some externally provided information for an algorithm is.

Chen et al. [9] gave bounds on the DELAYED $\mathcal{F}$-NODE-DELETION PROBLEM and DELAYED $\mathcal{F}$-EDGE-DELETION PROBLEM for several restricted families $\mathcal{F}$, dividing the problem by restricting whether $\mathcal{F}$ may contain only connected graphs and whether $\mathcal{F}$ consists of a single graph H or arbitrarily many. Tables 1 and 2 show our contributions: Up to some minor gap due to encoding details, we close the remaining gap for the DELAYED H-NODE-DELETION PROBLEM that was left open by Chen et al. [9], give tight bounds for the DELAYED H-EDGE-DELETION PROBLEM, and provide a variety of (tight) bounds for the DELAYED $\mathcal{F}$-NODE-DELETION PROBLEM depending on the nature of $\mathcal{F}$.

The problem for families of *connected* graphs $\mathcal{F}$ has one nice characteristic that one can exploit when designing algorithms solving node- or edge-deletion problems over such families. Intuitively, one can build an adversarial instance by presenting some copy of an $H \in \mathcal{F}$, extend this H depending on the behavior of an algorithm to force some specific deletion and continue by presenting the next copy of H. The analysis can then focus on each individual copy of H to determine the advice complexity. For families of *disconnected* graphs, this is not always so simple: Remnants of a previous copy of some H together with parts of another copy of H may themselves be a copy of some other $H' \in \mathcal{F}$. Thus, while constructing gadgets for some singular copy of $H \in \mathcal{F}$ is usually simple and can force an algorithm to distinguish the whole edge set of H, this generally breaks down once the next copy of H is presented in the same instance.

The rest of this work is structured as follows. We first give formal definitions of the problems that we analyze in this work and introduce necessary notation. We then give tight bounds on the DELAYED $\mathcal{F}$-NODE-DELETION PROBLEM for completely connected and completely disconnected families $\mathcal{F}$ and analyze the advice complexity of the Delayed H-Edge-Deletion Problem and Delayed H-Node-Deletion Problem. We then take a closer look at the general Delayed $\mathcal{F}$-Node-Deletion Problem, showing that its advice complexity is heavily dependent on the concrete obstruction set $\mathcal{F}$. To this end, we show that depending on $\mathcal{F}$, constant, logarithmic or linear advice can be necessary and sufficient to optimally solve the Delayed $\mathcal{F}$-Node-Deletion Problem. A full version that contains all the proofs is available via arXiv [2].

1.1 Notation and Problem Definitions

We use standard graph notation in this work. Given an undirected graph $G = (V, E)$, $G[V']$ with $V' \subseteq V$ denotes the graph induced by the node set V'. We use $|G|$ to denote $|V(G)|$ and $||G||$ to denote $|E(G)|$. We use the notation $\overline{G}$ for the complement graph of G. K_n and $\overline{K_n}$ denote the n-clique and the n-independent set respectively. P_n denotes the path on n nodes. The neighborhood of a vertex v in a graph G consists of all vertices adjacent to v and is denoted by $N^G(v)$. A vertex v is called universal, if $N^G(v) = V \backslash \{v\}$. We measure the advice complexity in *opt*, which denotes the size of the optimal solution of a given problem.

We adapt some of the notation of Chen et al. [9]. We use $H \trianglelefteq_\varphi G$ for graphs H and G iff there exists an isomorphism φ such that $\varphi(H)$ is an induced subgraph of G. A graph G is called *$\mathcal{F}$-free* if there is no $H \trianglelefteq_\varphi G$ for any $H \in \mathcal{F}$. Furthermore, a *gluing* operation works as follows: Given two graphs G and G', identify a single node from G and a single node from G'. For example, if we glue together with at the gray nodes, the resulting graph is 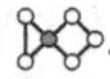 .

Definition 1. *Let $\mathcal{F}$ be a fixed family of graphs. Given an online graph G induced by its nodes $V(G) = \{v_1, \ldots, v_n\}$, ordered by their occurrence in an online instance. The* Delayed $\mathcal{F}$-Node-Deletion Problem *is for every i to select a set $S_i \subseteq \{v_1, \ldots, v_i\}$ such that $G[\{v_1, \ldots, v_i\}] - S_i$ is $\mathcal{F}$-free. Furthermore, it has to hold that $S_1 \subseteq \ldots \subseteq S_n$, where $|S_n|$ is to be minimized.*

The definition of the Delayed $\mathcal{F}$-Edge-Deletion Problem is identical, with S being a set of edges of G instead of nodes. If $\mathcal{F} = \{H\}$ we speak of an H-Deletion Problem instead of an {H}-Deletion Problem.

For an obstruction set $\mathcal{F}$ we assume that there exist no distinct $H_1, H_2 \in \mathcal{F}$ with $H_1 \trianglelefteq_\varphi H_2$ for some isomorphism φ, as each online graph containing H_2 also contains H_1, making H_2 redundant. Furthermore, we assume in the case of the Delayed $\mathcal{F}$-Edge-Deletion Problem that $\mathcal{F}$ contains no $\overline{K_n}$ for any n. This assumption is arguably reasonable as no algorithm that can only delete edges is able to remove a set of isolated nodes from a graph.

2 Essentially Tight Advice Bounds

One can easily construct a naive online algorithm for the Delayed $\mathcal{F}$-Node-Deletion Problem that is provided a complete optimal solution on the advice tape and deletes nodes accordingly. The online algorithm does not make any decisions itself, but strictly follows the solution provided by the oracle. The resulting trivial upper bounds on the advice complexity of the problem is summarized in the following theorem.

Theorem 1. *Let $\mathcal{F}$ be an arbitrary family of graphs, and let $H \in \mathcal{F}$ be a maximum order graph. Then there is an optimal online algorithm for the* Delayed $\mathcal{F}$-Node-Deletion Problem *with advice that reads at most $opt \cdot \log|H| + O(\log opt)$ bits of advice.*

Proof. The online algorithm reads *opt* from the tape using self-delimiting encoding (see [12]). Then it reads $\lceil opt \cdot \log|H| \rceil$ bits and interprets them as *opt* numbers $d_1, ..., d_{opt} \leq |H|$. Whenever some forbidden induced graph is detected, the algorithm deletes the d_ith node. □

With the same idea we can easily get a similar upper bound for the Delayed H-Edge-Deletion Problem.

Theorem 2. *There is an optimal online algorithm with advice for the* Delayed H-Edge-Deletion Problem *that reads at most $opt \cdot \log \|H\| + O(\log opt)$ bits of advice.*

In this section we show that for the Delayed $\mathcal{F}$-Node-Deletion Problem for connected or disconnected $\mathcal{F}$, as well as for the Delayed H-Edge-Deletion Problem this naive strategy is already the best possible. More formally, we meet these trivial upper bounds by essentially tight lower bounds for the aforementioned problems. We call lower bounds of the form $opt \cdot \log|H|$, or $opt \cdot \log\|H\|$ *essentially tight*, because they only differ from the trivial upper bound by some logarithmic term in *opt*. This additional term stems from the fact that the advisor must encode *opt* onto the advice tape in order for the online algorithm to correctly interpret the advice. If the online algorithm knew *opt* in advance, we would have exactly tight bounds.

2.1 Connected and Disconnected $\mathcal{F}$-Node Deletion Problems

Chen et al. [9] previously proved essentially tight bounds on the advice complexity of the Delayed $\mathcal{F}$-Node-Deletion Problem for the case that all graphs in $\mathcal{F}$ are connected. They found a lower bound of $opt \cdot \log|H|$ where H is a maximum order graph in $\mathcal{F}$. Additionally, they proved a lower bound on the advice complexity of the Delayed H-Node-Deletion Problem for disconnected H that was dependent on a maximum order connected component C_{max} of H: $opt \cdot \log|C_{max}| + O(\log opt)$. We improve this result and provide a lower bound on the advice complexity of the Delayed $\mathcal{F}$-Node-Deletion Problem for families $\mathcal{F}$ of disconnected graphs, that essentially matches the trivial upper bound from Theorem 1.

Lemma 1. *Let $\mathcal{F}$ be an arbitrary obstruction set, and let $\overline{\mathcal{F}} := \{\overline{H} \mid H \in \mathcal{F}\}$ be the family of complement graphs. Then the advice complexity of the* Delayed $\mathcal{F}$-Node-Deletion Problem *is the same as for the* Delayed $\overline{\mathcal{F}}$-Node-Deletion Problem.

Proof. We provide an advice preserving reduction from DELAYED $\mathcal{F}$-NODE-DELETION PROBLEM to DELAYED $\overline{\mathcal{F}}$-NODE-DELETION PROBLEM: If G is an online instance for the $\mathcal{F}$-problem, the complement graph of G is revealed for $\overline{\mathcal{F}}$-problem. An optimal online algorithm has to delete the same nodes in the same time steps in both instances. This proves that the advice complexity for the $\mathcal{F}$-problem is at most the advice complexity for the $\overline{\mathcal{F}}$-problem. The same reduction in the other direction yields equality. □

From this follows immediately the desired lower bound on the advice complexity of the DELAYED $\mathcal{F}$-NODE-DELETION PROBLEM for disconnected $\mathcal{F}$.

Theorem 3. *Let $\mathcal{F}$ be a family of disconnected graphs, and $H \in \mathcal{F}$ a maximum order graph. Then any optimal online algorithm for the* DELAYED $\mathcal{F}$-NODE-DELETION PROBLEM *needs $opt \cdot \log|H|$ bits of advice.*

Proof. Since all graphs in $\mathcal{F}$ are disconnected, $\overline{\mathcal{F}}$ is a family of connected graphs. For $\overline{\mathcal{F}}$, the lower bound proven by Chen et al. of $opt \cdot \log|H|$ holds. The claim follows from Lemma 1. □

In summary, this lower bound of $opt \cdot \log(\max_{H \in \mathcal{F}} |H|)$ holds for all families of graphs $\mathcal{F}$ that contain either only connected or only disconnected graphs. In particular, these results imply a tight lower bound on the advice complexity of the DELAYED H-NODE-DELETION PROBLEM for arbitrary graphs H, which Chen et al. raised as an open question.

Corollary 1. *Let H be an arbitrary graph. Then any online algorithm for the* DELAYED H-NODE-DELETION PROBLEM *requires $opt \cdot \log|H|$ bits of advice to be optimal.*

We want to briefly reiterate the main steps of the lower bound proof by Chen et al. for connected $\mathcal{F}$. Let H be a maximum order graph in $\mathcal{F}$. The idea is to construct $|H|^{opt}$ different instances with optimal solution size opt such that no two of these instances can be handled optimally with the same advice string. These instances consist of the disjoint unions of opt gadgets where each gadget is constructed by gluing two copies of H at an arbitrary node. This way, each gadget needs at least one node deletion, and deleting the glued node is the only optimal way to make a gadget $\mathcal{F}$-free. Since in each of the opt gadgets we have $|H|$ choices of where to glue the two copies together, we in total construct $|H|^{opt}$ instances. This procedure of constructing instances that have to be handled by different advice strings is a standard method of proving lower bounds on the advice complexity.

As the proof of Theorem 3 uses this result by Chen et al. one can examine closer the instances that are constructed implicitly in this proof. As a "dual" approach to the disjoint gadget constructions, we will use the *join* of graphs.

Definition 2. *Given two graphs G_1, G_2. The* join graph *$G = G_1 \nabla G_2$ is constructed by connecting each vertex of G_1 with each vertex of G_2 with an edge, i.e. $V(G) = V(G_1) \cup V(G_2)$, $E(G) = E(G_1) \cup E(G_2) \cup \{ v_1v_2 \mid v_1 \in V(G_1), v_2 \in V(G_2) \}$.*

First of all we look at how the gadgets for disconnected H look now. Let H be a maximum order graph of $\mathcal{F}$ where $\mathcal{F}$ consists of disconnected graphs. Then a gadget of H is the complement graph of a gadget of the connected graph $\overline{H}$. Therefore, a gadget of H is constructed by gluing two copies of H at some arbitrary vertex and then joining them everywhere else. For example, the graph ∘ ∘–∘ has the following three possible gadgets: • , , and 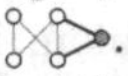 . Here the gray vertices were used for gluing, and the upper copy was joined with the lower copy everywhere else. Since in the connected case the instance consisted of the disjoint union of gadgets, in the "dual" case of disconnected forbidden graphs we join them. Thus, the constructed instances are join graphs of gadgets where each gadget is the join of two copies of H glued at an arbitrary vertex. Just as in the proof by Chen et al. we can construct $|H|^{opt}$ such instances which all need different advice strings in order to be handled optimally, and therefore the lower bound of $opt \cdot \log |H|$ holds also for disconnected $\mathcal{F}$.

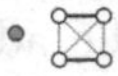

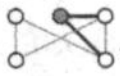

Similar constructions even work for the Delayed H-Edge-Deletion Problem and result in an essentially tight lower bound on its advice complexity as we will see in the next subsection.

2.2 H-Edge Deletion Problem

Chen et al. previously proved a lower bound of $opt \cdot \log \|H\|$ on the advice complexity of the Delayed H-Edge-Deletion Problem for connected graphs H which essentially matches the trivial upper bound from Theorem 2. We show that the same bound even holds if H is disconnected. For the node-deletion problem with disconnected $\mathcal{F}$ we constructed instances that make extensive use of the join operation. We will see that similar constructions can be used for the edge-deletion case. It will be insightful to understand why exactly the join operation behaves nicely with disconnected graphs. The most important observation is summarized in the following lemma.

Lemma 2. *Let H be a disconnected graph and let G_1, G_2 be two other graphs. Then $G_1 \nabla G_2$ is H-free iff G_1 and G_2 are H-free.*

We introduce the notion of an e-extension of a graph H. Intuitively, an e-extension is a graph $U_H(e)$ that extends H in such a way that the unique optimal way to make $U_H(e)$ H-free is to delete the edge e.

Definition 3. *For a disconnected graph H and an edge $e \in E(H)$ we call a graph $U_H(e)$ an* e-extension *of H if it satisfies*

(E.1) $H \trianglelefteq U_H(e)$,
(E.2) $H \not\trianglelefteq_{\varphi} U_H(e) - e$, *and*
(E.3) $H \trianglelefteq_{\varphi} U_H(e) - f$ *for all* $f \in E(U_H(e)) \setminus \{e\}$.

We call H edge-extendable *if for every $e \in E(H)$ there is such an e-extension.*

It turns out that extendability is a sufficient condition for the desired lower bound to hold.

Theorem 4. *Let H be an edge-extendable graph. Then any optimal online algorithm for the* Delayed H-Edge-Deletion Problem *needs* $opt \cdot \log \|H\|$ *bits of advice.*

Proof. Let $m \in \mathbf{N}$ be arbitrary. We construct a family of instances with optimal solution size m such that any optimal online algorithm needs advice to distinguish these instances. Take m disjoint copies $H^{(1)}, ..., H^{(m)}$ of H. We denote the vertices of $H^{(i)}$ by $v^{(i)}$. Furthermore, let $e_1, ..., e_m$ be arbitrary edges such that $e_i \in E(H^{(i)})$. We construct the instance $G(e_1, ..., e_m)$ in m phases. In the ith phase we reveal $H^{(i)}$ and join it with the already revealed graph from previous phases. Then we extend $H^{(i)}$ to $U_{H^{(i)}}(e_i)$ and again join the newly added vertices with the already revealed graph from the previous phases. If $G(e_1, ..., e_{i-1})$ is the graph after phase $i-1$, after phase i we have revealed a graph isomorphic to $G(e_1, ..., e_{i-1}) \nabla U_{H^{(i)}}(e_i)$. Thus, $G := G(e_1, ..., e_m) \simeq U_{H^{(1)}}(e_1) \nabla ... \nabla U_{H^{(m)}}(e_m)$. We claim that $X := \{e_1, ..., e_m\}$ is the unique optimal solution for the H-Edge Deletion problem on G. Deleting all e_i from G yields a graph isomorphic to $(U_{H^{(1)}}(e_1) - e_1) \nabla ... \nabla (U_{H^{(m)}}(e_m) - e_m)$. By definition of an e-extension, and Lemma 2 this graph is H-free. Thus X is a solution. It is also optimal because G contains m edge-disjoint copies of H. Finally, if in one of the $U_{H^{(i)}}(e_i)$ we delete any other edge than e_i, by definition we need to delete at least one more edge to make $U_{H^{(i)}}(e_i)$ H-free. Hence, X is the unique optimal solution.

We can construct $\|H\|^m$ such instances that pairwise only differ in the choice of the edges $e_1, ..., e_m$. Any online algorithm needs advice to distinguish these instances, and therefore requires $m \cdot \log \|H\|$ bits to be optimal on all of them. Since $m = opt$, the claim is proven. □

We prove constructively that each disconnected graph H without isolated vertices is edge-extendable. We then handle the case that H has isolated vertices.

Lemma 3. *Let H be a disconnected graph without isolated vertices. Then H is edge-extendable.*

Proof (Sketch). We create two copies of H and identify two edges, one of each H. We join the remaining nodes of the two copies. Figure 1 shows an example. □

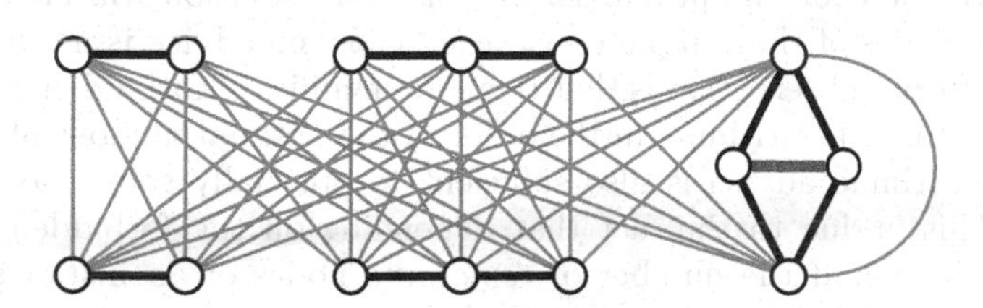

Fig. 1. Example for the e-extension of the graph $P_2 \cup P_3 \cup K_3$ as constructed in Lemma 3. The edge e is depicted in orange. (Color figure online)

The results from Chen et al. together with Theorem 3 and 4 yield the following corollary.

Corollary 2. *Let H be a graph without isolated vertices. Then any optimal online algorithm for the* DELAYED H-EDGE-DELETION PROBLEM *needs* $opt \cdot \log \|H\|$ *bits of advice.*

We now turn finally to the case where H has isolated nodes. We prove via a simple advice-preserving reduction from the known case to the case where H has isolated vertices that the same lower bound holds.

Theorem 5. *Let H be a graph with $k > 0$ isolated vertices. Then any optimal online algorithm for the* DELAYED H-EDGE-DELETION PROBLEM *needs at least* $opt \cdot \log \|H\|$ *bits of advice.*

Proof. Let H' be the graph we obtain from H by deleting all isolated vertices. Of course, $\|H\| = \|H'\|$. For any graph G' we have: G' is H'-free iff $\overline{K_k} \cup G'$ is H-free. Let G' be an online instance for the DELAYED H'-EDGE-DELETION PROBLEM. We construct an online instance G for the DELAYED H-EDGE-DELETION PROBLEM that presents $\overline{K_k}$ in the first k time steps and then continues to present G' node by node. Note that $|G| = |G'| + k$. The deletions in G' can be translated in to deletions in G by shifting k time steps. The optimal solutions of G and G' coincide up to this shifting by k time steps, and of course $opt_H(G) = opt_{H'}(G') = opt$. Thus, the advice complexity for the DELAYED H-EDGE-DELETION PROBLEM is at least the advice complexity for the DELAYED H'-EDGE-DELETION PROBLEM. For this problem, however, we already have a lower bound from Corollary 2. Thus, the same lower bound applies to the case where H has isolated vertices. □

3 The Delayed $\mathcal{F}$-Node-Deletion Problem

We have seen that for obstruction sets, in which all or none of the graphs are connected, the advice complexity is linear in the number of optimal deletions. This is not always the case when considering general families of graphs $\mathcal{F}$ as obstruction sets. An easy example is the following: consider the family $\mathcal{F}_4$ that contains all graphs over four nodes. Clearly, whenever any fourth node of an online graph is revealed, a node has to be deleted. Yet which concrete node is deleted is arbitrary for an optimal solution as every solution will have to delete all but three nodes of the complete instance. Thus, no advice is needed.

A more curious observation is that there are families of forbidden graphs that need advice that is logarithmic in the order of the optimal number of deletions. Further, logarithmic advice is also sufficient to optimally solve most of these problems. This is due to the fact that depending on the forbidden family of graphs we can bound the number of *remaining* nodes of an instance after an optimal number of deletions has been made.

We start by observing that when an $\mathcal{F}$ contains both an independent set and a clique, the size of the biggest graph that contains no $H \in \mathcal{F}$ is bounded.

Lemma 4. *Let $\mathcal{F}$ be an arbitrary family of graphs. Then there exists a minimal $R \in \mathbf{N}$ such that all graphs of size at least R are not $\mathcal{F}$-free iff $K_n, \overline{K_m} \in \mathcal{F}$ for some $n, m \in \mathbf{N}$.*

Proof. Ramsey's Theorem guarantees the existence of R if $K_n, \overline{K_m} \in \mathcal{F}$. Conversely, if $\mathcal{F}$ contains no clique (independent set), then arbitrarily big cliques (independent sets) are $\mathcal{F}$-free. □

We can use this observation to construct an algorithm that is mainly concerned with the remaining graph after all deletions have been made. As the size of this graph is bounded by a constant, we can simply tell an online algorithm when it should *not* delete every node of an H that it sees and which one that is.

Theorem 6. *If $K_n, \overline{K_m} \in \mathcal{F}$ for some $n, m \in \mathbf{N}$, then there is an optimal online algorithm with advice for the* Delayed $\mathcal{F}$-Node-Deletion Problem *that uses $O(\log opt)$ bits of advice.*

Proof. Let R be as in Lemma 4, and k be the size of the biggest graph in $\mathcal{F}$. Algorithm 1 uses at most $\lceil \log(R-1) \rceil + \lceil (R-1) \cdot \log(opt \cdot k) \rceil = (R-1)\log(opt) + O(1)$ bits of advice. We assume that the algorithm is given *opt* beforehand, which can be encoded using $O(\log opt)$ bits of advice using self-delimiting encoding as in [3]. The advisor computes an optimal offline solution. After deleting the nodes from G, a graph with at most $R-1$ nodes must remain, otherwise it would not be $\mathcal{F}$-free by Lemma 4. Let $u \leq R-1$ be the number of nodes that are considered by the online algorithm below and that will *not* be deleted. The advisor writes u onto the tape. Next, for all those u nodes $(v_i)_{i \leq u}$, the advisor computes in which round the algorithm considers this node for the first time. A node is considered by the algorithm if it is part of an $H \in \mathcal{F}$ that at some point is recognized. The node (v_i) can thus be identified by a pair $(r_i, a_i) \in \{1, \ldots, opt\} \times \{1, \ldots, k\}$. Then the algorithm encodes all these pairs $((r_i, a_i))_{i \leq u}$ onto the tape.

The algorithm starts by reading u and these pairs from the tape. Then it sets its *round counter* r to 1, and the set of *fixed nodes* (i.e. the set of enounctered nodes that will not be deleted) F to $\emptyset$. Whenever the algorithm finds a forbidden induced subgraph it checks in the list (r_i, a_i) which of its nodes it must not delete, and adds them to F. Then it deletes any other vertex from $W \backslash F$. □

This proof implies that, given some $\mathcal{F}$, if we can always bound the size of the graph remaining after deleting an optimal number of nodes by a constant, we can construct an algorithm that solves the Delayed $\mathcal{F}$-Node-Deletion Problem with advice logarithmic in *opt*. Under certain conditions we also get a lower bound logarithmic in *opt* as we will see in the following two theorems.

Theorem 7. *Let $K_n, \overline{K_m} \in \mathcal{F}$, and let D be a graph that is $\mathcal{F}$-free, $|D| = R-1$, and $\|D\|$ is maximal among such graphs. If D has no universal vertex, then any optimal online algorithm for the* Delayed $\mathcal{F}$-Node-Deletion Problem *needs $\Omega(\log opt)$ bits of advice.*

With a similar construction for independent sets instead of cliques we get the following sufficient condition for the necessity of logarithmic advice.

Theorem 8. *Let $K_n, \overline{K_m} \in \mathcal{F}$, and let D be a graph that is $\mathcal{F}$-free, $|D| = R-1$, and $\|D\|$ is minimal among such graphs. If D has no isolated vertex, then any optimal online algorithm for the* Delayed $\mathcal{F}$-Node-Deletion Problem *needs $\Omega(\log opt)$ bits of advice.*

Algorithm 1. Optimal Online Algorithm with Logarithmic Advice

1: Read $\lceil \log(R-1) \rceil$ bits of advice, interpret as number $u \in \{1, \dots, R-1\}$
2: Read $\lceil u \cdot \log(k \cdot opt) \rceil$ bits of advice, interpret as u pairs $((r_i, a_i))_{i \leq u} \subseteq \{1, \dots, opt\} \times \{1, \dots, k\}$
3: $r \leftarrow 1, F \leftarrow \emptyset$
4: **for all** $t = 1, \dots, T$ **do**
5: $\quad G_t \leftarrow$ reveal next node
6: $\quad$ **while** $G_t[W] \simeq H \in \mathcal{F}$ for some $W \subseteq V(G_t)$ **do**
7: $\quad\quad$ **for all** $i = 1, \dots, u$ **do**
8: $\quad\quad\quad$ **if** $r == r_i$ **then**
9: $\quad\quad\quad\quad v_i \leftarrow a_i$'th vertex of W
10: $\quad\quad\quad\quad F \leftarrow F \cup \{v_i\}$
11: $\quad\quad$ Delete any vertex from $W \backslash F$ (or all at once)
12: $\quad\quad r \leftarrow r + 1$

4 Further Work

While we were able to shed further light on the bigger picture of online node and edge deletion problems with advice, the most general problems of their kind are still not solved. For node deletion problems, the case of an obstruction set with both connected and disconnected graphs proves to be much more involved, with the advice complexity being heavily dependent on the obstruction set, as we have seen in the previous section.

The logarithmic bounds of this paper cannot be directly transferred to the Delayed $\mathcal{F}$-Edge-Deletion Problem, as independent sets cannot be part of the obstruction set. There are, of course, families $\mathcal{F}$ for which no advice is necessary, e.g., $\mathcal{F} = \{\circ\!\!-\!\!\circ\}$, but it seems hard to find non-trivial families for which less than linear advice is both necessary and sufficient. An additional difficulty is that forbidden graphs may be proper (non-induced) subgraphs of one another, which makes it difficult to count deletions towards individual copies of forbidden graphs. Chen et al. [9] proposed a recursive way to do so, but it is unclear if their analysis can be generalized to arbitrary families of forbidden graphs $\mathcal{F}$.

References

1. Ahmadian, S., Esfandiari, H., Mirrokni, V.S., Peng, B.: Robust load balancing with machine learned advice. In: Naor, J.S., Buchbinder, N. (eds.) Proceedings of the 2022 ACM-SIAM Symposium on Discrete Algorithms, SODA 2022, Virtual Conference / Alexandria, 9 January–12 January 2022, pp. 20–34. SIAM (2022). https://doi.org/10.1137/1.9781611977073.2
2. Berndt, N., Lotze, H.: Advice complexity bounds for online delayed $\mathcal{F}$-node-, H-node- and H-edge-deletion problems. CoRR abs/2303.17346 (2023)
3. Böckenhauer, H., Komm, D., Královič, R., Královič, R.: On the advice complexity of the k-server problem. J. Comput. Syst. Sci. **86**, 159–170 (2017). https://doi.org/10.1016/j.jcss.2017.01.001

4. Böckenhauer, H., Komm, D., Královič, R., Královič, R., Mömke, T.: Online algorithms with advice: the tape model. Inf. Comput. **254**, 59–83 (2017). https://doi.org/10.1016/j.ic.2017.03.001
5. Borodin, A., El-Yaniv, R.: Online Computation and Competitive Analysis. Cambridge University Press, Cambridge (1998)
6. Boyar, J., Eidenbenz, S.J., Favrholdt, L.M., Kotrbcík, M., Larsen, K.S.: Online dominating set. In: Pagh, R. (ed.) 15th Scandinavian Symposium and Workshops on Algorithm Theory, SWAT 2016, June 22–24, 2016, Reykjavik, Iceland. LIPIcs, vol. 53, pp. 21:1–21:15. Schloss Dagstuhl - Leibniz-Zentrum für Informatik (2016). https://doi.org/10.4230/LIPIcs.SWAT.2016.21
7. Boyar, J., Favrholdt, L.M., Kotrbčík, M., Larsen, K.S.: Relaxing the irrevocability requirement for online graph algorithms. In: WADS 2017. LNCS, vol. 10389, pp. 217–228. Springer, Cham (2017). https://doi.org/10.1007/978-3-319-62127-2_19
8. Boyar, J., Favrholdt, L.M., Kudahl, C., Larsen, K.S., Mikkelsen, J.W.: Online algorithms with advice: a survey. ACM Comput. Surv. **50**(2), 19:1–19:34 (2017). https://doi.org/10.1145/3056461
9. Chen, L.-H., Hung, L.-J., Lotze, H., Rossmanith, P.: Online node- and edge-deletion problems with advice. Algorithmica **83**(9), 2719–2753 (2021). https://doi.org/10.1007/s00453-021-00840-9
10. Dobrev, S., Královič, R., Pardubská, D.: Measuring the problem-relevant information in input. ITA **43**(3), 585–613 (2009). https://doi.org/10.1051/ita/2009012
11. Hromkovič, J., Královič, R., Královič, R.: Information complexity of online problems. In: Hliněný, P., Kučera, A. (eds.) MFCS 2010. LNCS, vol. 6281, pp. 24–36. Springer, Heidelberg (2010). https://doi.org/10.1007/978-3-642-15155-2_3
12. Komm, D.: An Introduction to Online Computation - Determinism, Randomization Advice. Texts in Theoretical Computer Science. An EATCS Series. Springer, Switzerland (2016). https://doi.org/10.1007/978-3-319-42749-2
13. Lindermayr, A., Megow, N.: Permutation predictions for non-clairvoyant scheduling. In: Agrawal, K., Lee, I.A. (eds.) SPAA 2022: 34th ACM Symposium on Parallelism in Algorithms and Architectures, Philadelphia, PA, USA, July 11–14, 2022, pp. 357–368. ACM (2022). https://doi.org/10.1145/3490148.3538579
14. Lindermayr, A., Megow, N., Simon, B.: Double coverage with machine-learned advice. In: Braverman, M. (ed.) 13th Innovations in Theoretical Computer Science Conference, ITCS 2022, January 31 - February 3, 2022, Berkeley, CA, USA. LIPIcs, vol. 215, pp. 99:1–99:18. Schloss Dagstuhl - Leibniz-Zentrum für Informatik (2022). https://doi.org/10.4230/LIPIcs.ITCS.2022.99
15. Lykouris, T., Vassilvitskii, S.: Competitive caching with machine learned advice. J. ACM **68**(4), 24:1–24:25 (2021). https://doi.org/10.1145/3447579
16. Sleator, D.D., Tarjan, R.E.: Amortized efficiency of list update rules. In: DeMillo, R.A. (ed.) Proceedings of the 16th Annual ACM Symposium on Theory of Computing, 30 April–2 May 1984, Washington, pp. 488–492. ACM (1984). https://doi.org/10.1145/800057.808718

Parameterized Algorithms for Eccentricity Shortest Path Problem

Sriram Bhyravarapu[1(✉)], Satyabrata Jana[1], Lawqueen Kanesh[2], Saket Saurabh[1,3], and Shaily Verma[1]

[1] The Institute of Mathematical Sciences, HBNI, Chennai, India
{sriramb,satyabrataj,saket,shailyverma}@imsc.res.in
[2] Indian Institute of Technology Jodhpur, Jodhpur, India
lawqueen@iitj.ac.in
[3] University of Bergen, Bergen, Norway

Abstract. Given an undirected graph $G = (V, E)$ and an integer ℓ, the Eccentricity Shortest Path (ESP) problem asks to check if there exists a shortest path P such that for every vertex $v \in V(G)$, there is a vertex $w \in P$ such that $d_G(v, w) \leq \ell$, where $d_G(v, w)$ represents the distance between v and w in G. Dragan and Leitert [Theor. Comput. Sci. 2017] studied the optimization version of this problem which asks to find the minimum ℓ for ESP and showed that it is NP-hard even on planar bipartite graphs with maximum degree 3. They also showed that ESP is W[2]-hard when parameterized by ℓ. On the positive side, Kučera and Suchý [IWOCA 2021] showed that ESP is fixed-parameter tractable (FPT) when parameterized by modular width, cluster vertex deletion set, maximum leaf number, or the combined parameters disjoint paths deletion set and ℓ. It was asked as an open question in the same paper, if ESP is FPT parameterized by disjoint paths deletion set or feedback vertex set. We answer these questions and obtain the following results:

1. ESP is FPT when parameterized by disjoint paths deletion set, or the combined parameters feedback vertex set and ℓ.
2. A $(1 + \epsilon)$-factor FPT approximation algorithm when parameterized by the feedback vertex set number.

Keywords: Shortest path · Eccentricity · Feedback vertex set · FPT · W[2]-hardness

1 Introduction

Given a graph $G = (V, E)$ and a path P, the *distance* from a vertex $v \in V(G)$ to P is $\min\{d_G(v, w) \mid w \in V(P)\}$, where $d_G(v, w)$ is the distance between v and w in G. Given a graph G and a path P, the *eccentricity* of P, denoted by $\mathsf{ecc}_G(P)$, with respect to G is defined as the maximum over all of the shortest distances between each vertex of G and P. Formally, $\mathsf{ecc}_G(P) = \max\{d_G(u, P) | u \in V(G)\}$. Dragan and Leitert [6] introduced the problem of finding a shortest path with minimum eccentricity, called the Minimum Eccentricity Shortest Path

S.-Y. Hsieh et al. (Eds.): IWOCA 2023, LNCS 13889, pp. 74–86, 2023.
https://doi.org/10.1007/978-3-031-34347-6_7

problem (for short MESP) in a given undirected graph. They found interesting connections between MESP and the MINIMUM DISTORTION EMBEDDING problem and obtained a better approximation algorithm for MINIMUM DISTORTION EMBEDDING. MESP may be seen as a generalization of the DOMINATING PATH Problem [7] that asks to find a path such that every vertex in the graph either belongs to the path or has a neighbor in the path. In MESP, the objective is to find a shortest path P in G such that the eccentricity of P is minimum. Throughout the paper, we denote the minimum value over the eccentricities of all the shortest paths in G as the *eccentricity of the graph* G, denoted by $\mathsf{ecc}(G)$. MESP has applications in transportation planning, fluid transportation, water resource management, and communication networks.

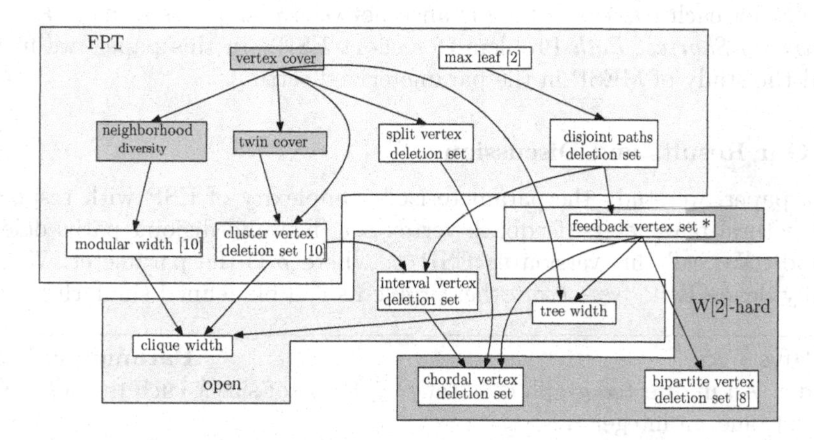

Fig. 1. The hierarchy of parameters explored in this work. Arrow pointing from parameter a to parameter b indicates $b \leq f(a)$, for some computable function f. Parameters in red are studied in this paper (see full version [1]). The symbol "$*$" attached to the feedback vertex set means it is FPT in combination with the desired eccentricity. The results for the parameters in grey boxes are subsumed by the existing results. (Color figure online)

Dragan and Leitert [5] demonstrated that fast algorithms for MESP imply fast approximation algorithms for MINIMUM LINE DISTORTION, and the existence of low eccentricity shortest paths in special graph classes will imply low approximation bounds for those classes. They also showed that MESP is NP-hard on planar bipartite graphs with maximum degree 3. In parameterized settings, they showed that MESP is W[2]-hard for general graphs and gave an XP algorithm for the problem when parameterized by eccentricity. Furthermore, they designed 2-approximation, 3-approximation, and 8-approximation algorithms for MESP running in time $O(n^3)$, $O(nm)$, and $O(m)$ respectively, where n and m represents the number of vertices and edges of the graph. The latter 8-approximation algorithm uses the double-BFS technique. In 2016, Birmelé et al. [2] showed that the algorithm is, in fact, a 5-approximation algorithm by a deeper analysis of the double-BFS procedure and further extended the idea to get a 3-approximation algorithm, which still runs in linear time. Furthermore, they study the link between MESP and the laminarity of graphs introduced by Volké et al. [9] in which the covering path is required to be a diameter and established some tight bounds between MESP and the laminarity parameters. Dragan and Leitert [5] showed that MESP can be solved in linear time on distance-hereditary graphs and in polynomial time on chordal and dually chordal graphs. Recently, Kučera and Suchý [8] studied MESP with respect to some structural parameters and provided FPT algorithms for the problem with respect to modular width, cluster vertex deletion (clvd), maximum leaf number, or the combined parameters disjoint paths deletion (dpd) and eccentricity (ecc), see Fig. 1. We call the decision version of MESP, which is to check if there exists a shortest path P such that for each $v \in V(G)$, the distance between v and P is at most ℓ, as the *Eccentricity Shortest Path* Problem (for short ESP). In this paper, we further extend the study of MESP in the parameterized setting.

1.1 Our Results and Discussion

In this paper, we study the parameterized complexity of ESP with respect to the structural parameters: feedback vertex set (fvs) and disjoint paths deletion set (dpd). We call this version as ESP/ρ, where ρ is the parameter. We now formally define ESP/fvs + ecc (other problems can be defined similarly).

> ESP/fvs + ecc **Parameter:** $k + \ell$
> **Input:** An undirected graph G, a set $S \subseteq V(G)$ of size k such that $G - S$ is a forest, and an integer ℓ.
> **Question:** Does there exist a shortest path P in G such that for each $v \in V(G)$, $dist_G(v, P) \leq \ell$?

First, we show an algorithm for ESP/fvs + ecc, in Sect. 2, that runs in $2^{\mathcal{O}(k \log k)} \ell^k n^{\mathcal{O}(1)}$ time where ℓ is the eccentricity of the graph and k is the size of a feedback vertex set. In Sect. 3, we design a $(1+\epsilon)$-factor FPT algorithm for ESP/fvs. Then, in Sect. 4 we design an algorithm for ESP/dpd running in time $2^{\mathcal{O}(k \log k)} \cdot n^{\mathcal{O}(1)}$.

Graph Notations. All the graphs considered in this paper are finite, unweighted, undirected, and connected. For standard graph notations, we refer to the graph theory book by R. Diestel [4]. For parameterized complexity terminology, we refer to the parameterized algorithms book by Cygan et al. [3]. For $n \in \mathbb{N}$, we denote the sets $\{1, 2, \cdots, n\}$ and $\{0, 1, 2, \cdots, n\}$ by $[n]$ and $[0, n]$ respectively. For a graph $G = (V, E)$, we use n and m to denote the number of vertices and edges of G. Given an integer ℓ, we say that a path P *covers* a vertex v if there exists a vertex $u \in V(P)$ such that the distance between the vertices u and v, denoted by, $d_G(v, u)$, is at most ℓ. A *feedback vertex set* of a graph G is a set $S \subseteq V(G)$ such that $G - S$ is acyclic.

The proofs of the results marked ($\star$) are presented in the full version of the paper [1].

2 Parameterized by Feedback Vertex Set and Eccentricity

The main theorem of this section is formally stated as follows.

Theorem 1. *There is an algorithm for* ESP/fvs + ecc *running in time* $\mathcal{O}(2^{\mathcal{O}(k \log k)}\ell^k n^{\mathcal{O}(1)})$.

Outline of the Algorithm. Given a graph G and a feedback vertex set S of size k, we reduce ESP/fvs + ecc to a "path problem" (which we call COLORFUL PATH-COVER) on an auxiliary graph G' (a forest) which is a subgraph of $G[V \setminus S]$, using some reduction rules and two intermediate problems called SKELETON TESTING and EXT-SKELETON TESTING. In Sect. 2.1, we show that ESP/fvs + ecc and SKELETON TESTING are FPT-equivalent. Next, in Sect. 2.3, we reduce SKELETON TESTING to EXT-SKELETON TESTING. Then in Sect. 2.4, we reduce EXT-SKELETON TESTING to COLORFUL PATH-COVER. Finally, in Sect. 2.5, we design a dynamic programming based algorithm for COLORFUL PATH-COVER that runs in $\mathcal{O}(\ell^2 2^{\mathcal{O}(k \log k)} n^{\mathcal{O}(1)})$ time. Together with the time taken for the reductions to the intermediate problems, we get our desired FPT algorithm. A flow chart for the steps of the algorithm is shown in Fig. 2.

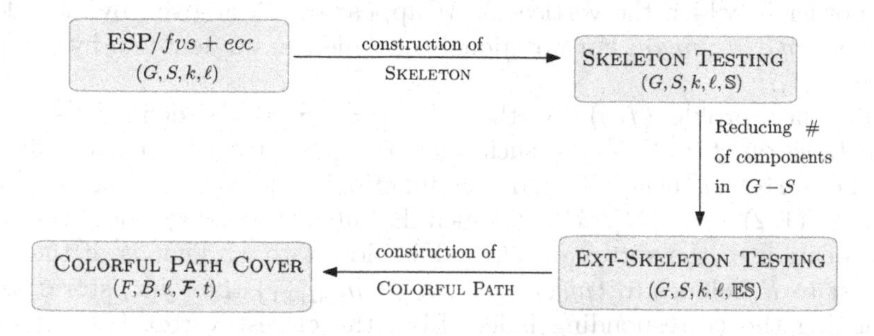

Fig. 2. Flow chart of the Algorithm for ESP/fvs + ecc.

2.1 Reducing to SKELETON TESTING

The input to the problem is an instance (G, S, k, ℓ) where $S \subseteq V(G)$ is a feedback vertex set of size k in G. Let (G, S, k, ℓ) be a yes instance, and P be a solution path which is a shortest path such that for each $v \in V(G)$, there exists $u \in V(P)$ such that $d_G(u, v) \leq \ell$. Our ultimate goal is to construct such a path P. Towards this, we try to get as much information as possible about P in time $f(k, \ell)n^{\mathcal{O}(1)}$. Observe that if S is an empty set, then we can obtain P by just knowing its end-points as there is a unique path in a tree between any two vertices. Generalizing this idea, given the set S, we define the notion of *skeleton* of P.

Definition 1 (Skeleton). A skeleton of P, denoted by $\mathbb{S}$, is the following set of information.

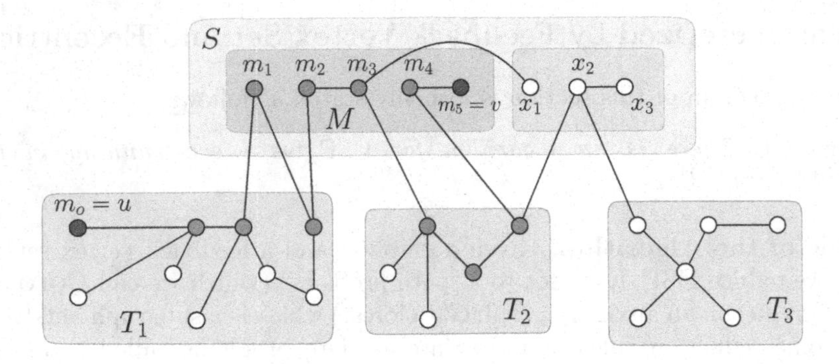

Fig. 3. Example of a skeleton of P. Here P is a shortest path (blue edges) between two red colored vertices u and v through green colored internal vertices. For the vertices x_1, x_2 and x_3, $f(x_1) = f(x_2) = 1$, $f(x_3) = 2$, $g(x_1) = 3$, $g(x_2) = g(x_3) = (3, 4)$. (Color figure online)

- End-vertices of P, say $u, v \in V(G)$.
- A subset of $S \backslash \{u, v\}$, say M, of vertices that appear on P. That is, $V(P) \cap (S \backslash \{u, v\}) = M$.
- The order in which the vertices of M appear on P, is given by an ordering $\pi = m_1, m_2, \ldots, m_{|M|}$. For notational convenience, we denote u by m_0 and v by $m_{|M|+1}$.
- A distance profile (f, g) for the set $X = S \backslash M$, is defined as follows: The function $f : X \to [\ell]$ such that $f(x)$ denotes the shortest distance of the vertex x from P, and the function $g : X \to \{0, 1, \cdots, |M| + 1, (0, 1), (1, 2), \cdots, (|M|, |M|+1)\}$ such that $g(x)$ stores the information about the location of the vertex on P, that is closest to x. That is, if the vertex closest to P belongs to $\{m_0, m_1, \ldots, m_{|M|}, m_{|M|+1}\}$ then $g(x)$ stores this by assigning the corresponding index. Else, the closest vertex belongs to the path segment between m_i, m_{i+1}, for some $0 \leq i \leq |M|$, which $g(x)$ stores by assigning $(i, i + 1)$.

An illustration of a skeleton is given in Fig. 3. By following the definition of skeletons, we get an upper bound on them.

Observation 1. *The number of skeletons is upper bounded by* $n^2 2^k k! \ell^k (2k+2)^k$.

We say that a path P *realizes* a skeleton $\mathbb{S}$ if the following holds.

1. $M = S \cap V(P)$, $X \cap V(P) = \emptyset$, the ordering of vertices in M in P is equal to π, endpoints of P are m_0 and $m_{|M|+1}$,
2. For each $v \in V(G)$, there exists a vertex $u \in V(P)$ such that $d_G(u, v) \leq \ell$,
3. For each $v \in X$, $d_G(v, w) \geq f(v)$ for all $w \in V(P)$ (where $f(v)$ is the shortest distance from v to any vertex on P in G), and
4. For each $v \in X$, if $g(v) = i$, where $i \in [0, |M| + 1]$, then $d_G(v, m_i) = f(v)$ and if $g(v) = (i, i + 1)$ where $i \in [0, |M|]$, then there exists a vertex u on a subpath m_i to m_{i+1} in P such that $u \notin \{m_i, m_{i+1}\}$ and $d_G(u, v) = f(v)$.

Now, given an input (G, S, k, ℓ) and a skeleton $\mathbb{S}$, our goal is to test whether the skeleton can be *realized* into a desired path P. This leads to the following problem.

SKELETON TESTING	**Parameter:** $k + \ell$
Input: A graph G, a set $S \subseteq V(G)$ of size k such that $G - S$ is a forest, an integer ℓ, and a skeleton $\mathbb{S}$.	
Question: Does there exist a shortest path P in G that realizes $\mathbb{S}$?	

Our next lemma shows a reduction from ESP/fvs + ecc to SKELETON TESTING problem.

Lemma 1 ($\star$). *(G, S, k, ℓ) is a yes instance of* ESP/fvs + ecc, *if and only if there exists a skeleton $\mathbb{S}$ such that $(G, S, k, \ell, \mathbb{S})$ is a yes instance of* SKELETON TESTING.

Observation 1 upper bounds the number of skeletons by $2^{\mathcal{O}(k(\log k + \log \ell))}n^2$. This together with Lemma 1, implies that ESP/fvs + ecc and SKELETON TESTING are FPT-equivalent. Thus, from now onwards, we focus on SKELETON TESTING.

2.2 Algorithm for SKELETON TESTING

Let $(G, S, k, \ell, \mathbb{S})$ be an instance of SKELETON TESTING, where $\mathbb{S} = (M, X, \pi, m_0, m_{|M|+1}, f, g)$. Our algorithm works as follows. First, the algorithm performs a simple sanity check by reduction rule. In essence, it checks whether the different components of the skeleton $\mathbb{S}$ are valid.

Reduction Rule 1 (Sanity Test 1). *Return that $(G, S, k, \ell, \mathbb{S})$ is a no instance of* SKELETON TESTING, *if one of the following holds:*

1. *For $i \in [0, |M|]$, $m_i m_{i+1}$ is an edge in G and $g^{-1}((i, i+1)) \neq \emptyset$. ($g$ is not valid.)*
2. *For a vertex $v \in X$, there exists a vertex $u \in M \cup \{m_0, m_{|M|+1}\}$ such that $d_G(u, v) < f(v)$. (f is not valid.)*
3. *For a vertex $v \in X$, $g(v) = i$ and $d_G(v, m_i) > f(v)$. (f is not valid.)*
4. *For an $i \in [0, |M|]$, $m_i m_{i+1}$ is not an edge in G, and there is either no m_i to m_{i+1} path in $G - (S \backslash \{m_i, m_{i+1}\})$ or the length of the path is larger than the shortest path length of m_i to m_{i+1} path in G. (π is not valid.)*
5. *For $i, j \in [0, |M|]$, $i < j$, there exists m_i to m_{i+1} shortest path P_i in $G - (S \backslash \{m_i, m_{i+1}\})$ and a m_j to m_{j+1} shortest path P_j in $G - (S \backslash \{m_j, m_{j+1}\})$ such that if $j = i + 1$, then $(V(P_i) \backslash \{m_{i+1}\}) \cap (V(P_j) \backslash \{m_j\}) \neq \emptyset$, otherwise $V(P_i) \cap V(P_j) \neq \emptyset$. ($\pi$ is not valid – shortest path claim will be violated.)*
6. *For $i \in [0, |M|]$ such that $m_i m_{i+1} \notin E(G)$, $g^{-1}((i, i+1)) \neq \emptyset$, and for every connected component C in $G - S$, and for every m_i to m_{i+1} path P in $G[V(C) \cup \{m_i, m_{i+1}\}]$ there exists a vertex $u \in g^{-1}((i, i+1))$ such that there is no vertex $v \in V(P) \backslash \{m_i, m_{i+1}\}$ for which $d_G(u, v) = f(u)$. (g is not valid.)*

Lemma 2 ($\star$). *Reduction rule 1 is safe.*

Reducing the Components of $G - S$: Now, we describe our marking procedure and reduction rules that are applied on the connected components in $G - S$. Let P_i be a path segment (subpath) of P, between m_i and m_{i+1}, with at least two edges. Further, let P_i^{int} be the subpath of P_i, obtained by deleting m_i and m_{i+1}. Then, we have that P_i^{int} is a path between two vertices in $G - S$ (that is, a path in the forest $G - S$). This implies that P is made up of S and at most $k + 1$ paths of forest $G - S$. Let these paths be $\mathbb{P} = P_1^{int}, \ldots, P_q^{int}$, where $q \leq k+1$. Next, we try to understand these $k+1$ paths of forest $G - S$. Indeed, if there exists a component C in $G - S$ such that it has a vertex that is far away from every vertex in S, then C must contain one of the paths in $\mathbb{P}$ (*essential components*). The number of such components can be at most $k + 1$. The other reason that a component contains a path from $\mathbb{P}$ is to select a path that helps us to satisfy constraints given by the g function (*g-satisfying components*). Next, we give a procedure that marks $\mathcal{O}(k)$ components, and later, we show that all unmarked components can be safely deleted.

Marking Procedure: Let $\mathcal{C}^*$ be the set of marked connected components of $G - S$. Initially, let $\mathcal{C}^* = \emptyset$.

- **Step 1.** If there exists a connected component C in $G - (S \cup V(\mathcal{C}^*))$, such that it contains a vertex v with $d_G(v, m_i) > \ell$, for all $m_i \in M$, and $d_G(v, u) > \ell - f(u)$, for all $u \in X$, then add C to $\mathcal{C}^*$. (Marking essential components)
- **Step 2.** For $i = 0$ to $|M|$ proceed as follows: Let C be some connected component in $G - (S \cup V(\mathcal{C}^*))$ such that there exists a m_i to m_{i+1} path P_i in $G[V(C) \cup \{m_i, m_{i+1}\}]$, which is a shortest m_i to m_{i+1} path in G and for every vertex $v \in g^{-1}((i, i+1))$, there exists a vertex $u \in V(P_i) \backslash \{m_i, m_{i+1}\}$ for which $d_G(u, v) = f(v)$. Then, add C to $\mathcal{C}^*$ and increase the index i. (Marking g-satisfying components)

Let $\mathcal{C}_1$ be the set of connected components added to $\mathcal{C}^*$ in Step 1. We now state a few reduction rules the algorithm applies exhaustively in the order in which they are stated.

Reduction Rule 2. *If $|\mathcal{C}_1| \geq k + 2$, then return that $(G, S, k, \ell, \mathbb{S})$ is a no instance of* Skeleton Testing.

Lemma 3. *Reduction rule 2 is safe.*

Proof. For each component C in $\mathcal{C}_1$, C contains a vertex v such that $d_G(v, m_i) > \ell$, for all $m_i \in M$ and $d_G(v, u) > \ell - f(u)$, for all $u \in X$, which implies we must add a path from component C in solution path as a subpath such that it contains a vertex that covers v. Observe that we can add at most $|M| + 1$ subpaths in the solution path. Therefore, $|\mathcal{C}_1| \leq |M| + 1$ if $(G, S, k, \ell, \mathbb{S})$ is a yes instance of Skeleton Testing. We obtain the required bound as $|M| \leq k$. □

Reduction Rule 3. *If there exists a connected component C in $G - S$ such that $C \notin \mathcal{C}^*$, then delete $V(C)$ from G. The resultant instance is $(G - V(C), S, k, \ell, \mathbb{S})$.*

Lemma 4 ($\star$). *Reduction rule 3 is safe.*

Observe that when Reduction rule 2 and Reduction rule 3 are no longer applicable, the number of connected components in $G-S$ is bounded by $2(k+1)$. This is because $|\mathcal{C}_1| \leq k+1$ and there exists a path (that is part of the solution) from each component in $\mathcal{C}^* - \mathcal{C}_1$ and therefore $|\mathcal{C}^* - \mathcal{C}_1| \leq k+1$. Otherwise, the given instance is a no instance of SKELETON TESTING. Notice that all our reduction rules can be applied in $n^{\mathcal{O}(1)}$ time.

2.3 Reducing SKELETON TESTING to EXT-SKELETON TESTING:

Let $(G, S, k, \ell, \mathbb{S})$ be a reduced instance of SKELETON TESTING. That is, an instance on which Reduction Rules 1, 2 and 3 are no longer applicable. This implies that the number of connected components in $G - S$ is at most $2k + 2$. Next, we enrich our skeleton by adding a function γ, which records an index of a component in $G - S$ that gives the m_i to m_{i+1} subpath in P or records that $m_i m_{i+1}$ is an edge in the desired path P, where $i \in [0, |M|]$.

Definition 2 (Enriched Skeleton). An *enriched skeleton* of a path P, denoted by $\mathbb{ES}$, contains $\mathbb{S}$ and a segment profile of paths between m_i and m_{i+1}, for $i \in [0, M]$. Let $C_1, C_2, \ldots, C_q$ be the connected components in $G - S$. Then, the segment profile is given by a function $\gamma : [0, |M|] \to [0, q]$. The function γ represents the following: For each $i \in [0, |M|]$, if $\gamma(i) = 0$, then the pair m_i, m_{i+1} should be connected by an edge in the solution path P, otherwise if $\gamma(i) = j$, then in P, the m_i to m_{i+1} subpath is contained in $G[V(C_j) \cup \{m_i, m_{i+1}\}]$. Also, $\mathbb{ES}$ is said to be enriching the skeleton $\mathbb{S}$.

Let $\mathbb{S}$ be a skeleton. The number of $\mathbb{ES}$, that enrich $\mathbb{S}$ is upper bounded by $(q+1)^{k+1}$. Thus, this is not useful for us unless q is bounded by a function of k, ℓ. Fortunately, the number of connected components in $G - S$ is at most $2k + 2$, and thus the number of $\mathbb{ES}$ is upper bounded by $2^{\mathcal{O}(k \log k)}$.

We say that a path P *realizes* an enriched skeleton $\mathbb{ES}$ enriching $\mathbb{S}$, if P realizes $\mathbb{S}$ and satisfies γ. Similar to SKELETON TESTING, we can define EXT-SKELETON TESTING, where the aim is to test if a path exists that realizes an enriched skeleton $\mathbb{ES}$. Further, it is easy to see that SKELETON TESTING and EXT-SKELETON TESTING are FPT-equivalent, and thus we can focus on EXT-SKELETON TESTING. Let $(G, S, k, \ell, \mathbb{ES})$ be an instance of EXT-SKELETON TESTING, where $G - S$ has at most $2k + 2$ components. Similarly, as SKELETON TESTING, we first apply some sanity testing on an instance of EXT-SKELETON TESTING.

Reduction Rule 4 (Sanity Test 2). *Return that $(G, S, k, \ell, \mathbb{ES})$ is a no instance of* EXT-SKELETON TESTING, *if one of the following holds:*

1. *$m_i m_{i+1}$ is an edge in G and $\gamma(i) \neq 0$, (or) $m_i m_{i+1}$ is not an edge in G and $\gamma(i) = 0$.*
2. *For an $i \in [|M|]$, $\gamma(i) = j \neq 0$ and there is,*
   - *No m_i to m_{i+1} path in $G[V(C_j) \cup \{m_i, m_{i+1}\}]$, (or)*
   - *No m_i to m_{i+1} path in $G[V(C_j) \cup \{m_i, m_{i+1}\}]$ which is also a shortest m_i to m_{i+1} path in G, (or)*

- *There does not exist a m_i to m_{i+1} path P_i in $G[V(C_j) \cup \{m_i, m_{i+1}\}]$ which is also a shortest m_i to m_{i+1} path in G and satisfies the property that for every vertex $v \in g^{-1}((i, i+1))$, there exists a vertex $u \in V(P_i) \setminus \{m_i, m_{i+1}\}$ for which $d_G(u, v) = f(v)$.*

The safeness of the above rule follows from Definition 2.

2.4 Reducing Ext-Skeleton Testing to Colorful Path-Cover

Let $(G, S, k, \ell, \mathbb{ES})$ be an instance of Ext-Skeleton Testing on which Reduction Rule 4 is no longer applicable. Further, let us assume that the number of components in $G - S$ is $k' \leq 2k + 2$ and $\gamma : [0, |M|] \to [0, k']$ be the function in $\mathbb{ES}$. Our objective is to find a path P that realizes $\mathbb{ES}$. Observe that for an $i \in [0, |M|]$, if $\gamma(i) = j \neq 0$, then the *interesting* paths to connect m_i, m_{i+1} pair are contained in component C_j in $G - S$. Moreover, among all the paths that connect m_i to m_{i+1} in C_j, only the shortest paths that satisfy the function g are the interesting paths. Therefore, we enumerate all the feasible paths for each m_i, m_{i+1} pair in a family $\mathcal{F}_i$ and focus on finding a solution that contains subpaths from this enumerated set of paths only. Notice that now our problem is reduced to finding a set of paths $\mathcal{P}$ in $G - S$ which contains exactly one path from each family of feasible paths and covers all the vertices in $G - S$ which are far away from S. In what follows, we formalize the above discussion. First, we describe our enumeration procedure.

For each $i \in [0, |M|]$ where $\gamma(i) = j \neq 0$, we construct a family $\mathcal{F}_i$ of *feasible paths* as follows. Let P_i be a path in $G[V(C_j) \cup \{m_i, m_{i+1}\}]$, such that (i) P_i is a shortest m_i to m_{i+1} path in G, (ii) for every vertex $v \in g^{-1}((i, i+1))$, $d_G(v, P_i - \{m_i, m_{i+1}\}) = f(v)$. Let m'_i, m'_{i+1} be the neighbours of m_i, m_{i+1}, respectively in P_i. Then we add m'_i to m'_{i+1} subpath to $\mathcal{F}_i$. Observe that a family $\mathcal{F}_i$ of feasible paths satisfies the following properties: (1) $V(\mathcal{F}_i) \cap V(\mathcal{F}_{i'}) = \emptyset$, for all $i, i' \in \gamma^{-1}(j), i \neq i'$, as item 5 of reduction Rule 1 is not applicable, and we add only shortest paths in families. (2) $\mathcal{F}_i$ contains paths from exactly one component in $G - S$ (by the construction). Let $\mathcal{F}$ be the collection of all the families of feasible paths.

The above discussion leads us to the following problem.

Colorful Path-Cover
Input: A forest F, a set $B \subseteq V(F)$, an integer ℓ, and a family $\mathcal{F} = \{\mathcal{F}_1, \mathcal{F}_2, \ldots, \mathcal{F}_t\}$ of t disjoint families of feasible paths.
Question: Is there a set $\mathcal{P}$ of t paths such that for each $\mathcal{F}_i$, $i \in [t]$, $|\mathcal{P} \cap \mathcal{F}_i| = 1$ and for every vertex $v \in B$, there exists a path $P \in \mathcal{P}$ and a vertex $u \in V(P)$, such that $d_F(u, v) \leq \ell$?

Let F be the forest obtained from $G - S$ by removing all the components C_j in $G - S$ such that $\gamma^{-1}(j) = \emptyset$, that is, components which do not contain any interesting paths. Notice that the number of components that contain interesting paths is at most $2k + 2$. We let $B \subseteq V(F)$ be the set of vertices which is not covered by vertices in S, that is, it contains all the vertices $v \in V(F)$ such that $d_G(v, m_i) > \ell$, for all $i \in [0, |M| + 1]$ and $d_G(v, u) > \ell - f(u)$, for all $u \in X$. We claim that it is sufficient to solve COLORFUL PATH-COVER on instance $(F, B, \ell, \mathcal{F})$ where F consists of at most $2k + 2$ trees. The following lemma shows a reduction formally and concludes that EXT-SKELETON TESTING parameterized by k and COLORFUL PATH-COVER problem parameterized by k, are FPT-equivalent.

Lemma 5 (⋆). *$(G, S, k, \ell, \mathbb{ES})$ is a yes instance of* EXT-SKELETON TESTING *if and only if $(F, B, \ell, \mathcal{F})$ is a yes instance of* COLORFUL PATH-COVER.

We design a dynamic programming-based algorithm for the COLORFUL PATH-COVER problem parameterized by k. Since the number of trees is at most $2k+2$, and the number of families of feasible paths is $|\mathcal{F}| = t$, we first guess the subset of families of feasible paths that comes from each tree in $\mathcal{F}$ in $\mathcal{O}(k^t)$ time. Now we are ready to work on a tree with its guessed family of feasible paths. We now present an overview of the algorithm in the next section.

Lemma 6 (⋆). COLORFUL PATH-COVER *can be solved in time $\mathcal{O}(\ell^2 \cdot 2^{\mathcal{O}(k \log k)} n^{\mathcal{O}(1)})$ when F is a forest with $\mathcal{O}(k)$ trees.*

2.5 Overview of the Algorithm for COLORFUL PATH-COVER

Consider an instance $(T, B, \ell, \mathcal{F} = \{\mathcal{F}_1, \mathcal{F}_2, \ldots, \mathcal{F}_t\})$ of COLORFUL PATH-COVER problem where T is a tree, $B \subseteq V(T)$, and $\ell \in \mathbb{N}$ and $\mathcal{F}$ is a disjoint family of feasible paths. The aim is to find a set $\mathcal{P}$ of t paths such that for each $\mathcal{F}_i$, $i \in [t]$, $|\mathcal{P} \cap \mathcal{F}_i| = 1$ and for every vertex $v \in B$, there exists a path $P \in \mathcal{P}$ and a vertex $u \in V(P)$, such that $d_T(u, v) \leq \ell$.

For a vertex $v \in V(T)$, the bottom-up dynamic programming algorithm considers subproblems for each child w of v which are processed from left to right. To compute a partial solution at the subtree rooted at a child of v, we distinguish whether there exists a path containing v that belongs to $\mathcal{P}$ or not. For this purpose, we define a variable that captures a path containing v in $\mathcal{P}$. If there exists such a path, we guess the region where the endpoints of the path belong, which includes the cases that the path contains: (i) only the vertex v, (ii) the parent of v and one of its endpoints belongs to the subtree rooted at w or v's child that is to the left of w or v's child that is to the right of w, (iii) both its endpoints belong to the subtrees of the children which are to the left or the right of w, and (iv) one of the endpoints belongs to the subtree rooted at w while the other belongs to the subtree of the child to the left or the right of w.

At each node v, we store the distance of the nearest vertex (say w') in the subtree of v, that is, on a path in $\mathcal{P}$, from v. We store this with the hope that w' can cover vertices of B that come in the future. In addition, we also store the farthest vertex (say w'') in the subtree of v that is not covered by any chosen paths of $\mathcal{P}$ in the subtree. Again, we store this with the hope that $w'' \in B$ can be covered by a future vertex, and the current solution leads to a solution overall.

At each node v, we capture the existence of the following: there exists a set of $t' \leq t$ paths Y, one from each $\mathcal{F}_i$, that either includes v or not on a path from Y in $\mathcal{P}$ satisfying the distances of the nearest vertex w' and the farthest vertex w'' (from v) that are on Y and already covered and not yet covered by Y, respectively. To conclude the existence of a colorful path cover at the root node, we check for the existence of an entry that consists of a set Y of t paths, one from each $\mathcal{F}_i$, and all the farthest distance of an uncovered vertex is zero.

Proof of Theorem 1. ESP/fvs + ecc and SKELETON TESTING are FPT-equivalent from Lemma 1. Observation 1 upper bounds the number of skeletons by $2^{\mathcal{O}(k(\log k+\log \ell))}n^2$. Then, we show that SKELETON TESTING and EXT-SKELETON TESTING are FPT-equivalent and for each skeleton we have at most $2^{\mathcal{O}(k \log k)}$ enriched skeletons. Finally, given an instance of EXT-SKELETON TESTING, we construct an instance of COLORFUL PATH-COVER in polynomial time. The COLORFUL PATH-COVER problem can be solved in $\ell^2 \cdot 2^{\mathcal{O}(k \log k)} n^{\mathcal{O}(1)}$ time, and this completes the proof of Theorem 1. □

3 (1+ϵ)-Factor Parameterized by Feedback Vertex Set

Theorem 2. *For any $\epsilon > 0$, there is an $(1+\epsilon)$-factor approximation algorithm for* ESP/fvs *running in time $2^{\mathcal{O}(k \log k)} n^{\mathcal{O}(1)}$.*

We make use of our algorithm in Theorem 1 that runs in $2^{\mathcal{O}(k \log k)} \ell^k n^{\mathcal{O}(1)}$ time. Notice that, ℓ^k comes because of the number of skeletons (Observation 1). Specifically, for the function $f : X \to [\ell]$ that maintains a distance profile of the set of vertices of S that do not appear on P. To design a $(1+\epsilon)$-factor FPT approximation algorithm, we replace the image set $[\ell]$ with a set of fixed size using ϵ such that we approximate the shortest distance for vertices of S that do not appear on P, with the factor $(1+\epsilon)$. The rest is similar to Theorem 1.

Let the function $f : X \to \{\epsilon\ell, \ell\}$ denote the approximate shortest distance of each vertex $x \in X$ from a hypothetical solution P of ESP/fvs. Formally,

$$f(v) = \begin{cases} \epsilon\ell & \text{if } d_G(v,P) < \epsilon\ell, \\ \ell & \text{if } \epsilon\ell \leq d_G(v,P) \leq \ell. \end{cases}$$

Correctness. Suppose that P^* is a shortest path, with eccentricity ℓ and the function f as defined in the proof of Theorem 1, returned by the algorithm in Theorem 1. We prove that for each vertex $v \in V(G)$, $d_G(v,P) \leq (1+\epsilon)\ell$. Observe that for a vertex $x \in X$, if $1 \leq d_G(x,P^*) < \epsilon\ell$, then for a correct guess of f, $f(x) = \epsilon\ell$ and $d_G(x,P^*) < \epsilon\ell$. Also if $\epsilon\ell \leq d_G(x,P^*) \leq \ell$, then for a correct guess of f, $f(x) = \ell$ and $d_G(x,P^*) \leq \ell$. Recall that, in the algorithm when we construct instances for a good function γ (reducing to instance of COLORFUL PATH-COVER), we remove such vertices to construct an instance of COLORFUL PATH-COVER. The assumption (or guess) we made was that the eccentricity requirement for v is satisfied using x. More explicitly, we use the following conditions: if $f(x) = \epsilon\ell$ (resp, $f(x) = \ell$), then the eccentricity requirement for the vertex v is satisfied using x if $d_G(v,x) \leq \ell$ (resp, $d_G(v,x) \leq \epsilon\ell$). Now consider a vertex $v \in V(G)\setminus S$. Suppose that there exists a shortest path from v to P^* containing no vertex from S, then by the description and correctness of algorithm of Theorem 1, we obtain that $d_G(v,P) \leq \ell$. Next, suppose that the shortest path from v to P^* contains a vertex $x \in X$, then $d_G(v,x) + d_G(x,P^*) \leq \ell$. Therefore, for such vertices, while $d_G(x,v) \leq \ell$ and $d_G(x,P) < \epsilon\ell$, we obtain that $d_G(v,P) \leq d_G(x,v) + d_G(x,P) \leq \ell + \epsilon\ell = (1+\epsilon)\ell$ and similarly, if $d_G(x,v) \leq \epsilon\ell$ and $d_G(x,P) \leq \ell$, then $d_G(v,P) \leq d_G(x,v) + d_G(x,P) \leq \epsilon\ell + \ell = (1+\epsilon)\ell$. This completes the correctness of the proof of Theorem 2.

4 Disjoint Paths Deletion Set

To eliminate the eccentricity parameter from the running time, we construct a set Q (in Lemma 7) of possible distance values of disjoint paths deletion set S to a solution path such that $|Q|$ is bounded by a function of $|S|$.

Theorem 3 ($\star$). *There is an algorithm for* ESP/dpd *running in time* $\mathcal{O}(2^{\mathcal{O}(k \log k)} n^{\mathcal{O}(1)})$.

Lemma 7 ($\star$). *Let (G,S,k) be a yes instance of* ESP/dpd, *and P be a hypothetical solution. Then there is a set $Q \subseteq [\ell]$ of size $\leq 2k^2$ such that for each $w \in S$, $d_G(w,P) \in Q$. Moreover, one can construct such a Q in $O(k^2n^2)$ time.*

Funding Information. The first author acknowledges SERB-DST for supporting this research via grant PDF/2021/003452.

References

1. Bhyravarapu, S., Jana, S., Kanesh, L., Saurabh, S., Verma, S.: Parameterized algorithms for eccentricity shortest path problem (2023). http://arxiv.org/abs/2304.03233arXiv:2304.03233
2. Birmelé, É., de Montgolfier, F., Planche, L.: Minimum eccentricity shortest path problem: an approximation algorithm and relation with the k-laminarity problem. In: Chan, T.-H.H., Li, M., Wang, L. (eds.) COCOA 2016. LNCS, vol. 10043, pp. 216–229. Springer, Cham (2016). https://doi.org/10.1007/978-3-319-48749-6_16
3. Cygan, M., et al.: Parameterized Algorithms. Springer, Cham (2015). https://doi.org/10.1007/978-3-319-21275-3
4. Diestel, R.: Graph Theory. GTM, vol. 173. Springer, Heidelberg (2017). https://doi.org/10.1007/978-3-662-53622-3
5. Dragan, F.F., Leitert, A.: Minimum eccentricity shortest paths in some structured graph classes. In: Mayr, E.W. (ed.) WG 2015. LNCS, vol. 9224, pp. 189–202. Springer, Heidelberg (2016). https://doi.org/10.1007/978-3-662-53174-7_14
6. Dragan, F.F., Leitert, A.: On the minimum eccentricity shortest path problem. Theoret. Comput. Sci. **694**, 66–78 (2017)
7. Faudree, R.J., Gould, R.J., Jacobson, M.S., West, D.B.: Minimum degree and dominating paths. J. Graph Theor. **84**(2), 202–213 (2017)

8. Kučera, M., Suchý, O.: Minimum eccentricity shortest path problem with respect to structural parameters. In: Flocchini, P., Moura, L. (eds.) IWOCA 2021. LNCS, vol. 12757, pp. 442–455. Springer, Cham (2021). https://doi.org/10.1007/978-3-030-79987-8_31
9. Völkel, F., Bapteste, E., Habib, M., Lopez, P., Vigliotti, C.: Read networks and k-laminar graphs. arXiv preprint arXiv:1603.01179 (2016)

A Polynomial-Time Approximation Scheme for Thief Orienteering on Directed Acyclic Graphs

Andrew Bloch-Hansen[1(✉)], Daniel R. Page[2,3], and Roberto Solis-Oba[1]

[1] Department of Computer Science, Western University, London, ON, Canada
ablochha@uwo.ca, solis@csd.uwo.ca

[2] PageWizard Games, Learning & Entertainment, Sunnyside, MB, Canada
drpage@pagewizardgames.com

[3] Department of Computer Science, University of Regina, Regina, SK, Canada

Abstract. We consider the scenario of routing an agent called a *thief* through a weighted graph $G = (V, E)$ from a start vertex s to an end vertex t. A set I of items each with weight w_i and profit p_i is distributed among $V \setminus \{s, t\}$. In the thief orienteering problem, the thief, who has a knapsack of capacity W, must follow a simple path from s to t within a given time T while packing in the knapsack a set of items, taken from the vertices along the path, of total weight at most W and maximum profit. The travel time across an edge depends on the edge length and current knapsack load.

The thief orienteering problem is a generalization of the orienteering problem and the 0–1 knapsack problem. We present a polynomial-time approximation scheme (PTAS) for the thief orienteering problem when G is directed and acyclic, and adapt the PTAS for other classes of graphs and special cases of the problem. In addition, we prove there exists no approximation algorithm for the thief orienteering problem with constant approximation ratio, unless $\mathsf{P} = \mathsf{NP}$.

Keywords: thief orienteering problem · knapsack problem · dynamic programming · approximation algorithm · approximation scheme

1 Introduction

The *thief orienteering problem* (ThOP) is defined as follows. Let $G = (V, E)$ be a weighted graph with n vertices, where two vertices $s, t \in V$ are designated the *start* and *end* vertices, and every edge $e = (u, v) \in E$ has a length $d_{u,v} \in \mathbb{Q}^+$. In addition, let there be a set I of items, where each item $i_j \in I$ has a non-negative

---

Andrew Bloch-Hansen and Roberto Solis-Oba were partially supported by the Natural Sciences and Engineering Research Council of Canada, grants 6636-548083-2020 and RGPIN-2020-06423, respectively. Daniel Page was partially supported publicly via Patreon.

S.-Y. Hsieh et al. (Eds.): IWOCA 2023, LNCS 13889, pp. 87–98, 2023.
https://doi.org/10.1007/978-3-031-34347-6_8

integer weight w_j and profit p_j. Each vertex $u \in V \setminus \{s,t\}$ stores a subset $S_u \subseteq I$ of items such that $S_u \cap S_v = \emptyset$ for all $u \neq v$ and $\bigcup_{u \in V \setminus \{s,t\}} S_u = I$. There is an agent called a thief that has a knapsack with capacity $W \in \mathbb{Z}^+$ and the goal is for the thief to travel a simple path from s to t within a given time $T \in \mathbb{Q}^+$ while collecting items in the knapsack, taken from the vertices along the path, of total weight at most W and maximum total profit. The amount of time needed to travel between two adjacent vertices u, v depends on the length of the edge connecting them and on the weight of the items in the knapsack when the edge is traveled; specifically, the travel time between adjacent vertices u and v is $d_{u,v}/\mathcal{V}$ where $\mathcal{V} = \mathcal{V}_{\max} - w(\mathcal{V}_{\max} - \mathcal{V}_{\min})/W$, w is the current weight of the items in the knapsack, and $\mathcal{V}_{\min}$ and $\mathcal{V}_{\max}$ are the minimum and maximum velocities at which the thief can travel.

The thief orienteering problem is a generalization of the orienteering problem [6] and the 0–1 knapsack problem, so it is NP-hard. The problem was formulated by Santos and Chagas [12] who provided the first two heuristics for ThOP: An iterated local search algorithm and a biased random-key genetic algorithm. In 2020, Faêda and Santos [4] presented a genetic algorithm for ThOP. Chagas and Wagner [2] designed a heuristic using an ant colony algorithm, and in 2022, Chagas and Wagner [3] further improved this algorithm.

While ThOP is a relatively new problem, the closely related family of travelling problems, such as the travelling thief problem [1] and some variants of orienteering [7], are well-studied and have applications in areas as diverse as route planning [5] and circuit design [1], among others.

In 2017, Polyakovskiy and Neumann [10] introduced the packing while travelling problem (PWTP). This is a problem similar to ThOP, but in PWTP the thief must follow a fixed path and the goal is to maximize the difference between the profit of the items collected in the knapsack and the transportation cost. Polyakovskiy and Neumann provided two exact algorithms for PWTP, one based on mixed-integer programming and another on branch-infer-and-bound. In 2019, Roostapour et al. [11] presented three evolutionary algorithms on variations of PWTP. Most recently, Neumann et al. [9] presented an exact dynamic programming algorithm and a fully polynomial-time approximation scheme (FPTAS) for PWTP. Their dynamic programming algorithm, unfortunately, cannot be applied to ThOP because in ThOP we must bound both the total weight of the items and the travelling time of the thief.

To the best of our knowledge, study on ThOP to this date has focused on the design of heuristics and no previous work has presented an approximation algorithm for it. In this paper, we present a dynamic programming-based polynomial-time approximation scheme (PTAS) for ThOP on directed acyclic graphs (DAGs). Our algorithm can be extended to other types of graphs like outerplanar, series-parallel, and cliques. Furthermore, variations of the algorithm yield several FPTAS on special versions of ThOP on arbitrary undirected graphs, like the case when $\mathcal{V}_{\min} = \mathcal{V}_{\max}$ and T is equal to the length L of a shortest path from s to t plus a constant K. We also show that ThOP on undirected graphs cannot be approximated within a constant factor unless P = NP.

There are several challenges in the design of our PTAS. To achieve polynomial running time at least two of the parameters of the problem (weight, profit, and travelling time) need to be rounded to keep the size of the dynamic programming table polynomial. Since travelling time depends on the weight of the items in the knapsack, it seems that the most natural parameters whose values need to be rounded are profit and weight. Rounding weights, however, needs to be done carefully because (1) rounding the weights of small weight items with high profit can yield solutions with profit much smaller than the optimum one, and (2) rounding the weights of items located far away from the destination vertex can cause large errors in the travelling times.

We solve the first problem through enumeration by ensuring that a constant number of the items with largest profit in an optimum solution belong to the solution computed by our algorithm. We solve the second problem by using actual item weights and not rounded weights when computing travelling times.

The rest of the paper is organized in the following way. In Sect. 2 we prove the inaproximability of ThOP. In Sect. 3 we present an exact algorithm for ThOP on DAGs using dynamic programming and we prove its correctness in Sect. 4. Our main result is a PTAS for ThOP when the input graph is a DAG, which we present in Sect. 5, and we provide its analysis in Sect. 6. In Sect. 7 we show our algorithm can be adapted to a restricted version of ThOP on undirected graphs.

2 Inapproximability

Theorem 1. *There is no approximation algorithm for ThOP with constant approximation ratio, unless* $\mathsf{P} = \mathit{NP}$*. Furthermore, for any* $\epsilon > 0$*, there is no approximation algorithm for ThOP with approximation ratio* $2^{O(\log^{1-\epsilon} n)}$ *unless* $\mathsf{NP} \subseteq \mathsf{DTIME}(2^{O(\log^{1-\epsilon} n)})$*. These hardness results hold even if the input graph has bounded degree, all edges have unit length, and each vertex stores only one item of unit weight and profit.*

Proof. The longest path problem [8] is a special case of ThOP, where s and t are the endpoints of a longest path in the input graph $G = (V, E)$, every edge has length 1, every vertex $u \in V \setminus \{s, t\}$ stores one item of weight 1 and profit 1, the capacity of the knapsack is $W = |V| - 2$, the bound on the time is $T = |V| - 1$, and $\mathcal{V}_{\min} = \mathcal{V}_{\max}$. Since there are $O(|V|^2)$ possible choices for the endpoints of a longest path in G then the inapproximability properties of the longest path problem [8] apply also to ThOP. □

Corollary 1. *The fractional version of ThOP, that allows the thief to select only a fraction of each item, cannot be approximated in polynomial time within any constant factor unless* $\mathsf{P} = \mathit{NP}$.

Proof. Consider the same reduction as in the proof of Theorem 1. Note than an optimal fractional solution must collect the whole items stored in the vertices of an optimal path. □

3 Algorithm for Thief Orienteering on DAGs

To simplify the description of our algorithm, for each vertex u that does not store any items we add to I an item of weight 0 and profit 0 and store it in u; hence, every vertex stores at least one item. We can assume that the minimum and maximum velocities $\mathcal{V}_{\min}$ and $\mathcal{V}_{\max}$ are δ and 1, respectively, where $\delta \in \mathbb{Q}^+$ and $\delta \leq 1$. Then, the *travel time* from u to v, when the knapsack's total weight is w just prior to leaving vertex u, is $d_{u,v}/\eta$, where $\eta = 1 - \frac{(1-\delta)w}{W}$.

Since DAGs have no cycles, every path from s to t is a simple path. We index the vertices using a topological ordering. As s is the start vertex, we delete from G all vertices that are unreachable from s, and since t is the end vertex, we delete from G all vertices that cannot reach t. For a vertex u, let $u.index$ be the index of the vertex as determined by the topological ordering.

Note that s and t have the lowest and highest indices, respectively. Additionally, observe that the indices of the vertices encountered along a simple path from s to any vertex u appear in increasing order. We consider the vertices in index order starting at s. We index the items stored in the vertices so that item indices are unique and items in vertex u_x have smaller indices than items in vertex u_y for all $x < y$. Let the items be $i_1, i_2, ..., i_{|I|}$.

We define the *parents* of a vertex u to be the vertices v such that (v, u) is a directed edge of G. Our algorithm for ThOP on DAGs is shown below.

Algorithm 1. ThOPDAG$(G, W, T, s, t, I, \delta)$

1: **Input:** DAG $G = (V, E)$, knapsack capacity W, time limit T, start vertex s, end vertex t, item assignments I, and minimum velocity δ.
2: **Output:** An optimum solution for the thief orienteering problem.
3: Delete from G vertices unreachable from s and vertices that cannot reach t.
4: Compute a topological ordering for G.
5: Index V by increasing topological ordering and index items as described above.
6: Let A be an empty profit table. Set $A[1] = (0, 0, 0, -1)$.
7: **for** $i = s.index$ **to** $t.index$ **do**
8: Let u be the vertex with index i.
9: Call *UpdateProfitTable*(W, T, δ, A, u).
10: **end for**
11: Return *BuildKnapsack*(A).

Due to space limitations, the *BuildKnapsack* algorithm has been omitted, but note that it simply retrieves the path and the items in the knapsack corresponding to the highest profit solution by backtracking through the profit table.

3.1 Profit Table

Let S be a subset of items. In the sequel, we define the *travel time* of S to a vertex u as the minimum time needed by the thief to collect all items in S

while travelling along a simple path p_{su} from s to u that includes all the vertices storing the items in S. Path p_{su} is called a *fastest path* of S to u. Additionally, we define the *total travel time* of S as the travel time of S to t.

Definition 1. *Let* $S_z = \{i_1, ..., i_z\}$ *for all* $z = 1, ..., |I|$. *A feasible subset* S *of* S_z *has weight* $w_S = \sum_{i_j \in S} w_j \leq W$ *and total travel time at most* T.

Our algorithm builds a *profit table* A where each entry $A[j]$, for $j = 1, ..., |I|$, corresponds to the item with index j, and $A[j]$ is a list of quadruples $(p, w, time, prev)$. Let u be the vertex that contains item i_j. A quadruple $(p, w, time, prev)$ in the list of $A[j]$ indicates that there is a subset S of S_j such that:

- the profit of the items in S is p, their weight is $w \leq W$, the travel time of S to u is $time \leq T$, and
- a fastest path of S to u includes a vertex storing i_{prev}. Note that item i_{prev} does not need to be in S.

A quadruple $(p, w, time, prev)$ *dominates* a quadruple $(p', w', time', prev')$ if $p \geq p'$, $w \leq w'$, and $time \leq time'$. We remove dominated quadruples from each list of A so that no quadruple in the list of each entry $A[j]$ dominates another quadruple in the same list. Therefore, we can assume each list $A[j]$ has the following properties: (i) the quadruples are sorted in non-decreasing order of their profits, (ii) there might be multiple quadruples with the same profit, and if so these quadruples are sorted in non-decreasing order of their weights, and (iii) if there were several quadruples in $A[j]$ with the same profit p and weight w, only the quadruple with the smallest value of $time$ is kept in $A[j]$.

3.2 UpdateProfitTable

Algorithm 2 shows how each vertex u updates the profit table. The start vertex $s \in V$ has no parents and holds a single item i_1 of weight and profit 0; therefore, we initialize $A[1]$ to store the quadruple $(0, 0, 0, -1)$.

When two or more different paths from s to t are routed through some intermediate vertex u, it is vital that subsets of items corresponding to each of the paths are recorded correctly; the entries in the profit table A must represent the item subsets from each path from s to u, but none of the quadruples in A should contain information from items stored in vertices from different paths.

When a vertex u's first item i_j is considered, two things must happen: (1) the profit table must correctly store the path information for each quadruple corresponding to i_j by storing the index of the previous item associated with the quadruple, and (2) the quadruples taken from all parents of u must have their travel time updated.

Observe that after the thief has reached vertex u and the list of quadruples corresponding to u's first item has been created, the travel times for the quadruples corresponding to the other items of u do not need to be further increased.

Algorithm 2. UpdateProfitTable(W, T, δ, A, u)

```
1: Input: Knapsack capacity W, time limit T, minimum velocity δ, profit table A,
   and vertex u.
2: Output: The entries of the profit table A corresponding to u's items are updated
   to represent the subsets of items found along all paths from s to u.
3: for each item i_j of u do
4:     Let p the profit and w be the weight of i_j.
5:     if i_j is u's first item then
6:         A[j] = ∅.
7:         for each parent v of u do
8:             Let id_v be the index of v's last item.
9:             for each(p', w', time', prev') ∈ A[id_v] do
10:                Let d_{u,v} be the distance from v to u.
11:                Let η = 1 − (1 − δ)w'/W.
12:                Let travel = d_{u,v}/η.
13:                if time' + travel ≤ T then
14:                    Append (p', w', time' + travel, id_v) to A[j].
15:                end if
16:            end for
17:        end for
18:    else
19:        A[j] = A[j − 1].
20:        For each quadruple (p', w', time', prev') in A[j] change prev' to j − 1.
21:    end if
22:    for each(p', w', time', prev') ∈ A[j] do
23:        if w + w' ≤ W then
24:            Append (p + p', w + w', time', prev') to A[j].
25:        end if
26:    end for
27:    Remove dominated quadruples from A[j].
28: end for
```

4 Algorithm Analysis

Recall that the items are indexed such that for two items with indices h and j where $h < j$, item i_h must belong to a vertex whose index is less than or equal to the index of the vertex containing item i_j.

To prove that our algorithm is correct we must prove that each entry $A[z]$ of the profit table is such that for every feasible subset S of S_z the entry $A[z]$ contains either (i) a quadruple $(p_S, w_S, time_S, prev)$, where $p_S = \sum_{i_j \in S} p_j$, $w_S = \sum_{i_j \in S} w_j$, and $time_S$ is the travel time of S to the vertex u storing i_z, or (ii) a quadruple $(p', w', time', prev')$ that dominates $(p_S, w_S, time_S, prev)$. This implies that A contains a quadruple representing a simple path from s to t whose vertices store a maximum profit set S^* of items of weight at most W and total travel time at most T.

Lemma 1. *Let* $1 \leq z \leq |I|$. *For each feasible subset* S *of* S_z *there is a quadruple* $(p, w, time, prev)$ *in the profit table* A *at entry* $A[z]$ *such that* $p \geq p_S = \sum_{i_j \in S} p_j$, $w \leq w_S = \sum_{i_j \in S} w_j$, *and* $time \leq time_S$, *where* $time_S$ *is the travel time of* S *to the vertex* u *storing* i_z.

Proof. We use a proof by induction on the number of entries of the profit table A. The base case is trivial as there are only two feasible subsets of S_1, the empty set and the set containing item i_1 with weight and profit 0, and $A[1]$ stores the quadruple $(0, 0, 0, -1)$.

Assume that the lemma holds true for every feasible subset S of S_q for all $q = 1, ..., z-1$. Let S be a feasible subset of S_z; we show that there is a quadruple $(p, w, time, prev) \in A[z]$ such that $p \geq p_S$, $w \leq w_S$, and $time \leq time_S$. Let u be the vertex storing i_z. We need to consider two cases:

- Case 1: Item $i_z \in S$. Let $S' = S - \{i_z\}$. Note that since the travel time of S to u is at most T, then the travel time of S' to u is also at most T.
  - Assume first that i_z is u's first item. Let u_α be the vertex before u in the fastest path of S to u; let i_α be the last item at u_α, and let $time_\alpha$ be the travel time of S' from u_α to u. Note that by the way we indexed the items, $\alpha < z$. By the induction hypothesis, there is a quadruple $(p', w', time', prev')$ in $A[\alpha]$ such that $p' \geq p_{S'}$ and $w' \leq w_{S'}$. Lines 7 to 16 in Algorithm 2 copy all the tuples in $A[\alpha]$ to $A[z]$ whose travel time, $time' + time_\alpha$, is at most T and update their travel times. Moreover, in lines 22 to 26 Algorithm 2 adds the quadruple $(p' + p_z, w' + w_z, time' + time_\alpha, prev')$ to $A[z]$ because $w' + w_z \leq w_{S'} + w_z = w_S \leq W$; furthermore, $p' + p_z \geq p_{S'} + p_z = p_S$. Therefore, there is a quadruple $(p, w, time, prev)$ in $A[z]$ such that $p \geq p_S$, $w \leq w_S$, and $time \leq time_S$.
  - Assume now that i_z is not u's first item. Note that then the item i_{z-1} is also located at u because of the way in which we indexed the items. By the induction hypothesis, there is a quadruple $(p', w', time', prev')$ in $A[z-1]$ such that $p \geq p_{S'}$, $w' \leq w_{S'}$, and $time' \leq time_S$. Lines 19 to 26 in Algorithm 2 add the quadruple $(p' + p_z, w' + w_z, time', prev')$ to $A[z]$, as $w' + w_z \leq w_{S'} + w_z = w_S \leq W$ and $p' + p_z \geq p_{S'} + p_z = p_S$. Therefore, there is a quadruple $(p, w, time, prev)$ in $A[z]$ such that $p \geq p_S$, $w \leq w_S$, and $time \leq time_S$.
- Case 2: Item $i_z \notin S$. Since S is a feasible subset, by the induction hypothesis, either (i) there is a quadruple $(p, w, time, prev)$ in $A[z-1]$ such that $p \geq p_S$, $w \leq w_S$, and $time \leq time_S$ that Algorithm 2 would have copied to $A[z]$ in lines 19–20 if i_z is not u's first item, or (ii) there is a quadruple $(p, w, time, prev)$ in $A[\alpha]$ (where i_α is the last item in vertex u_α defined above) such that $p \geq p_S$, $w \leq w_S$, and $time \leq time_S$ that Algorithm 2 would have copied to $A[z]$ in lines 7–17 if i_z is u's first item. Therefore, there is a quadruple $(p, w, time, prev)$ in $A[z]$ such that $p \geq p_S$, $w \leq w_S$, $time \leq time_S$. □

5 PTAS for Thief Orienteering on DAGs

Algorithm 1 might not run in polynomial time because the size of the profit table might become too large. We can convert Algorithm 1 into a PTAS by carefully rounding down the profit and rounding up the weight of each item.

Note that if we simply use rounded weights in the profit table described in Sect. 3.1, then we might introduce a very large error to the travel times computed by Algorithm 2. To prevent this, we modify the profit table so that every entry $A[j]$ holds a list of quintuples $(p, w_r, w_t, time, prev)$, where each quintuple indicates that there is a subset S of S_j in which the sum of their rounded profits is p, the sum of their rounded weights is w_r, the sum of their true weights is w_t, the travel time of S to the vertex u holding item i_j is $time$, and the fastest path of S to u includes the vertex which contains i_{prev}. Let P_{max} be the maximum profit of an item that can be transported within the allotted time T from the vertex that initially stored it to the destination vertex. Our PTAS is described in Algorithm 3.

Algorithm 3. ThOPDAGPTAS$(G, W, T, s, t, I, \delta, \epsilon)$

1: **Input:** DAG $G = (V, E)$, knapsack capacity W, time limit T, start vertex s, end vertex t, item assignments I, minimum velocity δ, and constant $\epsilon > 0$.
2: **Output:** A feasible solution for the thief orienteering problem of profit at least $(1 - 3\epsilon)OPT$, where OPT is the value of an optimum solution.
3: Delete from G vertices unreachable from s and vertices that cannot reach t.
4: Compute a topological ordering for G.
5: Index V by increasing topological ordering and items as described in Section 3.
6: Let $K = \frac{1}{\epsilon}$.
7: Let $\mathbb{S}$ be the set of all feasible subsets S of $S_{|I|}$ such that $|S| \leq K$.
8: **for** each $S \in \mathbb{S}$ **do**
9: Let A be an empty profit table. Set $A[1] = (0, 0, 0, -1)$.
10: Let $W' = W - \sum_{i_j \in S} w_j$.
11: Round down the profit of each item in $I - S$ to the nearest multiple of $\frac{\epsilon P_{max}}{|I|}$.
12: Round up the weight of each item in $I - S$ to the nearest multiple of $\frac{\epsilon W'}{|I|^2}$.
13: **for** $i = s.index$ *to* $t.index$ **do**
14: Let u be the vertex with index i.
15: Call *UpdateProfitTable**(W, T, δ, A, u, S).
16: **end for**
17: **end for**
18: Select the profit table A^* storing the quintuple with maximum profit.
19: Return *BuildKnapsack*(A^*).

Algorithm *UpdateProfitTable** is a slight modification of Algorithm 2 that computes the travel time of each tuple using the true weights, but it uses the rounded weights when determining if an item subset fits in the knapsack or when discarding dominated tuples. Additionally, for each $S \in \mathbb{S}$, Algorithm *UpdateProfitTable** includes the items from S in the knapsack when it processes the vertices storing these items.

6 PTAS Analysis

Since the weights of some items are rounded up, the solution produced by Algorithm 3 might have unused space where the algorithm could not fit any rounded items. However, an optimal solution would not leave empty space if there were items that could be placed in the knapsack while still travelling from s to t in at most T time, so we need to bound the maximum profit lost due to the rounding of the weights and profits.

Lemma 2. *For any constant $\epsilon > 0$, Algorithm 3 computes a feasible solution to the thief orienteering problem on DAGs with profit at least $(1 - 3\epsilon)OPT$, where OPT is the profit of an optimum solution.*

Proof. Let S_{OPT} be the set of items in an optimum solution with maximum profit and let $OPT = \sum_{i_j \in S_{OPT}} p_j$. If $|S_{OPT}| \leq K$ then Algorithm 3 computes an optimum solution. Hence, for the rest of the proof we assume that $|S_{OPT}| > K$. Let S_K be the set of the K items with largest profit from S_{OPT}, where $K = \frac{1}{\epsilon}$, and let $W' = W - \sum_{i_j \in S_K} w_j$. Let S_A be the set of items in the solution selected by our algorithm, and let $SOL = \sum_{i_j \in S_A} p_j$.

Recall that our algorithm tries including every possible feasible subset of at most K items in the knapsack. Therefore, our algorithm must have included S_K in the knapsack in one of the iterations and filled the remainder of the knapsack using the profit table. Let S_A^* be the solution computed by our algorithm in the iteration where it chose to include the items of S_K in the knapsack, and let $SOL^* = \sum_{i_j \in S_A^*} p_j$. Since our algorithm returns the best solution that it found over all iterations, then $SOL \geq SOL^*$.

To compare SOL^* to OPT, we round up the weight of each item in $S_{OPT} - S_K$ to the nearest multiple of $\frac{\epsilon W'}{|I|^2}$; note that the weights and profits of the items in S_K are not rounded. In the sequel, we will use w'_j and p'_j to refer to the rounded weight and profit of item i_j, and w_j and p_j to refer to the true weight and profit of item i_j. Given a set X of items let $weight(X) = \sum_{i_j \in X} w_j$, $weight'(X) = \sum_{i_j \in X} w'_j$, $profit(X) = \sum_{i_j \in X} p_j$, and $profit'(X) = \sum_{i_j \in X} p'_j$. We let $profit'(S_K) = profit(S_K)$.

Rounding up the weight of a single item increases the weight of that item by at most $\frac{\epsilon W'}{|I|^2}$, so $weight'(S_{OPT}) \leq weight(S_{OPT}) + \frac{\epsilon W'}{|I|} \leq W + \frac{\epsilon W'}{|I|}$ as $|S_{OPT}| \leq |I|$. Let A_{OPT} be a subset of $S_{OPT} - S_K$ with $weight'(A_{OPT}) \leq W'$ and maximum rounded profit and such that $A_{OPT} \cup S_K$ is a feasible set of items. Note that our algorithm must have included a quintuple $(p^*, w_r^*, w_t^*, time^*, prev^*)$ in the profit table such that $w_r^* \leq weight'(A_{OPT} \cup S_K)$, $time^*$ is at most the time to transport from s to t the items in $A_{OPT} \cup S_K$, and $p^* \geq profit'(A_{OPT} \cup S_K)$. Hence,

$$SOL^* \geq profit'(A_{OPT} \cup S_K) \tag{1}$$

We now bound $profit'(A_{OPT} \cup S_K)$. Let $S_{OPT}^- = S_{OPT} - S_K - A_{OPT}$: These are the items in the optimum solution whose profit is not included in the right hand side of (1). Note that if S_{OPT}^- is empty, then $S_K \cup A_{OPT} = S_{OPT}$ and so

$SOL^* \geq profit'(S_{OPT})$. If S^-_{OPT} is not empty, we show that $weight'(S^-_{OPT}) \leq \frac{\epsilon W'}{|I|} + w_L$, where w_L is the largest weight of the items in S^-_{OPT}. To see this, recall that $weight'(S_{OPT}) \leq W + \frac{\epsilon W'}{|I|}$ and note that $W - w_L < weight(S_K) + weight'(A_{OPT}) \leq W$, because by the way in which A_{OPT} was defined the empty space that $S_K \cup A_{OPT}$ leave in the knapsack is not large enough to fit the rounded weight of another item from S^-_{OPT}. Therefore,

$$\begin{aligned} weight'(S^-_{OPT}) &= weight'(S_{OPT}) - (weight(S_K) + weight'(A_{OPT})) \\ &< W + \frac{\epsilon W'}{|I|} - (W - w_L) = \frac{\epsilon W'}{|I|} + w_L \end{aligned} \tag{2}$$

Now we bound $profit'(S^-_{OPT})$. If S^-_{OPT} consists of only one item i_j, then since the least profitable item in S_K has profit at most $\frac{1}{K}OPT$ and i_j is not in S_K, then $p_j \leq \frac{1}{K}OPT$, which means that $profit'(S^-_{OPT}) \leq \frac{1}{K}OPT = \epsilon OPT$. If S^-_{OPT} consists of two or more items, we can partition S^-_{OPT} into the singleton $\{i_L\}$ consisting of the item with largest weight w_L and the set i_* of remaining items, which by (2) has $weight'(i_*) \leq \frac{\epsilon W'}{|I|}$.

Item i_L is not in S_K, so it has profit $p_L \leq \frac{1}{K}OPT$. As for i_* we show that $profit'(i_*) \leq \frac{1}{K}OPT$:

- $W' - weight'(A_{OPT}) < \frac{\epsilon W'}{|I|} = \frac{W'}{K|I|}$, as otherwise the items i_* would have been included in A_{OPT}, and so

$$weight'(A_{OPT}) = \sum_{i_j \in A_{OPT}} w'_j > W' - \frac{W'}{K|I|} > \frac{K-1}{K}W', \text{ as } |I| \geq 1 \tag{3}$$

- There is at least one item i_ψ in A_{OPT} with $w'_\psi \geq \frac{W'}{K|I|}$. To see this, note that if all of the items in A_{OPT} had weight strictly less than $\frac{W'}{K|I|}$, then $\sum_{i_j \in A_{OPT}} w'_j < \frac{W'}{K|I|}|A_{OPT}| < \frac{1}{K}W'$, as $|A_{OPT}| \leq |I|$, which contradicts (3).
- By definition, A_{OPT} includes items from $S_{OPT} - S_K$ with $weight'(A_{OPT}) \leq W'$ and maximum profit, and since the items i_* are not in A_{OPT} then $profit'(i_*) \leq p'_\psi$, as otherwise the items i_* would be in A_{OPT} instead of i_ψ. Since item i_ψ is not in S_K then $p'_\psi \leq \frac{1}{K}OPT$ and so $profit'(i_*) \leq \frac{1}{K}OPT$.

Therefore, $profit'(S^-_{OPT}) = profit'(i_L) + profit'(i_*) \leq \frac{2}{K}OPT$. Since $S_{OPT} = S_K \cup A_{OPT} \cup S^-_{OPT}$, then we know $profit'(A_{OPT}) + profit'(S_K) = profit'(S_{OPT}) - profit'(S^-_{OPT}) \geq profit'(S_{OPT}) - \frac{2}{K}OPT$. Since $SOL \geq SOL^*$, by (1),

$$\begin{aligned} SOL \geq SOL^* \geq profit'(A_{OPT}) + profit'(S_K) &\geq profit'(S_{OPT}) - \frac{2}{K}OPT \\ &\geq profit(S_{OPT}) - \frac{\epsilon P_{max}}{|I|}|S_{OPT}| - \frac{2}{K}OPT \\ &\geq OPT - \epsilon P_{max} - 2\epsilon OPT \geq (1 - 3\epsilon)OPT \end{aligned}$$

□

Theorem 2. *There is a PTAS for the thief orienteering problem on DAGs.*

Proof. As shown by Lemma 2, Algorithm 3 computes solutions with profit at least $(1-3\epsilon)OPT$. The profit table A has $|I|$ entries, and each entry $A[j]$ can have at most $O(P_rW_r)$ quintuples, where P_r is the number of different values for the rounded profit of any subset of items and W_r is the number of different values for the rounded weight of any subset of items of total weight at most W. Since profits are rounded down to the nearest multiple of $\frac{\epsilon P_{max}}{|I|}$ then P_r is $O(\frac{|I|^2}{\epsilon})$. Since weights of items not in the selected feasible subsets S were rounded up to the nearest multiple of $\frac{\epsilon W'}{|I|^2}$, then W_r is $O(\frac{|I|^2}{\epsilon})$.

Algorithm 3 iterates through the list at each entry $A[j]$ exactly once if i_j is not the last item in a vertex u; if i_j is the last item in a vertex u, then our algorithm iterates through the list at entry $A[j]$ once for each outgoing edge of u. Thus, for each feasible subset S with at most K items the algorithm loops through $|I|$ rows in the profit table, iterates over a particular row at most $n = |V|$ times, and each row can have at most $O(\frac{|I|^4}{\epsilon^2})$ quintuples in it; since the number of feasible subsets S is $O(|I|^{\frac{1}{\epsilon}})$ the running time is $O(n|I|^{5+\frac{1}{\epsilon}})$. □

7 Thief Orienteering with $\mathcal{V}_{\min} = \mathcal{V}_{\max}$

We consider the case when $G = (V, E)$ is undirected, every edge has length at least 1, $\mathcal{V}_{\min} = \mathcal{V}_{\max}$, and T is equal to the length L of a shortest path from s to t plus a constant K. If $\mathcal{V}_{\min} = \mathcal{V}_{\max}$ the travel time for any edge is equal to the length of the edge and independent of the weight of the items in the knapsack.

For a vertex u, let $u.dist$ be the length of a shortest path from u to t, and let σ be a path from s to t with length at most $L + K$. If the length of σ is larger than L, then σ contains a set of *detours*; a detour is a subpath $\sigma_{u,v}$ of σ from u to v such that the length of $\sigma_{u,v}$ is larger than $|u.dist - v.dist|$. Note that σ contains a set of detours formed by at most $2K$ vertices.

Theorem 3. *There is a FPTAS for ThOP where* $\mathcal{V}_{\min} = \mathcal{V}_{\max}$ *and the time* T *is equal to the length* L *of a shortest path from* s *to* t *plus a constant* K.

Proof. Graph G can be transformed into a set of DAGs that contain all simple paths from s to t of length at most $L+K$. To do this, we enumerate all possible sets of detours D containing at most $2K$ vertices. For each set D of detours we build a DAG G_D by removing edges from G until every path from s to t travels through all the detours in D and directing the remaining edges such that (i) edges belonging to a detour of D form a simple directed path from the start of the detour to the end of the detour and (ii) edges not belonging to detours are directed towards t.

Each DAG G_D is an instance of ThOP that can be solved using Algorithm 3. We finally choose the path with the most profit among those computed for all DAGs G_D.

Since $\mathcal{V}_{\min} = \mathcal{V}_{\max}$, the item weights do not need to be rounded up and, furthermore, the number of possible sets of detours is polynomial with respect to the constant K. It is not hard to see that a simple modification of Algorithms 2 and 3 yield a FPTAS. □

In this paper we presented a PTAS for ThOP for the case when the input graph is a DAG. Additionally, we showed that ThOP cannot be approximated within a constant factor, unless P = NP. While our paper focused on DAGs, our PTAS can be used to solve ThOP on other classes of graphs, such as on chains, trees, and cycle graphs. Finally, our ideas can be used to design algorithms for ThOP on other kinds of graphs like outerplanar, series-parallel, and cliques.

References

1. Bonyadi, M., Michalewicz, Z., Barone, L.: The travelling thief problem: the first step in the transition from theoretical problems to realistic problems. In: IEEE Congress on Evolutionary Computation (CEC), pp. 1037–1044. IEEE (2013)
2. Chagas, J., Wagner, M.: Ants can orienteer a thief in their robbery. Oper. Res. Lett. **48**(6), 708–714 (2020)
3. Chagas, J., Wagner, M.: Efficiently solving the thief orienteering problem with a max-min ant colony optimization approach. Opt. Lett. **16**(8), 2313–2331 (2022)
4. Faêda, L., Santos, A.: A genetic algorithm for the thief orienteering problem. In: 2020 IEEE Congress on Evolutionary Computation (CEC), pp. 1–8. IEEE (2020)
5. Freeman, N., Keskin, B., Çapar, İ: Attractive orienteering problem with proximity and timing interactions. Eur. J. Oper. Res. **266**(1), 354–370 (2018)
6. Golden, B., Levy, L., Vohra, R.: The orienteering problem. Naval Res. Logistics (NRL) **34**(3), 307–318 (1987)
7. Gunawan, A., Lau, H.C., Vansteenwegen, P.: Orienteering problem: a survey of recent variants, solution approaches and applications. Eur. J. Oper. Res. **255**(2), 315–332 (2016)
8. Karger, R., Motwani, R., Ramkumar, G.: On approximating the longest path in a graph. Algorithmica **18**(1), 82–98 (1997). https://doi.org/10.1007/BF02523689
9. Neumann, F., Polyakovskiy, S., Skutella, M., Stougie, L., Wu, J.: A fully polynomial time approximation scheme for packing while traveling. In: Disser, Y., Verykios, V.S. (eds.) ALGOCLOUD 2018. LNCS, vol. 11409, pp. 59–72. Springer, Cham (2019). https://doi.org/10.1007/978-3-030-19759-9_5
10. Polyakovskiy, S., Neumann, F.: The packing while traveling problem. Eur. J. Oper. Res. **258**(2), 424–439 (2017)
11. Roostapour, V., Pourhassan, M., Neumann, F.: Analysis of baseline evolutionary algorithms for the packing while travelling problem. In: 1st ACM/SIGEVO Conference on Foundations of Genetic Algorithms, pp. 124–132 (2019)
12. Santos, A., Chagas, J.: The thief orienteering problem: formulation and heuristic approaches. In: IEEE Congress on Evolutionary Computation, pp. 1–9. IEEE (2018)

Deterministic Performance Guarantees for Bidirectional BFS on Real-World Networks

Thomas Bläsius and Marcus Wilhelm(✉)

Karlsruhe Institute of Technology (KIT), Karlsruhe, Germany
{thomas.blaesius,marcus.wilhelm}@kit.edu

Abstract. A common technique for speeding up shortest path queries in graphs is to use a bidirectional search, i.e., performing a forward search from the start and a backward search from the destination until a common vertex on a shortest path is found. In practice, this has a massive impact on performance in some real-world networks, while it seems to save only a constant factor in other types of networks. Although finding shortest paths is a ubiquitous problem, only few studies have attempted to explain the apparent asymptotic speedups on some networks using average case analysis on certain models of real-world network.

In this paper we provide a new perspective on this, by analyzing deterministic properties that allow theoretical analysis and that can be easily checked on any particular instance. We prove that these parameters imply sublinear running time for the bidirectional breadth-first search in several regimes, some of which are tight. Furthermore, we perform experiments on a large set of real-world networks and show that our parameters capture the concept of practical running time well.

Keywords: scale-free networks · bidirectional BFS · bidirectional shortest paths · distribution-free analysis

1 Introduction

A common way to speed up the search for a shortest path between two vertices is to use a bidirectional search strategy instead of a unidirectional one. The idea is to explore the graph from both, the start and the destination vertex, until a common vertex somewhere in between is discovered. Even though this does not improve upon the linear worst-case running time of the unidirectional search, it leads to significant practical speedups on some classes of networks. Specifically, Borassi and Natale [6] found that bidirectional search seems to run asymptotically faster than unidirectional search on scale-free real-world networks. This does, however, not transfer to other types of networks like for example transportation networks, where the speedup seems to be a constant factor [1].

There are several results aiming to explain the practical run times of the bidirectional search, specifically of the balanced bidirectional breadth-first search (short: bidirectional BFS). These results have in common that they analyze

S.-Y. Hsieh et al. (Eds.): IWOCA 2023, LNCS 13889, pp. 99–110, 2023.
https://doi.org/10.1007/978-3-031-34347-6_9

the bidirectional BFS on probabilistic network models with different properties. Borassi and Natale [6] show that it takes $O(\sqrt{n})$ time on Erdös-Rényi-graphs [7] with high probability. The same holds for slightly non-uniform random graphs as long as the edge choices are independent and the second moment of the degree distribution is finite. For more heterogeneous power-law degree distributions with power-law exponent in $(2, 3)$, the running time is $O(n^c)$ for $c \in [1/2, 1)$. Note that this covers a wide range of networks with varying properties in the sense that it predicts sublinear running times for homogeneous as well as heterogeneous degree distributions. However, the proof for these results heavily relies on the independence of edges, which is not necessarily given in real-world networks. Bläsius et al. [3] consider the bidirectional BFS on network models that introduce dependence of edges via an underlying geometry. Specifically, they show sublinear running time if the underlying geometry is the hyperbolic plane, yielding networks with a heterogeneous power-law degree distribution. Moreover, if the underlying geometry is the Euclidean plane, they show that the speedup is only a constant factor.

Summarizing these theoretical results, one can roughly say that the bidirectional BFS has sublinear running time unless the network has dependent edges and a homogeneous degree distribution. Note that this fits to the above observation that bidirectional search works well on many real-world networks, while it only achieves a constant speedup on transportation networks. However, these theoretical results only give actual performance guarantees for networks following the assumed probability distributions of the analyzed network models. Thus, the goal of this paper is to understand the efficiency of the bidirectional BFS in terms of deterministic structural properties of the considered network.

Intuition. To present our technical contribution, we first give high-level arguments and then discuss where these simple arguments fail. As noted above, the bidirectional BFS is highly efficient unless the networks are homogeneous and have edge dependencies. In the field of network science, it is common knowledge that these are the networks with high diameter, while other networks typically have the small-world property. This difference in diameter coincides with differences in the expansion of search spaces. To make this more specific, let v be a vertex in a graph and let $f_v(d)$ be the number of vertices of distance at most d from v. In the following, we consider two settings, namely the setting of *polynomial expansion* with $f_v(d) \approx d^2$ and that of *exponential expansion* with $f_v(d) \approx 2^d$ for all vertices $v \in V$. Now assume we use a BFS to compute the shortest path between vertices s and t with distance d.

To compare the unidirectional with the bidirectional BFS, note that the former explores the $f_s(d)$ vertices at distance d from s, while the latter explores the $f_s(d/2) + f_t(d/2)$ vertices at distance $d/2$ from s and t. In the polynomial expansion setting, $f_s(d/2) + f_t(d/2)$ evaluates to $2(d/2)^2 = d^2/2 = f_s(d)/2$, yielding a constant speedup of 2. In the exponential expansion setting, $f_s(d/2) + f_t(d/2)$ evaluates to $2 \cdot 2^{d/2} = 2\sqrt{f_s(d)}$, resulting in a polynomial speedup.

With these preliminary considerations, it seems like exponential expansion is already the deterministic property explaining the asymptotic performance improvement of the bidirectional BFS on many real-world networks. However, though this property is strong enough to yield the desired theoretic result, it is too strong to actually capture real-world networks. There are two main reasons for that. First, the expansion in real-world networks is not that clean, i.e., the actual increase of vertices varies from step to step. Second, and more importantly, the considered graphs are finite and with exponential expansion, one quickly reaches the graph's boundary where the expansion slows down. Thus, even though search spaces in real-world networks are typically expanding quickly, it is crucial to consider the number of steps during which the expansion persists. To actually capture real-world networks, weaker conditions are needed.

Contribution. The main contribution of this paper is to solve this tension between wanting conditions strong enough to imply sublinear running time and wanting them to be sufficiently weak to still cover real-world networks. We solve this by defining multiple parameters describing expansion properties of vertex pairs. These parameters address the above issues by covering a varying amount of expansion and stating requirements on how long the expansion lasts. We refer to Sect. 2 and Sect. 3.1 for the exact technical definitions, but intuitively we define the *expansion overlap* as the number of steps for which the exploration cost is growing exponentially in both directions. Based on this, we give different parameter settings in which the bidirectional search is sublinear. In particular, we show sublinear running time for logarithmically sized expansion overlap (Theorem 1) and for an expansion overlap linear in the distance between the queried vertices (Theorem 2, the actual statement is stronger). For a slightly more general setting we also prove a tight criterion for sublinear running time in the sense that the parameters either guarantee sublinear running time or that there exists a family of graphs that require linear running time (Theorem 3). Note that the latter two results also require the relative difference between the minimum and maximum expansion to be constant. Finally, we demonstrate that our parameters do indeed capture the behavior actually observed in practice by running experiments on more than 3 k real-world networks.

Due to space constraints, some proofs and explanations have been shortened or omitted. They can be found in the full version of this paper [4].

Related Work. Our results fit into the more general theme of defining distribution-free [10] properties that capture real-world networks and analyzing algorithms based on these deterministic properties.

Borassi, Crescenzi, and Trevisan [5] analyze heuristics for graph properties such as the diameter and radius as well as centrality measures such as closeness. The analysis builds upon a deterministic formulation of how edges form based on independent probabilities and the birthday paradox. The authors verify their properties on multiple probabilistic network models as well as real-world networks.

Fox et al. [8] propose a parameterized view on the concept of triadic closure in real-world networks. This is based on the observation that in many networks, two vertices with a common neighbor are likely to be adjacent. The authors thus call a graph c-closed if every pair of vertices u, v with at least c common neighbors is adjacent. They show that enumerating all maximal cliques is in FPT for parameter c and also for a weaker property called weak c-closure. The authors also verify empirically that real-world networks are weakly c-closed for moderate values of c.

2 Preliminaries

We consider simple, undirected, and connected graphs $G = (V, E)$ with $n = |V|$ vertices and $m = |E|$ edges. For vertices $s, t \in V$ we write $d(s, t)$ for the *distance* of s and t, that is the number of edges on a shortest path between s and t. For $i, j \in \mathbb{N}$, we write $[i]$ for the set $\{1, \ldots, i\}$ and $[i, j]$ for $\{i, \ldots, j\}$. In a (unidirectional) *breadth-first search* (*BFS*) from a vertex s, the graph is explored layer by layer until the target vertex $t \in V$ is discovered. More formally, for a vertex $v \in V$, the i-th *BFS layer around* v (short: *layer*), $\ell_G(v, i)$, is the set of vertices that have distance exactly i from v. Thus, the BFS starts with $\ell_G(s, 0) = \{s\}$ and then iteratively computes $\ell_G(s, i)$ from $\ell_G(s, i-1)$ by iterating through the neighborhood of $\ell_G(s, i-1)$ and ignoring vertices contained in earlier layers. We call this the i-th *exploration step* from s. We omit the subscript G from the above notation when it is clear from context.

In the *bidirectional* BFS, layers are explored both from s and t until a common vertex is discovered. This means that the algorithm maintains layers $\ell(s, i)$ of a *forward search* from s and layers $\ell(t, j)$ of a *backward search* from t and iteratively performs further exploration steps in one of the directions. The decision about which search direction to progress in each step is determined according to an *alternation strategy*. Note that we only allow the algorithm to switch between the search directions after fully completed exploration steps. If the forward search performs k exploration steps and the backward search the remaining $d(s, t) - k$, then we say that the search *meets* at layer k.

In this paper, we analyze a particular alternation strategy called the *balanced* alternation strategy [6]. This strategy greedily chooses to continue with an exploration step in either the forward or backward direction, depending on which is cheaper. Comparing the anticipated cost of the next exploration step requires no asymptotic overhead, as it only requires summing the degrees of vertices in the preceding layer. The following lemma gives a running time guarantee for balanced BFS relative to any other alternation strategy. This lets us consider arbitrary alternation strategies in our mathematical analysis, while only costing a factor of $d(s, t)$, which is typically at most logarithmic.

Lemma 1 (**[3, Theorem 3.2]**). *Let G be a graph and (s, t) a start–destination pair with distance $d(s, t)$. If there exists an alternation strategy such that the bidirectional BFS between s and t explores $f(n)$ edges, then the balanced bidirectional search explores at most $d(s, t) \cdot f(n)$ edges.*

The forward and backward search need to perform a total of $d(s, t)$ exploration steps. To ease the notation, we say that exploration step i (of the bidirectional search between s and t) is either the step of finding $\ell(s, i)$ from $\ell(s, i-1)$

in the forward search or the step of finding $\ell(t, d(s,t)+1-i)$ from $\ell(t, d(s,t)-i)$ in the backward search. For example, exploration step 1 is the step in which either the forward search finds the neighbors of s or in which s is discovered by the backwards search. Also, we identify the i-th exploration step with its index i, i.e., $[d(s,t)]$ is the set of all exploration steps. We often consider multiple consecutive exploration steps together. For this, we define the interval $[i,j] \subseteq [d(s,t)]$ to be a *sequence* for $i, j \in [d(s,t)]$. The *exploration cost* of exploration step i from s equals the number of visited edges with endpoints in $\ell(s, i-1)$), i.e., $c_s(i) = \sum_{v \in \ell(s,i-1)} \deg(v)$. The exploration cost for exploration step i from t is $c_t(i) = \sum_{v \in \ell(t,d(s,t)-i)} \deg(v)$. For a sequence $[i,j]$ and $v \in \{s,t\}$, we define the cost $c_v([i,j]) = \sum_{k \in [i,j]} c_v(k)$. Note that the notion of exploration cost is an independent graph theoretic property and also valid outside the context of a particular run of the bidirectional BFS in which the considered layers are actually explored.

For a vertex pair s,t we write $c_{\text{bi}}(s,t)$ for the cost of the bidirectional search with start s and destination t. Also, as we are interested in polynomial speedups, i.e., $\mathcal{O}(m^{1-\varepsilon})$ vs. $\mathcal{O}(m)$, we use $\tilde{\mathcal{O}}$-notation to suppress poly-logarithmic factors.

3 Performance Guarantees for Expanding Search Spaces

We now analyze the bidirectional BFS based on expansion properties. In Sect. 3.1, we introduce expansion, including the concept of expansion overlap, state some basic technical lemmas and give an overview of our results. In the subsequent sections, we then prove our results for different cases of the expansion overlap.

3.1 Expanding Search Spaces and Basic Properties

We define *expansion* as the relative growth of the search space between adjacent layers. Let $[i,j]$ be a sequence of exploration steps. We say that $[i,j]$ is *b-expanding from s* if for every step $k \in [i,j)$ we have $c_s(k+1) \geq b \cdot c_s(k)$. Analogously, we define $[i,j]$ to be *b-expanding from t* if for every step $k \in (i,j]$ we have $c_t(k-1) \geq b \cdot c_t(k)$. Note that the different definitions for s and t are completely symmetrical. With this definition layed out, we investigate its relationship with logarithmic distances.

Lemma 2. *Let $G = (V,E)$ be a graph and let $s,t \in V$ be vertices such that the sequence $[1, c \cdot d(s,t)]$ is b-expanding from s for constants $b > 1$ and $c > 0$. Then $d(s,t) \leq \log_b(2m)/c$.*

Proof Sketch. Follows via simple derivation, see full version. □

Note that this lemma uses s and t symmetrically and also applies to expanding sequences from t. Together with Lemma 1, this allows us to consider arbitrary alternation strategies that are convenient for our proofs. Next, we show that the total cost of a b-expanding sequence of exploration steps is constant in the cost of the last step, which often simplifies calculations.

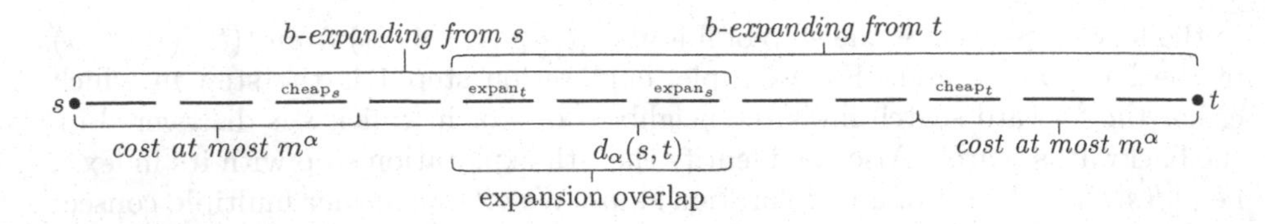

Fig. 1. Visualization of cheap_v, expan_v and related concepts. Each line stands for an exploration step between s and t. Additionally, certain steps and sequences relevant for Theorem 1 and Theorem 2 are marked.

Lemma 3. *For $b > 1$ let $f : \mathbb{N} \mapsto \mathbb{R}$ be a function with $f(i) \geq b \cdot f(i-1)$ and $f(1) = c$ for some constant c. Then $f(n)/\sum_{i=1}^{n} f(i) \geq \frac{b-1}{b}$.*

Proof Sketch. Follows via simple derivation, see full version. □

We define four specific exploration steps depending on two constant parameters $0 < \alpha < 1$ and $b > 1$. First, $\text{cheap}_s(\alpha)$ is the latest step such that $c_s([1, \text{cheap}_s(\alpha)]) \leq m^\alpha$. Moreover, $\text{expan}_s(b)$ is the latest step such that the sequence $[1, \text{expan}_s(b)]$ is b-expanding from s. Analogously, we define $\text{cheap}_t(\alpha)$ and $\text{expan}_t(b)$ to be the smallest exploration steps such that $c_t([\text{cheap}_t(\alpha), d(s,t)]) \leq m^\alpha$ and $[\text{expan}_t(b), d(s,t)]$ is b-expanding from t, respectively. If $\text{expan}_t(b) \leq \text{expan}_s(b)$, we say that the sequence $[\text{expan}_t(b), \text{expan}_s(b)]$ is a *b-expansion overlap* of size $\text{expan}_s(b) - \text{expan}_t(b) + 1$. See Fig. 1 for a visualization of these concepts. Note that the definition of expan_s (reps. expan_t) cannot be relaxed to only require expansion behind cheap_s (resp. cheap_t), as in that case an existing expansion overlap no longer implies logarithmic distance between s and t. This allows for the construction of instances with linear running time. To simplify notation, we often omit the parameters α and b as well as the subscript s and t if they are clear from the context. Note that cheap_s or cheap_t is undefined if $c_s(1) > m^\alpha$ or $c_t(d(s,t)) > m^\alpha$. Moreover, in some cases expan_v may be undefined for $v \in \{s,t\}$, if the first exploration step of the corresponding sequence is not b-expanding. Such cases are not relevant in the remainder of this paper.

Overview of Our Results. Now we are ready to state our results. Our first result (Theorem 1) shows that for $b > 1$ we obtain sublinear running time if the expansion overlap has size at least $\Omega(\log m)$. Note that this already motivates why the two steps expan_s and expan_t and the resulting expansion overlap are of interest.

The logarithmic expansion overlap required for the above result is of course a rather strong requirement that does not apply in all cases where we expect expanding search spaces to speed up bidirectional BFS. For instance, the expansion overlap is at most the distance between s and t, which might already be too small. This motivates our second result (Theorem 2), where we only require an expansion overlap of sufficient relative length, as long as the maximum expansion is at most a constant factor of the minimum expansion b. Additionally, we make use of the fact that cheap_s and cheap_t can give us initial steps of the search that are cheap. Formally, we define the *(α-)relevant distance* as $d_\alpha(s,t) = \text{cheap}_t - \text{cheap}_s - 1$ and require expansion overlap linear in $d_\alpha(s,t)$, i.e., we obtain sublinear running time if the expansion overlap is at least $c \cdot d_\alpha(s,t)$ (see also Fig 1) for some constant c.

Finally, in our third result (Theorem 3), we relax the condition of Theorem 2 further by allowing expansion overlap that is sublinear in $d_\alpha(s,t)$ or even non-existent. The latter corresponds to non-positive expansion overlap, when extending the above definition to the case $\text{expan}_t > \text{expan}_s$. Specifically, we define $S_1 = \text{expan}_s$, $S_2 = \text{cheap}_t - \text{expan}_s - 1$, $T_1 = d(s,t) - \text{expan}_t + 1$, and $T_2 = \text{expan}_t - \text{cheap}_s - 1$ (see Fig. 2) and give a bound for which values of

$$\rho = \frac{\max\{S_2, T_2\}}{\min\{S_1, T_1\}},$$

sublinear running time can be guaranteed. We write $\rho_{s,t}(\alpha, b)$ if these parameters are not clear from context. This bound is tight (see Lemma 7), i.e., for all larger values of ρ we give instances with linear running time.

3.2 Large Absolute Expansion Overlap

We start by proving sublinear running time for a logarithmic expansion overlap.

Theorem 1. *For parameter $b > 1$ let $s, t \in V$ be a start–destination pair with a b-expansion overlap of size at least $c \log_b(m)$ for a constant $c > 0$. Then $c_{\text{bi}}(s,t) \leq 8 \log_b(2m) \cdot \frac{b^2}{b-1} \cdot m^{1-c/2}$.*

Proof Sketch. We assume the search meets in the middle of the expansion overlap and note that there are at least $\frac{c}{2} \log_b(m)$ more b-expanding layers behind the last explored layer. The cost of these layers is high enough that this lets us derive the stated sublinear bound on the cost of the last layer compared to the number of edges of the graph. See the full version for a complete proof. □

3.3 Large Relative Expansion Overlap

Note that Theorem 1 cannot be applied if the length of the expansion overlap is too small. We resolve this in the next theorem, in which the required length of the expansion overlap is only relative to α-relevant distance between s and t, i.e., the distance without the first few cheap steps around s and t. Additionally, we say that b^+ is the *highest expansion between s and t* if it is the smallest number, such that there is no sequence of exploration steps that is more than b^+-expanding from s or t.

Theorem 2. *For parameters $0 \leq \alpha < 1$ and $b > 1$, let $s, t \in V$ be a start–destination pair with a b-expansion overlap of size at least $c \cdot d_\alpha(s,t)$ for some constant $c > 0$ and assume that $b^+ \geq b$ is the highest expansion between s and t. Then $c_{\text{bi}}(s,t) \in \tilde{\mathcal{O}}\left(m^{1-\varepsilon}\right)$ for $\varepsilon = \frac{c(1-\alpha)}{\log_b(b^+)+c} > 0$.*

Proof Sketch. The proof works via a case distinction on the size of the expansion overlap. Either, it is large enough to apply Theorem 1. Otherwise, the number of layers is small enough that even if their cost grows with a factor b^+ per step, the overall cost will be sublinear. See the full version of this paper for the complete proof. □

Note that Theorem 2 does not require $\text{expan}_t > \text{cheap}_s$ or $\text{expan}_s < \text{cheap}_t$, i.e., the expansion overlap may intersect the cheap prefix and suffix. Before extending this result to an even wider regime, we want to briefly mention a simple corollary of the theorem, in which we consider vertices with an expansion overlap region and polynomial degree.

Corollary 1. *For parameter $b > 1$, let $s, t \in V$ be a start–destination pair with a b-expansion overlap of size at least $c \cdot d(s,t)$ for a constant $0 < c \leq 1$. Further, assume that $\deg(t) \leq \deg(s) \leq m^\delta$ for a constant $\delta \in (0,1)$ and that b^+ is the highest expansion between s and t. Then $c_{\text{bi}}(s,t) \in \tilde{\mathcal{O}}\left(m^{1-\frac{c(1-\delta)}{\log_b(b^+)+c}}\right)$.*

This follows directly from Theorem 2, using $\text{cheap}_s(\delta)$ and $\text{cheap}_t(\delta)$.

3.4 Small or Non-existent Expansion Overlap

Theorem 2 is already quite flexible, as it only requires an expansion overlap with constant length relative to the distance between s and t, minus the lengths of a cheap prefix and suffix. In this section, we weaken these conditions further, obtaining a tight criterion for polynomially sublinear running time. In particular, we relax the length requirement for the expansion overlap as far as possible. Intuitively, we consider the case in which the cheap prefix and suffix cover almost all the distance between start and destination. Then, the cost occurring between prefix and suffix can be small enough to stay sublinear, regardless of whether there still is an expansion overlap or not.

In the following we first examine the sublinear case, before constructing a family of graphs with linear running time for the other case and putting together the complete dichotomy in Theorem 3. We begin by proving an upper bound for the length of low-cost sequences, such as $[1, \text{cheap}_s]$ and $[\text{cheap}_t, d(s,t)]$.

Lemma 4. *Let v be a vertex with a b-expanding sequence S starting at v with cost $c_v(S) \leq C$. Then $|S| \leq \log_b(C) + 1$.*

Proof Sketch. Follows via a similar derivation as for Lemma 2. □

This statement is used in the following technical lemma that is needed to prove sublinear running times in the case of small expansion overlap. Recall from Sect. 3.1 that $\rho_{s,t}(\alpha, b) = \frac{\max\{S_2, T_2\}}{\min\{S_1, T_1\}}$; also see Fig. 2.

Lemma 5. *For parameters $0 \leq \alpha < 1$ and $b > 1$, let $s, t \in V$ be a start–destination pair and assume that b^+ is the highest expansion between s and t and $\rho_{s,t}(\alpha, b) < \frac{1-\alpha}{1-\alpha+\alpha\log_b(b^+)}$. There are constants $c > 0$ and k such that if the size of the b-expansion overlap is less than $c \cdot \log_b(m) - k$, then there is a constant $x < 1$ such that $c_s([1, \text{cheap}_s + T_2]) \leq 2^{1-\alpha} \cdot m^x$ and $c_t([\text{cheap}_t - S_2, d(s,t)]) \leq 2^{1-\alpha} \cdot m^x$.*

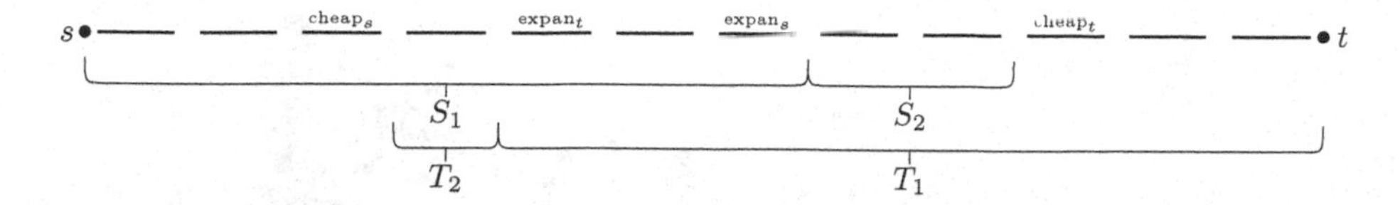

Fig. 2. Visualization of exploration steps and (lengths of) sequences relevant for Lemmas 5 and 6

Proof Sketch. Following the definitions, we can write the size of the expansion overlap as $d_{\text{overlap}} = S_1 - T_2 - \text{cheap}_s$. We use this to derive a suitable upper bound for S_2 and T_2 that we use to give a sublinear upper bound for the cost of $c_s(\text{cheap}_s + T_2)$. See the full version of this paper for the complete proof. □

This lets us prove the sublinear upper bound.

Lemma 6. *For parameters $0 \leq \alpha < 1$ and $b > 1$, let $s, t \in V$ be a start-destination pair and assume that b^+ is the highest expansion between s and t. If $\rho_{s,t}(\alpha, b) < \frac{1-\alpha}{1-\alpha+\alpha\log_b(b^+)}$, then $c_{\text{bi}}(s,t) \in \tilde{O}(m^{1-\varepsilon})$ for a constant $\varepsilon > 0$.*

Proof Sketch. Similarly to the proof of Theorem 2, we obtain sublinear cost if the size of the expansion overlap is large, via Theorem 1. Otherwise, Lemma 5 gives sublinear cost for exploration sequences around s and t. If these sequences intersect, sublinear cost follows immediately. Otherwise, sublinear cost follows via direct application of Theorem 2, as the expansion steps between the considered sequences are b-expanding both from s and from t. See the full version of this paper for the complete proof. □

The following lemma covers the other side of the dichotomy, by proving a linear lower bound on the running time for the case where the conditions on ρ in Lemma 6 are not met. The rough idea is the following. We construct symmetric trees of depth d around s and t. The trees are b-expanding for $(1-\rho)d$ steps and b^+-expanding for subsequent ρd steps and are connected at their deepest layers.

Lemma 7. *For any choice of the parameters $0 < \alpha < 1$, $b^+ > b > 1$, $\rho_{s,}(\alpha, b) \geq \frac{1-\alpha}{1-\alpha+\alpha\log_b(b^+)}$ there is an infinite family of graphs with two designated vertices s and t, such that in the limit $\text{cheap}_s(\alpha)$, $\text{cheap}_t(\alpha)$, $\text{expan}_s(b)$, and $\text{expan}_t(b)$ fit these parameters, b^+ is the highest expansion between s and t and $c_{\text{bi}}(s,t) \in \Theta(m)$.*

This lets us state a complete characterization of the worst case running time of bidirectional BFS depending on $\rho_{s,t}(\alpha, b)$. It follows directly from Lemma 6 and Lemma 7.

Theorem 3. *Let an instance (G, s, t) be a graph with two designated vertices, let b^+ be the highest expansion between s and t and let $0 < \alpha < 1$ and $b > 1$ be parameters. For a family of instances we have $c_{\text{bi}}(s,t) \in \mathcal{O}(m^{1-\varepsilon})$ for some constant $\varepsilon > 0$ if $\rho_{s,t}(\alpha, b) < \frac{1-\alpha}{1-\alpha+\alpha\log_b(b^+)}$ and $c_{\text{bi}}(s,t) \in \Theta(m)$ otherwise.*

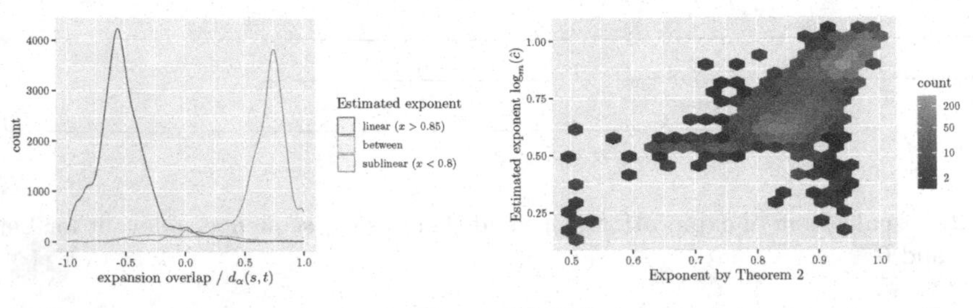

(a) Distribution of parameter c of Theorem 2 for $b = 2$ and $\alpha = 0.1$ for graphs with different estimated exponents.

(b) Relationship between estimated exponent and asymptotic exponent predicted by Theorem 2 for $b = 2$.

Fig. 3. Empirical validation of Theorem 2.

4 Evaluation

We conduct experiments to evaluate how well our concept of expansion captures the practical performance observed on real-world networks. For this, we use a collection of 3006 networks selected from Network Repository [2,9]. The data-set was obtained by selecting all networks with at most 1 M edges and comprises networks from a wide range of domains such as social-, biological-, and infrastructure-networks. Each of these networks was reduced to its largest connected component and multi-edges, self-loops, edge directions and weights were ignored. Finally, only one copy of isomorphic graphs was kept. The networks have a mean size of 12386 vertices (median 522.5) and are mostly sparse with a median average degree of 5.6.

4.1 Setup and Results

For each graph, we randomly sample 250 start–destination pairs s, t. We measure the cost of the bidirectional search as the sum of the degrees of explored vertices. For each graph we can then compute the average cost $\hat{c}$ of the sampled pairs. Then, assuming that the cost behaves asymptotically as $\hat{c} = m^x$ for some constant x, we can compute the *estimated exponent* as $x = \log_m \hat{c}$.

We focus our evaluation on the conditions in Theorem 2 and Theorem 3. For this, we compute $\text{expan}_s(b)$, $\text{expan}_t(b)$, $\text{cheap}_s(\alpha)$, and $\text{cheap}_t(\alpha)$ for each sampled vertex pair for all values of α, by implicitly calculating the values of α corresponding to cheap sequences of different length.

By Theorem 2 a vertex pair has asymptotically sublinear running time, if the length of the expansion overlap is a constant fraction of the relevant distance $d_\alpha(s,t)$. We therefore computed this fraction for every pair and then averaged over all sampled pairs of a graph. Note that for any graph of fixed size, there is a value of α, such that $\text{cheap}_s(\alpha) \geq \text{cheap}_t(\alpha)$. We therefore set $\alpha \leq 0.1$ in order to not exploit the asymptotic nature of the result. Also we set the minimum base of the expansion b to 2. Outside of extreme ranges, the exact choice of

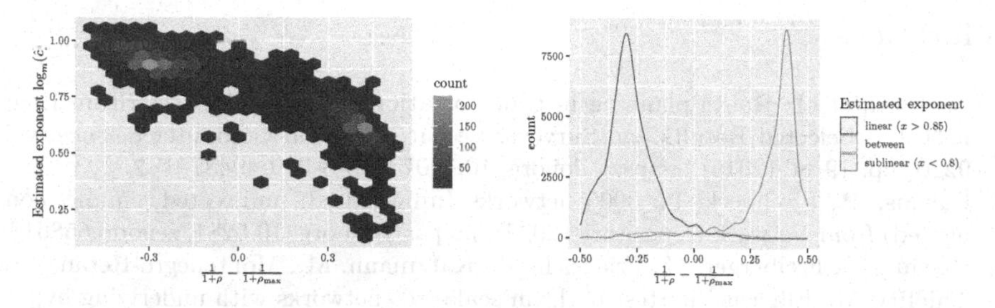

Fig. 4. Relationship between estimated exponent and $\delta_\rho = 1/(1+\rho_{s,t}(\alpha,b) - 1/(1+\rho_{\max})$ for $b=2$. Theorem 3 predicts sublinear running time for all points with $\delta_\rho > 0$.

these parameters makes only little difference. Figure 3a shows the distribution of the relative length of the expansion overlap for different values of the estimated exponent. It separates the graphs into three categories; graphs with estimated exponent $x > 0.85$ ((almost) linear), with $x < 0.8$ (sublinear) and the graphs in between. We note that the exact choice of these break points makes little difference.

Note that Theorem 2 states not only sublinear running time but actually gives the exponent as $1 - \frac{c(1-\alpha)}{2(\log_b(b^+)+c)}$. Figure 3b shows the relationship between this exponent (averaged over the (s,t)-pairs) and the estimated exponent. For each sampled pair of vertices we chose α optimally to minimize the exponent. This is valid even for individual instances of fixed size, because even while higher values of α increase the fraction of the distance that is included in the cheap prefix and suffix, this increases the predicted exponent.

Finally, Theorem 3 proves sublinear running time if $\rho_{s,t}(\alpha,b) \leq \frac{1-\alpha}{1-\alpha+\alpha\log_b(b^+)}$. To evaluate how well real-world networks fit this criterion, we computed $\rho_{s,t}(\alpha,b)$ for each sampled pair (s,t) as well as the upper bound $\rho_{\max} := \frac{1-\alpha}{1-\alpha+\alpha\log_b(b^+)}$. Again, choosing large values for α does not exploit the asymptotic nature of the statement, as $\rho_{\max}$ tends to 0 for large values of α. For each vertex pair, we therefore picked the optimal value of α, minimizing $\rho_{\max} - \rho_{s,t}(\alpha,b)$ and recorded the average over all pairs for each graph. Figure 4 shows the difference between $1/(1+\rho_{s,t}(\alpha,b))$ and $1/(1+\rho_{\max})$. This limits the range of these values to $[0,1]$ and is like dividing S_2 by S_1+S_2 instead of S_2 by S_1 in the definition of ρ.

4.2 Discussion

Both Fig. 4 and Fig. 3a show that our notion of expansion not only covers some real networks, but actually gives a good separation between networks where the bidirectional BFS performs well and those where it requires (close to) linear running time. With few exceptions, exactly those graphs that seem to have sublinear running time satisfy our conditions for asymptotically sublinear running time. Furthermore, although the exponent stated in Theorem 2 only gives an asymptotic worst-case guarantee, Fig. 3b clearly shows that the estimated exponent of the running time is strongly correlated with the exponent given in the theorem.

References

1. Bast, H., et al.: Route planning in transportation networks. In: Algorithm Engineering - Selected Results and Surveys, Lecture Notes in Computer Science, vol. 9220, pp. 19–80 (2016). https://doi.org/10.1007/978-3-319-49487-6_2
2. Bläsius, T., Fischbeck, P.: 3006 networks (unweighted, undirected, simple, connected) from Network Repository (2022). https://doi.org/10.5281/zenodo.6586185
3. Bläsius, T., Freiberger, C., Friedrich, T., Katzmann, M., Montenegro-Retana, F., Thieffry, M.: Efficient shortest paths in scale-free networks with underlying hyperbolic geometry. ACM Trans. Algorithms 18(2) (2022). https://doi.org/10.1145/3516483
4. Bläsius, T., Wilhelm, M.: Deterministic performance guarantees for bidirectional bfs on real-world networks. Computing Research Repository (CoRR) abs/2209.15300 (2022). https://doi.org/10.48550/arXiv.2209.15300
5. Borassi, M., Crescenzi, P., Trevisan, L.: An axiomatic and an average-case analysis of algorithms and heuristics for metric properties of graphs. In: Proceedings of the 28th Annual ACM-SIAM Symposium on Discrete Algorithms, SODA 2017, pp. 920–939. SIAM (2017). https://doi.org/10.1137/1.9781611974782.58
6. Borassi, M., Natale, E.: KADABRA is an adaptive algorithm for betweenness via random approximation. ACM J. Exp. Algorithmics 24(1), 1.2:1–1.2:35 (2019). https://doi.org/10.1145/3284359
7. Erdős, P., Rényi, A.: On random graphs I. Publicationes Mathematicae 6, 290–297 (1959). https://www.renyi.hu/~p_erdos/1959-11.pdf
8. Fox, J., Roughgarden, T., Seshadhri, C., Wei, F., Wein, N.: Finding cliques in social networks: a new distribution-free model. SIAM J. Comput. **49**(2), 448–464 (2020). https://doi.org/10.1137/18M1210459
9. Rossi, R.A., Ahmed, N.K.: The network data repository with interactive graph analytics and visualization. In: AAAI (2015). https://networkrepository.com
10. Roughgarden, T., Seshadhri, C.: Distribution-free models of social networks, Chap. 28, pp. 606–625. Cambridge University Press (2021). https://doi.org/10.1017/9781108637435.035

A Polyhedral Perspective on Tropical Convolutions

Cornelius Brand[1(✉)], Martin Koutecký[2], and Alexandra Lassota[3]

[1] Department of Computer Science, Vienna University of Technology, Vienna, Austria
cbrand@ac.tuwien.ac.at
[2] Computer Science Institute, Charles University, Prague, Czech Republic
koutecky@iuuk.mff.cuni.cz
[3] Institute of Mathematics, EPFL, Lausanne, Switzerland
alexandra.lassota@epfl.ch

Abstract. Tropical (or min-plus) convolution is a well-studied algorithmic primitive in fine-grained complexity. We exhibit a novel connection between polyhedral formulations and tropical convolution, through which we arrive at a dual variant of tropical convolution. We show this dual operation to be equivalent to primal convolutions. This leads us to considering the geometric objects that arise from dual tropical convolution as a new approach to algorithms and lower bounds for tropical convolutions. In particular, we initiate the study of their extended formulations.

1 Introduction

Given two sequences $a = (a_0, \ldots, a_n) \in \mathbb{R}^{n+1}$ and $b = (b_0, \ldots, b_n) \in \mathbb{R}^{n+1}$, their *tropical convolution* $a * b$ is defined as the sequence $c = (c_0, \ldots, c_{2n}) \in \mathbb{R}^{2n+1}$, where

$$c_k = \min_{i+j=k} a_i + b_j \text{ for } k = 0, \ldots, 2n. \quad (1)$$

Other names for this operation include $(\min, +)$-convolution, inf-convolution or epigraphic sum. It can be equivalently formulated with a max in lieu of the min by negating a and b point-wise.

In its continuous guise, it is a classic object of study in convex analysis (see e.g. the monograph of Fenchel [7]). The discrete variant above has been studied algorithmically for more than half a century, going back at least to the classic work of Bellman and Karush [2].

C. Brand was supported by the Austrian Science Fund (FWF, Project Y1329: ParAI). M. Koutecký was partially supported by Charles University project UNCE/SCI/004 and by the project 22-22997S of GA ČR. A. Lassota was supported by the Swiss National Science Foundation within the project *Complexity of integer Programming* (207365).
The original version of the chapter has been revised. The first reference in this chapter has been corrected. A correction to this chapter can be found at
https://doi.org/10.1007/978-3-031-34347-6_34

S.-Y. Hsieh et al. (Eds.): IWOCA 2023, LNCS 13889, pp. 111–122, 2023.
https://doi.org/10.1007/978-3-031-34347-6_10

This article initiates the study of tropical convolution from the point of view of polyhedral geometry and linear programming. This leads to a novel dual variant of the operation. We prove that this dual convolution is equivalent to the usual, primal tropical convolution under subquadratic reductions. This motivates our subsequent study of the associated polyhedra.

Tropical Convolution. There is a trivial quadratic-time procedure to compute the tropical convolution of two sequences, and significant research efforts have been directed at improving over this upper bound. While there are $o(n^2)$-time algorithms for the problem [3,14], there is no known $n^{1.99}$-time algorithm. More efficient algorithms are known only in restricted special cases or relaxed models of computation [3–5]. This has led researchers to considering the quadratic lower bound on tropical convolution as one of the various hardness assumptions in fine-grained complexity [6]. To wit, Cygan et al. [6] and, independently, Künnemann et al. [11] formally conjectured the following statement to hold true:

Conjecture 1. (MinConv-Conjecture). There is no algorithm computing Eq. (1) with a, b having integral entries of absolute value at most d in time $n^{2-\varepsilon} \cdot \text{polylog}(d)$, for any $\varepsilon > 0$.

This conjecture joins the ranks of several similar conjectures (or hypotheses) made in the same spirit. These are all part of a line of research that is often referred to as the study of "hardness in P," and has seen a tremendous amount of progress in recent years.

For example, a classic hardness assumption in computational geometry is that the well-known 3-Sum problem cannot be solved in truly subquadratic time. More recent examples include the All-Pairs-Shortest-Paths (APSP) conjecture (where the assumption is that no truly subcubic algorithms exist) [16] or the Orthogonal-Vectors problem. The latter gives a connection of these polynomial-time hardness assumptions to the world of exact exponential-time hardness and its main hypothesis, the Strong Exponential-Time Hypothesis (SETH) [10,15].

Polyhedral Formulations. It is a fundamental technique of combinatorial optimization to model the solution space of a problem as polyhedra (which are solution sets of finite systems of linear inequalities over the reals or the rationals), or rather, polytopes (bounded polyhedra). In this way, polytopes become a model of computation, which necessitates endowing them with a complexity measure. This is usually done using their *extension complexity*. For a polytope P, this is the minimum of the number of inequalities needed to describe any polytope that maps to P under linear projections (so-called *extended formulations*). There is a large body of work on extended formulations of various fundamental problems of combinatorial optimization, including recent breakthroughs on lower bounds [9,13]. On the other hand, surprisingly small extended formulations do exist for polyhedral formulations of some combinatorial objects, for instance for spanning trees in minor-closed graph classes, as shown by Aprile et al. [1] (see also [8]).

Our Contribution. While for many combinatorial optimization problems, there is an obvious associated polytope such that optimizing over the polytope corresponds to solving the combinatorial problem, this is not the case for tropical convolutions. It is a natural question to ask whether there is even a way to come up with such a polytope, and if so, how the resulting optimization problem relates to the usual tropical convolution.

In this article, we answer this question affirmatively by exhibiting a novel connection between polyhedral geometry and tropical convolutions. This connection can be seen from a variety of perspectives.

On the one hand, from the point of view of fine-grained complexity, we introduce a new, dual variant of tropical convolution and show its equivalence to primal convolutions under subquadratic reductions. This can be interpreted as a new avenue for designing fast algorithms and lower bounds for primal convolutions.

On the other hand, the polyhedral formulations we use to derive this new problem may provide structural insights into tropical convolutions, and point towards interesting special cases worth studying. In particular, extended formulations for these polyhedra might shed light on the complexity of tropical convolutions.

Finally, we find the geometric objects that arise from dual tropical convolution to be interesting in their own right.

Proofs of the claims in this article and all formal details will be made available in the full version.

2 Polyhedral Formulations and Dual Convolutions

There is a straightforward way of formulating the solution of the defining Eq. (1) as a linear program in variables $x_0, \ldots, x_{2n}$, given a and b:

$$\max \sum_{i=0}^{2n} x_i \quad \text{subject to} \quad x_{i+j} \leq a_i + b_j \text{ for } i, j = 0, \ldots, n. \tag{2}$$

As it is, this program captures very little of the original problem. It is not helpful algorithmically, as writing it down amounts to a brute-force computation of all terms considered in Eq. (1). It is also not interesting geometrically, since the polytope described by these inequalities is just a translated orthant with a vertex at $a * b$.

A Better Formulation. In order to model the geometric structure of Eq. (1) more faithfully, we can generalize this program from fixed to variable a and b. Consider the polyhedron P_n defined on (x, y, z)-space as

$$P_n = \{(x, y, z) \in \mathbb{R}^{2n+1} \times \mathbb{R}^{n+1} \times \mathbb{R}^{n+1} \mid x_{i+j} \leq y_i + z_j \text{ for } i, j = 0, \ldots, n\}. \tag{3}$$

While this formulation is more expressive geometrically, it is still not clear how to make use of P_n algorithmically: The optimization problem of maximizing $\sum_{i=0}^{2n} x_i$ over P_n is unbounded, and in order to incorporate an instance of Eq. (1), we need to consider the intersection of P_n with the hyperplanes defined by $y_i = a_i$ and $z_i = b_i$, which brings us back to the previous formulation.

The Dual Approach. However, the dual of the linear program resulting from the intersection of P_n with these hyperplanes sheds more light on the situation. It can be written down in the variables $\mu_i, \nu_j, \lambda_{i,j}$ with $i, j = 0, \ldots, n$ as

$$\begin{aligned} &\min \sum_{i=0}^{n} a_i \mu_i + \sum_{j=0}^{n} b_j \nu_j \\ &\text{subject to } \lambda \geq 0 \text{ and } \sum_{i+j=k} \lambda_{i,j} = 1 \qquad \text{for all } k = 0, \ldots, 2n, \\ &\mu_i = \sum_{j=0}^{n} \lambda_{i,j},\ \nu_j = \sum_{i=0}^{n} \lambda_{i,j}, \qquad \text{for all } i, j = 0, \ldots, n. \end{aligned} \tag{4}$$

Note that a point in the polyhedron defined by system (4) is fully determined by its λ-coordinates. It is therefore sensible geometrically to just discard μ and ν whenever considering the system (4) as a polytope. Let us collect the objects we have encountered so far in a definition.

Definition 1 (Convolution polyhedra). *The polyhedron* $P_n \subseteq \mathbb{R}^{4n+1}$ *is called the* primal convolution polyhedron of order n. *The projection of the polyhedron defined by the inequalities in* (4) *to* λ*-space is called the* dual convolution *polyhedron of order* n *and is denoted as* $D_n \subseteq \mathbb{R}^{(n+1)\times(n+1)}$.

We prove later that dual convolution polyhedra are bounded and integral. This justifies understanding the dual variable $\lambda_{i,j}$ as a decision variable that indicates whether or not the k-th entry c_k of the result in the tropical convolution is decomposed by the solution into $c_k = a_i + b_j$ with $i + j = k$. Correspondingly, μ_i indicates how often a_i is chosen as part of *any* sum $c_{i+j} = a_i + b_j$ for some j, and similarly, ν_j records the multiplicity of b_j as part of a sum $c_{i+j} = a_i + b_j$ for some i.

Projections to Multiplicities. Now, the program (4) has the desirable property that the inputs a and b are reflected only in the target functional, and not in the definition of the underlying polyhedron. Unfortunately, this comes at the price of introducing a quadratic number of variables $\lambda_{i,j}$ into the program. However, none of the λ-variables directly enter into the objective function, but only via the coordinates μ and ν. Optimizing over D_n as in (4) is therefore equivalent to optimizing over the *projection* of the dual convolution polyhedron to (μ, ν)-space.

Definition 2 (Multiplicity polytope). *The projection of the dual convolution polyhedron to* μ *and* ν *is called the* multiplicity polytope *of order* n. *We denote it with* $M_n \subseteq \mathbb{R}^{2n+2}$.

Note that it is a well-known fact of polyhedral geometry that the projection of a polyhedron is a polyhedron. Observe further that multiplicity polytopes are bounded and integral, as dual convolution polyhedra are bounded and integral, as we are yet to see. Hence, multiplicity polytopes are indeed polytopes.

Dual Tropical Convolution. This allows us to formulate a dual optimization variant of tropical convolution:

Definition 3 (Dual tropical convolution). *Given $a, b \in \mathbb{R}^n$,* dual tropical convolution *is the problem of solving the following linear optimization problem:*

$$\min a^T \mu + b^T \nu \text{ subject to } (\mu, \nu) \in M_n. \tag{5}$$

We refer to μ and ν as multiplicities.

In more combinatorial terms, we want to select, for each a_i and each b_j, how often it has to appear in order to obtain an optimal solution for the tropical convolution $a * b$.

Fine-Grained Complexity. Since M_n is a projection of D_n, every pair of multiplicities (that is, every point in M_n) corresponds to some point in D_n, which projects to this pair. Now, suppose we could solve the dual tropical convolution in time $n^{1.9}$. Does this imply, say, an $n^{1.999}$-time algorithm for tropical convolution, that is: can we efficiently lift the multiplicities in M_n to a solution in D_n, and hence, a solution for Eq. (1)? We show below that this is indeed the case. As our main contribution to the fine-grained complexity of tropical convolution, we show:

Theorem 1. *If there is an algorithm that solves dual tropical convolution on numbers of absolute value at most d in time $n^{2-\delta} \cdot \text{polylog}(d)$ for some $\delta > 0$, then there is an algorithm that solves primal tropical convolution on numbers of absolute value at most d in time $n^{2-\varepsilon} \cdot \text{polylog}(d)$ for some $\varepsilon > 0$.*

In order to prove Theorem 1, we give a fine-grained Turing reduction that works, roughly speaking, as follows. Given an instance (a, b) of primal tropical convolution, we use the purported algorithm for solving dual convolutions as an oracle. We first massage the instance into a more useful form, where every entry of a and b is triplicated and then carefully perturbed to ensure uniqueness of solutions. We then split this modified instance into small pieces, and we query the oracle to dual convolution on instances obtained by replacing one of these pieces by infinite (rather, sufficiently large) values. This way, we can determine successively which pairs of a and b are matched together to produce the final solution $c = a * b$.

We note that the other direction, that is, reading off the multiplicities from a solution to primal tropical convolution as witnessed by pairs i, j decomposing each k as $k = i + j$, is trivial. However, in fine-grained complexity, tropical convolution is usually considered just as the task of outputting the vector $a * b$, not giving the decomposition of indices. Even in this setting, we show that a converse of Theorem 1 holds:

Theorem 2. *If there is an algorithm that outputs $a * b$ given $a, b \in \mathbb{Z}^{n+1}$ containing numbers of absolute value at most d in time $n^{2-\varepsilon} \cdot \text{polylog}(d)$ for some $\varepsilon > 0$, then there is an algorithm that solves dual tropical convolution on numbers of absolute value at most d in time $n^{2-\delta} \cdot \text{polylog}(d)$ for some $\delta > 0$.*

Theorems 1 and 2 in turn motivate the study of M_n as a geometric object. If it turned out that M_n can be described in some simple manner that allows the corresponding linear program to be solved in time $n^{1.99}$, this would imply a truly subquadratic algorithm for tropical convolution.

An Equivalent Approach. We now turn to another formulation of tropical convolutions in terms of convex optimization, which we show to be equivalent to the approach laid out above. It is clear that a solution of Eq. (1) is encoded, for each k, by a choice of i, j such that $i + j = k$. It is apparent from the definition (and made explicit in Eq. (2)) that in order to decide optimality of a candidate solution (that is, a set of pairs (i, j)), it is enough to decide whether the sum of the entries decomposed in this way is optimal. Formally, this defines a function

$$\sigma_n : \mathbb{R}^n \times \mathbb{R}^n \to \mathbb{R}, (a, b) \mapsto \sum_{k=0}^{2n} c_k, \text{ where } c = a * b. \tag{6}$$

We can now ask if there is a polytope such that optimizing in direction of (a, b) over this polytope yields $\sigma_n(a, b)$. Naively, this approach for computing σ_n can be encoded using the following set:

Definition 4. *We denote with Σ_n the following infinite intersection of half-spaces:*

$$\Sigma_n = \bigcap_{a,b \in \mathbb{R}^n} \{(x, y) \in \mathbb{R}^{n+1} \times \mathbb{R}^{n+1} \mid a^T x + b^T y \geq \sigma_n(a, b)\}. \tag{7}$$

Since it is an intersection of half-spaces, Σ_n is a convex set. Furthermore, by construction, it has the property that

$$\min_{(x,y) \in \Sigma_n} a^T x + b^T y \geq \sigma_n(a, b) \tag{8}$$

holds for all $a, b \in \mathbb{R}^{n+1}$. However, if we were agnostic about the approach via P_n and D_n from before, it is neither clear whether or not equality holds in Eq. (8), nor if Σ_n can be written as a finite intersection of half-spaces. In fact, both statements hold. Even more, we show the following:

Theorem 3. *Σ_n is equal to the multiplicity polytope of order n, that is, $\Sigma_n = M_n$ for all n.*

Given the properties of M_n, Theorem 3 can be shown by using strong duality and the fact that polytopes are determined by their so-called supporting functions. However, in order to actually describe how one would arrive at Theorem 3 from first principles, we proceed to show $M_n = \Sigma_n$ via a decomposition of D_n into a Minkowski sum of simplices of varying dimensions, and then arguing about the support functions of appropriate projections.

Theorem 3 means that optimizing over Σ_n is the same as optimizing over M_n, which is a polytope (hence all minima exist). As an immediate consequence, we obtain:

Corollary 1. *For all $a, b \in \mathbb{R}^n$, the following holds:*

$$\min_{(x,y)\in\Sigma_n} a^T x + b^T y = \sigma_n(a, b). \tag{9}$$

Reading Theorem 3 as a statement about M_n gives us a representation of M_n as an intersection of half-spaces, albeit infinite. Conceptually, this justifies understanding M_n as the natural polytope capturing tropical convolutions in yet a different way than the primal-dual approach above.

2.1 Extended Formulations

In light of Theorem 1, it is a natural approach to study the structure of M_n in order to obtain insights into the complexity of tropical convolutions. In particular, we are interested in determining the complexity of the polytope M_n as a surrogate quantity for the complexity of tropical convolutions. Indeed, the complexity measure used on polytopes, as mentioned above, is usually their *extension complexity*. In the literature, this has been used as a formal model of computation that corresponds to dynamic programming based approaches to combinatorial problems [12]. Insofar, lower bounds on the complexity M_n translate to lower bounds in restricted computational models on tropical convolution.

Yannakakis' Algebraic Bound. Recall that the extension complexity of a polytope P is the smallest number of inequalities needed to describe any polytope Q that maps to P under linear projections, and is usually denoted as $\text{xc}(P)$. Such polytopes Q are called *extended formulations* of P. While this quantity by itself may seem rather intangible, the seminal work of Yannakakis [17] provides an algebraic approach to lower-bounding $\text{xc}(P)$. It is based on *slack matrices* of the polytope P, which we define in a general form as follows:

Definition 5 (Slack matrix). *Let $V = \{v_1, \ldots, v_t\} \subseteq \mathbb{R}^N$, and let $L = \{\ell_1, \ldots, \ell_m\}$ be a set of affine linear functionals $\ell_j : \mathbb{R}^N \to \mathbb{R}$ such that $\ell_i(v_j) \geq 0$ holds for all i, j. Then, the matrix $S \in \mathbb{R}^{m\times t}$ defined through $S_{i,j} = \ell_i(v_j)$ is called the* slack matrix *of V with respect to L.*

The crucial quantity related to these matrices studied by Yannakakis is the following:

Definition 6 (Non-negative rank). *For a matrix with non-negative entries $M \in \mathbb{R}^{m\times t}$, its* non-negative rank $\text{rk}_+(M)$ *is defined as the smallest number r such that there exist non-negative column vectors $u_1, \ldots, u_r \in \mathbb{R}^m_+$ and $v_1, \ldots, v_r \in \mathbb{R}^t_+$ such that $M = \sum_{i=1}^r u_i v_i^T$ holds.*

An *extreme point* (or *vertex*) of a polytope is one that cannot be written as a convex combination of other points of the polytope. One connection that Yannakakis showed is the following:

Theorem 4 ([17]). *Let V be the set of vertices of a polytope P. Let S be the slack matrix of V with respect to some functionals L. Then, the following inequality holds:*

$$\text{rk}_+(S) \leq \text{xc}(P). \tag{10}$$

For an appropriate choice of V and L, equality can be shown to hold in Eq. (10). Since we are interested in lower bounds, this is not relevant to us.

For a polytope P, the set L is usually given through *slack functionals* associated with a description of P as a system of linear inequalities. Concretely, suppose that P is represented as $P = \{x \mid Ax \leq b\} \subseteq \mathbb{R}^N$ for some matrix $A \in \mathbb{R}^{m \times N}$ and $b \in \mathbb{R}^m$, and let a_i be the i-th row of A. Then, the linear functionals ℓ_i are given through $\ell_i : x \mapsto b_i - a_i^T x$. Thus, by definition, $\ell_i(p) \geq 0$ for all $p \in P$. In particular, this holds when p is a vertex of P.

To lower-bound $\mathrm{rk}_+(S)$, Yannakakis suggested an approach based on nondeterministic communication complexity, which only features in this article in its form as rectangle-coverings as follows.

Definition 7 (Rectangle-covering number). *For a matrix with non-negative entries $M \in \mathbb{R}^{m \times t}$, let* $\mathrm{supp}(M) \subseteq [m] \times [t]$ *be the set of indices where M is non-zero. The* rectangle covering number $\mathrm{rc}(M)$ *of M is defined as the smallest number r such that there are sets $L_1, \ldots, L_r \subseteq [m]$ and $R_1, \ldots, R_r \subseteq [t]$ that satisfy* $\mathrm{supp}(M) = \bigcup_{i=1}^{r} L_i \times R_i$. *The cartesian products of sets $L_i \times R_i$ are referred to as* combinatorial rectangles.

Non-negativity of the matrix M then implies:

$$\mathrm{rc}(M) \leq \mathrm{rk}_+(M), \tag{11}$$

which together with Eq. (10) implies a lower bound of $\mathrm{rc}(S)$ on $\mathrm{xc}(P)$.

It took some twenty years until this elegant insight of Yannakakis came to bear fruit. Indeed, in their celebrated work, Fiorini et al. [9] were first to prove exponential rectangle-covering lower bounds on the slack matrix of a polytope corresponding to a problem in combinatorial optimization, namely the Traveling Salesman Problem. Their proof goes via the so-called *correlation polytope.*

Simultaneously, a central open question at the time revolved around the extension complexity of the polytope corresponding to matchings in the complete graph on n vertices. One can show quite easily that the rectangle-covering bound for all slack matrices of this polytope is polynomially bounded in n. However, this does not imply polynomial extension complexity, and the existence of extended formulations of polynomial size for the matching polytope was not known.

In a breakthrough result, Rothvoss [13] finally settled this issue, proving an exponential lower bound on the extension complexity of the matching polytope. The central ingredient of his proof is a direct lower bound for the non-negative rank of slack matrices instead of rectangle-coverings.

Rectangle Coverings for the Multiplicity Polytope. This article studies the extension complexity of M_n through its rectangle-covering bounds. In fact, we give results that suggest that arguments along the lines of Rothvoss' proof may be necessary if we were to prove lower bounds on $\mathrm{xc}(M_n)$ (or rather, the related polytope S_n that we encounter below).

By its definition as a projection of D_n, we have:

Proposition 1. *The extension complexity of the multiplicity polytope M_n is bounded as* $\mathrm{xc}(M_n) \leq O(n^2)$.

Furthermore, we see later that the "hard part" of M_n is not to formulate which μ or ν appear in *some* pair in M_n, but when a pair (μ, ν) appears *together*:

Proposition 2. *Let N_n be the projection of M_n to ν-coordinates. That is, $N_n = \{\nu \mid (\mu, \nu) \in M_n$ for some $\mu\}$. Then, $\mathrm{xc}(N_n) \leq O(n)$.*

The proof proceeds by first decomposing M_n into a Minkowski sum of two polytopes that correspond to a truncated version of tropical convolution. For these truncated polytopes, one can then show a linear upper bound on their extension complexity. This goes by proving that all ν projections of the truncated polytopes arise through a sequence of elementary operations from some initial point.

One major obstacle to proving bounds on $\mathrm{xc}(M_n)$ is that we have no useful combinatorial description of the extreme points of M_n available (or the integral points of M_n, for that matter). We therefore have to transform M_n to a closely related polytope:

Definition 8 (Polar convolution polytopes). *Let $\{(e_i, e_j, e_k)\}_{i,j,k}$ be the standard basis of $\mathbb{R}^{n+1} \oplus \mathbb{R}^{n+1} \oplus \mathbb{R}^{2n+1}$. We define $s_{i,j} = (e_i, e_j, e_{i+j})$ for all $i, j = 0, \ldots, n$. The* polar convolution polytope *S_n is defined as*

$$S_n = \mathrm{conv}\{s_{i,j}\}_{i \neq j} \subseteq \mathbb{R}^{n+1} \oplus \mathbb{R}^{n+1} \oplus \mathbb{R}^{2n+1}. \tag{12}$$

Here, $\mathrm{conv}\, X$ *designates the convex hull of the set X. The* projected polar convolution polytope *$S_n = \pi(S_n) \subseteq \mathbb{R}^{2n+1}$ is the image of S_n under the projection*

$$\pi : (e_i, e_j, e_k) \mapsto e_i + e_j + e_k. \tag{13}$$

The name of S_n is derived from some formal semblance it carries to the so-called polar dual of M_n. However, this is not a direct correspondence, and when arguing about S_n, we do not make use of results from the theory of polarity.

Theorem 5. *For the extension complexity of M_n and S_n, the following holds:*

$$\mathrm{xc}(M_n) \leq \mathrm{xc}(S_n) + O(n). \tag{14}$$

Theorem 5 can be seen by realizing M_n as an intersection of a radial cone of S_n with translates of the coordinate hyperplanes.

While the problem with determining $\mathrm{xc}(M_n)$ was that we have no description of its vertices available, the problem with S_n is now, dually, that we do not have a description of its facets (that is, an irredundant set of inequalities describing S_n), although we know its vertices. We give some evidence for the possibility that, contrary to what one might expect from the MinConv-conjecture, $\mathrm{xc}(S_n) \leq o(n^2)$ could hold.

Valid Inequalities and a Slack Matrix. As a bit of notation for the following, we call an inequality that holds for all points of a polyhedron *valid* for the polyhedron. If, furthermore, the inequality is tight for some point of the polyhedron, we call this inequality *face-defining.* In order to lower-bound $\mathrm{xc}(S_n)$, the first step is to exhibit face-defining valid inequalities for S_n that allow us to prove lower bounds on the non-negative rank of the corresponding slack matrix.

We take inspiration from the inequalities defined by Fiorini et al. [9] for the correlation polytope: They start with trivially valid *quadratic* inequalities instead of linear inequalities. These quadratic inequalities are then reduced modulo equations on the vertices of the correlation polytope, which then yield linearized valid inequalities.

The quadratic inequalities used by Fiorini et al. are an encoding of those pairs (A, B) of subsets of $[n]$ such that $|A \cap B| \neq 1$. Their exponential lower bound then follows from the fact that this relation (interpreted as the support of a non-negative matrix) has high rectangle-covering number. In the same vein, we can encode the property that two pairs $(i, j), (k, \ell)$ of size two of $[n]$ satisfy $|(i, j) \cap (k, \ell)| \neq 1$ (which is to be read as meaning (i, j) and (k, ℓ) agree in either none or both of the coordinates) as a quadratic inequality on the variables of S_n. namely as:

$$(x_k + y_\ell - 1)^2 \geq 0. \tag{15}$$

Here, we assume the linear functionals on $\mathbb{R}^{n+1} \oplus \mathbb{R}^{n+1} \oplus \mathbb{R}^{2n+1}$ to be endowed with a standard dual basis in coordinates x, y, z for each of the respective direct summands, respectively.

Obviously, Ineq. (15) is valid for S_n (after all, it is valid on all of $\mathbb{R}^{n+1} \oplus \mathbb{R}^{n+1} \oplus \mathbb{R}^{2n+1} \supset S_n$). It is tight precisely at those vertices $s_{i,j}$ of S_n that agree in either both or no coordinates.

Since S_n is a 0/1-polytope, the vertices of S_n satisfy $x_k^2 = x_k$ and $y_\ell^2 = x_\ell$. By definition of S_n, the vertices furthermore satisfy $x_k y_\ell \leq z_{k+\ell}$. Therefore, we may linearize the left-hand side of the quadratic inequality as

$$x_k^2 + y_\ell^2 + 1 + 2x_k y_\ell - 2x_k - 2y_\ell \leq 2z_{k+\ell} - x_k - y_\ell + 1$$

and obtain

$$2z_{k+\ell} \geq x_k + y_\ell - 1 \tag{16}$$

as a valid linear inequality for S_n, which is again tight at $s_{i,j}$ if and only if (i, j) and (k, ℓ) agree in both or no coordinates, and is hence face-defining.

One may now suspect that the exponential lower bound for subsets of *all* sizes translates to quadratic lower bounds for subsets of size two. We see later, however, that this is not true in terms of rectangle coverings. For $k, \ell = 0, \ldots, n$, let

$$\lambda_{k,\ell} : \mathbb{R}^{n+1} \oplus \mathbb{R}^{n+1} \oplus \mathbb{R}^{2n+1} \to \mathbb{R}, \ (x, y, z) \mapsto 2z_{k+\ell} - x_k - y_\ell + 1$$

be the corresponding slack functional.

Theorem 6. *The slack matrix Λ_n of S_n with respect to the slack functionals $\lambda_{k,\ell}$ has*

$$\mathrm{rc}(\Lambda_n) \leq O(n).$$

The proof uses a folklore bound on the disjointness relation on sets of fixed size, and the fact that the $(n, 2)$-Kneser graph has clique-covers of linear size.

2.2 Open Questions

With this article, we establish a new connection between fine-grained complexity and polyhedral geometry. While we have initiated the study of its main protagonists, this connection leads to an abundance of further questions that, despite our best efforts, could not be fully resolved.

The most pertinent question is to give a combinatorial description of the facets and vertices of M_n and S_n, and use this description to give either a lower bound or a better upper bound on $\mathrm{xc}(M_n)$. We remain agnostic as to whether $\mathrm{xc}(M_n)$ is indeed lower bounded by $\Omega(n^2)$. The only data we *do* have on $\mathrm{xc}(M_n)$ is a linear upper bound on the rectangle covering number of some slack matrix of the related polytope S_n, which, if anything, points in the opposite direction. On the other hand, the fact that the quadratic lower bound on the computational complexity of tropical convolution is a widely believed conjecture lends credibility to this bound also holding in the polyhedral model.

Furthermore, beyond mere extension complexity, various questions about the complexity of optimizing over M_n arise. In particular, it might be interesting to study the extension complexity of the radial cones of M_n as well as the structure of the circuits of M_n, which are quantities directly related to the complexity of optimizing over M_n.

References

1. Aprile, M., Fiorini, S., Huynh, T., Joret, G., Wood, D.R.: Smaller extended formulations for spanning tree polytopes in minor-closed classes and beyond. Electron. J. Comb. **28**(4), 4–47 (2021)
2. Bellman, R., Karush, W.: Mathematical programming and the maximum transform. J Soc. Ind. Appl. Math. **10**(3), 550–567 (1962)
3. Bremner, D., et al.: Necklaces, convolutions, and X+Y. Algorithmica **69**(2), 294–314 (2014)
4. Bussieck, M.R., Hassler, H., Woeginger, G.J., Zimmermann, U.T.: Fast algorithms for the maximum convolution problem. Oper. Res. Lett. **15**(3), 133–141 (1994)
5. Chan, T.M., Lewenstein, M.: Clustered Integer 3SUM via Additive Combinatorics. In: Servedio, A.R., Rubinfeld, R. (eds.) ACM, pp. 31–40 (2015)
6. Cygan, M., Mucha, M., Wegrzycki, K., Wlodarczyk. , M.: On problems equivalent to (min, +)-convolution. ACM Trans. Algorithms **15**(1), 14:1–14:25 (2019)
7. Fenchel, W., Blackett, W.D.: Convex cones, sets, and functions. Dept. Math. Logistics Res. Proj. (1953). Princeton University

8. Fiorini, S., Huynh, T., Joret, G., Pashkovich, K.: Smaller extended formulations for the spanning tree polytope of bounded-genus graphs. Discrete Comput. Geom. **57**(3), 757–761 (2017). https://doi.org/10.1007/s00454-016-9852-9
9. Fiorini, S., Massar, S., Pokutta, S., Tiwary, H. R., De Wolf, R.: Exponential Lower Bounds for Polytopes in Combinatorial Optimization. J. ACM **62**(2), 17:1–17:23 (2015)
10. Impagliazzo, R., Paturi, R., Zane, F.: Which problems have strongly exponential complexity? J. Comput. Syst. Sci. **63**(4), 512–530 (2001)
11. Künnemann, M., Paturi, R., Schneider, S.: On the fine-grained complexity of one-dimensional dynamic programming. In: Chatzigiannakis, i., Indyk, P., Kuhn, F., Muscholl, A. (eds.) 44th International Colloquium on Automata, Languages, and Programming, ICALP 2017, 10–14 July 2017, Warsaw, Poland, vol. 80. LIPIcs. Schloss Dagstuhl - Leibniz-Zentrum für Informatik, pp. 21:1–21:15 (2017)
12. Martin, R.K., Rardin, R.L., Campbell, B.A.: Polyhedral characterization of discrete dynamic programming. Oper. Res. **38**(1), 127–138 (1990)
13. Rothvoss, T.: The matching polytope has exponential extension complexity. J. ACM 64(6), 41:1–41:19 (2017)
14. Williams, R.R.: Faster all-pairs shortest paths via circuit complexity. SIAM J. Comput. **47**(5), 1965–1985 (2018)
15. Williams, R.: A new algorithm for optimal 2-constraint satisfaction and its implications. Theor. Comput. Sci. **348**(2–3), 357–365 (2005)
16. Williams, V.V., Williams, R.R.: Subcubic equivalences between path, matrix, and triangle problems. J. ACM 65(5), 27:1–27:38 (2018)
17. Yannakakis, M.: Expressing combinatorial optimization problems by linear programs. J. Comput. Syst. Sci. **43**(3), 441–466 (1991)

Online Knapsack with Removal and Recourse

Hans-Joachim Böckenhauer[1], Ralf Klasing[2], Tobias Mömke[3], Peter Rossmanith[4], Moritz Stocker[1(✉)], and David Wehner[1]

[1] Department of Computer Science, ETH Zürich, Zürich, Switzerland
{hjb,moritz.stocker,david.wehner}@inf.ethz.ch
[2] CNRS, LaBRI, Université de Bordeaux, Talence, France
ralf.klasing@labri.fr
[3] Institute of Computer Science, University of Augsburg, Augsburg, Germany
moemke@informatik.uni-augsburg.de
[4] Department of Computer Science, RWTH Aachen University, Aachen, Germany
rossmani@cs.rwth-aachen.de

Abstract. We analyze the proportional online knapsack problem with removal and limited recourse. The input is a sequence of item sizes; a subset of the items has to be packed into a knapsack of unit capacity such as to maximize their total size while not exceeding the knapsack capacity. In contrast to the classical online knapsack problem, packed items can be removed and a limited number of removed items can be re-inserted to the knapsack. Such re-insertion is called *recourse*. Without recourse, the competitive ratio is known to be approximately 1.618 (Iwama and Taketomi, ICALP 2002). We show that, even for only one use of recourse for the whole instance, the competitive ratio drops to 3/2. We prove that, with a constant number of $k \geq 2$ uses of recourse, a competitive ratio of $1/\left(\sqrt{3}-1\right) \leq 1.367$ can be achieved and we give a lower bound of $1 + 1/(k+1)$ for this case. For an extended use of recourse, i.e., allowing a constant number of $k \geq 1$ uses per step, we derive tight bounds for the competitive ratio of the problem, lying between $1 + 1/(k+2)$ and $1 + 1/(k+1)$. Motivated by the observation that the lower bounds heavily depend on the fact that the online algorithm does not know the end of the input sequence, we look at a scenario where an algorithm is informed when the instance ends. We show that with this information, the competitive ratio for a constant number of $k \geq 2$ uses of recourse can be improved to strictly less than $1 + 1/(k+1)$. We also show that this information improves the competitive ratio for one use of recourse per step and give a lower bound of ≥ 1.088 and an upper bound of 4/3 in this case.

Keywords: online knapsack · proportional knapsack · recourse · semi-online algorithm

R. Klasing—Partially supported by the ANR project TEMPOGRAL (ANR-22-CE48-0001).

T. Mömke—Partially supported by DFG Grant 439522729 (Heisenberg-Grant) and DFG Grant 439637648 (Sachbeihilfe).

S.-Y. Hsieh et al. (Eds.): IWOCA 2023, LNCS 13889, pp. 123–135, 2023.
https://doi.org/10.1007/978-3-031-34347-6_11

1 Introduction

In the classical knapsack problem, we are given a knapsack of capacity B and a set of items, each of which has a size and a value. The goal is to pack a subset of items such that the total size does not exceed the capacity B, maximizing the value of the packed items. In the *proportional* variant of the problem (also called *unweighted* or *simple* knapsack problem), the size and the value of each item coincide. Stated as an online problem, B is given upfront, but the items are revealed one by one in a sequence of requests. The online algorithm has to decide whether the requested item is packed or discarded before the subsequent item arrives and cannot revoke the decision. To measure the quality of the solution, we use the competitive ratio, i.e., the value attainable by an optimal offline algorithm divided by the value of the solution computed by the online algorithm in the worst case. It is well-known that, even in the proportional variant, no online algorithm for the online knapsack problem achieves a bounded competitive ratio [20]. This indicates that the classical notion of online algorithms is overly restrictive for the knapsack problem. Unless otherwise stated, we focus on the proportional variant in this paper.

In the literature, several ways of relaxing the online requirements have been considered for various online problems, leading to so-called *semi-online* problems [7,10]. Such semi-online problems enable to study the effect of different degrees of online behavior on the hardness of computational problems. Semi-online problems can be roughly divided into two classes: On the one hand, one can equip the algorithm with some extra information about the instance, e.g., the size of an optimal solution. On the other hand, one can relax the irrevocability of the algorithm's decisions. Several models from the second class have already been considered for the online knapsack problem: In the model of *delayed decisions* [23], one grants the online algorithm the right to postpone its decisions until they are really necessary. This means that the algorithm is allowed to temporarily pack items into the knapsack as long as its capacity allows and to remove them later on to avoid overpacking of the knapsack [17]. In the *reservation model*, the algorithm has the option to reserve some items in an extra storage at some extra cost, with the possibility of packing them into the knapsack later [2].

In this paper, we consider a semi-online model called *recourse*. In the model of recourse, the algorithm is allowed to withdraw a limited number of its previous decisions. Recourse has mainly been studied for the Steiner tree problem, MST, and matchings [1,6,12,13,16,21,22]. In case of the knapsack problem, we distinguish two types of recourse: (i) An item that has been selected previously is discarded (to make space for a new item); and (ii) an item was previously discarded and is added afterwards. The second type of recourse is costlier than the first type, as an unlimited number of items has to stay at disposal.

Applying the first type of recourse directly to the classical online knapsack problem does not get us too far: The same hard example as for the problem without recourse, i.e., one instance consisting of the items ε and 1 (for some arbitrarily small $\varepsilon > 0$) and another one consisting of the ε only, also proves

Table 1. Our results on the competitive ratio for online knapsack with removal and limited recourse. Here, $f(k) = 2/(\sqrt{k^2+6k+5} - k - 1)$.

	Upper bound		Lower bound	
1 recourse in total	$\frac{3}{2}$	(Theorem 2)	$\frac{3}{2}$	(Theorem 1)
$k \geq 2$ recourses in total	$\frac{1}{\sqrt{3}-1} \leq 1.367$	(Theorem 3)	$1 + \frac{1}{k+1}$	(Theorem 1)
k recourses per step	$f(k) \leq 1 + \frac{1}{k+1}$	(Theorem 4)	$f(k) \geq 1 + \frac{1}{k+2}$	(Theorem 5)

Table 2. Our results on the competitive ratio for online knapsack with removal and limited recourse, given information on the end of an instance. Here, $f(k) = 2/(\sqrt{k^2+6k+5} - k - 1)$.

	Upper bound		Lower bound	
$k \geq 2$ recourses in total	$f(k) \leq 1 + \frac{1}{k+1}$	(Theorem 6)		
1 recourse per step	$\frac{4}{3}$	(Theorem 8)	$\frac{18}{5+\sqrt{133}} \geq 1.088$	(Theorem 7)

an unlimited competitive ratio here since discarding the first item makes the instance stop and does not leave room for any recourse.

We combine the option of unlimited removal with limited recourse, that is, a limited number of re-packings of discarded items. The resulting upper and lower bounds on the competitive ratio are shown in Table 1. Classically, in the online model, upon arrival of an item, the algorithm does not know whether this will be the last item in the sequence or not. Besides analyzing this standard model, we additionally consider various ways of communicating information about the end of the sequence to the algorithm. The problem exhibits a surprisingly rich structure with respect to this parameter. Our respective bounds are shown in Table 2. Due to space constraints, some proofs are omitted in this paper.

1.1 Preliminaries

In the online knapsack problem OnlineKP used in this paper, an instance is given as a sequence of items $I = (x_1, \ldots, x_n)$. For convenience of notation, we identify an item with its size. In each step $1 \leq i \leq n$, an algorithm receives the item $x_i > 0$. At this point, the algorithm has no knowledge of the items $(x_{i+1}, \ldots, x_n)$ and no knowledge of the length n of the instance. It maintains a *knapsack* $S \subseteq I$ such that, in each step, the items in the knapsack do not exceed the capacity of the knapsack. We normalize this size to 1 and thus assume that $x_i \in [0, 1]$ for all i. All items x_j for $j < i$ that are not in S are considered to be in a *buffer* of infinite size.

In the framework of this paper, given the item x_i, an algorithm first adds it to the knapsack S, potentially exceeding the size limit. It may then *remove* any number of items from S, moving them to the buffer, and possibly return a certain number of items from the buffer to S afterwards, expending one use of *recourse* for each returned item. For simplicity, we say that the algorithm *packs* an item if it is kept in the knapsack upon its arrival, and that it *discards* an item if it is immediately removed in the step of its arrival. The number of such uses available to the algorithm will vary between various scenarios. After this process, the condition $\sum_{x_i \in S} x_i \leq 1$ must hold, i.e., the selected items respect the capacity of the knapsack. In our algorithms, we frequently classify the items by their sizes: For a bound b with $1/2 < b < 1$, we call an item *small* if it is of size at most $1 - b$, *medium* if its size is greater than $1 - b$ but smaller than b, and *large* if its size is at least b.

The *gain* $\text{gain}_{\text{ALG}}(I)$ of the algorithm ALG on the instance I is given as the sum $\sum_{x_i \in S} x_i$ of the item sizes in S after the last step n. Its *strict competitive ratio* $\rho_{\text{ALG}}(I)$ on I is defined as $\rho_{\text{ALG}}(I) = \text{gain}_{\text{OPT}}(I)/\text{gain}_{\text{ALG}}(I)$. The smaller the ratio, the better the algorithm performs. The strict competitive ratio ρ_{ALG} of the algorithm is then defined as the worst case over all possible instances, $\rho_{\text{ALG}} = \sup_I \rho_{\text{ALG}}(I)$. This definition is often generalized to the *competitive ratio* of an algorithm, which allows for a constant additive term to be added to $\text{gain}_{\text{ALG}}(I)$ in the definition of $\rho_{\text{ALG}}(I)$. However, since in ONLINEKP the optimal solution for any instance is bounded by 1, this relaxation does not add any benefit to the analysis of this problem. We will therefore only consider the strict version and refer to it simply as *competitive ratio*.

1.2 Related Work

Online problems with recourse date back to Imase and Waxman [16] who have studied the online Steiner tree problem and utilized the benefit of a limited number of rearrangements. The number of required recourse steps was subsequently reduced by a sequence of papers [12,13,22]. Recently, recourse was considered for further problems, in particular online matching [1,6,21].

Many different kinds of semi-online problems have been considered in the literature; Boyar et al. [7] give an overview of some of these. In particular, many results on semi-online algorithms focus on makespan scheduling problems with some extra information, starting with the work by Kellerer et al. [18]; see the survey by Dwibety and Mohanty [10] for a recent overview of this line of research.

Many semi-online settings assume the availability of some extra information, e.g., the total makespan in scheduling problems. In the model of *advice complexity*, one tries to measure the performance of an online algorithm in the amount of any information conveyed by some oracle that knows the whole input in advance. This very general approach to semi-onlineness provides a powerful tool for proving lower bounds. The model was introduced by Dobrev et al. [9] in 2008 and shortly afterwards revised by Emek et al. [11], Böckenhauer et al. [4], and Hromkovič et al. [15].

Since then, it has been applied to many different online problems; for a survey, see the work by Boyar et al. [7] and the textbook by Komm [19].

In this paper, we consider a slightly different kind of semi-online problems, where the online condition is not relaxed by giving some extra information to the algorithm, but by relaxing the irrevocability of its decisions. In one approach, the online algorithm is allowed to delay its decisions until there is a real need for it. For instance, in the knapsack problem, the algorithm is allowed to pack all items into the knapsack until it is overpacked, and only then has to decide which items to remove. Iwama and Taketomi [17] gave the first results for online knapsack with removal. A version in which the removal is not completely for free, but induces some extra cost was studied by Han et al. [14]. *Delayed decisions* were also studied for other online problems by Rossmanith [23] and by Chen et al. [8]. Böckenhauer et al. [2] analyzed another semi-online version of online knapsack which gives the algorithm a third option besides packing or rejecting an item, namely to reserve it for possible later packing at some reservation cost. The advice complexity of online knapsack was analyzed by Böckenhauer et al. [5] in the normal model and later in the model with removal [3].

2 Number of Uses of Recourse Bounded per Instance

In this section, we analyze the scenario in which an algorithm can only use recourse a limited number of $k \geq 1$ times in total. Even if we just allow one use of recourse per instance, the upper bound of $(\sqrt{5}+1)/2 \approx 1.618$ proven by Iwama and Taketomi [17] for the online knapsack problem with removal (but without recourse) can be improved, as we show in the following. We find a lower bound of $1 + 1/(k+1)$ for the competitive ratio of any algorithm. In the case $k = 1$ where the algorithm can use recourse exactly once, we present an algorithm that matches this lower bound of $3/2$. In the case $k > 1$, we find an upper bound of $1/(\sqrt{3}-1) \leq 1.367$ that does not improve as k gets larger.

2.1 Lower Bound

Theorem 1. *Any algorithm that uses recourse at most $k \geq 1$ times in total cannot have a competitive ratio of less than $1 + 1/(k+1)$.*

Proof. We present a family of instances dependent on $\varepsilon > 0$, such that any algorithm that uses at most k recourses in total cannot have a competitive ratio of less than $(k+2)/(k+1)$ for at least one of these instances in the limit $\varepsilon \to 0$. These instances are given in Table 3; they all start with k copies of the item $x_1 = \frac{1}{k+2} + \varepsilon$. For the proof to work as intended, ε is chosen such that $0 < \varepsilon < \frac{1}{(k+2)^2}$. Note that any deterministic algorithm must act identically on these instances up to the point where they differ.

Now, let Alg be any algorithm that uses at most k recourses in total and assume that it has a competitive ratio strictly less than $(k+2)/(k+1)$.

1. The algorithm must pack item x_2 in each instance, removing all previously packed copies of x_1: otherwise, its competitive ratio on instance I_1 is at least

$$\rho_{\text{ALG}} \geq \frac{(k+1)/(k+2) + (k+2)\varepsilon}{k \cdot (1/(k+2) + \varepsilon)} \to \frac{k+1}{k} > \frac{k+2}{k+1} \text{ as } \varepsilon \to 0 .$$

2. The algorithm must then pack the item x_3, remove x_2 and use its entire recourse to retrieve the k copies of x_1 in instances I_2 to I_5:

Table 3. A family of instances showing that no algorithm that uses at most k recourses in total can achieve a competitive ratio better than $(k+2)/(k+1)$. All instances start with k items of size x_1, where $x_1 = x_3 = \frac{1}{k+2} + \varepsilon$ and $x_2 = \frac{k+1}{k+2} + (k+2)\varepsilon$.

	k copies					
I_1	x_1	x_2				
I_2	x_1	x_2	x_3			
I_3	x_1	x_2	x_3	$y_3 = \frac{1}{k+2} - (k+1)\varepsilon$		
I_4	x_1	x_2	x_3	$x_4 = \frac{k+1}{k+2} + \varepsilon$	$y_4 = \frac{1}{k+2} - \varepsilon$	
I_5	x_1	x_2	x_3	$x_4 = \frac{k+1}{k+2} + \varepsilon$	$x_5 = \frac{1}{k+2}$	$y_5 = \frac{k+1}{k+2} - \varepsilon$
I_6	x_1	x_2	x_3	$x_4 = \frac{k+1}{k+2} + \varepsilon$	$x_5 = \frac{1}{k+2}$	

- If it packs item x_3 but only uses its recourse to retrieve $m < k$ copies of x_1, its competitive ratio on instance I_2 is at least

$$\rho_{\text{ALG}} \geq \frac{\frac{k+1}{k+2} + (k+2)\varepsilon}{(m+1)(\frac{1}{k+2} + \varepsilon)} \geq \frac{\frac{k+1}{k+2} + (k+2)\varepsilon}{k(\frac{1}{k+2} + \varepsilon)} \to \frac{k+1}{k} > \frac{k+2}{k+1} \text{ as } \varepsilon \to 0\,.$$

- If it does not pack item x_3 and keeps x_2, its competitive ratio on instance I_3 is at least

$$\rho_{\text{ALG}} \geq \frac{1}{\frac{k+1}{k+2} + (k+2)\varepsilon} \to \frac{k+2}{k+1} \text{ as } \varepsilon \to 0\,.$$

So, from here on, the algorithm cannot use any further recourse.

3. The algorithm must then pack item x_4 in instances I_4 to I_6, removing x_3 and all copies of x_1: otherwise, its competitive ratio on instance I_4 is at least

$$\rho_{\text{ALG}} \geq \frac{1}{\frac{k+1}{k+2} + (k+2)\varepsilon} \to \frac{k+2}{k+1} \text{ as } \varepsilon \to 0\,.$$

4. The algorithm must then pack item x_5 and remove x_4 in instances I_5 and I_6: otherwise, its competitive ratio on instance I_5 is at least

$$\rho_{\text{ALG}} \geq \frac{1}{\frac{k+1}{k+2} + \varepsilon} \to \frac{k+2}{k+1} \text{ as } \varepsilon \to 0\,.$$

5. However, in this situation, its competitive ratio on instance I_6 is at least

$$\rho_{\text{ALG}} \geq \frac{\frac{k+1}{k+2} + (k+2)\varepsilon}{\frac{1}{k+2}} \to k+1 > \frac{k+2}{k+1} \text{ as } \varepsilon \to 0\,.$$

Hence, ALG has a competitive ratio of at least $(k+2)/(k+1) = 1 + 1/(k+1)$. □

2.2 Upper Bound

Upper Bound for $k = 1$. We present an algorithm ALG_1 that uses recourse at most once and that achieves a competitive ratio of $3/2$. We set $b = 2/3$ and distinguish between small, medium, and large items with respect to b as described in Subsect. 1.1. The algorithm ALG_1

- packs any large item immediately, to this end removes some items from the knapsack if necessary, and discards all other items from there on;
- packs any small item immediately. If a small item does not fit or has to be discarded to fit a medium item, it discards all other items from there on;
- always keeps the largest medium item. If it encounters a medium item x_i that fits together with a previously encountered medium item x_j, it removes the currently packed one, uses its recourse to retrieve x_j and discards all other items from there on.

Theorem 2. *The algorithm* ALG_1 *has a competitive ratio of at most* $3/2$.

Proof. We prove that ALG_1 is either optimal or achieves a gain of at least $2/3$ and thus a competitive ratio of at most $3/2$. If there is a large item in the instance, the algorithm packs it, leading to a gain of at least $2/3$. We can therefore assume that the instance contains no large items.

If the algorithm discards a small item at any point due to size limits, its gain at that point must be at least $1 - 1/3 = 2/3$. We can therefore assume that the algorithm never discards a small item.

Now, consider the number of medium items in the optimal solution on the instance, which cannot be more than two. If it contains exactly two, the algorithm can pack two medium items, leading to a gain of at least $1/3 + 1/3 = 2/3$. If the optimal solution contains zero or one medium item, the algorithm packs all small items and (if there is one) the largest medium item. Since all small items in the instance are packed by assumption, the algorithm is optimal in this case. □

Upper Bound for $k > 1$. We define b as the unique positive solution of the equation $b^2 + 2b - 2 = 0$, so $b = \sqrt{3} - 1 \approx 0.73$. We present an algorithm ALG_2 with a competitive ratio of at most $1/b \leq 1.367$ that uses recourse at most twice. The algorithm again distinguishes between small, medium, and large items with respect to b.

- The algorithm ALG_2 treats any small or large item the same as ALG_1.
- The algorithm only keeps the largest medium item, until it is presented with a medium item that will fit with one that has already been encountered. If that is the case, it will spend one use of its recourse to retrieve that item if it is in the buffer. From there on, it will keep the two smallest medium items encountered so far. (i) If it encounters a third medium item that will fit with these two, it will pack it and discard everything else from there on. (ii) If, however, at any point the algorithm encounters a medium item x_i, such that

there is a previously encountered medium item x_j with $b \leq x_i + x_j \leq 1$, it packs x_i, spends one use of recourse to retrieve x_j if it is in the buffer and discards everything else from there on.
- If at any point a medium item does not fit because of small items that have already been packed, the algorithm removes small items one by one until the medium item can be packed and discards everything else from there on.

Theorem 3. *The algorithm* ALG$_2$ *has a competitive ratio of at most* $1/(\sqrt{3}-1)$.

3 Number of Uses of Recourse Bounded per Step

In this scenario, an algorithm could use recourse a limited number of k times per step. We give sharp bounds on the competitive ratio of an optimal algorithm in this case, tending to 1 when k tends to infinity. Let b_k be the unique positive root of the quadratic equation $b_k^2 + (k+1) \cdot b_k - (k+1) = 0$. Then it is easy to check that $1 + 1/(k+1) \leq 1/b_k \leq 1 + 1/(k+2)$.

3.1 Upper Bound

Let $k \in \mathbb{N}$. We define $b_k = (\sqrt{k^2+6k+5} - k - 1)/2$ as the unique positive solution of the quadratic equation $b_k^2 + (k+1) \cdot b_k - (k+1) = 0$. We construct an algorithm ALG$_k$ with a competitive ratio of at most $1/b_k$ that uses recourse at most k times per step, again distinguishing between small, medium, and large items with respect to b_k.

- The algorithm ALG$_k$ treats any small or large item the same as ALG$_1$ in Subsect. 2.2.
- As long as at most k medium items fit, ALG$_k$ packs them optimally, using its recourse to do so. As soon as it is presented with a medium item that will fit with k previous ones, it will use its recourse to retrieve any of these that are in the buffer. From there on, it will keep the $k+1$ smallest medium items encountered so far. (i) If it encounters an additional medium item that will fit with these $k+1$, it will pack it and discard everything else from there on. (ii) If however at any point the algorithm encounters a medium item that fits with at most k previously encountered ones, such that their sum is at least b_k, it packs it, uses its recourse to retrieve any of the others that might be in the buffer and discards everything else from there on.
- If at any point a medium item does not fit because of small items that have already been packed, ALG$_k$ removes small items one by one until the medium item can be packed and discards everything else from there on.

In the case $k = 1$, this is the algorithm ALG$_2$ that is used in Subsect. 2.2 for $k > 1$ uses of recourse in total. The algorithm not only uses recourse at most twice but never spends both uses in the same step.

Theorem 4. *The algorithm* ALG$_k$ *has a competitive ratio of at most* $1/b_k \leq 1 + \frac{1}{k+1}$.

3.2 Lower Bound

We now prove that the algorithm provided in Subsect. 3.1 is the best possible. Let b_k be defined as in Subsect. 3.1.

Theorem 5. *Any algorithm that uses recourse at most k times per step cannot have a competitive ratio of less than* $1/b_k \geq 1 + \frac{1}{k+2}$.

Proof (Sketch). We define $a_k = (1 - b_k) + \varepsilon$ and present two instances $I_1 = (a_k, \dots, a_k, b_k + (k+2)\varepsilon)$ and $I_2 = (a_k, \dots, a_k, b_k + (k+2)\varepsilon, 1 - (k+1)a_k)$. Both instances start with $k+1$ items of size a_k. The competitive ratio of any algorithm must be at least $1/b_k$ on at least one of these instances. □

4 Information on the End of the Instance

Previously, all problems were defined in a way where the algorithm had no information on whether a certain item was the last item of the instance or not. It might be natural to allow an algorithm access to this information. In the situation where no recourse is allowed, this distinction does not matter: Any instance could be followed by a final item of size 0, in which case removing any items would not lead to a better solution. With recourse however, this might change.

4.1 Two Different Ways for Handling End of Instance

There appear to be two different ways in which the information that the instance ends might be encoded. One hand, the instance might be given in the form $(x_1, \dots, x_n, \perp)$ where $\perp$ informs the algorithm that the previous item was the last one, allowing it to perform one last round of removal and recourse. On the other hand, the instance could be given in the form $(x_1, \dots, x_{n-1}, (x_n, \perp))$, where the algorithm is informed that an item is the last of the instance in the same step that the item is given. It can be shown that an algorithm will always perform at least as well if it receives the information in the former variant than in the latter. However, in the scenario where the size of the recourse is bounded by k uses in total, the chosen variant does not matter.

4.2 Upper Bound for k Uses of Recourse in Total Given Information on the End of the Instance

When information about the end of an instance is available, k uses of recourse in total are at least as useful as k uses of recourse per step without that information. Since the way that information is received does not matter as mentioned in Subsect. 4.1 we will assume that the instance is given in the form $(x_1, \dots, x_n, \perp)$.

We can now adapt the algorithm ALG_k from Subsect. 3.1 to this situation, where b_k is again defined as the unique positive solution of the quadratic equation $b_k^2 + (k+1) \cdot b_k - (k+1) = 0$ and define small, medium and large items accordingly. The obtained algorithm $\text{ALG}_{(k,\perp)}$ works as follows.

- It treats any small or large item the same as ALG_1 in Subsect. 2.2.
- It only keeps the smallest medium item unless it can achieve a gain of at least b_k using only medium items. In this case, as in the algorithm in Subsect. 3.1, it retrieves at most k items and discards everything from there on.
- If the algorithm receives the item $\perp$ and has not yet decided to discard everything else, it computes the optimal solution on the entire previous instance, retrieves all medium items in that solution using its recourse and discards all small items not contained in that solution.

Theorem 6. *The algorithm* $\text{ALG}_{(k,\perp)}$ *has a competitive ratio of at most* $1/b_k$.

4.3 Lower Bound for One Use of Recourse per Step Given Information on the End of the Instance

We will assume that the algorithm handles instances of the form $(x_1, \ldots, (x_n, \perp))$.

Theorem 7. *No algorithm that uses recourse at most once per step and that recognizes the last item in an instance can have a competitive ratio of less than* $18/(5+\sqrt{133}) \geq 1.088$.

Proof (Sketch). We define b as the unique positive root of the equation $27b^2 - 5b - 1 = 0$, so $b = (5+\sqrt{133})/54 \approx 0.3062$. We further define the additional item $a = (1-b)/3+\varepsilon \approx 0.2313+\varepsilon$ and present two instances $I_1 = (a, a, a, b, b, b, 1-3a)$ and $I_2 = (a, a, a, b, b, b, \varepsilon)$. The competitive ratio of any algorithm must be at least $1/(3b)$ on at least one of these instances in the limit $\varepsilon \to 0$. □

4.4 Upper Bound for One Use of Recourse per Step Given Information on the End of the Instance

We assume that the instance is given in the form $(x_1, \ldots, (x_n, \perp))$ and present an algorithm $\text{ALG}_\perp$ with a competitive ratio of $4/3 \approx 1.33$, which is strictly better than the optimal algorithm without that information in Subsect. 3.1. We define $b = 3/4$ and distinguish between small, medium and large items, depending on b as before. The algorithm $\text{ALG}_\perp$ then works as follows.

- It treats any small or large item the same as ALG_1 in Subsect. 2.2.
- It packs medium items as follows. (i) As long as only one medium item fits, it keeps the smallest one of these. (ii) As soon as two medium items fit, it computes the optimal sum of two medium items and keeps the larger of these two, as well as the smallest medium item encountered so far. (iii) When it encounters a medium item that fits with two previously encountered ones (one being w.l.o.g. the smallest medium item, currently packed), it packs it, retrieves the third of these three and discards everything from there on.

– If it encounters the item $(x_n, \perp)$ and has not yet decided to discard everything else, it computes the optimal solution on the entire instance. (i) If this solution contains only one medium item, it retrieves it and returns the optimal solution. (ii) If this solution contains two medium items, one of these must be either x_n or already packed. The algorithm retrieves the other one and returns the optimal solution. (iii) If this solution contains three medium items, x_n must be a medium item and the algorithm proceeds as if it was not the last one.

Theorem 8. *The algorithm* $\text{ALG}_\perp$ *has a competitive ratio of at most* $4/3$.

5 Conclusion

Besides closing the gap between the upper and lower bounds for $k \geq 2$ uses of recourse in total without knowing the end of the instance and proving a lower bound in the case of knowing the end, it is an interesting open problem to consider a model in which the use of removal or recourse is not granted for free, but incurs some cost for the algorithm.

In the *general online knapsack problem*, the items have both a size and a value, and the goal is to maximize the value of the packed items while obeying the knapsack bound regarding their sizes. If we consider the general online knapsack problem with removal and one use of recourse per step, we can easily see that the competitive ratio is unbounded: Suppose there is an online algorithm with competitive ratio c. The adversary then presents first an item of size and value 1, if the algorithm does not take this item, the instance ends. Otherwise, it presents up to $(c+1)/\varepsilon$ many items of size $\varepsilon/(c+1)$ and value ε. If the online algorithm decides at some point to take such an item of value ε and uses its recourse to fetch another of these from the buffer, the instance ends and the competitive ratio is $1/(2\varepsilon)$. A similar argument shows that a recourse of size k per step cannot avoid an unbounded competitive ratio either. But, again, these arguments heavily depend on the algorithm's unawareness of the end of the instance. It remains as an open problem to extend the results for known instance lengths to the general case.

References

1. Angelopoulos, S., Dürr, C., Jin, S.: Online maximum matching with recourse. J. Comb. Optim. **40**(4), 974–1007 (2020). https://doi.org/10.1007/s10878-020-00641-w
2. Böckenhauer, H.-J., Burjons, E., Hromkovič, J., Lotze, H., Rossmanith, P.: Online simple knapsack with reservation costs. In: Bläser, M., Monmege, B., (eds.), 38th International Symposium on Theoretical Aspects of Computer Science, STACS 2021, 16–19 March 2021, Saarbrücken, Germany (Virtual Conference), vol. 187 of LIPIcs, pp. 16:1–16:18. Schloss Dagstuhl - Leibniz-Zentrum für Informatik (2021)

3. Böckenhauer, H.-J., Dreier, J., Frei, F., Rossmanith, P.: Advice for online knapsack with removable items. CoRR, abs/2005.01867 (2020)
4. Böckenhauer, H.-J., Komm, D., Královič, R., Královič, R., Mömke, T.: On the advice complexity of online problems. In: Dong, Y., Du, D.-Z., Ibarra, O. (eds.) ISAAC 2009. LNCS, vol. 5878, pp. 331–340. Springer, Heidelberg (2009). https://doi.org/10.1007/978-3-642-10631-6_35
5. Böckenhauer, H.-J., Komm, D., Královič, R., Rossmanith, P.: The online knapsack problem: Advice and randomization. Theor. Comput. Sci. **527**, 61–72 (2014)
6. Boyar, J., Favrholdt, L.M., Kotrbcík, M., Larsen, K.S.: Relaxing the irrevocability requirement for online graph algorithms. Algorithmica **84**(7), 1916–1951 (2022)
7. Boyar, J., Favrholdt, L.M., Kudahl, C., Larsen, K.S., Mikkelsen, J.W.: Online algorithms with advice: a survey. ACM Comput. Surv., **50**(2), 19:1–19:34 (2017)
8. Chen, L., Hung, L., Lotze, H., Rossmanith, P.: Online node- and edge-deletion problems with advice. Algorithmica **83**(9), 2719–2753 (2021)
9. Dobrev, S., Královič, R., Pardubská, D.: How much information about the future is needed? In: Geffert, V., Karhumäki, J., Bertoni, A., Preneel, B., Návrat, P., Bieliková, M, (eds.) SOFSEM 2008: theory and practice of computer science. In: 34th Conference on Current Trends in Theory and Practice of Computer Science, Nový Smokovec, Slovakia, 19-25 January 2008, Proceedings, vol. 4910 of Lecture Notes in Computer Science, pp. 247–258. Springer (2008)
10. Dwibedy, D., Mohanty, R.: Semi-online scheduling: a survey. Comput. Oper. Res. **139**, 105646 (2022)
11. Emek, Y., Fraigniaud, P., Korman, A., Rosén, A.: Online computation with advice. Theor. Comput. Sci. **412**(24), 2642–2656 (2011)
12. Gu, A., Gupta, A., Kumar, A.: The power of deferral: maintaining a constant-competitive steiner tree online. SIAM J. Comput. **45**(1), 1–28 (2016)
13. Gupta, A., Kumar, A.: Online steiner tree with deletions. In: SODA, pp. 455–467. SIAM (2014)
14. Han, X., Kawase, Y., Makino, K.: Online unweighted knapsack problem with removal cost. Algorithmica **70**(1), 76–91 (2014)
15. Hromkovič, J., Královič, R., Královič, R.: Information complexity of online problems. In: Hlinený, P., Kucera, A., (eds.), Mathematical Foundations of Computer Science 2010, 35th International Symposium, MFCS 2010, Brno, Czech Republic, 23-27 August 2010. Proceedings, vol. 6281 of Lecture Notes in Computer Science, pp. 24–36. Springer (2010)
16. Imase, M., Waxman, B.M.: Dynamic steiner tree problem. SIAM J. Discret. Math. **4**(3), 369–384 (1991)
17. Iwama, K., Taketomi, S.: Removable online knapsack problems. In: Widmayer, P., Eidenbenz, S., Triguero, F., Morales, R., Conejo, R., Hennessy, M. (eds.) ICALP 2002. LNCS, vol. 2380, pp. 293–305. Springer, Heidelberg (2002). https://doi.org/10.1007/3-540-45465-9_26
18. Kellerer, H., Kotov, V., Speranza, M.G., Tuza, Z.: Semi on-line algorithms for the partition problem. Oper. Res. Lett. **21**(5), 235–242 (1997)
19. Komm, D.: An introduction to online computation - determinism, randomization, advice. Texts in Theoretical Computer Science. An EATCS Series. Springer (2016)
20. Marchetti-Spaccamela, A., Vercellis, C.: Stochastic on-line knapsack problems. Math. Program. **68**, 73–104 (1995)
21. Megow, N., Nölke, L.: Online minimum cost matching with recourse on the line. In: APPROX-RANDOM, Vol. 176 of LIPIcs, pages 37:1–37:16. Schloss Dagstuhl - Leibniz-Zentrum für Informatik, (2020)

22. Megow, N., Skutella, M., Verschae, J., Wiese, A.: The power of recourse for online MST and TSP. SIAM J. Comput. **45**(3), 859–880 (2016)
23. Rossmanith, P.: On the advice complexity of online edge- and node-deletion problems. In: Böckenhauer, H.-J., Komm, D., Unger, W. (eds.) Adventures Between Lower Bounds and Higher Altitudes. LNCS, vol. 11011, pp. 449–462. Springer, Cham (2018). https://doi.org/10.1007/978-3-319-98355-4_26

Minimum Surgical Probing with Convexity Constraints

Toni Böhnlein[1(✉)], Niccolò Di Marco[2], and Andrea Frosini[2]

[1] Weizmann Institute of Science, Rehovot, Israel
toni.bohnlein@weizmann.ac.il
[2] Department of Mathematics and Informatics, University of Florence, Florence, Italy
{niccolo.dimarco,andrea.frosini}@unifi.it

Abstract. We consider a tomographic problem on graphs, called Minimum Surgical Probing, introduced by Bar-Noy et al. [2]. Each vertex $v \in V$ of a graph $G = (V, E)$ is associated with an (unknown) *label* ℓ_v. The outcome of *probing* a vertex v is $\mathcal{P}_v = \sum_{u \in N[v]} \ell_u$, where $N[v]$ denotes the closed neighborhood of v. The goal is to uncover the labels given *probes* $\mathcal{P}_v$ for all $v \in V$. For some graphs, the labels cannot be determined (uniquely), and the use of *surgical probes* is permitted but must be minimized. A surgical probe at vertex v returns ℓ_v.

In this paper, we introduce convexity constraints to Minimum Surgical Probing. For binary labels, convexity imposes constraints such as if $\ell_u = \ell_v = 1$, then for all vertices w on a shortest path between u and v, we must have that $\ell_w = 1$.

We show that convexity constraints reduce the number of required surgical probes for several graph families. Specifically, they allow us to recover the labels without using surgical probes for *trees* and *bipartite graphs* where otherwise $\lfloor |V|/2 \rfloor$ surgical probes might be needed. Our analysis is based on restricting the size of cliques in a graph using the concept of K_h-free graphs (forbidden induced subgraphs). Utilizing this approach, we analyze *grid graphs*, the *King's graph*, and *(maximal-) outerplanar graphs*.

Keywords: Reconstruction Problem · Discrete Tomography · Graph Convexity · Path Convexity · K_h-free Graphs

1 Introduction

Tomography is rooted in the task of retrieving spatial information about an object where direct observation is restricted or impossible. Instead, it is only possible to inspect the object indirectly using *projections*, e.g., taking aggregate measurements over spatial regions. The spatial data to be *reconstructed* from the projections depends on the specific application. Often, one wants to learn about the shape and position of one or several objects in order to produce an image. Tomography finds applications, for example, in the medical domain providing remarkable imaging technologies, like CT and MRI, which are able to

S.-Y. Hsieh et al. (Eds.): IWOCA 2023, LNCS 13889, pp. 136–147, 2023.
https://doi.org/10.1007/978-3-031-34347-6_12

create detailed internal images of the human body in a non-invasive manner. Various other engineering and scientific applications feature such indirect-access constraints constituting tomography as a fundamental field of study.

We are interested in *discrete tomography* (cf. [12]) where an exemplary problem asks to reconstruct the entries of a binary matrix given their row and column sums (horizontal and vertical projections). We investigate a line of research initiated by Frosini and Nivat [9] and further developed in [1,2,5]. The microscopic image reconstruction (MRI) problem, introduced in [9], asks to reconstruct the entries of a binary matrix given two-dimensional window projections or *probes*. Probing an entry of the matrix returns the sum of the entries within a window (of fixed size) around it (see Fig. 1a).

To deal with a wider range of topologies the setting was generalized to graphs where the local concept of probing windows generalizes naturally to neighborhoods of vertices. The model is defined in [2] as follows. Let $G = (V, E)$ be a connected and simple graph where $V = \{1, 2, \dots, n\}$. Each vertex $i \in V$ is associated with a *label* $\ell_i \in \mathbb{R}$, and the result of *probing* vertex i is

$$\mathcal{P}_i = \textstyle\sum_{j \in N[i]} \ell_j, \tag{1}$$

where $N[i]$ is the closed neighborhood of i. *Label vector* $\ell = (\ell_1, \ell_2, \dots, \ell_n)$ and *probe vector* $\mathcal{P} = (\mathcal{P}_1, \mathcal{P}_2, \dots, \mathcal{P}_n)$ are *consistent* if Eq. (1) is satisfied for all $i \in V$. The reconstruction problem asks to uncover a consistent *label vector* ℓ, given G and $\mathcal{P}$ (see Fig. 1b). We note that topologies specific to the MRI problem are realized if G is a grid graph.

0	0	1	0	0
0	2	3	3	1
0	3	5	5	2
0	2	4	4	2
0	1	2	2	1

(a) The colors of the 5 × 5 checkerboard encode a binary matrix: gray squares are 0 and yellow squares are 1. The numeral entries result from probing with a square 3 × 3 window.

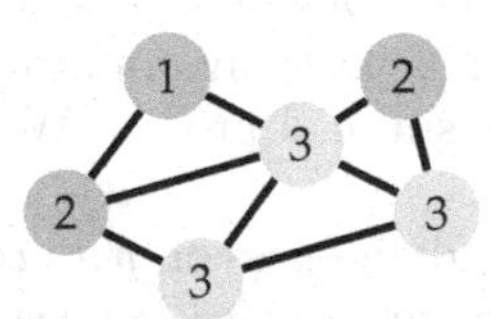

(b) A graph with binary labels on its vertices indicated by their color. Gray vertices have label 0, and yellow vertices have label 1. The numeral value inside each vertex is the result of probing it.

Fig. 1. Reconstruction problems on matrices and graphs. (Color figure online)

Attempting to find a consistent label vector raises the question of uniqueness resp. of counting consistent label vectors[1]. For example, consider the graph with two adjacent vertices, and observe that the label vectors $\ell = (1, 0)$ and $\ell' = (0, 1)$ are both consistent with the probe vector $\mathcal{P} = (1, 1)$.

[1] It is assumed that there is at least one consistent label vector.

In [2], it was shown that the number of solutions is tied to the rank of the adjacency matrix of G. To find a specific solution with respect to a label vector, the use of *surgical probes* is proposed. A surgical probe at vertex i returns ℓ_i.

The reconstruction problem is refined to the MINIMUM SURGICAL PROBING (MSP) problem where, given G and $\mathcal{P}$, the goal is to uncover ℓ uniquely using the minimum number of surgical probes. The problem can be solved in polynomial time using techniques from linear algebra. For this solution, it is paramount that the label vectors have real values.

The motivation for our paper is to analyze MSP with *convexity constraints.* Convexity constraints have been studied for various tomography problems as they are well-motivated by applications. For example, in medical imaging, many objects of interest have a convex shape naturally.

Consider the exemplary problem of reconstructing a binary matrix from its row and column sums again. In this context, *hv−convexity* implies that the 1 entries of the binary matrix are connected in each row (horizontally) and column (vertically). Note that in Fig. 1a, the 1 entries are *hv*-convex. Given a row of length n with k many 1 entries, there are $\mathcal{O}(n)$ possible configurations assuming *hv*-convexity, otherwise there are $\Omega(n^k)$.

Several studies show that convexity constraints help solve tomography problems, e.g., find unique solutions or speed up computation (see [3,4,7,10]).

Our Contribution. MSP with convexity constraints is not only motivated by the tomography literature but also because the graph theory literature covers several concepts of convexity (cf. [6,8,13]).

To introduce convexity on graphs, we assume that the label vector $\ell \in \{0,1\}^n$ is binary. To the best of our knowledge, MSP with binary labels has not been studied. The work by Gritzmann et al. [11] suggests that it is NP-hard. But the result of [2] is partially applicable and provides an upper bound on the minimum number of surgical probes. We call a vertex v a 1*-vertex* if $\ell_v = 1$, otherwise a 0*-vertex.*

Our study focuses on *path convexities*, which we introduce formally in Sect. 2. Informally, given two 1-vertices, path convexity implies that all vertices on particular path systems connecting them must also be 1-vertices.

Specifically, we consider three different path convexities: (i) g-convexity based on *shortest paths,* (ii) m-convexity based on *chordless paths,* and (iii) t-convexity based on *triangle paths.* Recall that a *chord* of a path is an edge incident to two non-consecutive vertices on the path. A path is chordless if it does not have a chord. A path is a triangle path if it does not have chords except *short chords,* which connect two vertices of distance 2 along the path. Figure 2 illustrates the differences between the three path convexities.

We define *convex* (shorthand for g-convex, m-convex or t-convex) subsets of vertices as those containing all vertices on the respective path systems between all pairs of its vertices. We say a label vector is convex if it's 1-vertices induce a convex set. Naturally, given G and $\mathcal{P}$, the goal of CONVEX MINIMUM SURGICAL PROBING (CMSP) is to uncover a convex label vector ℓ uniquely using the minimum number of surgical probes.

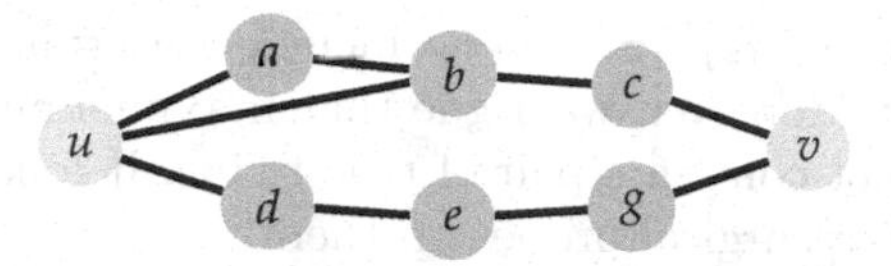

Fig. 2. For vertices u and v, let $I_g(u,v)$ be the set of all vertices on a shortest path between them (analogously, we define $I_m(u,v)$ and $I_t(u,v)$). For the example graph, verify that $I_g(u,v) = \{u,b,c,v\}$, $I_m(u,v) = I_g(u,v) \cup \{d,e,g\}$, and that $I_t(u,v) = I_m(u,v) \cup \{a\}$.

Considering convex label vectors, our first observation is that a 1-vertex must have a neighboring 1-vertex (assuming that there is another). This leads to our first result showing that if $\mathcal{P}_v \leq 1$, then $\ell_v = 0$, for any vertex v.

Interestingly, the converse holds for t-convex label vectors. The addition of short chords to a system of chordless paths, arguably, does not change its characteristics by much. But it simplifies CMSP drastically. We show that for t-convex labels, the 1-vertices can be identified as those vertices v where $\mathcal{P}_v \geq 2$.

We structure our analysis of g-convex and m-convex label vectors by bounding the size of cliques in a graph. Observe that in a complete graph of order n, denoted K_n, any subset of vertices is convex and that the notion of convexity is immaterial. Already in [2], it was shown that K_n requires a maximum number of $n-1$ surgical probes to uncover a non-convex label vector.

K_h-Free Graphs. In Sect. 3, we use the concept of H-free graphs to bound the size of the largest clique. We denote $G[U]$ as the subgraph of G *induced* by $U \subseteq V$. Graph G is *H-free* if H is not an induced subgraph of G. It follows that the largest clique in a K_h-free graph is of size at most $h-1$.

K_3-free graphs have essentially no cliques. We verify our intuition that CMSP is not a difficult problem on K_3-free graphs, showing that vertex v is a 1-vertex if and only if $\mathcal{P}_v \geq 2$. Hence, the 1-vertices are identified entirely by their own probe value, and without using surgical probes. The result holds for g-convex and m-convex label vectors. We remark that trees and bipartite graphs are important families of K_3-free graphs. Moreover, there are trees requiring $\lfloor n/2 \rfloor$ surgical probes to uncover a non-convex label vector (cf. [2]).

K_4-free graphs can contain triangles but no larger cliques. Worst-case examples show that $\mathcal{O}(n)$ many surgical probes may be necessary to uncover a g-convex or m-convex label vector. Our main result for K_4-free graphs relates the number of required surgical probes to the number of *true twins* in a graph. Recall that a pair of vertices are true twins if they are adjacent and have identical neighborhoods. A pair of true twins requires using one surgical probe if their and neighboring vertices' probes have specific values. We show that for m-convex label vectors, all necessary surgical probes are due to such true twins. For g-convex label vectors, additional structures incur the use of surgical probes. We leave their characterization as an open problem.

Interesting families of K_4-free graphs include outerplanar graphs, and our result applies directly. Moreover, we argue that maximal outerplanar graphs (of order 5 or more) cannot contain a pair of true twins. It follows that CMSP can be solved without using surgical probes on them.

Finally, we remark that our paper is an initial study of CMSP with great potential for future research. Different directions for MSP were considered in [1,5], which are also interesting with convexity constraints.

Our results are mostly "stand alone". Therefore, it would be compelling to derive deeper implications for CMSP from results that already exist in the graph theory literature on convexities. Note that the extended abstract omits proofs but they are presented in the full version of the paper.

2 Minimum Surgical Probing with Convex Labels

We start this section by introducing graph convexities. Let $G = (V, E)$ be a simple graph where $V = \{1, \dots, n\}$. A binary *label vector* $\ell \in \{0,1\}^n$ assigns a label to each vertex. The *support* of label vector ℓ is $\sum_{i \in V} \ell_i$.

2.1 Convexity on Graphs

Convexity on graphs is a well-studied concept. We introduce standard definitions following Pelayo [13]. Let $\mathcal{C} \subseteq 2^{|V|}$ be a family of subsets of V. The pair $(V, \mathcal{C})$ is a *convexity space* if

The elements of $\mathcal{C}$ are the *convex sets*. For $U \subseteq V$, the *convex hull* $\langle U \rangle_{\mathcal{C}}$ is the smallest convex set containing U. The pair $(G, \mathcal{C})$ is a *graph convexity space* if $(V, \mathcal{C})$ is a convexity space and $G[U]$ is connected for any $U \in \mathcal{C}$.

Arguably, the most natural graph convexities are *path convexities* which are defined by *interval functions* $I : V \times V \mapsto 2^V$. Interval functions extend to subsets $U \subseteq V$ as follows: $I(U) = \bigcup_{u,v \in U} I(u, v)$. For path convexities, the convex hull $\langle U \rangle_{\mathcal{C}}$ is given by $I(U)$. Set U is convex if $I(U) = U$, i.e., U is *closed* under the operator I. By definition $u, v \in I(u, v)$ and $I(v, v) = \{v\}$.

We consider convexities based on *shortest paths* and *chordless paths*. Recall that a *chord* of a path P is an edge that connects two non-consecutive vertices in P, and that a chordless path does not have chords. We define the following interval functions for vertices $u, v \in V$.

$$I_g^G(u, v) = \{w \in V \mid w \text{ is on a shortest path between } u \text{ and } v \text{ in } G\}.$$

The shortest paths between two vertices are called the *geodesics*. A set $U \subseteq V$ is *g-convex* if $U = I_g^G(U)$, and $\mathcal{C}_g = \{U \mid U = I_g^G(u)\}$ is the *geodesic convexity*.

$$I_m^G(u, v) = \{w \in V \mid w \text{ is on a chordless path between } u \text{ and } v \text{ in } G\}.$$

Chordless paths are induced paths and are called *monophonics*. A set $U \subseteq V$ is *m-convex* if $U = I_m^G(U)$, and $\mathcal{C}_m = \{U \mid U = I_m^G(U)\}$ is the *monophonic convexity*.

A *short chord* of a path P is a chord that connects two vertices with distance 2 along P , i.e., (u, v) is a short chord if (u, w) and (v, w) are path edges, for some vertex w. Note that u, v, w form a triangle. A *triangle path* is a path that has only short chords.

$$I_t^G(u,v) = \{w \in V \mid w \text{ is on a triangle path between } u \text{ and } v \text{ in } G\}.$$

A set $U \subseteq V$ is *t-convex* if $U = I_t^G(U)$, and $\mathcal{C}_t = \{U \mid U = I_t^G(U)\}$ is the *triangle path convexity*.

If the graph G is clear from the context, we drop the super-script from the interval functions. Figure 2 illustrates the differences between the path convexities. Observe that $I_g(u,v) \subseteq I_m(u,v) \subseteq I_t(u,v)$, for any $u, v \in V$. Finally, we remark that the sets $I(u,v)$ (for each path convexity) can be computed in polynomial time.

2.2 CONVEX MINIMUM SURGICAL PROBING

In this section, we introduce MSP with convexity constraints. Probe vector $\mathcal{P}$ is determined by Eq. (1). We use the term *convex*, shorthand, for g-convex, m-convex, or t-convex, and define that a label vector ℓ is *convex* if $\mathbb{K}_G^{\ell} = \{i \in V \mid \ell_i = 1\}$ is a convex subset of V in G.

Definition 1 (Convex Minimum Surgical Probing (CMSP)). *Given a connected graph G and a probe vector $\mathcal{P}$, determine the minimum number of surgical probes required to uniquely uncover a label vector ℓ that is convex on G.*

If a graph is not connected, we consider its components as independent instances of CMSP. Our first observation shows that the labels of some vertices can be uncovered readily. It also holds for non-convex label vectors.

Observation 1. *If $\mathcal{P}_v = 0$, then $\ell_u = 0$, for all $u \in N[v]$.*

Path convexities imply that a 1-vertex necessarily has another 1-vertex in its neighborhood (if the label vector has support 2 or more). This observation may identify additional 0-vertices. We formalize it in the next lemma.

Lemma 1. *Let $G = (V, E)$ be a connected graph with a convex label vector ℓ of support at least 2, and let $\mathcal{P}$ be the corresponding probe vector. If $\mathcal{P}_v \leq 1$, then $\ell_v = 0$, for $v \in V$.*

Proof. Let G, ℓ, and $\mathcal{P}$ be as in the lemma. Let $v \in V$ such that $\mathcal{P}_v \leq 1$. Towards a contradiction, suppose that $\ell_v = 1$.

Since ℓ has support at least 2, there is another vertex $u \in V$ such that $\ell_u = 1$. If v and u are adjacent, it follows that $\mathcal{P}_v \geq 2$. Otherwise, v and u are not adjacent, and since G is connected, there is a vertex $w \in I_g(u,v)$ (resp. $I_m(u,v)$ and $I_t(u,v)$) that is adjacent to v and $\ell_w = 1$. It follows that $\mathcal{P}_v \geq 2$. We reach the desired contradiction, and $\ell_v = 0$ holds. □

We treat label vectors with support 1 as a corner case since the notion of convexity is immaterial in this case. In the full paper, we show that they can be recognized efficiently. To close this section, we define the *core* $C_G = \bigcap_{i \in V} N_G[i]$ of G. It is the subset of vertices that are adjacent to all vertices. For this reason, it's easy to see that C_G is a clique in G.

2.3 Triangle Path Convexity

In this section, we show that short chords simplify CMSP dramatically. Namely, we prove that the converse of Lemma 1 is true if the label vector is t-convex.

Theorem 1. *Let $G = (V, E)$ be a connected graph with t-convex label vector ℓ of support at least 2, and let $\mathcal{P}$ be the corresponding probe vector. Then, the labels ℓ can be uncovered without using surgical probes. Moreover, $\Bbbk_G^\ell = \{i \in V \mid \mathcal{P}_i \geq 2\}$.*

Proof. Let G, ℓ, and $\mathcal{P}$ be as in the theorem. We show that $\ell_v = 1$ if and only if $\mathcal{P}_v \geq 2$, for $v \in V$. Sufficiency follows with Lemma 1.

To show necessity, assume that $\mathcal{P}_v \geq 2$, for $v \in V$. Either $\ell_v = 1$ or v has two neighbors $u, w \in N(v)$ such that $\ell_u = \ell_w = 1$. In the latter case, verify that $v \in I_t(u, w)$. Since ℓ is t-convex, $\ell_v = 1$, and the claim follows. □

2.4 Reduction Algorithm

In this section, we describe Algorithm 1 which removes vertices with known labels from graph G and *updates* a consistent probe vector $\mathcal{P}$.

Algorithm 1. REDUCE$(G, \mathcal{P}, \ell_i, i)$

$\mathcal{P}' \leftarrow \mathcal{P} \setminus \mathcal{P}_i$ and $G' \leftarrow G \setminus i$
for all $j \in N_G(i)$ **do**
  $\mathcal{P}'_j \leftarrow \mathcal{P}'_j - \ell_i$
end for
return $G', \mathcal{P}'$

Assume that ℓ_i is known, for $i \in V$. Let G' and $\mathcal{P}'$ be the result of running REDUCE$(G, \mathcal{P}, \ell, i)$. We denote the result of removing ℓ_i from ℓ by ℓ^{-i}. The next observation follows readily.

Observation 2. *Probe vector $\mathcal{P}$ is consistent with G and ℓ if and only if $\mathcal{P}'$ is consistent with G' and ℓ^{-i}.*

One needs to be careful when applying Algorithm 1 as it does not preserve convexity, and G' may not be connected. For example, consider the cycle graph of order n, for $n \geq 7$, with a g-convex label vector ℓ of support 3. The three 1-vertices are consecutive along the cycle. If the reduction algorithm is applied to the central 1-vertex, the resulting graph is a path where only the two vertices at the ends (having degree one) are 1-vertices. Clearly, the labels are not g-convex.

On the positive side, Algorithm 1 does preserve g-convexity if 0-vertices are removed as shown by the next lemma.

Lemma 2. *Let $G = (V, E)$ be a graph and ℓ a g-convex label where $\ell_i = 0$ for $i \in V$. Then, ℓ^{-i} is a g-convex label vector on the components of $G' = G \setminus i$.*

Proof. Let G, G', and i be as in the lemma. Towards a contradiction, suppose that ℓ^{-i} is not g-convex on a connected component of G'. Hence, there are vertices $u, v \in V'$ such that $\ell_v = \ell_u = 1$, and a vertex $w \in I_g^{G'}(u, v)$ such that $\ell_w = 0$. Since ℓ is g-convex on G, we have that $w \notin I_g^G(u, v)$. It follows that vertices u, v are connected on a path via vertex i that is shorter than the path via vertex w. Consequently, $i \in I_g^G(u, v)$, and $\ell_i = 1$. We reach the desired contradiction, and the claim follows. □

The claim of Lemma 2 applies to 1-vertices if ℓ is m-convex. Adding a vertex to a graph and connecting it arbitrarily to existing vertices cannot add a chord to a path and only increase the number of vertices in $I_m(u, v)$.

Observation 3. *For any $v, u \in V'$, we have that $I_m^{G'}(u, v) \subseteq I_m^G(u, v)$.*

Lemma 3. *Let $G = (V, E)$ be a graph, $i \in V$, and ℓ a m-convex label. Then, ℓ^{-i} is a m-convex label vector on the components of $G' = G \setminus i$.*

Proof. Let G, G', and i be as in the lemma. Suppose that ℓ^{-i} is not m-convex on a component of G'. Hence, there are vertices $u, v \in V'$ such that $\ell_v = \ell_u = 1$, and a vertex $w \in I_m^{G'}(u, v)$ such that $\ell_w = 0$. With Observation 3 it follows that $w \in I_m^G(u, v)$. We reach a contradiction since ℓ is m-convex on G. □

3 Graphs with Small Maximum Cliques

In this section, we analyze CMSP for graphs where the size of a clique is bounded by considering H-free graphs. As most of the results hold for m-convex and g-convex label vectors, we say *convex*, shorthand, for g-convex or m-convex.

3.1 K_3-Free Graphs

Triangle-free or K_3-free graphs essentially do not have cliques (except two adjacent vertices). This allows us to show that the converse of Lemma 1 is true.

Lemma 4. *Let $G = (V, E)$ be a connected and K_3-free graph with a convex label vector ℓ of support at least 2, and let $\mathcal{P}$ be a consistent probe vector. For $v \in V$, we have that $\ell_v = 1$ if and only if $\mathcal{P}_v \geq 2$.*

Proof. Let $G = (V, E)$, ℓ, and $\mathcal{P}$ be as in the lemma. Moreover, let $v \in V$. Sufficiency follows with Lemma 1.

To show necessity, we assume that $\mathcal{P}_v \geq 2$. Towards a contradiction, suppose $\ell_v = 0$. It follows that there are two neighbors $x, y \in N(v)$ such that $\ell_x = \ell_y = 1$. Next, verify that $(x, y) \notin E$. Otherwise the vertices v, x, y induce K_3 in G, which contradicts that G is K_3-free.

It follows that the path (x, v, y) is a shortest and chordless path between x and y. Since ℓ is convex, we have that $\ell_v = 1$. We reach the desired contradiction, and the claim follows. □

Lemma 4 allows us to uncover ℓ without using surgical probes if ℓ has support at least 2. In Sect. 1 we already observed, in a small example, that K_2 requires one surgical probe to uncover ℓ if (and only if) it has support 1.

As it turns out this is the only exception when considering K_3-free graphs.

Theorem 2. *Let G be a connected and K_3-free graph with a convex label vector ℓ, and let $\mathcal{P}$ be a consistent probe vector. Then, the labels ℓ can be uncovered without using surgical probes and $\Bbbk_G^\ell = \{v \in V \mid \mathcal{P}_v \geq 2\}$, except if $G = K_2$ and ℓ has support* 1*. In which case, one surgical probe is necessary and sufficient to uncover ℓ.*

3.2 K_4-Free Graphs

In this section, we study K_4-free graphs where the largest clique is a triangle. We start by showing that the local argument of Lemma 4 for K_3-free graphs also applies to K_4-free graphs (implying a weaker proposition).

Lemma 5. *Let $G = (V, E)$ be a connected K_4-free graph with a convex label vector ℓ, and let $\mathcal{P}$ be a consistent probe vector. If $\mathcal{P}_v \geq 3$, then $\ell_v = 1$, for $v \in V$.*

Lemma 5 and 1 together allow us to determine the labels of vertices whose probe is not equal to 2. However, we cannot close this gap and uncover all labels without using surgical probes. The following example shows that there are K_4-free graphs and probe vectors such that uncovering a convex label vector requires $\lfloor n/2 \rfloor$ surgical probes.

For an even integer $h \geq 1$, consider the K_4-free graph $G_h = (V, E)$ such that $V = \bigcup_{i=1}^{h} v_i \cup x$, and $E = \bigcup_{i=1}^{h} (x, v_i) \cup \bigcup_{i=1}^{h/2} (v_{2i-1}, v_{2i})$. To complete the CMSP instance, we define probe vector $\mathcal{P}$ as follows: $\mathcal{P}_x = 1 + h/2$, and $\mathcal{P}_{v_i} = 2$ for $i = 1, \ldots, h$. The graph G_4 with probes is depicted in Fig. 3.

The label of the central vertex x can be uncovered by Lemma 5 (if $h \geq 2$). The remaining vertices with unknown labels form adjacent pairs, e.g., v_1 and v_2. Observe that, for each pair, there are two convex label vectors $(0, 1)$ and $(1, 0)$ that are consistent with the probe vector. To distinguish between them, we need to use 1 surgical probe. It follows that $h/2$ surgical probes are required to uncover all labels.

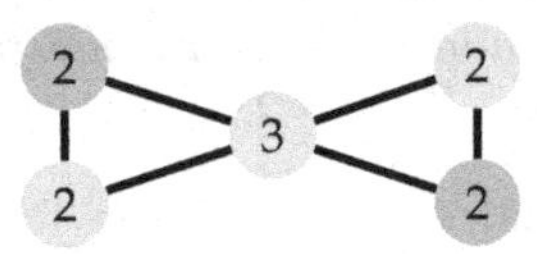

Fig. 3. A K_4-free graph with g-convex labels indicated by the vertices' color (yellow vertices have label 1). Numeral values are the result of probing. To uncover the labels 2 surgical probes are necessary. (Color figure online)

Note that to utilize Lemma 5, the support of the label vector ℓ has to be at least 3. Otherwise, there is no vertex v such that $\mathcal{P}_v \geq 3$. We treat label vectors with support 2 as a special case in the full paper. In the following, we assume that label vectors have support 3 or more.

For our main result of this section, we relate the number of necessary and sufficient surgical probes to the number of *true twins* in G. Recall that a pair of vertices u, v are true twins if $N[u] = N[v]$, i.e., u and v are adjacent and have the same open neighborhood. For a graph $G = (V, E)$ and probe vector $\mathcal{P}$, we define the set $\mathrm{TT}_{(G,\mathcal{P})} \subseteq E$ where $(u, v) \in \mathrm{TT}_{(G,\mathcal{P})}$ if

(i) $N[u] = N[v]$,
(ii) $\mathcal{P}_v = \mathcal{P}_u = 2$,
(iii) there is $w \in N[v]$ such that $\mathcal{P}_w \geq 3$, and
(iv) for $x \in N[v] \setminus \{u, v, w\}$, we have that $\mathcal{P}_x = 1$.

We note that $\mathrm{TT}_{(G,\mathcal{P})}$ can be computed in polynomial time. First, we show that $|\mathrm{TT}_{(G,\mathcal{P})}|$ surgical probes are necessary. This lower bound holds for g-convex and m-convex label vectors.

Lemma 6. *Let G be a connected K_4-free graph with a convex label vector ℓ of support 3 or more, and let $\mathcal{P}$ be a consistent probe vector. Then, $|\mathrm{TT}_{(G,\mathcal{P})}|$ surgical probes are necessary to uncover ℓ.*

However, that $|\mathrm{TT}_{(G,\mathcal{P})}|$ surgical probes are sufficient is only true for m-convex label vectors. Towards proof, we first describe Algorithm 2 which uncovers 1-vertices based on Lemma 5 and removes them by calling Algorithm 1. Its correctness follows readily for m-convex label vectors.

Algorithm 2. 4REDUCEALL-$K_4(G, \mathcal{P})$

1: **for all** $v \in V$ such that $\mathcal{P}_v \geq 3$ **do**
2:     mark vertex v ▷ based on Lemma 5
3: **end for**
4: **for all** marked vertices v **do**
5:     $G, \mathcal{P} \leftarrow$ REDUCE$(G, \mathcal{P}, 1, v)$
6: **end for**
7: **return** $G, \mathcal{P}$

Theorem 3. *Let G be a connected K_4-free graph with a m-convex label vector ℓ of support 3 or more, and let $\mathcal{P}$ be a consistent probe vector. Then, $|\mathrm{TT}_{(G,\mathcal{P})}|$ surgical probes are necessary and sufficient to uncover ℓ.*

Proof (Sketch). Let $G = (V, E)$, $\mathcal{P}$, and ℓ be as in the lemma. Necessity follows with Lemma 6. To show sufficiency, let $H, \mathcal{P}'$ be the result of applying Algorithm 2 to $G, \mathcal{P}$. As graph H may not be connected, let $H_1, H_2, \ldots, H_t$ be the subgraphs induced by the connected components of H.

In the following, we analyze the structure of a component H_i, for $i \in [t]$. We show that either $u, v \in V(H_i)$ and $(u, v) \in \mathrm{TT}_{(G,\mathcal{P})}$, or that the labels of vertices in $V(H_i)$ can be uncovered without using surgical probes.

Algorithm 2 uses Algorithm 1 to remove vertices from G and update $\mathcal{P}$ correctly. Note that, for $v \in V(H_i)$, we have $\mathcal{P}'_v \leq 2$. Let ℓ' be the ℓ restricted to $V(H_i)$. Due to Lemma 3, ℓ' is m-convex on $V(H_i)$.

We assume that there is a vertex $x \in V(H_i)$ such that $\ell_x = 1$. Otherwise, for all $v \in V(H_i)$, $\mathcal{P}_v = 0$ and the labels of vertices can be uncovered without using surgical probes. With Lemma 1 it follows that $\mathcal{P}_x = 2$. Moreover, define $\bar{U} = \{v \in V(H_i) \mid \mathcal{P}_v = 2\}$.

Claim. If $|\bar{U}| = 1$, then the labels of the vertices in $V(H_i)$ can be uncovered without using a surgical probe.

In the following, we assume that $|\bar{U}| \geq 2$. Next, define $\bar{V} = \{v \in V \mid \mathcal{P}_v \geq 3\}$, and verify that $\bar{V}$ is non-empty since ℓ has support at least 3.

Claim. There is a vertex $z \in \bar{V}$ such that $(x, z) \in E$.

Proof. As $\bar{V}$ is not empty, let $y \in \bar{V}$ and note that $\ell_y = 1$. The claim follows immediately if $(x, y) \in E$.

Assume that $(x, y) \notin E$. Since G is connected, there is a chordless path $(x, v_1, v_2, \ldots, v_s, y)$ connecting x and y in G where $s \geq 1$. Since ℓ is m-convex, it follows that $\ell_{v_j} = 1$, for $j \in [s]$. Then, $\mathcal{P}_{v_1} \geq 3$, and consequently, $v_1 \in \bar{V}$. Vertex $v_1 = z$ has the desired properties, and the claim follows. □

Since $\mathcal{P}_x = 2$ and $\ell_x = 1$, vertex z is the *only* neighbor of x in $\bar{V}$. It follows that $\mathcal{P}'_x = 1$.

Claim. For all $v \in V(H_i) \setminus \{x\}$, we have that $\ell_v = 0$.

The previous claim shows that ℓ' has support 1 on H_i. It follows that, for all $v \in V(H_i) \setminus \{x\}$, we have that $\mathcal{P}'_v \leq 1$ and that $(v, x) \in E$ if and only if $\mathcal{P}'_v = 1$. We conclude that x is contained in the core $C_{\bar{W}}$ where $\bar{W} = \{v \in V(H_i) \mid \mathcal{P}'_v = 1\}$. Note that $\bar{U} \subseteq \bar{W}$ and that $x \in \bar{U} \cap C_{\bar{W}}$.

Next, we use that G is K_4-free to restrict the size of $C_{\bar{W}}$. Consider a vertex $v \in \bar{U}$. Either $\mathcal{P}'_v = 0$ and v is adjacent to two vertices in $\bar{V}$, or $\mathcal{P}'_v = 1$ and v is adjacent to x and to one vertex in $\bar{V}$ in G. Verify that if vertex $v \in \bar{U} \cap \bar{W}$, then $(v, z) \in E$.

Claim. If $|\bar{U}| \geq 3$, then $|C_{\bar{W}}| = 1$ and the labels of vertices in $V(H_i)$ can be uncovered without using surgical probes.

In the following, we assume that $\bar{U} = \{x, u\}$. To finish the proof, we argue that $(u, x) \in \mathrm{TT}_{(G,\mathcal{P})}$. By definition, $\mathcal{P}_x = \mathcal{P}_u = 2$, and we showed that $z \in N(x)$ where $\mathcal{P}_z \geq 3$.

Claim. Either $N[x] = N[u]$, or ℓ' can be uncovered without using surgical probes.

In case, $N[x] = N[u]$, the neighborhood $N[x]$ may contain more vertices than x, u, z. Verify that if $v \in N[x] \setminus \{x, u, z\}$, then $\mathcal{P}_v = 1$. It follows that $(u, x) \in \mathrm{TT}_{(G,\mathcal{P})}$ and the theorem follows. □

An important family of K_4-free graphs are outerplanar graphs. Recall that an outerplanar graph can be embedded in the plane such that edges do not intersect and all vertices lie on the outer face. Moreover, in a maximal outerplanar graph each vertex appears exactly once on the outer face implying that it cannot have a pair of true twins. We conclude that if G is maximal outerplanar, then $\mathrm{TT}_{(G,\mathcal{P})}$ is empty and that ℓ can be uncovered without using surgical probes if its support is at least 3.

Finally, we observe that Lemma 5 generalizes to K_h-free graphs, for a fixed integer h. The proof is analog to that of Lemma 5.

Theorem 4. *Let $G = (V, E)$ be a connected K_h-free graph with a convex label vector ℓ, and let $\mathcal{P}$ be a consistent probe vector. If $\mathcal{P}_v \geq h$, then $\ell_v = 1$, for $v \in V$.*

References

1. Bar-Noy, A., Böhnlein, T., Lotker, Z., Peleg, D., Rawitz, D.: Weighted microscopic image reconstruction. In: Bureš, T., Dondi, R., Gamper, J., Guerrini, G., Jurdziński, T., Pahl, C., Sikora, F., Wong, P.W.H. (eds.) SOFSEM 2021. LNCS, vol. 12607, pp. 373–386. Springer, Cham (2021). https://doi.org/10.1007/978-3-030-67731-2_27
2. Bar-Noy, A., Böhnlein, T., Lotker, Z., Peleg, D., Rawitz, D.: The generalized microscopic image reconstruction problem. Dis. Appl. Math. **321**, 402–416 (2022)
3. Barcucci, E., Lungo, A.D., Nivat, M., Pinzani, R.: Reconstructing convex polyominoes from horizontal and vertical projections. Theor. Comput. Sci. **155**(2), 321–347 (1996)
4. Di Marco, N., Frosini, A.: Properties of sat formulas characterizing convex sets with given projections. In: Baudrier, É., Naegel, B., Krähenbühl, A., Tajine, M. (eds.) DGMM 2022. LNCS, vol. 13493, pp. 153–166. Springer, Cham (2022). https://doi.org/10.1007/978-3-031-19897-7_13
5. Di Marco, N., Frosini, A.: The generalized microscopic image reconstruction problem for hypergraphs. In: Barneva, R.P., Brimkov, V.E., Nordo, G. (eds.) IWCIA 2022. LNCS, vol. 13348, pp. 317–331. Springer, Cham (2023). https://doi.org/10.1007/978-3-031-23612-9_20
6. Dourado, M.C., Protti, F., Rautenbach, D., Szwarcfiter, J.L.: On the convexity number of graphs. Graphs Comb. **28**(3), 333–345 (2012)
7. Dulio, P., Frosini, A., Rinaldi, S., Tarsissi, L., Vuillon, L.: Further steps on the reconstruction of convex polyominoes from orthogonal projections. J. Comb. Optim. **44**(4), 2423–2442 (2021)
8. Farber, M., Jamison, R.E.: On local convexity in graphs. Discret. Math. **66**(3), 231–247 (1987)
9. Frosini, A., Nivat, M.: Binary matrices under the microscope: a tomographical problem. Theor. Comput. Sci. **370**(1–3), 201–217 (2007)
10. Gerard, Y.: Regular switching components. Theor. Comput. Sci. **777**, 338–355 (2019)
11. Gritzmann, P., Langfeld, B., Wiegelmann, M.: Uniqueness in discrete tomography: three remarks and a corollary. SIAM J. Discret. Math. **25**(4), 1589–1599 (2011)
12. Herman, G.T., Kuba, A.: Discrete Tomography: Foundations, Algorithms, and Applications. Springer, Heidelberg (2012)
13. Pelayo, I.M.: Geodesic Convexity in Graphs. Springer, Heidelberg (2013). https://doi.org/10.1007/978-1-4614-8699-2

A Linear Algorithm for Radio k-Coloring Powers of Paths Having Small Diameter

Dipayan Chakraborty[1(✉)], Soumen Nandi[2], Sagnik Sen[3], and D. K. Supraja[2,3(✉)]

[1] Université Clermont-Auvergne, CNRS, Mines de Saint-Étienne, Clermont-Auvergne-INP, LIMOS, 63000 Clermont-Ferrand, France
dipayan.chakraborty@uca.fr
[2] Netaji Subhas Open University, Kolkata, India
[3] Indian Institute of Technology Dharwad, Dharwad, India
dksupraja95@gmail.com

Abstract. The radio k-chromatic number $rc_k(G)$ of a graph G is the minimum integer λ such that there exists a function $\phi : V(G) \rightarrow \{0, 1, \cdots, \lambda\}$ satisfying $|\phi(u) - \phi(v)| \geq k + 1 - d(u, v)$, where $d(u, v)$ denotes the distance between u and v. To date, several upper and lower bounds of $rc_k(\cdot)$ is established for different graph families. One of the most notable works in this domain is due to Liu and Zhu [SIAM Journal on Discrete Mathematics 2005] whose main results were computing the exact values of $rc_k(\cdot)$ for paths and cycles for the specific case when k is equal to the diameter.

In this article, we find the exact values of $rc_k(G)$ for powers of paths where the diameter of the graph is strictly less than k. Our proof readily provides a linear time algorithm for providing such labeling. Furthermore, our proof technique is a potential tool for solving the same problem for other graph classes with "small" diameter.

Keywords: radio coloring · radio k-chromatic number · Channel Assignment Problem · power of paths

1 Introduction and Main Results

The theory of radio coloring and its variations are popular and well-known mathematical models of the Channel Assignment Problem (CAP) in wireless networks [1,2]. The connection between the real-life problem and the theoretical model has been explored in different bodies of works. In this article, we focus on the theoretical aspects of a particular variant, namely, the radio k-coloring. All the graphs considered in this article are undirected simple graphs and we refer to the book "Introduction to graph theory" by West [14] for all standard notations and terminologies used.

S.-Y. Hsieh et al. (Eds.): IWOCA 2023, LNCS 13889, pp. 148–159, 2023.
https://doi.org/10.1007/978-3-031-34347-6_13

A *λ-radio k-coloring* of a graph G is a function $\phi : V(G) \rightarrow \{0, 1, \cdots, \lambda\}$ satisfying $|\phi(u) - \phi(v)| \geq k + 1 - d(u, v)$. For every $u \in V(G)$, the value $\phi(u)$ is generally referred to as the *color* of u under ϕ. Usually, we pick λ in such a way that it has a preimage under ϕ, and then, we call λ to be the span of ϕ, denoting it by $span(\phi)$. The *radio k-chromatic number*[1] $rc_k(G)$ is the minimum $span(\phi)$, where ϕ varies over all radio k-colorings of G.

In particular, the radio 2-chromatic number is the most well-studied restriction of the parameter (apart from the radio 1-chromatic number, which is equivalent to studying the chromatic number of graphs). There is a famous conjecture by Griggs and Yeh [6] that claims $rc_2(G) \leq \Delta^2$ where Δ is the maximum degree of G. The conjecture has been resolved for all $\Delta \geq 10^{69}$ by Havet, Reed and Sereni [7].

As one may expect, finding the exact values of $rc_k(G)$ for a general graph is an NP-complete problem [6]. Therefore, finding the exact value of $rc_k(G)$ for a given graph (usually belonging to a particular graph family) offers a huge number of interesting problems. Unfortunately, due to a lack of general techniques for solving these problems, not many exact values are known till date. One of the best contributions in this front remains the work of Liu and Zhu [12] who computed the exact value of $rc_k(G)$ where G is a path or a cycle and $k = diam(G)$.

As our work focuses on finding radio k-chromatic number of powers of paths, let us briefly recall the relevant related works. For a detailed overview of the topic, we encourage the reader to consult Chap. 7.5 of the dynamic survey on this topic maintained in the Electronic Journal of Combinatorics by Gallian [5] and the survey by Panigrahi [13]. For small paths P_n, that is, with $diam(P_n) < k$, Kchikech et al. [8] had established an exact formula for $rc_k(P_n)$; whereas, recall that, for paths of diameter equal to $k \geq 2$, Liu and Zhu [12] gave an exact formula for the radio number $rc_k(P_k)$. Moreover, a number of studies on the parameter $rc_k(P_n)$ depending on how k is related to $diam(P_n)$, or n alternatively, have been done by various authors [3,8–10]. So far as works on powers of paths are concerned, the only notable work we know is an exact formula for the radio number $rn(P_n^2)$ of the square of a path P_n by Liu and Xie [11]. Hence the natural question to ask is whether the results for the paths can be extended to paths of a general power m, where $1 \leq m \leq n$.

Progressing along the same line, in this article we concentrate on powers of paths having "small diameters", that is, $diam(P_n^m) < k$ and compute the exact value of $rc_k(P_n^m)$, where P_n^m denotes the m-th power graph of a path P_n on $(n+1)$ vertices. In other words, the graph P_n^m is obtained by adding edges between the vertices of P_n that are at most m distance apart, where $m \leq n$. Notice that, the so-obtained graph is, in particular, an interval graph. Let us now state our main theorem.

Theorem 1. *For all $k > diam(P_n^m)$ and $m \leq n$, we have*

[1] In the case that $diam(G) = k, k+1$ or $k+2$, the radio k-chromatic number is alternatively known as the *radio number* denoted by $rn(G)$, the *radio antipodal number* denoted by $ac(G)$ and the *nearly antipodal number* denoted by $ac'(G)$, respectively.

$$rc_k(P_n^m) = \begin{cases} nk - \frac{n^2-m^2}{2m} & \text{if } \lceil \frac{n}{m} \rceil \text{ is odd and } m|n, \\ nk - \frac{n^2-s^2}{2m} + 1 & \text{if } \lceil \frac{n}{m} \rceil \text{ is odd and } m \nmid n, \\ nk - \frac{n^2}{2m} + 1 & \text{if } \lceil \frac{n}{m} \rceil \text{ is even and } m|n, \\ nk - \frac{n^2-(m-s)^2}{2m} + 1 & \text{if } \lceil \frac{n}{m} \rceil \text{ is even and } m \nmid n, \end{cases}$$

where $s \equiv n \pmod m$ *and* $1 \leq s < m$.

In this article, we develop a robust graph theoretic tool for the proof. Even though the tool is specifically used to prove our result, it can be adapted to prove bounds for other classes of graphs. Thus, we would like to remark that, the main contribution of this work is not only in proving an important result that captures a significant number of problems with a unified proof, but also in devising a proof technique that has the potential of becoming a standard technique to attack similar problems. We will prove the theorem in the next section.

Moreover, our proof of the upper bound is by giving a prescribed radio k-coloring of the concerned graph, and then proving its validity, while the lower bound proof establishes its optimality. Therefore, as a corollary to Theorem 1, we can say that our proof provides a linear time algorithm radio k-color powers of paths, optimally.

Theorem 2. *For all* $k > diam(P_n^m)$ *and* $m \leq n$, *one can provide an optimal radio* k*-coloring of the graph* P_n^m *in* $O(n)$ *time.*

We prove Theorem 1 in the next section.

2 Proofs of Theorems 1 and 2

This section is entirely dedicated to the proofs of Theorems 1 and 2. The proofs use specific notations and terminologies developed for making it easier for the reader to follow. The proof is contained in several observations and lemmas and uses a modified and improved version of the DGNS formula [4].

As seen from the theorem statement, the graph P_n^m that we work on is the mth power of the path on $(n+1)$ vertices. One crucial aspect of this proof is the naming of the vertices of P_n^m. In fact, for convenience, we shall assign *two* names to each of the vertices of the graph and use them as required depending on the context. Such a naming convention will depend on the parity of the diameter of P_m^n.

Observation 1. *The diameter of the graph* P_n^m *is* $diam(P_n^m) = \lceil \frac{n}{m} \rceil$.

For the rest of this section, let $q = \lfloor \frac{diam(P_n^m)}{2} \rfloor$.

2.1 The Naming Conventions

We are now ready to present the first naming convention for the vertices of P_n^m. For convenience, let us suppose that the vertices of P_n^m are placed (embedded) on the X-axis having co-ordinates $(i, 0)$ where $i \in \{0, 1, \cdots, n\}$ and two (distinct) vertices are adjacent if and only if their Euclidean distance is at most m.

We start by selecting the layer L_0 consisting of the vertex, named c_0, say, positioned at $(qm, 0)$ for even values of $diam(P_n^m)$. On the other hand, for odd values of $diam(P_n^m)$, the layer L_0 consists of the vertices $c_0, c_1, \cdots, c_m$, say, positioned at $(qm, 0), (qm+1, 0), \cdots, (qm+m, 0)$, respectively, and inducing a maximal clique of size $(m+1)$. The vertices of L_0 are called the *central vertices*, and those positioned to the left and the right side of the central vertices are naturally called the *left vertices* and the *right vertices*, respectively.

After this, we define the layer L_i as the set of vertices that are at a distance i from L_0. Observe that the layer L_i is non-empty for all $i \in \{0, 1, \cdots, q\}$. Moreover, notice that, for all $i \in \{1, 2, \cdots, q\}$, L_i consists of both left and right vertices. In particular, for $i \geq 1$, the left vertices of L_i are named $l_{i1}, l_{i2}, \cdots, l_{im}$, sorted according to the increasing order of their Euclidean distances from L_0. Similarly, for $i \in \{1, 2, \cdots, q-1\}$, the right vertices of L_i are named $r_{i1}, r_{i2}, \cdots, r_{im}$, sorted according to the increasing order of their Euclidean distance from L_0. However, the right vertices of L_q are $r_{q1}, r_{q2}, \cdots, r_{qs}$, where $s = (n+1) - (2q-1)m - |L_0|$ (observe that this s is the same as the s mentioned in the statement of Theorem 1), again sorted according to the increasing order of their Euclidean distances from L_0. That is, if $m \nmid n$, then there are $s = (n+1) - (2q-1)m - |L_0|$ right vertices in L_q. Besides L_q, every layer L_i, for $i \in \{1, 2, \cdots, q-1\}$, has exactly m left vertices and m right vertices. This completes our first naming convention.

Now, we move to the second naming convention. This depends on yet another observation.

Observation 2. *For $k \geq diam(P_n^m)$, let ϕ be a radio k-coloring of P_n^m. Then $\phi(x) \neq \phi(y)$ for all distinct $x, y \in V(P_n^m)$.*

Let ϕ be a radio k-coloring of P_n^m. Thus, due to Observation 2, it is possible to sort the vertices of P_n^m according to the increasing order of their colors. That is, our second naming convention which names the vertices of P_n^m as $v_0, v_1, \cdots, v_n$ satisfying $\phi(v_0) < \phi(v_1) < \cdots < \phi(v_n)$. Clearly, the second naming convention depends only on the coloring ϕ, which, for the rest of this section, will play the role of any arbitrary radio k-coloring of P_n^m.

2.2 The Lower Bound

Next, we shall proceed to establish the lower bound of Theorem 1 by showing it to be a lower bound of $span(\phi)$. To do so, however, we need to introduce yet another notation. Let $f : V(P_n^m) \to \{0, 1, \cdots, q\}$ be the function which indicates the layer of a vertex, that is, $f(x) = i$ if $x \in L_i$. With this notation, we initiate the lower bound proof with the following result.

Lemma 1. *For any* $i \in \{0, 1, \cdots, n-1\}$, *we have*

$$\phi(v_{i+1}) - \phi(v_i) \geq \begin{cases} k - f(v_i) - f(v_{i+1}) + 1 & \text{if } diam(P_n^m) \text{ is even,} \\ k - f(v_i) - f(v_{i+1}) & \text{if } diam(P_n^m) \text{ is odd.} \end{cases}$$

Proof. If $diam(P_n^m)$ is even, then L_0 consists of the single vertex c_0. Observe that, as v_i is in $L_{f(v_i)}$, it is at a distance $f(v_i)$ from c_0. Similarly, v_{i+1} is at a distance $f(v_{i+1})$ from c_0. Hence, by the triangle inequality, we have

$$d(v_i, v_{i+1}) \leq d(v_i, c_0) + d(c_0, v_{i+1}) = f(v_i) + f(v_{i+1}).$$

Therefore, by the definition of radio k-coloring,

$$\phi(v_{i+1}) - \phi(v_i) \geq k - f(v_i) - f(v_{i+1}) + 1.$$

If $diam(P_n^m)$ is odd, then L_0 is a clique. Thus, by the definition of layers and the function f, there exist vertices c_j and $c_{j'}$ $(j \neq j')$ in L_0 satisfying $d(v_i, c_j) = f(v_i)$ and $d(v_{i+1}, c_{j'}) = f(v_{i+1})$. Hence, by triangle inequality again, we have

$$d(v_i, v_{i+1}) \leq d(v_i, c_j) + d(c_j, c_{j'}) + d(c_{j'}, v_{i+1}) = f(v_i) + 1 + f(v_{i+1}).$$

Therefore, by the definition of radio k-coloring,

$$\phi(v_{i+1}) - \phi(v_i) \geq k - f(v_i) - f(v_{i+1}).$$

Hence we are done. □

Notice that it is not possible to improve the lower bound of the inequality presented in Lemma 1. Motivated by this fact, whenever we have

$$\phi(v_{i+1}) - \phi(v_i) = \begin{cases} k - f(v_i) - f(v_{i+1}) + 1 & \text{if } diam(P_n^m) \text{ is even,} \\ k - f(v_i) - f(v_{i+1}) & \text{if } diam(P_n^m) \text{ is odd.} \end{cases}$$

for some $i \in \{0, 1, \cdots, n-1\}$, we say that the pair (v_i, v_{i+1}) is *optimally colored* by ϕ. Moreover, we can naturally extend this definition to a sequence of vertices of the type $(v_i, v_{i+1}, \cdots, v_{i+i'})$ by calling it an *optimally colored sequence* by ϕ if (v_{i+j}, v_{i+j+1}) is optimally colored by ϕ for all $j \in \{0, 1, \cdots, i'-1\}$. Furthermore, a *loosely colored sequence* $(v_i, v_{i+1}, v_{i+2}, \cdots, v_{i+i'})$ is a sequence that does not contain any optimally colored sequence as a subsequence.

An important thing to notice is that the sequence of vertices $(v_0, v_1, \cdots, v_n)$ can be written as a concatenation of maximal optimally colored sequences and loosely colored sequences. That is, it is possible to write

$$(v_0, v_1, \cdots, v_n) = Y_0 X_1 Y_1 X_2 \cdots X_t Y_t$$

where Y_is are loosely colored sequences and X_js are maximal optimally colored sequences. Here, we allow the Y_is to be empty sequences as well. In fact, for

$1 \leq i \leq t-1$, a Y_i is empty if and only if there exist two consecutive vertices $v_{s'}$ and $v_{s'+1}$ of P_n^m in the second naming convention such that $(v_{s'}, v_{s'+1})$ is loosely colored and that $X_i = (v_s, v_{s+1}, \cdots, v_{s'})$ and $X_{i+1} = (v_{s'+1}, v_{s'+2}, \cdots, v_{s''})$ for some $s \leq s' < s''$. Moreover, Y_0 (resp. Y_t) is empty if and only if the pair (v_0, v_1) (resp. (v_{n-1}, v_n)) is optimally colored. By convention, empty sequences are always loosely colored and a sequence having a singleton vertex is always optimally colored. From now onward, whenever we mention a radio k-coloring ϕ of P_n^m, we shall also suppose an associated concatenated sequence using the same notation as mentioned above.

Let us now prove a result which plays an instrumental role in the proof of the lower bound.

Lemma 2. *Let ϕ be a radio-k coloring of P_n^m such that*

$$(v_0, v_1, \cdots, v_n) = Y_0 X_1 Y_1 X_2 \cdots X_t Y_t.$$

Then, for even values of $diam(P_n^m)$, we have

$$span(\phi) \geq \left[n(k+1) - 2\sum_{i=1}^{q} i|L_i| \right] + \left[f(v_0) + f(v_n) + \sum_{i=0}^{t} |Y_i| + t - 1 \right]$$

and, for odd values of $diam(P_n^m)$, we have

$$span(\phi) \geq \left[nk - 2\sum_{i=1}^{q} i|L_i| \right] + \left[f(v_0) + f(v_n) + \sum_{i=0}^{t} |Y_i| + t - 1 \right],$$

where $|Y_i|$ denotes the length of the sequence Y_i.

As we shall calculate the two additive components of Lemma 2 separately, we introduce short-hand notations for them for the convenience of reference. So, let

$$\alpha_1 = \begin{cases} n(k+1) - 2\sum_{i=1}^{q} i|L_i| & \text{if } diam(P_n^m) \text{ is even,} \\ nk - 2\sum_{i=1}^{q} i|L_i| & \text{if } diam(P_n^m) \text{ is odd,} \end{cases}$$

and

$$\alpha_2(\phi) = f(v_0) + f(v_n) + \sum_{i=0}^{t} |Y_i| + t - 1.$$

Observe that α_1 and α_2 are functions of a number of variables and factors such as, n, m, k, ϕ, etc. However, to avoid clumsy and lengthy formulations, we have avoided writing α_1 and α_2 as multivariate functions, as their definitions are not ambiguous in the current context. Furthermore, as k and P_n^m are assumed to be fixed in the current context and, as α_1 does not depend on ϕ (follows from its definition), it is treated and expressed as a constant as a whole. On the other hand, α_2 is expressed as a function of ϕ.

Now we shall establish lower bounds for α_1 and $\alpha_2(\phi)$ separately to prove the lower bound of Theorem 1. Let us start with α_1 first.

Lemma 3. *We have*

$$\alpha_1 = \begin{cases} nk - \frac{n^2+m^2-s^2}{2m} & \text{if } diam(P_n^m) \text{ is even,} \\ nk - \frac{n^2-s^2}{2m} & \text{if } diam(P_n^m) \text{ is odd,} \end{cases}$$

where $s = (n+1) - (2q-1)m - |L_0|$.

Next, we focus on $\alpha_2(\phi)$. We shall handle the cases with odd $diam(P_n^m)$ first.

Lemma 4. *We have*

$$\alpha_2(\phi) \geq \begin{cases} 0 & \text{if } diam(P_n^m) \text{ is odd and } m|n, \\ 1 & \text{if } diam(P_n^m) \text{ is odd and } m \nmid n. \end{cases}$$

Next, we consider the cases with even $diam(P_n^m)$. Before starting with it though, we are going to introduce some terminologies to be used during the proofs. So, let X_i be an optimally colored sequence. As X_i cannot have two consecutive left (resp., right) vertices as elements, the number of left vertices can be at most one more than the number of right vertices and the central vertex, the latter two combined together.

Lemma 5. *We have*

$$\alpha_2(\phi) \geq \begin{cases} 1 & \text{if } diam(P_n^m) \text{ is even and } m|n, \\ m-s+1 & \text{if } diam(P_n^m) \text{ is even and } m \nmid n, \end{cases}$$

where $s \equiv n \pmod{m}$.

Combining Lemmas 2, 3, 4 and 5, therefore, we have the following lower bound for the parameter $rc_k(P_n^m)$.

Lemma 6. *For all* $k \geq diam(P_n^m)$ *and* $m \leq n$, *we have*

$$rc_k(P_n^m) \geq \begin{cases} nk - \frac{n^2-m^2}{2m} & \text{if } \lceil \frac{n}{m} \rceil \text{ is odd and } m|n, \\ nk - \frac{n^2-s^2}{2m} + 1 & \text{if } \lceil \frac{n}{m} \rceil \text{ is odd and } m \nmid n, \\ nk - \frac{n^2}{2m} + 1 & \text{if } \lceil \frac{n}{m} \rceil \text{ is even and } m|n, \\ nk - \frac{n^2-(m-s)^2}{2m} + 1 & \text{if } \lceil \frac{n}{m} \rceil \text{ is even and } m \nmid n, \end{cases}$$

where $s \equiv n \pmod{m}$ *and* $1 \leq s < m$.

Remark 1. Our lower bound technique can be applied to a graph G of diameter more than k also. This can be achieved by taking a subgraph H of G induced on $\bigcup_{i=0}^{q} L_i$, where $q = \lfloor \frac{k}{2} \rfloor$ and $diam(H) \leq k$. Thus, a lower bound for H serves as a lower bound for G as well.

2.3 The Upper Bound

Now let us prove the upper bound. We shall provide a radio k-coloring ψ of P_n^m and show that its span is the same as the value of $rc_k(P_n^m)$ stated in Theorem 1. To define ψ, we shall use both naming conventions. That is, we shall express the ordering $(v_0, v_1, \cdots, v_n)$ of the vertices of P_n^m with respect to ψ in terms of the first naming convention.

Let us define some ordering for the right (and similarly for the left) vertices:

(1) $r_{ij} \prec_1 r_{i'j'}$ if either (i) $j < j'$ or (ii) $j = j'$ and $(-1)^{j-1}i < (-1)^{j'-1}i'$;
(2) $r_{ij} \prec_2 r_{i'j'}$ if either (i) $j < j'$ or (ii) $j = j'$ and $(-1)^{m-j}i < (-1)^{m-j'}i'$;
(3) $r_{ij} \prec_3 r_{i'j'}$ if either (i) $j < j'$ or (ii) $j = j'$ and $i > i'$; and
(4) $r_{ij} \prec_4 r_{i'j'}$ if either (i) $j < j'$ or (ii) $j = j'$ and $(-1)^{j}i < (-1)^{j'}i'$.

Observe that, the orderings are based on comparing the second co-ordinate of the indices of the right (resp., left) vertices, and if they happen to be equal, then comparing the first co-ordinate of the indices with conditions on their parities. Moreover, all the above four orderings define total orders on the set of all right (resp., left) vertices. Thus, there is a unique increasing (resp., decreasing) sequence of right (or the left) vertices with respect to $\prec_1$, $\prec_2$, $\prec_3$, and $\prec_4$. Based on these orderings, we are going to construct a sequence of vertices of the graph and then greedy color the vertices to provide our labeling.

The sequences of the vertices are given as follows:

(1) An *alternating chain* as a sequence of vertices of the form $(a_1, b_1, a_2, b_2, \cdots, a_p, b_p)$ such that $(a_1, a_2, \cdots, a_p)$ is the increasing sequence of right vertices with respect to $\prec_1$ and $(b_1, b_2, \cdots, b_p)$ is the decreasing sequence of left vertices with respect to $\prec_2$.
(2) A *canonical chain* as a sequence of vertices of the form $(a_1, b_1, a_2, b_2, \cdots, a_p, b_p)$ such that $(a_1, a_2, \cdots, a_p)$ is the increasing sequence of right vertices with respect to $\prec_3$ and $(b_1, b_2, \cdots, b_p)$ is the decreasing sequence of left vertices with respect to $\prec_3$;
(3) A *special alternating chain* as a sequence of vertices of the form $(a_1, b_1, a_2, b_2, \cdots, a_p, b_p)$ such that $(a_1, a_2, \cdots, a_p)$ is the increasing sequence of right vertices with respect to $\prec_2$ and $(b_1, b_2, \cdots, b_p)$ is the decreasing sequence of left vertices with respect to $\prec_1$; and
(4) A *special canonical chain* as a sequence of vertices of the form $(a_1, b_1, a_2, b_2, \cdots, a_p, b_p)$ such that $(a_1, a_2, \cdots, a_p)$ is the increasing sequence of right vertices with respect to $\prec_4$ and $(b_1, b_2, \cdots, b_p)$ is the decreasing sequence of left vertices with respect to $\prec_4$.

Notice that the special alternating chains, the reverse alternating chain and the canonical chains can exist only when the number of right and left vertices are equal. Of course, when $m|n$, both the chains exist. Otherwise, we shall modify the names of the vertices a little to make them exist.

We are now ready to express the sequence $(v_0, v_1, \cdots, v_n)$ by splitting it into different cases which are depicted in Figs. 1, 2, 3 and 4 for example. In the figures, the both naming conventions for each of the vertices are given so that the reader may cross verify the correctness for that particular instance for each case. For convenience, also recall that $q = \lfloor \frac{diam(P_n^m)}{2} \rfloor$.

Case 1: when $diam(P_n^m)$ is even, $m|n$ and $k > diam(P_n^m)$. First of all, $(v_0, v_1, \cdots, v_{2qm-1})$ is the alternating chain. Moreover, $v_n = c_0$.

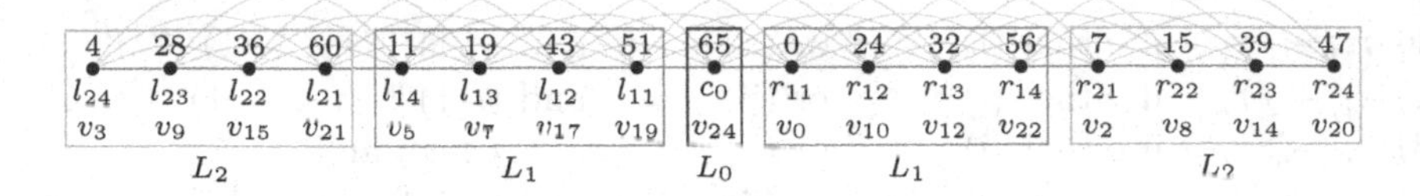

Fig. 1. *Case 1.* $n = 16$, $m = 4$, $diam(P_{24}^4) = 4$, $k = 6$.

Case 2: when $diam(P_n^m)$ is odd, $m|n$ and $k > diam(P_n^m)$. Let the ordering of the vertices be $(v_0, v_1, \cdots, v_{2qm+m})$. Now, $v_{j(2q+1)} = c_j$ for all $0 \leq j \leq m$. The remaining vertices follow the canonical chain.

18 41 64 87 9 32 55 78 0 23 46 69 92 14 37 60 83 5 28 51 74
l_{24} l_{23} l_{22} l_{21} l_{14} l_{13} l_{12} l_{11} c_0 c_1 c_2 c_3 c_4 r_{11} r_{12} r_{13} r_{14} r_{21} r_{22} r_{23} r_{24}
v_4 v_9 v_{14} v_{19} v_2 v_7 v_{12} v_{17} v_0 v_5 v_{10} v_{15} v_{20} v_3 v_8 v_{13} v_{18} v_1 v_6 v_{11} v_{16}
L_2 L_1 L_0 L_1 L_2

Fig. 2. *Case 2.* $n = 20$, $m = 4$, $diam(P_{20}^4) = 5$, $k = 7$.

Case 3: when $diam(P_n^m)$ is odd, $m \nmid n$ and $k > diam(P_n^m)$. For any set A, let $A^\star$ represent an ordered sequence of the elements of A. Let $G \cong P_n^m$ and $S = V(G) = \{v_0, v_1, v_2, \cdots, v_{2qm+s}\}$. Then $S^\star$ is defined as described. First, define

$$T = \{v_t : 0 \leq t \leq s(2q+1)\} - \{v_{j(2q+1)} : 0 \leq j \leq s\}.$$

Order $T^\star$ as a canonical chain. Also, define $v_{j(2q+1)} = c_j$ for all $0 \leq j \leq s$. Assume G' to be the subgraph of G induced by the subset $S - \{r_{q1}, r_{q2}, \cdots, r_{qs}\}$ of S. Then $G' \cong P_{n'}^m$, $m|n'$ and $diam(G') = \frac{n'}{m}$ is even, where $n' = n - s$. Define

$$v_n = l_{11} \text{ and } U = \{v_t : s(2q+1) + 1 \leq t < n\}.$$

Note that $U \subset V(G')$. Order $U^\star$ (as vertices of G') by the following.

(i) Special alternating chain when m and s have the same parity.
(ii) Alternating chain when m is even and s is odd.
(iii) Special canonical chain when m is odd and s is even.

Fig. 3. *Case 3.* $n = 19$, $m = 4$, $diam(P_{19}^4) = 5$, $k = 7$, $s = 3$.

Case 4: when $diam(P_n^m)$ is even, $m \nmid n$ and $k > diam(P_n^m)$. Notice that, in this case, the left vertices are $(m - s)$ more than the right vertices. Also, L_0 has only one vertex in this case. We shall discard some vertices from the set of left vertices, and then present the ordering. To be specific, we disregard the subset $\{l_{11}, l_{12}, \cdots, l_{1(m-s)}\}$, temporarily, from the set of left vertices and consider the alternating chain. First of all, $(v_0, v_1, \cdots, v_{2qm-2m+2s-1})$ is the alternating chain. Additionally, $(v_{2qm-2m+2s}, v_{2qm-2m+2s+1}, v_{2qm-2m+2s+2}, \cdots, v_{2qm-m+s}) = (c_0, l_{11}, l_{12}, \cdots, l_{1(m-s)})$.

Fig. 4. *Case 4.* $n = 14$, $m = 4$, $diam(P_{22}^4) = 4$, $k = 6$, $s = 2$.

Thus, we have obtained a sequence $(v_0, v_1, \cdots, v_n)$ in each case under consideration. Now, we define, $\psi(v_0) = 0$ and $\psi(v_{i+1}) = \psi(v_i) + k + 1 - d(v_i, v_{i+1})$, recursively, for all $i \in \{0, 1, 2, \cdots, n-1\}$. Next, we note that ψ is a radio k-coloring.

Lemma 7. *The function ψ is a radio k-coloring of P_n^m.*

This brings us to the upper bound for $rc_k(P_n^m)$.

Lemma 8. *For all $k > diam(P_n^m)$ and $m \leq n$, we have*

$$rc_k(P_n^m) \leq \begin{cases} nk - \frac{n^2 - m^2}{2m} & \text{if } \lceil \frac{n}{m} \rceil \text{ is odd and } m \mid n, \\ nk - \frac{n^2 - s^2}{2m} + 1 & \text{if } \lceil \frac{n}{m} \rceil \text{ is odd and } m \nmid n, \\ nk - \frac{n^2}{2m} + 1 & \text{if } \lceil \frac{n}{m} \rceil \text{ is even and } m \mid n, \\ nk - \frac{n^2 - (m-s)^2}{2m} + 1 & \text{if } \lceil \frac{n}{m} \rceil \text{ is even and } m \nmid n, \end{cases}$$

where $s \equiv n \pmod{m}$ and $1 \leq s < m$.

Proof. Observe that, $rc_k(P_n^m) \leq span(\psi)$. So, to prove the upper bound, it is enough to show that for all $k > diam(P_n^m)$ and $s \equiv n \pmod m$,

$$span(\psi) = \begin{cases} nk - \frac{n^2-m^2}{2m} & \text{if } \lceil \frac{n}{m} \rceil \text{ is odd and } m|n, \\ nk - \frac{n^2-s^2}{2m} + 1 & \text{if } \lceil \frac{n}{m} \rceil \text{ is odd and } m \nmid n, \\ nk - \frac{n^2}{2m} + 1 & \text{if } \lceil \frac{n}{m} \rceil \text{ is even and } m|n, \\ nk - \frac{n^2-(m-s)^2}{2m} + 1 & \text{if } \lceil \frac{n}{m} \rceil \text{ is even and } m \nmid n. \end{cases}$$

As ψ is explicitly known, it is possible to calculate it and prove the above. However, we omit the rest of the proof due to space constraint. □

2.4 The Proofs

Finally we are ready to conclude the proofs.

Proof of Theorem 1. The proof follows directly from Lemmas 6 and 8. □

Proof of Theorem 2. Notice that the proof of the upper bound for Theorem 1 is given by prescribing an algorithm (implicitly). The algorithm requires ordering the vertices of the input graph, and then providing the coloring based on the ordering. Each step runs in linear order of the number of vertices in the input graph. Moreover, we have theoretically proved the tightness of the upper bound. Thus, we are done. □

For the full version of the paper, please go to https://homepages.iitdh.ac.in/~sen/Supraja_IWOCA.pdf.

Acknowledgements. This work is partially supported by the following projects: "MA/IFCAM/18/39", "SRG/2020/001575", "MTR/2021/000858", and "NBHM/RP-8 (2020)/Fresh". Research by the first author is partially sponsored by a public grant overseen by the French National Research Agency as part of the "Investissements d'Avenir" through the IMobS3 Laboratory of Excellence (ANR-10-LABX-0016) and the IDEX-ISITE initiative CAP 20–25 (ANR-16-IDEX-0001).

References

1. Chartrand, G., Erwin, D., Harary, F., Zhang, P.: Radio labelings of graphs. Bull. Inst. Combin. Appl. **33**, 77–85 (2001)
2. Chartrand, G., Erwin, D., Zhang, P.: A graph labeling problem suggested by FM channel restrictions. Bull. Inst. Combin. Appl. **43**, 43–57 (2005)
3. Chartrand, G., Nebeský, L., Zhang, P.: Radio k-colorings of paths. Discussiones Math. Graph Theory **24**(1), 5–21 (2004)
4. Das, S., Ghosh, S.C., Nandi, S., Sen, S.: A lower bound technique for radio k-coloring. Discret. Math. **340**(5), 855–861 (2017)
5. Gallian, J. A.: A dynamic survey of graph labeling. Electron. J. Comb. **1**(DynamicSurveys), DS6 (2018)

6. Griggs, J.R., Yeh, R.K.: Labelling graphs with a condition at distance 2. SIAM J. Discret. Math. **5**(4), 586–595 (1992)
7. Havet, F., Reed, B., Sereni, J.S.: L(2,1)-labelling of graphs. In: ACM-SIAM Symposium on Discrete Algorithms (SODA 2008), pp. 621–630 (2008)
8. Kchikech, M., Khennoufa, R., Togni, O.: Linear and cyclic radio k-labelings of trees. Discussiones Math. Graph Theory **27**(1), 105–123 (2007)
9. Khennoufa, R., Togni, O.: A note on radio antipodal colourings of paths. Math. Bohem. **130**(3), 277–282 (2005)
10. Kola, S.R., Panigrahi, P.: Nearly antipodal chromatic number $ac'(P_n)$ of the path P_n. Math. Bohem. **134**(1), 77–86 (2009)
11. Liu, D.D.F., Xie, M.: Radio number for square paths. Ars Combin. **90**, 307–319 (2009)
12. Liu, D.D.F., Zhu, X.: Multilevel distance labelings for paths and cycles. SIAM J. Discret. Math. **19**(3), 610–621 (2005)
13. Panigrahi, P.: A survey on radio k-colorings of graphs. AKCE Int. J. Graphs Comb. **6**(1), 161–169 (2009)
14. West, D.B.: Introduction to Graph Theory, 2nd edn. Prentice Hall, Hoboken (2001)

Capacity-Preserving Subgraphs of Directed Flow Networks

Markus Chimani and Max Ilsen(✉)

Theoretical Computer Science, Osnabrück University, Osnabrück, Germany
{markus.chimani,max.ilsen}@uos.de

Abstract. We introduce and discuss the MINIMUM CAPACITY-PRESERVING SUBGRAPH (MCPS) problem: given a directed graph and a retention ratio $\alpha \in (0, 1)$, find the smallest subgraph that, for each pair of vertices (u, v), preserves at least a fraction α of a maximum u-v-flow's value. This problem originates from the practical setting of reducing the power consumption in a computer network: it models turning off as many links as possible while retaining the ability to transmit at least α times the traffic compared to the original network.

First we prove that MCPS is NP-hard already on directed acyclic graphs (DAGs). Our reduction also shows that a closely related problem (which only considers the arguably most complicated core of the problem in the objective function) is NP-hard to approximate within a sublogarithmic factor already on DAGs. In terms of positive results, we present a simple linear time algorithm that solves MCPS optimally on directed series-parallel graphs (DSPs). Further, we introduce the family of laminar series-parallel graphs (LSPs), a generalization of DSPs that also includes cyclic and very dense graphs. Not only are we able to solve MCPS on LSPs in quadratic time, but our approach also yields straightforward quadratic time algorithms for several related problems such as MINIMUM EQUIVALENT DIGRAPH and DIRECTED HAMILTONIAN CYCLE on LSPs.

Keywords: Maximum flow · Minimum equivalent digraph · Series-parallel graphs · Inapproximability

1 Introduction

We present the MINIMUM CAPACITY-PRESERVING SUBGRAPH (MCPS) problem. Interestingly, despite it being very natural, simple to formulate, and practically relevant, there seems to have been virtually no explicit research regarding it. We may motivate the problem by recent developments in Internet usage and routing research: Not only does Internet traffic grow rapidly [22], current Internet usage shows distinct traffic peaks in the evening (when people, e.g., are streaming videos) and lows at night and in the early morning [19]. This has sparked research into the reduction of power consumption in backbone Internet providers (Tier 1) by turning off unused resources [4,23]: One natural way is to turn off as many connections between servers as possible, while still retaining

Supported by the German Research Foundation (DFG) grant CH 897/7-1.

S.-Y. Hsieh et al. (Eds.): IWOCA 2023, LNCS 13889, pp. 160–172, 2023.
https://doi.org/10.1007/978-3-031-34347-6_14

the ability to route the occurring traffic. Typically, one assumes that (a) the original routing network is suitably dimensioned and structured for the traffic demands at peak times, and (b) the traffic demands in the low times are mostly similar to the peak demands but "scaled down" by some ratio.

Graph-theoretically, we are given a directed graph $G = (V, E)$, a capacity function $cap\colon E \to \mathbb{R}^+$ on its edges, and a retention ratio $\alpha \in (0, 1)$. All graphs are simple (i.e., contain no self-loops nor parallel edges). For every pair of vertices $(s, t) \in V^2$, let $c_G(s, t)$ denote the value of a maximum flow (or equivalently, a minimum cut) from s to t in G according to the capacity function cap. Thus, in the following we unambiguously refer to $c_G(s, t)$ as the *capacity* of the vertex pair (s, t) in G, which intuitively represents how much flow can be sent from s to t in G. The lowest capacity among all vertex pairs corresponds to the size $\mathfrak{c}(G)$ of the global minimum cut. One may ask for an edge-wise minimum subgraph $G' = (V, E')$, $E' \subseteq E$, such that $\mathfrak{c}(G') \geq \alpha \cdot \mathfrak{c}(G)$. We call this problem Minimum Global Capacity-Preserving Subgraph (MGCPS):

Observation 1. MGCPS *is NP-hard, both on directed and undirected graphs, already with unit edge capacities.*

Proof. Identifying a Hamiltonian cycle in a directed strongly-connected (or undirected 2-edge-connected) graph G is NP-hard [12]. Consider an optimal MGCPS solution for G with unit edge capacities and $\alpha = {}^1\!/\mathfrak{c}(G)$ $({}^2\!/\mathfrak{c}(G))$: every vertex pair is precisely required to have a capacity of at least $\lceil \alpha \cdot \mathfrak{c}(G) \rceil = 1$ $(\lceil \alpha \cdot \mathfrak{c}(G) \rceil = 2)$. Hence, an α-MGCPS of G must also be strongly-connected (2-edge-connected, respectively) and is a Hamiltonian cycle if and only if one exists in G. □

However, in our practical scenario, MGCPS is not so interesting. Thus, we rather consider the problem where the capacities $c_G(u, v)$ have to be retained for each vertex pair (u, v) individually: In the Minimum Capacity-Preserving Subgraph (MCPS) problem, we are given a directed graph $G = (V, E)$ including edge capacities cap and a retention ratio $\alpha \in (0, 1)$. We ask for a set of edges $E' \subseteq E$ with minimum size $|E'|$ yielding the subgraph $G' = (V, E')$, such that $c_{G'}(s, t) \geq \alpha \cdot c_G(s, t)$ for all $(s, t) \in V^2$. For an MCPS instance (G, α), we will call a vertex pair (s, t) (or edge st) *covered* by an edge set E' if the graph $G' = (V, E')$ satisfies $c_{G'}(s, t) \geq \alpha \cdot c_G(s, t)$. In the following, we discuss the special setting where the capacity function cap assigns 1 to every edge—in this setting, $c_G(s, t)$ equals the maximum number of edge-disjoint s-t-paths in G.

Related Work. Capacity-preserving subgraphs are related to the research field of sparsification. There, given a graph G, one is typically interested in an upper bound on the size of a graph H that preserves some of G's properties up to an error margin ε. Graph H may be a minor of G, a subgraph of G, or a completely new graph on a subset of G's vertices (in which case it is called a vertex sparsifier [15]). Such research does not necessarily yield approximation algorithms w.r.t. minimum sparsifier size as the obtained upper bound may not

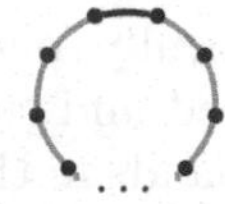

Fig. 1. Two examples of subgraphs in red whose stretch (left: $|V|-1$, right: 2) differs greatly from their retention ratio (left: $\frac{1}{2}$, right: $\frac{1}{|V|-1}$). All edge lengths and edge capacities are 1, showing that using the reciprocals of the edge lengths as edge capacities does not lead to a direct relation between stretch and capacity either.

be easily correlated to the instance-specific smallest possible H (e.g., the general upper bound may be $|E(H)| = \mathcal{O}(\frac{|V|\log|V|}{\varepsilon^2})$ whereas there are instances where an optimal H is a path); however, sparsifiers often can be used as a black box in other approximation algorithms. A majority of cut/flow sparsification research only concerns undirected graphs (for more details, see [5]).

Closely related to sparsifiers are spanners (see, e.g., [1] for a survey on this rich field). These are subgraphs that preserve the length of a shortest path within a given ratio (stretch factor) between each pair of vertices. However, even the most basic results in this line of work cannot be applied to MCPS due to fundamental differences between shortest paths and minimum cuts (as Fig. 1 illustrates).

When the capacity is equal to the number of edge-disjoint paths (i.e. for unit edge capacities), MCPS is a special case of the DIRECTED SURVIVABLE NETWORK DESIGN (DSND) problem, where one asks for the smallest subgraph of a directed graph that satisfies given edge-connectivity requirements for each vertex pair. Dahl [6,7] studied DSND from a polyhedral point of view and presented an ILP approach that can easily be adapted to solve MCPS. But algorithmically, DSND has not received as much attention as its undirected counterpart [13] (for which a 2-approximation algorithm exists [11]).

Lastly, MCPS can be seen as a generalization of the well-established MINIMUM EQUIVALENT DIGRAPH (MED) problem [2,14,21]: Given a directed graph $G = (V, E)$, one asks for a cardinality-wise minimum edge set $E' \subseteq E$ such that the subgraph $G' = (V, E')$ preserves the reachability relation for every pair of vertices. We may think of the MED as a directed version of the MINIMUM SPANNING TREE (despite not being tree-like)—the latter contains an undirected path from each vertex to every other reachable vertex, the former contains a directed one. MED has been shown to be NP-hard via a reduction from DIRECTED HAMILTONIAN CYCLE [10,12,18]. The NP-hardness of MCPS follows from a simple observation:

Observation 2. MED *is the special case of* MCPS *with* $\alpha = \min_{(s,t)\in V^2} {}^1/_{c_G(s,t)}$.

There exist several polynomial approximation algorithms for MED, which are all based on the contraction of cycles [14,24]; the currently best ratio is 1.5 [21]. Moreover, MED can be solved optimally in linear time on graphs whose *shadow*—the underlying undirected graph obtained by ignoring edge orientations—is series-parallel [17], and in quadratic time on DAGs [2]. The latter algorithm simply deletes all those edges uv for which there exists another u-v-path.

Our Contribution. In this paper, we introduce the natural problem MCPS, which we assume may be of wider interest to the algorithmic community.

Based on the fact that MED is simple on DAGs, one might expect to similarly find a polynomial algorithm for MCPS on DAGs as well. However, in Sect. 2 we show that the arguably most complex core of MCPS cannot even be approximated within a sublogarithmic factor, already on DAGs, unless P = NP.

In Sect. 3 we introduce the class of *laminar series-parallel graph (LSPs)*—a generalization of directed series-parallel graphs (DSPs) that also allows, e.g., cycles and dense subgraphs. LSPs have the potential to allow simple algorithms and proofs for a wider array of problems, not just MCPS, due to their structural relation to DSPs. For example, the MCPS-analogue of a well-known spanner property [3] holds on LSPs but not on general graphs: if the retention constraint is satisfied for all edges, it is also satisfied for every vertex pair (see Theorem 11).

In Sect. 4, we complement the hardness result by a linear-time algorithm for MCPS on DSPs, and a quadratic one on LSPs. While the latter algorithm's proof requires the one of the former, both algorithms themselves are independent and surprisingly simple. Lastly, we show a further use of LSPs: our algorithms can directly be applied to other related problems, and we prove that the algorithm for MED on DAGs described in [2] in fact also works on general LSPs (Sect. 4.3).

2 Inapproximability on DAGs

Since a capacity-preserving subgraph always contains an MED (which might be quite large), MCPS can be approximated on sparse graphs by simply returning an arbitrary feasible solution (i.e., a set of edges such that the corresponding subgraph satisfies the capacity requirement for every vertex pair).

Observation 3. *Every feasible solution for a connected* MCPS *instance is an* $m/_{n-1}$*-approximation.*

Proof. Every feasible MCPS solution must contain at least as many edges as an MED to ensure a capacity of 1 for all vertex pairs (u, v) where v is reachable from u. An MED of a connected graph is also connected. Thus, an optimal MCPS solution contains at least $n-1$ of all m original edges. □

Hence, it seems sensible to consider a slightly altered version MCPS* of the MCPS with a tweaked objective function $|E'| - m_{\text{MED}}$, which does not take into account the number of edges m_{MED} in an MED but aims at focusing on the problem's core complexity beyond the MED. We show that it is NP-hard to approximate MCPS* on DAGs to within a sublogarithmic factor using a reduction from the decision variant of SET COVER (SC): given a universe U and a family of sets $\mathcal{S} = \{S_i \subseteq U \mid i = 1, \ldots, k\}$ with $k \in \mathcal{O}(\text{poly}(|U|))$, one asks for a subfamily $\mathcal{C} \subseteq \mathcal{S}$ with minimum size $|\mathcal{C}|$ such that $\bigcup_{S \in \mathcal{C}} S = U$. For an SC instance $(U, \mathcal{S})$, let $f(u) := |\{S \in \mathcal{S} \mid S \ni u\}|$ denote u's *frequency*, i.e., the number of sets that contain u, and $f := \max_{u \in U} f(u)$ the maximum frequency.

Theorem 4. *Any polynomial algorithm can only guarantee an approximation ratio in $\Omega(\log |E|)$ for* MCPS*, *unless P=NP. This already holds on DAGs with maximum path length 4.*

Proof. We give a reduction from SC to MCPS* on DAGs such that any feasible solution for the new MCPS* instance can be transformed into a feasible solution for the original SC instance with an equal or lower objective function value in linear time. The size $|E|$ of our MCPS* instance is linear in the size $N \in \mathcal{O}(|U| \cdot k) = \mathcal{O}(|U|^r)$ of the SC instance, i.e., $|E| = c \cdot |U|^r$ for some constants c, r: if it was possible to approximate MCPS* on DAGs within a factor in $o(\log |E|) = o(\log(c \cdot |U|^r)) = o(\log |U|)$, one could also approximate SC within $o(\log |U|)$. However, it is NP-hard to approximate SC within a factor of $\varepsilon \ln(|U|)$ for any positive $\varepsilon < 1$ [8,16]. To create the MCPS* instance (G, α), let $\alpha := \frac{1}{2}$ and construct G as follows (see Fig. 2 for a visualization of G):

$$G := (V_U \cup V_{\mathcal{S}}^U \cup V_{\mathcal{S}} \cup V_t^{\mathcal{S}} \cup \{t\},\ E_U \cup E_{\mathcal{S}} \cup E_{\mathcal{G}} \cup E_{\mathcal{R}})$$

$$V_U := \{v_u \mid \forall u \in U\} \qquad V_{\mathcal{S}} := \{v_S \mid \forall S \in \mathcal{S}\}$$

$$V_{\mathcal{S}}^U := \{x_S^u, y_S^u \mid \forall S \in \mathcal{S}, u \in S\} \qquad V_t^{\mathcal{S}} := \{z_S \mid \forall S \in \mathcal{S}\}$$

$$E_U := \{v_u x_S^u, v_u y_S^u, x_S^u v_S, y_S^u v_S \mid \forall S \in \mathcal{S}, u \in S\} \qquad E_{\mathcal{S}} := \{v_S z_S, z_S t \mid s \in \mathcal{S}\}$$

$$E_{\mathcal{G}} := V_{\mathcal{S}} \times \{t\} \qquad E_{\mathcal{R}} := V_U \times \{t\}$$

As G is a DAG, its MED is unique [2]. This MED is formed by $E_U \cup E_{\mathcal{S}}$ and already covers all vertex pairs except $V_U \times \{t\}$. Let $(v_u, t) \in V_U \times \{t\}$: Its capacity in G is $2f(u) + 1$ and this pair thus requires a capacity of $\lceil \frac{1}{2} \cdot (2f(u) + 1) \rceil = f(u) + 1$ in the solution. Since the MED already has a v_u-t-capacity of $f(u)$, only one additional edge is needed to satisfy the capacity requirement: either the *corresponding red edge* $v_u t \in E_{\mathcal{R}}$, or one of the *corresponding green edges* $\{v_S t \in E_{\mathcal{G}} \mid S \ni u\}$. After choosing a corresponding red or green edge for each item $u \in U$, the number of these edges is the value of the resulting solution.

Given an SC solution $\mathcal{C}$, we can construct an MCPS* solution $E_U \cup E_{\mathcal{S}} \cup \{v_S t \in E_{\mathcal{G}} \mid S \in \mathcal{C}\}$ with the same value. Since every item is covered by the sets in $\mathcal{C}$, the constructed MCPS* solution includes at least one corresponding green edge for each item, ensuring its feasibility.

To turn a feasible solution E' for the MCPS* instance into a feasible solution for the original SC instance with an equal or lower value, we remove all red edges $v_u t \in E_{\mathcal{R}}$ from the MCPS* solution and replace each of them—if necessary—by one green edge $v_S t \in E_{\mathcal{G}}$ with $S \ni u$. Since the MCPS* solution has at least one corresponding green edge for each item $u \in U$, the resulting SC solution $\{S \mid v_S t \in E_{\mathcal{G}} \cap E'\}$ also contains at least one covering set for each item. □

The same reduction shows the NP-hardness of MCPS on DAGs: an optimal SC solution $\{S \mid v_S t \in E_G \cap E'\}$ can be easily obtained from an optimal solution E' for the MCPS instance (G, α), $\alpha = \frac{1}{2}$. Moreover, the largest capacity between any two vertices in G is $2f + 1$. Since SC is already NP-hard for $f = 2$ (in the form of Vertex Cover [10,12]), we arrive at the following corollary:

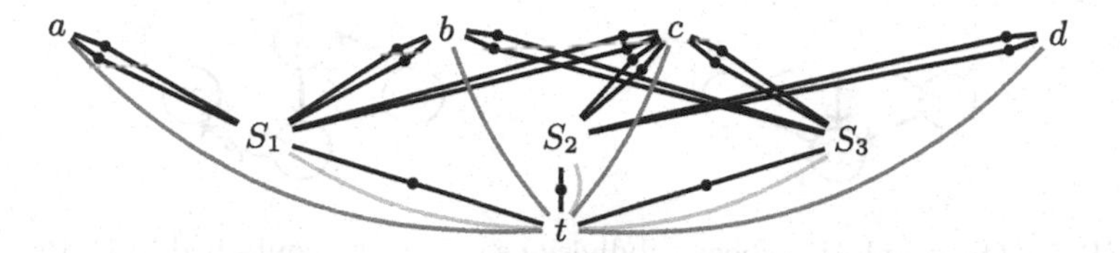

Fig. 2. MCPS instance constructed from the SC instance $(U, \mathcal{S})$ with $U = \{a, b, c, d\}$, $\mathcal{S} = \{\{a, b, c\}, \{c, d\}, \{b, c\}\}$. An optimal solution contains the MED (drawn in black) as well as one corresponding red or green edge for each $u \in U$. Edges are directed from upper to lower vertices. (Color figure online)

Corollary 5. MCPS *is NP-hard already on DAGs G with maximum path length 4 and* $\max_{(u,v)\in V^2} c_G(u, v) = 5$.

The above reduction for MCPS* with $\alpha = \frac{1}{2}$ can be generalized to MCPS* for every $\alpha = \frac{p}{p+1}$ with $p \in \mathbb{N}_{>0}$. This only requires a small change in the construction: E_U must contain $p + 1$ v_u-v_S-paths of length 2 for all $(v_u, v_S) \in V_U \times V_S$, and E_S must contain p v_S-t-paths of length 2 for all $v_S \in V_S$.

3 Laminar Series-Parallel Graphs

In this section, we introduce laminar series-parallel graph (LSPs)—a rich graph family that not only includes the directed series-parallel graph (DSPs) but also cyclic graphs and graphs with multiple sources and sinks. A (directed) graph G is *(directed) s-t-series-parallel (s-t-(D)SP)* if and only if it is a single edge st or there exist two (directed, resp.) s_i-t_i-series-parallel graphs G_i, $i \in \{1, 2\}$, such that G can be created from their disjoint union by one of the following operations [9]:

1. P-composition: Identify s_1 with s_2 and t_1 with t_2. Then, $s = s_1$ and $t = t_1$.
2. S-composition: Identify t_1 with s_2. Then, $s = s_1$ and $t = t_2$.

There also exists a widely known forbidden subgraph characterization of DSPs:

Theorem 6 (see [20]). *A directed graph $G = (V, E)$ is a DSP if and only if it is acyclic, has exactly one source, exactly one sink, and G does not contain a subgraph that is a subdivision of W (displayed in Fig. 3(left)).*

Given a directed graph $G = (V, E)$, for every vertex pair $(u, v) \in V^2$, let $G\langle u, v\rangle$ be the graph induced by the edges on u-v-paths. Note that such a path-induced subgraph may contain cycles but a single path may not. If $e = uv$ is an edge, we call $G\langle u, v\rangle$ an *edge-anchored subgraph (EAS)* and may use the shorthand notation $G\langle e\rangle$. Based on these notions, we can define LSPs:

Definition 7 (Laminar Series-Parallel Graph). *A directed graph* $G = (V, E)$ *is a* laminar series-parallel graph (LSP) *if and only if it satisfies:*

P1. *For every $(s, t) \in V^2$, $G\langle s, t\rangle$ is either an s-t-DSP or contains no edges; and*

Fig. 3. (Left) The graph W, whose subdivisions cannot be contained in DSPs. (Right) Graph W with two added paths of length 2, see Observation 12. The u-v-capacity is 3. For $\alpha = \frac{1}{2}$, all edges of the original graph are covered by the MED (black edges) but the vertex pair (u, v) is not. Observe that the graph is not a DSP but its shadow is s-t-series-parallel.

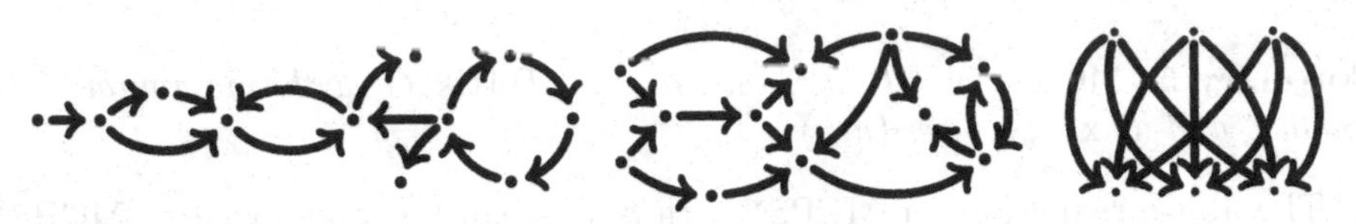

Fig. 4. Examples of LSPs. Every edge represents a DSP of arbitrary size.

P2. $\{E(G\langle e\rangle)\}_{e\in E}$ *form a laminar set family, i.e., for all edges* $e_1, e_2 \in E$ *we have*

$$G\langle e_1\rangle \subseteq G\langle e_2\rangle \;\vee\; G\langle e_2\rangle \subseteq G\langle e_1\rangle \;\vee\; E(G\langle e_1\rangle) \cap E(G\langle e_2\rangle) = \emptyset.$$

Figure 4 shows some example LSPs. LSPs not only include the graphs whose biconnected components are all DSPs but also some cyclic graphs, e.g., *cyclic DSPs* constructed by identifying the source and sink of a DSP. Moreover, there exist very dense LSPs, e.g., the natural orientations of complete bipartite graphs. Below, we present some interesting properties of LSPs; for proofs see [5, App. A].

Theorem 8. *A directed graph* $G = (V, E)$ *satisfies P1 if and only if* G *does not contain a subgraph that is a subdivision of* W *(displayed in Fig. 3(left)).*

Theorem 9. *Every directed graph* G *has a subdivision* $\bar{G}$ *that satisfies P2.*

Theorem 10. *Every DSP* G *is an LSP.*

4 Efficient Algorithms

We first present an algorithm that finds an optimal MCPS solution for DSPs in linear time (Sect. 4.1) and then a quadratic-time algorithm for LSPs (Section 4.2), which can also be applied to several related problems (Sect. 4.3).

4.1 MCPS on Directed Series-Parallel Graphs

Our linear-time algorithm will exploit a useful property of capacities in P1-graphs: if every edge is covered, then *all* vertex pairs are covered.

Theorem 11. *Given a retention ratio* $\alpha \in (0,1)$, *let* $G = (V, E)$ *be a P1-graph and* $G' = (V, E')$, $E' \subseteq E$, *with* $c_{G'}(u,v) \geq \alpha \cdot c_G(u,v)$ *for all edges* $uv \in E$. *Then,* E' *is a feasible* α*-MCPS solution, i.e.,* $c_{G'}(u,v) \geq \alpha \cdot c_G(u,v)$ *for all vertex pairs* $(u,v) \in V^2$.

Proof. Consider any vertex pair $(u,v) \in V^2$ with $u \neq v$ (as vertex pairs with $u = v$ are trivially covered). We can assume w.l.o.g. that a maximum flow from u to v passes only over edges of $H := G\langle u,v\rangle$: for any maximum flow from u to v that uses cycles outside of H, we can find one of the same value in H by simply removing these cycles. As G is a P1-graph, H is a u-v-DSP. We use induction over the series-parallel composition of H to prove that (u,v) is covered. If $uv \in E$, the edge is already covered as asserted in the hypothesis of the theorem; this includes the base case of H containing a single edge.

Let H be created from the disjoint union of two smaller u_i-v_i-DSPs H_i, $i \in \{1,2\}$, for which the theorem holds. Further, let $X' := (V(X), E(X) \cap E(G'))$ for any subgraph $X \in \{H, H_1, H_2\}$ of G. If H is constructed from an S-composition, i.e. $u = u_1$, $v = v_2$, and $v_1 = u_2$, each maximum u-v-flow in H (H') passes through both H_1 and H_2 (H'_1 and H'_2, resp.): $c_{H'}(u,v) = \min\{c_{H'_1}(u,v_1), c_{H'_2}(v_1,v)\} \geq \min\{\alpha \cdot c_{H_1}(u,v_1), \alpha \cdot c_{H_2}(v_1,v)\} = \alpha \cdot c_H(u,v)$. If H is constructed from a P-composition, i.e. $u = u_1 = u_2$ and $v = v_1 = v_2$, its u-v-capacity in H (H') is the sum of u-v-capacities in H_1 and H_2 (H'_1 and H'_2, resp.): $c_{H'}(u,v) = c_{H'_1}(u,v) + c_{H'_2}(u,v) \geq \alpha \cdot c_{H_1}(u,v) + \alpha \cdot c_{H_2}(u,v) = \alpha \cdot c_H(u,v)$. □

Observation 12. *Theorem 11 does not even hold for the smallest graph without property P1 when only two paths of length 2 are added to it, see Fig. 3(right).*

We give a simple linear-time algorithm to solve MCPS in DSPs. The algorithm requires the series-parallel decomposition tree to be *clean*, i.e., if there are multiple P-compositions of several s-t-DSPs $H_0, \ldots, H_k$ where $E(H_0) = \{st\}$ is a single edge, we first compose $H_1, \ldots, H_k$ into a common s-t-DSP H before composing H with H_0. Standard decomposition algorithms can easily achieve this property; the proof below also describes an independent linear-time method to transform a non-clean decomposition tree into a clean one.

Theorem 13. *Algorithm 1 solves* MCPS *on DSPs in* $\mathcal{O}(|V|)$ *time.*

Algorithm 1. Compute an optimal MCPS in DSPs.

Input: DSP $G = (V, E)$, retention ratio $\alpha \in (0,1)$.
1: $E' \leftarrow E$
2: $T \leftarrow$ clean series-parallel decomposition tree for G
3: **for each** tree node σ in a bottom-up traversal of T
4: $\quad (H, (s,t)) \leftarrow$ graph and terminal pair corresponding to σ
5: $\quad$ **if** σ is a P-composition **and** $st \in E(H)$ **and** (s,t) is covered by $(E' \cap E(H)) \setminus \{st\}$
6: $\qquad$ remove st from E'
7: **return** E'

Proof. We use induction over the clean series-parallel decomposition tree T of G, maintaining the following invariants: at the end of each for-loop iteration with $(H, (s, t))$ as the graph and terminal pair for the respective tree node, $E' \cap E(H)$ is an optimal solution for H, and all optimal solutions of H have equal s-t-capacity.

Let σ be a leaf of T: A graph H with a single edge st only allows one feasible (and hence optimal) solution consisting of its only edge. The edge is added to E' during the algorithm's initialization and is not removed from it before σ has been processed. Now observe S-compositions and those P-compositions where no edge st exists in any of the components: They produce no additional paths between the endpoints of any edge (which are the only vertex pairs that have to be covered, see Theorem 11). Thus, the feasible (optimal) solutions of H are exactly those that can be created by taking the union of one feasible (optimal, respectively) solution for each respective component. The algorithm proceeds accordingly by keeping E' unchanged. Since the components' respective optimal solutions all have the same source-sink-capacity (per induction hypothesis), this also holds true for their unions, i.e., the optimal solutions of H.

Now consider a P-composition with $st \in E(H)$. As T is clean, there are two components H_1 and H_2 with $E(H_2) = \{st\}$, and $G\langle s, t\rangle = H$. All edges $e \in H_1$ are covered optimally by $E' \cap E(H_1)$ both in H_1 and in H since $(s, t) \notin E(H\langle e\rangle)$.

Case 1: If one optimal solution for H_1 already covers st in H, then all optimal solutions for H_1 do so (as they all have the same s-t-capacity per induction hypothesis). Then, the optimal solutions for H_1 are exactly the optimal solutions for H, and the algorithm finds one of them by keeping its solution for H_1 intact and removing st from E'. Note that this removal does not affect the feasibility of E' for subgraphs of $G \setminus G\langle s, t\rangle$ that have already been processed.

Case 2: If st is not yet covered by our optimal solution for H_1, it is not covered by any optimal solution for H_1. Our algorithm chooses the edge st by keeping its optimal solutions for both H_1 and H_2. An optimal solution S for H must contain an optimal solution for H_1: $S' := S \setminus \{st\}$ covers all edges of H_1. If S' were not optimal, there would exist another solution S'' that covers all edges and thus vertex pairs of H_1 with $|S''| < |S'|$. But $S''' := S'' \cup \{st\}$ is feasible for H because the capacity requirements for vertex pairs in H and H_1 only differ by at most one. We arrive at $|S'''| = |S''| + 1 < |S'| + 1 \leq |S|$, a contradiction. — In addition to an optimal solution for H_1, we need exactly one more edge to increase the s-t-capacity and cover st in H: this additional edge is either st itself or another edge from H_1. Assume that adding an additional edge $e_1 \in E(H_1)$ (instead of st) increases the capacity for st or a later source-sink-pair by 1, then st by construction does so as well. Thus, adding st instead of e_1 is never worse; furthermore, all optimal solutions for H have the same s-t-capacity.

For the running time, note that a (clean) series-parallel decomposition tree T can be computed and traversed in linear time [20]. If T were not clean, it is trivial to establish this property in linear time: Traverse T bottom-up; whenever a leaf λ is the child of a P-node, ascend the tree as long as the parents are P-nodes. Let ϱ be the final such P-node, and γ the other child of ϱ that was not part of the ascent. Swap λ with γ. Observe that the ascents for different leafs are disjoint, and thus this operation requires overall only $\mathcal{O}(|T|) = \mathcal{O}(|V|)$ time.

In each step during the traversal in line 3, we can compute the capacity of the current source and sink—for both the current solution and G overall—in constant time using the values computed in previous steps: a single edge is assigned a capacity of 1, and an S-composition (P-composition) is assigned the minimum (sum, respectively) of the capacities of its two components. □

4.2 MCPS on Laminar Series-Parallel Graphs

An intuitive approach to solve MCPS on an LSP $G = (V, E)$ is based on the observation that every LSP can be partitioned into a set of edge-disjoint DSPs: Consider the *maximal edge-anchored subgraphs (MEASs)*, i.e., those $G\langle e\rangle$ for $e \in E$ such that there is no other edge $e' \in E \setminus \{e\}$ with $E(G\langle e\rangle) \subseteq E(G\langle e'\rangle)$. Since LSPs are P1-graphs, each of these MEAS must be a DSP, and it suffices to cover its edges (see Theorem 11). Further, the EAS $G\langle e''\rangle$ for each edge $e'' \in E$ is contained in a single MEAS (as LSPs are P2-graphs). Hence, one could identify the MEASs and run Algorithm 1 on each of them to obtain an optimal MCPS solution. We give a more straightforward but functionally equivalent Algorithm 2.

Theorem 14. *Algorithm 2 solves* MCPS *on LSPs in* $\mathcal{O}(|E|^2)$ *time.*

Proof. We prove by induction over μ_e that for each DSP (and hence for each MEAS), Algorithm 2 returns the same result as Algorithm 1: Edges e with $\mu_e = 1$ (i.e., MED-edges) are added to E' by Algorithm 2 in order to cover themselves. Similarly, Algorithm 1 will add such edges during its initialization and never remove them: edge e would only be removed if e connected the source and sink of a subgraph constructed with a P-composition, a contradiction to $\mu_e = 1$.

Now assume that the edges with a μ-value smaller than i for some $i > 1$ are already processed equivalently to Algorithm 1. Consider any edge $e = uv$ with $\mu_e = i$. Since G is a P1-graph, $H := G\langle e\rangle$ is a u-v-DSP. As $e \in E(H)$, H can be constructed with a P-composition from two graphs H_1 and H_2 where $E(H_2) = \{e\}$. All edges in H_1 have already been processed (they have a μ-value smaller than i), and the solutions of Algorithm 2 and Algorithm 1 thus coincide on H_1. Hence, both algorithms produce the same solution for H as they both contain e if and only if e is not already covered by the current solution for H_1.

Algorithm 2. Compute an optimal MCPS in LSPs.

Input: LSP $G = (V, E)$, retention ratio $\alpha \in (0, 1)$.
1: $E' \leftarrow \emptyset$
2: **for each** edge $e \in E$: $\mu_e \leftarrow$ number of edges in $G\langle e\rangle$
3: sort all edges $e \in E$ by non-descending μ_e
4: **for each** edge $e = uv \in E$ in order: **if** uv is not covered by E', add e to E'
5: **return** E'

For each MEAS, Algorithm 1 and, as we have now shown, Algorithm 2 both find the smallest subgraph that covers all of its edges. As LSPs are P1-graphs, this suffices to guarantee an optimal MCPS solution for the MEAS by Theorem 11. Further, we can consider the different MEASs in isolation since their solutions are independent of each other: LSPs are P2-graphs, and thus, for any edge $uv \in E(G)$, all u-v-paths are completely contained in a single MEAS.

It remains to argue the running time. For each $e = uv \in E$, we compute μ_e by starting a depth-first search at u and counting tree- and cross-edges when backtracking from v. Overall, this results in $\mathcal{O}(|E|^2)$ time. The sorting of the edges can be done in $\mathcal{O}(|E|)$ time as the domain are integer values between 1 and $|E|$. Lastly, to check whether an edge uv is covered, we precompute the s-t-capacity for every edge $st \in E$ on $G\langle s,t\rangle$ and then, when needed, compute the u-v-capacity on the graph $G[E' \cap E(G\langle u,v\rangle)]$ for the current solution E'. Note that both of these subgraphs are DSPs as G is a P1-graph. This allows us to compute a series-parallel decomposition tree in $\mathcal{O}(|E|)$ time and traverse it bottom-up to obtain the capacity (cf. the proof of Theorem 13). Doing so twice for every edge takes $\mathcal{O}(|E|^2)$ time overall. □

4.3 Applications of Algorithm 2 to Other Problems

Consider the MINIMUM STRONGLY CONNECTED SUBGRAPH (MSCS) problem [14,21], the special case of MED on strongly connected graphs, i.e., graphs where every vertex is reachable from every other vertex. Since there are straightforward reductions from DIRECTED HAMILTONIAN CYCLE to MSCS to MED to MCPS that all use the original input graph of the instance, Algorithm 2 can be adapted to solve these problems as well: To solve MED, one simply has to set $\alpha = \min_{(s,t)\in V^2} {}^1/_{c_G(s,t)}$ and then run the algorithm on the input graph. However, with α set this way, Algorithm 2 does precisely the same as the algorithm for finding the MED on DAGs [2]: it returns all those edges for which there is only one path between their endpoints (namely the edge itself). Hence, our new insight with regards to the MED is that the aforementioned approach does not only solve MED optimally on DAGs, but on arbitrary LSPs as well. Moreover, if the input graph is strongly connected (Hamiltonian), the returned MED forms the MSCS (directed Hamiltonian cycle, respectively).

Corollary 15. *There are quadratic time algorithms that return optimal solutions for* DIRECTED HAMILTONIAN CYCLE, MINIMUM STRONGLY CONNECTED SUBGRAPH *and* MINIMUM EQUIVALENT DIGRAPH *on any LSP.*

5 Conclusion and Open Questions

We have laid the groundwork for research into capacity-preserving subgraphs by not only showing the NP-hardness of MCPS on DAGs but also presenting a first inapproximability result as well as two algorithmically surprisingly simple algorithms for MCPS on DSPs and LSPs. Several questions remain, for example:

Is MCPS on undirected graphs (which is a generalization of MINIMUM SPANNING TREE) NP-hard? Is it NP-hard to approximate MCPS within a sublogarithmic factor? Is there a linear-time algorithm for MCPS on LSPs? Moreover, one may investigate MCPS with non-unit edge capacities, or other problems on LSPs.

References

1. Ahmed, A.R., et al.: Graph spanners: a tutorial review. Comput. Sci. Rev. **37**, 100253 (2020)
2. Aho, A.V., Garey, M.R., Ullman, J.D.: The transitive reduction of a directed graph. SIAM J. Comput. **1**(2), 131–137 (1972)
3. Althöfer, I., Das, G., Dobkin, D., Joseph, D., Soares, J.: On sparse spanners of weighted graphs. Discrete Comput. Geom. **9**(1), 81–100 (1993)
4. Chiaraviglio, L., Mellia, M., Neri, F.: Reducing power consumption in backbone networks. In: Proceedings of ICC 2009, pp. 1–6. IEEE (2009)
5. Chimani, M., Ilsen, M.: Capacity-preserving subgraphs of directed flow networks (2023). extended version of this paper including appendix: arXiv:2303.17274 [cs.DS]
6. Dahl, G.: The design of survivable directed networks. Telecommun. Syst. **2**(1), 349–377 (1993)
7. Dahl, G.: Directed Steiner problems with connectivity constraints. Discret. Appl. Math. **47**(2), 109–128 (1993)
8. Dinur, I., Steurer, D.: Analytical approach to parallel repetition. In: Proceedings of STOC 2014, pp. 624–633. ACM (2014)
9. Duffin, R.J.: Topology of series-parallel networks. J. Math. Anal. Appl. **10**(2), 303–318 (1965)
10. Garey, M.R., Johnson, D.S.: Computers and Intractability: A Guide to the Theory of NP-Completeness. W. H. Freeman (1979)
11. Jain, K.: A factor 2 approximation algorithm for the generalized Steiner network problem. Combinatorica **21**(1), 39–60 (2001)
12. Karp, R.M.: Reducibility among combinatorial problems. In: Proceedings of COCO 1972. The IBM Research Symposia Series, pp. 85–103. Plenum Press, New York (1972)
13. Kerivin, H., Mahjoub, A.R.: Design of survivable networks: a survey. Networks **46**(1), 1–21 (2005)
14. Khuller, S., Raghavachari, B., Young, N.E.: Approximating the minimum equivalent digraph. SIAM J. Comput. **24**(4), 859–872 (1995)
15. Moitra, A.: Approximation algorithms for multicommodity-type problems with guarantees independent of the graph size. In: Proceedings of FOCS 2009, pp. 3–12. IEEE Computer Society (2009)
16. Moshkovitz, D.: The projection games conjecture and the np-hardness of ln n-approximating set-cover. Theory Comput. **11**, 221–235 (2015)
17. Richey, M.B., Parker, R.G., Rardin, R.L.: An efficiently solvable case of the minimum weight equivalent subgraph problem. Networks **15**(2), 217–228 (1985)
18. Sahni, S.: Computationally related problems. SIAM J. Comput. **3**(4), 262–279 (1974)
19. Schüller, T., Aschenbruck, N., Chimani, M., Horneffer, M., Schnitter, S.: Traffic engineering using segment routing and considering requirements of a carrier IP network. IEEE/ACM Trans. Netw. **26**(4), 1851–1864 (2018)

20. Valdes, J., Tarjan, R.E., Lawler, E.L.: The recognition of series parallel digraphs. SIAM J. Comput. **11**(2), 298–313 (1982)
21. Vetta, A.: Approximating the minimum strongly connected subgraph via a matching lower bound. In: Proceedings of SODA 2001, pp. 417–426. ACM/SIAM (2001)
22. Wong, R.T.: Telecommunications network design: technology impacts and future directions. Networks **77**(2), 205–224 (2021)
23. Zhang, M., Yi, C., Liu, B., Zhang, B.: GreenTE: power-aware traffic engineering. In: Proceedings of ICNP 2010, pp. 21–30. IEEE Computer Society (2010)
24. Zhao, L., Nagamochi, H., Ibaraki, T.: A linear time 5/3-approximation for the minimum strongly-connected spanning subgraph problem. Inf. Process. Lett. **86**(2), 63–70 (2003)

Timeline Cover in Temporal Graphs: Exact and Approximation Algorithms

Riccardo Dondi[1(✉)] and Alexandru Popa[2]

[1] Università degli Studi di Bergamo, Bergamo, Italy
riccardo.dondi@unibg.it

[2] Department of Computer Science, University of Bucharest, Bucharest, Romania
alexandru.popa@fmi.unibuc.ro

Abstract. In this paper we study a variant of vertex cover on temporal graphs that has been recently introduced for timeline activities summarization in social networks. The problem has been proved to be NP-hard, even in restricted cases. In this paper, we present algorithmic contributions for the problem. First, we present an approximation algorithm of factor $O(T \log n)$, on a temporal graph of T timestamps and n vertices. Then, we consider the restriction where at most one temporal edge is defined in each timestamp. For this restriction, which has been recently shown to be NP-hard, we present a $4(T-1)$ approximation algorithm and a parameterized algorithm when the parameter is the cost (called span) of the solution.

1 Introduction

Novel representations of entity interactions have been considered in network science and graph literature, in order to take into account their dynamics and heterogeneity. This has led to the definition of novel graph models, a notable example being *temporal networks* [14,15,19]. Temporal networks or temporal graphs represent how interactions (or edges) evolve in a discrete time domain for a given set of entities (or vertices) [14,17]. The time domain consists of a sequence of timestamps and, for each timestamp of the considered time domain, a temporal graph defines a static graph (also called snapshot) on the same vertex set. Thus a temporal graph can be seen as sequence of static graphs, one for each timestamp, over the same vertex set, while the edge sets can change from one timestamp to the other. In particular, an edge observed in a timestamp, is called a *temporal edge*.

The main aspects of temporal graphs considered in the literature have been finding paths and studying their connectivity [2,7,10,17,18,22–24], but other problems have been studied, for example dense subgraph discovery [9,20]. A fundamental problem in computer science that has been recently considered for temporal graphs is Vertex Cover [3,21]. In this contribution we will consider a

This work was supported by a grant of the Ministry of Research, Innovation and Digitization, CNCS-UEFISCDI, project PN-III-P1-1.1-TE-2021-0253, within PNCDI III.

S.-Y. Hsieh et al. (Eds.): IWOCA 2023, LNCS 13889, pp. 173–184, 2023.
https://doi.org/10.1007/978-3-031-34347-6_15

variant of Vertex Cover on temporal graphs that has been introduced for interaction timeline summarization, a problem called MinTCover [21]. The problem considers a sequence of observed interactions between entities (for example, users of a social platform) and aims to explain these interactions with intervals of entity activities that, following a parsimony approach, have minimum length. From a graph theory perspective, the MinTCover problem is defined over a temporal graph, whose vertices represents entities and whose temporal edges represent interactions between entities. MinTCover then asks, for each vertex, for the definition of a temporal interval where the vertex is considered to be active such that each temporal edge e is covered, that is at least one of its endpoints has an interval activity that includes the timestamp where e is present. The length of an interval activity of a vertex is called the *span* of the vertex. The objective function of MinTCover asks to minimize the sum of the vertex spans.

MinTCover is NP-hard [21], also for some restrictions: when each timestamp contains at most one temporal edge [8], a restriction we denote by 1-MinTCover, when each vertex has degree at most two in each timestamp and the temporal graph is defined over three timestamps [8], and when the temporal graph is defined over two timestamps [11].

MinTCover has been considered also in the parameterized complexity framework. MinTCover admits a parameterized algorithm when parameterized by the span of a solution, when the temporal graph is defined over two timestamps [11].

Two results given in [11] can be applied also for the approximation complexity of MinTCover. A lower bound on the approximability of the problem can be proved by observing that the reduction from the Odd Cycle Transversal problem[1] to MinTCover on two timestamps presented in [11] is indeed an approximation preserving reduction. Since assuming the Unique Games conjecture Odd Cycle Transversal is known to be not approximable within constant factor [5], the following result holds.

Theorem 1 [11]. MinTCover *cannot be approximated within constant factor, assuming the Unique Games Conjecture, even on temporal graphs defined over two timestamps.*

The authors of [11] give a parameterized algorithm, when the problem is parameterized by the span of a solution and the temporal graph is defined over two timestamps. This parameterized algorithm is based on a parameterized reduction from MinTCover on a time domain of two timestamps to the Almost 2-SAT problem[2]. This reduction can be easily modified so that it is an approximation preserving reduction from MinTCover on two timestamps to Almost 2-SAT. Since Almost 2-SAT is approximable within factor $O(\sqrt{\log n})$, for a graph having n vertices [1], the following result holds.

---

[1] We recall that the Odd Cycle Transversal problem, given a graph, asks for the removal of the minimum number of edges such that the resulting graph is bipartite.

[2] We recall that the Almost 2-SAT, given a formula consisting of clauses on two literals, asks for the removal of the minimum number of clauses so that the resulting formula is satisfiable.

Theorem 2 [11]. MinTCover *on two timestamps can be approximated within factor* $O(\sqrt{\log n})$, *for a temporal graph having* n *vertices.*

Related Works. In [21] different variants of the MinTCover problem have been introduced, considering the cases that vertex activity is defined in one interval (as for MinTCover) or in more than one interval (but in a bounded number), and that the objective function asks for the minimization of (1) the sum of vertex spans (as for MinTCover) or (2) the maximum activity span of a vertex.

The computational complexity of the variants of the MinTCover problem has been analyzed recently. Unlike MinTCover, when the vertex activity is defined in one interval and the objective function is the minimization of the maximum activity span of a vertex, the problem admits a polynomial-time algorithm [21]. The variants where the vertex activity is defined as $k \geq 2$ intervals are NP-hard for both objective functions considered [21] and they also do not admit any polynomial-time approximation algorithm, as deciding whether there exists a solution of cost equal to 0 is already an NP-complete problem [21].

In [11] the parameterized complexity of the variants of MinTCover has been explored, considering as parameters the number of vertices of the temporal graph, the length of the time domain, the number of intervals of vertex activity and the span of a solution.

Two other variants of Vertex Cover in temporal graphs have been considered in [3,12]. A first variant, given a temporal graph, asks for the minimum number of pairs (u, t), where u is a vertex and t is a timestamp, such that each non-temporal edge $e = \{a, b\}$ is temporally covered, that is there exists a timestamp t where e is present and at least one of (a, t) and (b, t) belongs to the cover. A second variant asks for each temporal edge to be temporally covered at least once for every interval of a given length Δ.

Our Results. In this paper, we give algorithmic contributions for MinTCover, in particular for its approximability, and for 1-MinTCover (the restriction of MinT-Cover where at most one temporal edge is present in each timestamp, a restriction also known to be NP-hard [8]). First, we present in Sect. 3 a randomized polynomial-time approximation algorithm for MinTCover of factor $O(T \log n)$, for a temporal graph defined over T timestamps and n vertices. Then in Sect. 4 we focus on 1-MinTCover and we provide a parameterized algorithm, where the parameter is the span of the solution, and a polynomial-time approximation algorithm of factor $4(T-1)$. In the next section we introduce the main concepts related to MinTCover and we formally define the MinTCover problem. Due to space restrictions some proofs are placed in the appendix.

2 Preliminaries

We start this section by introducing the definition of discrete time domain over which a temporal graph is defined. A temporal graph is defined over a sequence $\mathcal{T}$ of timestamps between 1 and T, denoted by $[1 \ldots, T]$ (hence in what follows T

denotes the number of timestamp of the temporal graph). An interval $I = [i, j]$, with $1 \leq i \leq j \leq T$, is the sequence of timestamps with value between i and j.

We present now the definition of temporal graph, where recall that the vertex set is not changing in the time domain.

Definition 1. *A temporal graph $G = (V, E, \mathcal{T})$ consists of a set V of n vertices, a time domain $\mathcal{T} = [1, 2, \ldots, T]$, a set $E \subseteq V \times V \times \mathcal{T}$ of m temporal edges, where a temporal edge of G is a triple $\{u, v, t\}$, with $u, v \in V$ and $t \in \mathcal{T}$.*

Given an interval I of $\mathcal{T}$, E_I denotes the set of active edges in the timestamps of I, that is: $E_I = \{\{u, v, t\} | \{u, v, t\} \in E \wedge t \in I\}$. E_t denotes the set of active edges in timestamp t.

Given a vertex $v \in V$, an activity interval of v is defined as an interval $I_v = [l_v, r_v]$ of the time domain where v is considered active, while in any timestamp not in I_v, v is considered inactive. Notice that if $I_v = [l_v, r_v]$ is an activity interval of v, there may exist temporal edges $\{u, v, t\}$, with $t < l_v$ or $t > r_v$ (see the example in Fig. 1). An activity timeline $\mathcal{A}$ is a set of activity intervals, defined as $\mathcal{A} = \{I_v : v \in V\}$.

Given a temporal graph $G = (V, E, \mathcal{T})$, a timeline $\mathcal{A}$ covers $G = (V, E, \mathcal{T})$ if for each temporal edge $\{u, v, t\} \in E$, t belongs to I_u or to I_v, where I_u (I_v, respectively) is the activity interval of u (of v, respectively) defined by $\mathcal{A}$.

The span $s(I_v)$ of an interval $I_v = [l_v, r_v]$, for some $v \in V$, is defined as $s(I_v) = |r_v - l_v|$. Notice that for an interval $I_v = [l_v, r_v]$ consisting of a single timestamp, that is where $l_v = r_v$, it holds that $s(I_v) = 0$. The overall span of an activity timeline $\mathcal{A}$ is equal to $s(\mathcal{A}) = \sum_{I_v \in \mathcal{A}} s(I_v)$.

Now, we are ready to define the problem we are interested into (see the example of Fig. 1).

Problem 1. (MinTCover)
Input: A temporal graph $G = (V, E, \mathcal{T})$.
Output: An activity timeline of minimum span that covers G.

Given a temporal graph $G = (V, E, \mathcal{T})$ and a vertex $v \in V$, the local degree of v in a timestamp t, denoted by $\Delta_L(v, t)$, is the number of temporal edges $\{v, u, t\}$, with $u \in V$. The local degree Δ_L of G is the maximum over v and t of $\Delta_L(v, t)$. The global degree $\Delta(v)$ of a vertex $v \in V$ is the number of temporal edges incident in v in the overall time domain, that is $\Delta(v) = \sum_{t=1}^{T} \Delta_L(v, t)$. The global degree Δ of G is the maximum over v of $\Delta(v)$.

Consider a set $V_1 \subseteq V$ of vertices and an activity timeline $\mathcal{A}_1$ for V_1. Given a set $V_2 \subseteq V$, an activity timeline $\mathcal{A}_2$ for V_2 is in *agreement* with $\mathcal{A}_1$ if, for each $v \in V_1 \cap V_2$, the activity interval of v in $\mathcal{A}_1$ is identical to the activity interval of v in $\mathcal{A}_2$. Furthermore, we define as $\mathcal{A} = \mathcal{A}_1 + \mathcal{A}_2$ the activity timeline of $V_1 \cup V_2$ that is in agreement with both $\mathcal{A}_1$ and $\mathcal{A}_2$.

Given a temporal graph $G = (V, E, \mathcal{T})$, it is possible to check in polynomial time whether there exists a solution of MinTCover on G of span 0.

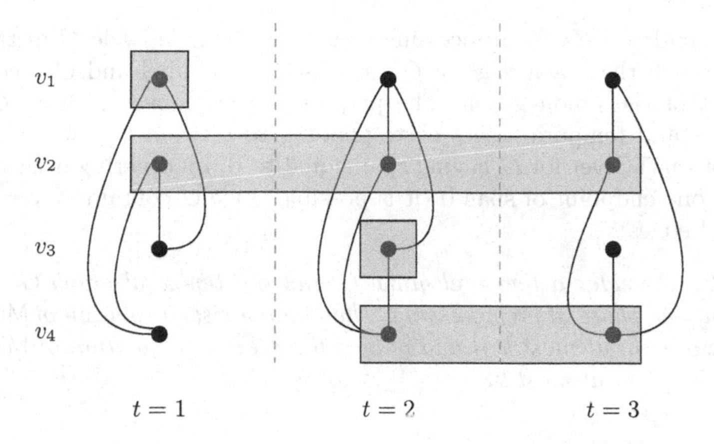

Fig. 1. An example of MinTCover on a temporal graph G consisting of four vertices (v_1, v_2, v_3, v_4) and three timestamps $(1, 2, 3)$. For each timestamp, the temporal edges defined in that timestamp are presented. The activity interval of each vertex is represented in gray. The overall span is equal to 3.

Lemma 1 ([21]). *Let G be an instance of* MinTCover. *We can compute in polynomial time whether there exists a solution of* MinTCover *on G that has span equal to* 0.

Given a temporal graph $G = (V, E, \mathcal{T})$, we can associate a labeled static graph, called *union graph*, where all the temporal edges are represented.

Definition 2. *Given a temporal graph $G = (V, E, \mathcal{T})$, the union graph associated with G is a labeled graph $G_U = (V, E_U, \lambda)$, where $E_U = \{\{u, v\} : \{u, v, t\} \in E$, for some $t \in \mathcal{T}\}$ and $\lambda \subseteq (E_U \times T)$ is a labeling of the edges in E_U defined as $\lambda(\{u, v\}) = \{t : \{u, v, t\} \in E$, for some $t \in \mathcal{T}\}$.*

In the paper we consider a variant of MinTCover, denoted by 1-MinTCover, when there exists at most one active temporal edge in each timestamp.

2.1 Preprocessing a Temporal Graph

We present a preprocessing procedure of a temporal graph and the corresponding union graph that allows to remove some easy to cover parts. The preprocessing consist of two phases. In the first phase, while there exists a vertex u with global degree 1, the preprocessing removes u and the single temporal edge $\{u, v, t\}$ incident in u from G (and G_U); u is defined to be active in t (that is u covers $\{u, v, t\}$) with span 0.

Lemma 2. *Consider a temporal graph G and the temporal graph G' obtained after the first phase of preprocessing. Then there exists a solution of* MinTCover *on G having span at most k if and only if there exists a solution of* MinTCover *on G' having span at most k.*

The second phase of preprocessing considers a simple cycle C in the union graph G_U such that each edge in C has exactly one label and C a connected component of the union graph. The preprocessing removes C from G_U (and the vertices and temporal edges corresponding to C from G) and computes a solution of MinTCover for C having span equal to 0, by covering each temporal edge with one endpoint of span 0 (it is possible since C contains z vertices and z temporal edges).

Lemma 3. *Consider a temporal graph G and the temporal graph G' obtained after the second phase of preprocessing. Then there exists a solution of* MinTCover *on G having span at most k if and only if there exists a solution of* MinTCover *on G' having span at most k.*

3 An Approximation Algorithm for MinTCover

In this section we present an approximation algorithm for MinTCover. First, we recall that by Lemma 1 it is possible to compute in polynomial time whether there exists a solution of MinTCover having span 0, so in what follows we assume that an optimal solution of MinTCover on instance G requires a span greater than 0. In this section we give an $O(T \log n)$ randomized approximation algorithm for the MinTCover problem (recall that n is the number of vertices of G and T is the number of timestamps).

The algorithm consists of two phases: (1) first it computes an approximated solution that covers the temporal edges having at least three occurrences, then (2) it computes a solution of the remaining part of graph.

3.1 Temporal Edges with at Least Three Occurrences

In the first phase, the approximation algorithm considers the subgraph $G'_U = (V, E'_U)$ of the underlying static graph G_U that contains edges representing temporal edges of G with at least three occurrences. E'_U is then defined as follows:

$$E'_U = \{\{u, v\} : \exists \{u, v, t_1\}, \{u, v, t_2\}, \{u, v, t_3\} \in E, \text{ with } 1 \le t_1 < t_2 < t_3 \le T\}.$$

Now, consider $G'_U = (V, E'_U)$ and compute in polynomial time a vertex cover $V'_U \subseteq V$ of G'_U, by applying a factor 2 approximation algorithm for Vertex Cover (for example with the approximation algorithm given in [16]). Define then the activity of vertices in V'_U as follows: $I_v = [1, T]$, for each $v \in V'_U$.

Let $\mathcal{A}_1$ be the activity of vertices in V'_U. We prove the following result on $\mathcal{A}_1$.

Lemma 4. *Consider the set of vertices V'_U and an optimal solution $\mathcal{A}^*$ of* MinTCover *on instance G. Then, it holds that (1) every temporal edge with an endpoint in V'_U is covered by $\mathcal{A}_1$ and (2) it holds that $s(\mathcal{A}_1) \le 2T s(\mathcal{A}^*)$.*

Now, the approximation algorithm removes from G the vertices in V'_U and the temporal edges covered by V'_U (hence all the temporal edges incident in one vertex of V'_U). We present now the second phase of the approximation algorithm.

3.2 Temporal Edges with at Most Two Occurrences

We present an approximation algorithm on graph G, where we assume that each pair of vertices is connected by at most two temporal edges. It follows that the number of temporal edges is bounded by $2\binom{n}{2} \leq n^2$.

The approximation algorithm is based on randomized rounding and it is inspired by the approximation algorithm for Set Cover [13]. First of all, we present an ILP formulation to model the following variant of the problem, called Minimum Non-Consecutive Timeline Cover (Min-NC-TCover), where: (1) each vertex can be active in several non necessarily consecutive timestamps and (2) if each vertex is active in x timestamps, $1 \leq x \leq |T|$, it has a span of $x - 1$, hence it contributes $x - 1$ to the objective function. Notice that since Min-NC-TCover is less restrictive than MinTCover, the optimum of Min-NC-TCover on a temporal graph G is not greater than the optimum of MinTCover on the same instance, since any solution of MinTCover on G is also a solution of Min-NC-TCover on the same instance. Furthermore, notice that Min-NC-TCover is an NP-hard problem, since on a temporal graph defined on two timestamps the two problems are identical and MinTCover is known to be NP-hard in this case [11].

We use a randomized rounding algorithm to find an $O(\log n)$ approximation solution to the Min-NC-TCover problem. Then, we transform the solution for the Min-NC-TCover into a solution for the MinTCover problem increasing the cost within a factor of at most T. Thus, we obtain a $O(T \log n)$ approximation for the MinTCover problem. The integer program formulation of the Min-NC-TCover problem is presented in Fig. 2.

$$\text{minimize} \sum_{v \in V} \left(\sum_{t=1}^{T} x_v^t - 1 \right) \quad (1)$$

$$\text{subject to} \sum_{t=1}^{T} x_v^t \geq 1 \qquad \forall v \in V \quad (2)$$

$$x_v^t + x_u^t \geq 1 \qquad \forall \{u, v, t\} \in E \quad (3)$$

$$x_v^t \in \{0, 1\} \qquad \forall v \in V,\ \forall t \in \{1, 2, \ldots, T\} \quad (4)$$

Fig. 2. ILP formulation for the timeline cover problem for the Min-NC-TCover problem

The $O(T \log n)$ approximation for the MinTCover problem is presented in Algorithm 1. We prove now the correctness and the approximation ratio of Algorithm 1.

Lemma 5. *Algorithm 1 is a $O(T \log n)$ approximation algorithm for the* MinTCover *problem on a temporal graph where each pair of vertices is connected by at most two temporal edges.*

Algorithm 1. $O(T \log n)$ approximation algorithm for the MinTCover problem.

1. Solve the LP relaxation of the ILP formulation from Fig. 2, where we relax constraint (4) to be $0 \leq x_v^t \leq 1$.
2. Define $c = 4n^2$.
3. For every variable x_v^t, define a boolean variable X_v^t, initialized to 0.
4. Repeat $\log_4 c$ times the following point (point 5)
5. For every variable x_v^t, assign X_v^t to 1 with probability x_v^t.
6. For every vertex v such that there exist at least two variables $X_v^t = X_v^t = 1$, let t_{min} be the smallest t such that $X_v^t = 1$ and t_{max} be the maximum t such that $X_v^t = 1$. We make the vertex v active in interval $[t_{min}, t_{max}]$.

We can prove now that the overall algorithm has an approximation factor $O(T \log n)$.

Theorem 3. MinTCover *can be approximated within factor* $O(T \log n)$ *in polynomial time.*

Proof. Let $\mathcal{A}_1$ be the activity assignment of the first phase of the algorithm and let $\mathcal{A}_2$ be the activity assignment of the second phase of the algorithm. Let $\mathcal{A}_f = \mathcal{A}_1 + \mathcal{A}_1$ be the solution consisting of the assignment of $\mathcal{A}_1$ and $\mathcal{A}_2$. Notice that, since all the vertices defined to be active by $\mathcal{A}_1$ are removed, $\mathcal{A}_f$ is well-defined.

Let $\mathcal{A}^*$ be an optimal solution of MinTCover on instance G, recall that by Lemma 4 it holds that $s(\mathcal{A}_1) \leq 2Ts(\mathcal{A}^*)$, while by Theorem 5 it holds that $s(\mathcal{A}_2) \leq O(T \log n \; s(\mathcal{A}^*)$.

Now, if $s(\mathcal{A}_1) \geq s(\mathcal{A}_2)$, it follows that

$$s(\mathcal{A}_f) = s(\mathcal{A}_1) + s(\mathcal{A}_2) \leq 2\, s(\mathcal{A}_1) \leq 4(T-1)s(\mathcal{A}^*) \leq O(T \log n)s(\mathcal{A}^*).$$

If $s(\mathcal{A}_1) < s(\mathcal{A}_2)$, it follows that

$$s(\mathcal{A}_f) = s(\mathcal{A}_1) + s(\mathcal{A}_2) \leq 2\, s(\mathcal{A}_2) \leq O(Ts(\mathcal{A}^*) \leq O(T \log n)s(\mathcal{A}^*).$$

thus concluding the proof. □

4 Algorithms for 1-MinTCover

In this section we study a variant of the timeline cover problem in which the graph from each timestamp contains a single edge. We give a fixed-parameter algorithm for this variant and we also show an approximation algorithm with approximation factor $4(T-1)$.

4.1 A Fixed-Parameter Algorithm

In this section we give a fixed-parameter algorithm for 1-MinTCover when parameterized by k (the span of the solution). Consider a solution $\mathcal{A}$ of 1-MinTCover on a temporal graph G. We assume that the instance has been preprocessed as described in Subsect. 2.1, hence the union graph G_U does not contain disjoint simple cycles and there is no vertex having global degree one. Moreover, we assume that the union graph G_U is connected, otherwise we can solve 1-MinTCover on each connected component independently.

Denote by $\mathcal{D} \subseteq V$ the set of vertices in G having global degree greater than two (that is those vertices having at least three incident temporal edges). We start by proving the following result on G.

Lemma 6. *Let G be an instance of* 1-MinTCover *that admits a solution of span k. Let $\mathcal{D} \subseteq V$ be the set of vertices in G having global degree greater than two. Then, (1) $|\mathcal{D}| \leq 2k$; (2) There are at most $6k$ temporal edges incident to vertices of $\mathcal{D}$.*

Now, we prove that the union graph G'_U obtained from G_U by removing the vertices in $\mathcal{D}$ consists of disjoint paths.

Lemma 7. *Let G' be the temporal graph obtained by removing the vertices in $\mathcal{D}$. Then, the union graph G'_U associated with G' consists of a set of disjoint paths.*

Based on Lemma 6 and Lemma 7, we present our fixed-parameter algorithm for 1-MinTCover (Algorithm 2). Informally, since $|\mathcal{D}| \leq 2k$ and the edges incident in some vertex of $\mathcal{D}$ are at most $6k$, we can iterate over the possible covers of temporal edges incident to $\mathcal{D}$ and then solve the problem on resulting graph.

Algorithm 2. The FPT Algorithm for the 1-MinTCover problem.

1. For each edge $e = \{u, v, t\}$, with $u \in \mathcal{D}$ or $v \in \mathcal{D}$, decide whether e is covered by u (which is then defined active by the activity timeline $\mathcal{A}$ in timestamp t) or by v (which is then defined active by activity timeline $\mathcal{A}$ in timestamp t);
2. Let G' be the temporal graph obtained by removing the vertices of $\mathcal{D}$: compute in polynomial time an optimal solution $\mathcal{A}'$ of 1-MinTCover on G', where $\mathcal{A}'$ is in agreement with $\mathcal{A}$;
3. return $\mathcal{A} + \mathcal{A}'$.

Algorithm 2 at step 2, starts from an activity timeline $\mathcal{A}$ and computes an optimal solution $\mathcal{A}'$ of 1-MinTCover on G' that is in agreement with $\mathcal{A}$. We describe in the following how $\mathcal{A}'$ is computed.

A timeline $\mathcal{A}'$ of minimum span in agreement with $\mathcal{A}$ is computed independently for each connected component (which is a path) P in G'_U. Notice that $\mathcal{A}$ may have already defined an endpoint of P active in a timestamp. Given a path P in G'_U having endpoints u and v, then the algorithm:

Case 1) If $\mathcal{A}$ defines at most one of u and v to be active in some timestamp t, then $\mathcal{A}'$ is defined to be an activity timeline for P in agreement with $\mathcal{A}$ having span 0.

Case 2) If $\mathcal{A}$ defines u active in timestamp t and v to be active timestamp t', then $\mathcal{A}'$ is computed as follows. For each vertex x of P, the algorithm computes an activity timeline $\mathcal{A}_x$ in agreement with $\mathcal{A}$ so that x covers each temporal edge of P incident in x, while each other vertex of P covers exactly one temporal edge and has a span of 0. Then $\mathcal{A}'$ is defined to be the timeline activity having minimum span among $\mathcal{A}_x$, with x a vertex of P.

The computation on paths requires polynomial time, thus leading to an algorithm of complexity $O^*(2^{6k})$.

Theorem 4. 1-MinTCover *can be solved in* $O^*(2^{6k})$ *time.*

4.2 Approximation Algorithm

In this subsection we present a polynomial time $4(T-1)$-approximation algorithm for the 1-MinTCover problem. We start by showing that starting from a union graph G_U where edges can have multiple labels, we can compute a corresponding simple union graph G'_U where each edge has a single label and such that G_U has a feedback vertex set of size k if and only if G'_U has a feedback vertex set of size k. Consider G_U, then compute G'_U by subdividing each edge of G_U with more than one label in a set of edges as follows: for each $\{u,v\} \in E_U$ labeled by t, a new vertex $e_{u,v,t}$ is added in G'_U connected to u and v.

Lemma 8. *Consider a multigraph* G_U*, then compute* G'_U *by subdividing each multiedge of* G_U*. Then* G_U *has a feedback vertex set of size* k *if and only if* G'_U *has a feedback vertex set of size* k*.*

Algorithm 3. $4(T-1)$-approximation algorithm for 1-MinTCover

---

1. Construct the union graph $G_U = (V, E)$ of G and the corresponding simple union graph G'_U.
2. Compute a 2-approximate feedback vertex set F of G'_U (and G_U) [4,6]
3. Make each vertex $v \in F$ active in the time interval $[1, T]$.
4. Cover the graph $G-F$ using the first reduction rule from Subsect. 2.1, since $G-F$ is acyclic.

The key to the analysis of the algorithm is the following structural lemma that relates the size of the optimal solution to the size of a minimum feedback vertex set.

Lemma 9. *Let* G *be a temporal graph, input of* 1-MinTCover *(each timestamp has precisely one edge). Let* $G_U = (V, E)$ *be the corresponding union graph and* G'_U *be the corresponding simple union graph. If the optimal solution of* 1-MinTCover *on* G *has a span of* k*, then a feedback vertex set of* G_U *and* G'_U *has at most* $2k$ *vertices.*

We analyze now the correctness and the performance of Algorithm 3.

Theorem 5. *Algorithm 3 is a polynomial* $4(T-1)$*-approximation algorithm for the* 1-MinTCover *problem.*

Proof. Algorithm 3 produces a valid cover of G, since F is a feedback vertex set of G'_U and $G'_U \setminus F$ is acyclic and thus, we can iteratively apply the first rule from Subsect. 2.1 and obtain a timeline activity of span 0 for $G - F$.

We analyze now the approximation ration of Algorithm 3. First, observe that the global span of the activity timeline returned by the algorithm is $ALG = 2(|T|-1)|F^*|$, where F^* is a minimum feedback vertex set in the graph G'_U. From Lemma 9 we have that $|F^*| \leq 2 \cdot OPT$, where OPT is the minimum span of any activity timeline on the graph G. Thus, it holds that $ALG = 2(T-1)|F^*| \leq 2(T-1)2OPT \leq 4(T-1)OPT$, concluding the proof. □

5 Conclusion

In this paper we have presented algorithmic contributions on MinTCover and on the 1-MinTCover restriction. There are several interesting open problems related to MinTCover. There is a gap in the approximation complexity of the problem. In particular is it possible to obtain an approximation algorithm whose factor does not depend on T? Another interesting open problem is whether it is possible to obtain a fixed-parameter tractable algorithm with parameter the span of the solution.

Acknowledgement. This work was supported by a grant of the Ministry of Research, Innovation and Digitization, CNCS - UEFISCDI, project number PN-III-P1-1.1-TE-2021-0253, within PNCDI III.

References

1. Agarwal, A., Charikar, M., Makarychev, K., Makarychev, Y.: O(sqrt(log n)) approximation algorithms for min UnCut, min 2CNF deletion, and directed cut problems. In: Gabow, H.N., Fagin, R. (eds.) Proceedings of the 37th Annual ACM Symposium on Theory of Computing, Baltimore, MD, USA, 22–24 May 2005, pp. 573–581. ACM (2005)
2. Akrida, E.C., Mertzios, G.B., Spirakis, P.G., Raptopoulos, C.L.: The temporal explorer who returns to the base. J. Comput. Syst. Sci. **120**, 179–193 (2021)
3. Akrida, E.C., Mertzios, G.B., Spirakis, P.G., Zamaraev, V.: Temporal vertex cover with a sliding time window. J. Comput. Syst. Sci. **107**, 108–123 (2020)
4. Bafna, V., Berman, P., Fujito, T.: A 2-approximation algorithm for the undirected feedback vertex set problem. SIAM J. Discret. Math. **12**(3), 289–297 (1999)
5. Bansal, N., Khot, S.: Optimal long code test with one free bit. In: 50th Annual IEEE Symposium on Foundations of Computer Science, FOCS 2009, 25–27 October 2009, Atlanta, Georgia, USA, pp. 453–462. IEEE Computer Society (2009)
6. Becker, A., Geiger, D.: Optimization of pearl's method of conditioning and greedy-like approximation algorithms for the vertex feedback set problem. Artif. Intell. **83**(1), 167–188 (1996)

7. Bumpus, B.M., Meeks, K.: Edge exploration of temporal graphs. In: Flocchini, P., Moura, L. (eds.) IWOCA 2021. LNCS, vol. 12757, pp. 107–121. Springer, Cham (2021). https://doi.org/10.1007/978-3-030-79987-8_8
8. Dondi, R.: Insights into the complexity of disentangling temporal graphs. In: Lago, U.D., Gorla, D. (eds.) Proceedings of the 23th Italian Conference on Theoretical Computer Science, ICTCS 2022. CEUR-WS.org (2022)
9. Dondi, R., Hosseinzadeh, M.M.: Dense sub-networks discovery in temporal networks. SN Comput. Sci. **2**(3), 158 (2021)
10. Erlebach, T., Hoffmann, M., Kammer, F.: On temporal graph exploration. J. Comput. Syst. Sci. **119**, 1–18 (2021)
11. Froese, V., Kunz, P., Zschoche, P.: Disentangling the computational complexity of network untangling. In: Raedt, L.D. (ed.) Proceedings of the Thirty-First International Joint Conference on Artificial Intelligence, IJCAI 2022, Vienna, Austria, 23–29 July 2022, pp. 2037–2043. ijcai.org (2022)
12. Hamm, T., Klobas, N., Mertzios, G.B., Spirakis, P.G.: The complexity of temporal vertex cover in small-degree graphs. CoRR, abs/2204.04832 (2022)
13. Hochbaum, D.S.: Approximation algorithms for the set covering and vertex cover problems. SIAM J. Comput. **11**(3), 555–556 (1982)
14. Holme, P.: Modern temporal network theory: a colloquium. Eur. Phys. J. B **88**(9), 1–30 (2015). https://doi.org/10.1140/epjb/e2015-60657-4
15. Holme, P., Saramäki, J.: A map of approaches to temporal networks. In: Holme, P., Saramäki, J. (eds.) Temporal Network Theory. CSS, pp. 1–24. Springer, Cham (2019). https://doi.org/10.1007/978-3-030-23495-9_1
16. Karakostas, G.: A better approximation ratio for the vertex cover problem. ACM Trans. Algorithms **5**(4), 41:1–41:8 (2009)
17. Kempe, D., Kleinberg, J.M., Kumar, A.: Connectivity and inference problems for temporal networks. J. Comput. Syst. Sci. **64**(4), 820–842 (2002)
18. Marino, A., Silva, A.: Königsberg sightseeing: Eulerian walks in temporal graphs. In: Flocchini, P., Moura, L. (eds.) IWOCA 2021. LNCS, vol. 12757, pp. 485–500. Springer, Cham (2021). https://doi.org/10.1007/978-3-030-79987-8_34
19. Michail, O.: An introduction to temporal graphs: an algorithmic perspective. Internet Math. **12**(4), 239–280 (2016)
20. Rozenshtein, P., Bonchi, F., Gionis, A., Sozio, M., Tatti, N.: Finding events in temporal networks: segmentation meets densest subgraph discovery. Knowl. Inf. Syst. **62**(4), 1611–1639 (2020)
21. Rozenshtein, P., Tatti, N., Gionis, A.: The network-untangling problem: from interactions to activity timelines. Data Min. Knowl. Discov. **35**(1), 213–247 (2021)
22. Wu, H., Cheng, J., Huang, S., Ke, Y., Lu, Y., Xu, Y.: Path problems in temporal graphs. Proc. VLDB Endow. **7**(9), 721–732 (2014)
23. Wu, H., Cheng, J., Ke, Y., Huang, S., Huang, Y., Wu, H.: Efficient algorithms for temporal path computation. IEEE Trans. Knowl. Data Eng. **28**(11), 2927–2942 (2016)
24. Zschoche, P., Fluschnik, T., Molter, H., Niedermeier, R.: The complexity of finding small separators in temporal graphs. J. Comput. Syst. Sci. **107**, 72–92 (2020)

Finding Small Complete Subgraphs Efficiently

Adrian Dumitrescu[1(✉)] and Andrzej Lingas[2]

[1] Algoresearch L.L.C., Milwaukee, WI 53217, USA
ad.dumitrescu@algoresearch.org

[2] Department of Computer Science, Lund University, Box 118, 22100 Lund, Sweden
Andrzej.Lingas@cs.lth.se

Abstract. (I) We revisit the algorithmic problem of finding all triangles in a graph $G = (V, E)$ with n vertices and m edges. According to a result of Chiba and Nishizeki (1985), this task can be achieved by a combinatorial algorithm running in $O(m\alpha) = O(m^{3/2})$ time, where $\alpha = \alpha(G)$ is the graph arboricity. We provide a new very simple combinatorial algorithm for finding all triangles in a graph and show that is amenable to the same running time analysis. We derive these worst-case bounds from first principles and with very simple proofs that do not rely on classic results due to Nash-Williams from the 1960s.

(II) We extend our arguments to the problem of finding all small complete subgraphs of a given fixed size. We show that the dependency on m and α in the running time $O(\alpha^{\ell-2} \cdot m)$ of the algorithm of Chiba and Nishizeki for listing all copies of K_ℓ, where $\ell \geq 3$, is asymptotically tight.

(III) We give improved arboricity-sensitive running times for counting and/or detection of copies of K_ℓ, for small $\ell \geq 4$. A key ingredient in our algorithms is, once again, the algorithm of Chiba and Nishizeki. Our new algorithms are faster than all previous algorithms in certain high-range arboricity intervals for every $\ell \geq 7$.

Keywords: triangle · subgraph detection/counting · graph arboricity · rectangular matrix multiplication

1 Introduction

The problem of deciding whether a given graph $G = (V, E)$ contains a complete subgraph on k vertices is among the most natural and easily stated algorithmic graph problems. If the subgraph size k is part of the input, this is the CLIQUE problem which is NP-complete [15]. For every fixed k, determining whether a given graph $G = (V, E)$ contains a complete subgraph on k vertices can be accomplished by a brute-force algorithm running in $O(n^k)$ time.

For $k = 3$, deciding whether a graph contains a triangle and finding one if it does, or counting all triangles in a graph, can be done in $O(n^\omega)$ time by the algorithm of Itai and Rodeh [14], where $\omega < 2.373$ is the exponent of matrix multiplication [1,6]. The algorithm compares M and M^2, where M is the graph adjacency matrix. Alternatively, this task can be done in $O(m^{2\omega/(\omega+1)}) = O(m^{1.408})$

S.-Y. Hsieh et al. (Eds.): IWOCA 2023, LNCS 13889, pp. 185–196, 2023.
https://doi.org/10.1007/978-3-031-34347-6_16

time by the algorithm of Alon, Yuster, and Zwick [2]. Itai and Rodeh [14] and also Papadimitriou and Yannakakis [26] as well as Chiba and Nishizeki [5] showed that triangles in planar graphs can be found in $O(n)$ time.

For $k = 4$, deciding whether a graph contains a K_4 and finding one if it does (or counting all K_4's in a graph) can be done in $O(n^{\omega+1}) = O(n^{3.373})$ time by the algorithm of Alon, Yuster, and Zwick [2], in $O(n^{3.252})$ time by the algorithm of Eisenbrand and Grandoni [10], and in $O(m^{(\omega+1)/2}) = O(m^{1.687})$ time by the algorithm of Kloks, Kratsch, and Müller [16].

In contrast to the problem of detecting the existence of subgraphs of a certain kind, the analogous problem of listing *all* such subgraphs has usually higher complexity, as expected. For example, finding all triangles in a given graph (each triangle appears in the output list) can be accomplished in $O(m^{3/2})$ time and with $O(n^2)$ space by an extended version of the algorithm of Itai and Rodeh [14]. Bar-Yehuda and Even [3] improved the space complexity of the algorithm from $O(n^2)$ to $O(n)$ by avoiding the use of the adjacency matrix. Chiba and Nishizeki [5] further refined the time complexity in terms of graph arboricity (the minimum number of edge-disjoint forests into which its edges can be partitioned); their algorithm lists all triangles in a graph in $O(m\alpha)$ time, where α is the arboricity. They also showed that $\alpha = O(\sqrt{m})$; consequently, the running time is $O(m^{3/2})$. If G is planar, $\alpha(G) \leq 3$ (see [13, p. 124]), so the algorithm runs in $O(m) = O(n)$ (i.e., linear) time on planar graphs. Since there are graphs G with $\alpha(G) = \Theta(m^{1/2})$, this does not improve the worst-case dependence on m (which, in fact, cannot be improved). More general, for every fixed $\ell \geq 3$, the same authors gave an algorithm for listing all copies of K_ℓ in $O(\alpha^{\ell-2} \cdot m) = O(m^{\ell/2})$ time.

We distinguish several variants of the general problem of finding triangles in a given undirected graph $G = (V, E)$: (i) the triangle *detection* (or *finding*) problem is that of finding a triangle in G or reporting that none exists; (ii) the triangle *counting* problem is that of determining the total number of triangles in G; (iii) the triangle *listing* problem is that of listing (reporting) all triangles in G, with each triangle appearing in the output list. Obviously any algorithm for listing all triangles can be easily transformed into one for triangle detection or into one for listing a specified number of triangles (by stopping after the required number of triangles have been output).

Our Results. We obtain the following results for the problem of finding small complete subgraphs of a given size.

(i) We provide a new combinatorial algorithm for finding all triangles in a graph running in $O(m\alpha) = O(m^{3/2})$ time, where $\alpha = \alpha(G)$ is the graph arboricity (Algorithm HYBRID in Sect. 2). We derive these worst-case bounds from first principles and with very simple proofs that do not rely on classic results due to Nash-Williams from the 1960s.

(ii) For every n, $b \leq n/2$ and a fixed $\ell \geq 3$, there exists a graph G of order n with m edges and $\alpha(G) \leq b$ that has $\Omega(\alpha^{\ell-2} \cdot m)$ copies of K_ℓ (Lemma 3 in Sect. 3). As such, the dependency on m and α in the running time $O(\alpha^{\ell-2} \cdot m)$ for listing all copies of K_ℓ, is asymptotically tight.

(iii) We give improved arboricity-sensitive running times for counting and/or detection of copies of K_ℓ, for small $\ell \geq 4$ (Sect. 4). A key ingredient in our algorithms is the algorithm of Chiba and Nishizeki. Our new algorithms beat all previous algorithms in certain high-range arboricity intervals for every $\ell \geq 7$. Up-to-date running times based on rectangular matrix multiplication times are included.

Preliminaries and Notation. Let $G = (V, E)$ be an undirected graph. The *neighborhood* of a vertex $v \in V$ is the set $N(v) = \{w : (v, w) \in E\}$ of all adjacent vertices, and its cardinality $\deg(v) = |N(v)|$ is called the *degree* of v in G. A *clique* in a graph $G = (V, E)$ is a subset $C \subset V$ of vertices, each pair of which is connected by an edge in E. The CLIQUE problem is to find a clique of maximum size in G. An *independent set* of a graph $G = (V, E)$ is a subset $I \subset V$ of vertices such that no two of them are adjacent in G.

Graph Parameters. For a graph G, its *arboricity* $\alpha(G)$ is the minimum number of edge-disjoint forests into which G can be decomposed [5]. For instance, it is known and easy to show that $\alpha(K_n) = \lceil n/2 \rceil$. A characterization of arboricity is provided by the following classic result [22,23,29]; see also [8, p. 60].

Theorem 1. (Nash-Williams 1964; Tutte 1961) *A multigraph $G = (V, E)$ can be partitioned into at most k forests if and only if every set $U \subseteq V$ induces at most $k(|U| - 1)$ edges.*

The *degeneracy* $d(G)$ of an undirected graph G is the smallest number d for which there exists an *acyclic orientation* of G in which all the out-degrees are at most d. The degeneracy of a graph is linearly related to its arboricity, i.e., $\alpha(G) = \Theta(d(G))$; more precisely $\alpha(G) \leq d(G) \leq 2\alpha(G) - 1$; see [2,11,21,31].

1.1 Triangle Finding Algorithms

To place our algorithm in Sect. 2 in a proper context, we first present a summary of previous work in the area of triangle finding and enumeration.

The Algorithms of Itai and Rodeh (1978). The authors gave three methods for finding triangles. Initially intended for triangle detection, the first algorithm [14] runs in $O(m^{3/2})$ time. It can be extended to list all triangles within the same overall time and works as follows:

ITAI-RODEH(G)

Input: an undirected graph $G = (V, E)$

1 Find a spanning tree for each connected component of G
2 List all triangles containing at least one tree edge
3 Delete the tree edges from G and go to Step 1

The second is a randomized algorithm that checks whether there is an edge contained in a triangle. It runs in $O(mn)$ worst-case time and $O(n^{5/3})$ expected time. The third algorithm relies on Boolean matrix multiplication and runs in $O(n^\omega)$ time.

The Algorithm of Chiba and Nishizeki (1985). The algorithm uses a vertex-iterator approach for listing all triangles in G. It relies on the observation that each triangle containing a vertex v corresponds to an edge joining two neighbors of v. The graph is represented with doubly-linked adjacency lists and mutual references between the two stubs of an edge ensure that each deletion takes constant time. A more compact version described by Ortmann and Brandes [25] is given below.

```
CHIBA-NISHIZEKI(G)
   Input: an undirected graph G = (V, E)
 1 Sort vertices such that deg(v_1) ≥ deg(v_2) ≥ ... deg(v_n)
 2 for u = v_1, v_2, ..., v_{n-2} do
 3     foreach vertex v ∈ N(u) do mark v
 4     foreach vertex v ∈ N(u) do
 5         foreach vertex w ∈ N(v) do
 6             if w is marked then output triangle uvw
 7         end
 8         unmark v
 9     end
10     G ← G − u
11 end
```

The authors [5] showed that their algorithm runs in $O(m\alpha)$ time. As a corollary, the number of triangles is $O(m\alpha)$ as well. The $O(m^{3/2})$ upper bound on the number of triangles in a graph is likely older than these references indicate. In any case, other proofs are worth mentioning [9,17], including algebraic ones [27]. We derive yet another one in Sect. 2.

Corollary 1 ([5,14]). *For any graph G of order n with m edges and arboricity α, G contains $O(m\alpha) = O(m^{3/2})$ triangles.*

Ortmann and Brandes [25] gave a survey of other approaches, including edge-iterators. Algorithms in this category iterate over all edges and intersect the neighborhoods of the endpoints of each edge. A straightforward neighborhood merge requires $O(\deg(u)+\deg(v))$ time per edge uv, but this is not good enough to list all triangles in $O(m^{3/2})$ time. Two variants developed by Shanks [28] use $O(m)$ extra space to represent neighborhoods in hash sets and obtain the intersection in $O(\min(\deg(u), \deg(v)))$ time, which suffices for listing all triangles in $O(m^{3/2})$ time.

The Algorithm of Alon, Yuster and Zwick (1997). The authors showed that deciding whether a graph contains a triangle and finding one if it does (or counting all triangles in a graph) can be done in $O(m^{2\omega/(\omega+1)}) = O(m^{1.408})$ time [2]. The idea is to find separately triangles for which at least one vertex has low degree (for an appropriately set threshold) and triangles whose all three vertices have high degree. Triangles of the latter type are handled using matrix multiplication in a smaller subgraph.

Recent Algorithms. Björklund et al. [4] obtained output-sensitive algorithms for finding (and listing) all triangles in a graph; their algorithms are tailored for dense and sparse graphs. Several approaches [12,18] provide asymptotic improvements by taking advantage of the bit-level parallelism offered by the word-RAM model. If the number of triangles is small, Zechner and Lingas [30] showed how to list all triangles in $O(n^\omega)$ time.

2 A Simple Hybrid Algorithm for Listing All Triangles

In this section we present a new algorithm for listing all triangles. While its general idea is not new, the specific hybrid data structure therein does not appear to have been previously considered. Using both an adjacency list representation and an adjacency matrix representation of the graph allows one to obtain a very time-efficient neighborhood merge (intersection) procedure. Let $V = \{1, 2, \ldots, n\}$ and let M be the adjacency matrix of G. A triangle ijk with $i < j < k$ is reported when edge ij is processed, and so each triangle is reported exactly once. For each edge ij, the algorithm scans the adjacency list of the endpoint of lower degree (among i and j) and for each neighbor it checks for the needed entry in the adjacency matrix M corresponding to the third triangle edge.

HYBRID(G)

Input: an undirected graph $G = (V, E)$
1 **foreach** edge $ij \in E$, $(i < j)$ **do**
2   **if** $\deg(i) \leq \deg(j)$ **then** $x \leftarrow i$, $y \leftarrow j$
3   **else** $x \leftarrow j$, $y \leftarrow i$
4   **foreach** $k \in ADJ(x)$ **do**
5   **if** $j < k$ and $M(y, k) = 1$ **then** report triangle ijk
6   **end**
7 **end**

We subsequently show that the algorithm runs in time $O(m\alpha)$. Whereas the space used is quadratic, the hybrid algorithm appears to win by running time and simplicity. In particular, no additional data structures nor hashing are used, no sorting (by degree or otherwise) and no doubly linked lists are needed, etc.

2.1 The Analysis

Define the following function on edges of G. For $uv = e \in E$, let

$$f(e) = \min(\deg(u), \deg(v)) \text{ and } F(G) = \sum_{e \in E} f(e). \tag{1}$$

There are at most $f(e)$ triangles based on edge $e = ij$ and overall at most $F(G)$ triangles in G. The runtime of the algorithm is proportional to this quantity; the space used is quadratic. A short and elegant decomposition argument by Chiba and Nishizeki [5, Lemma 2] shows that

$$F(G) \leq 2\,m\alpha, \tag{2}$$

thus our algorithm runs in time $O(m\alpha)$. The same analysis applies to the algorithm of Chiba and Nishizeki [5]. By Theorem 1, the above authors deduced that $\alpha(G) \leq \lceil (2m+n)^{1/2}/2 \rceil$, which implies that $\alpha(G) = O(m^{1/2})$ for a connected graph G. As such, both algorithms run in $O(m^{3/2})$ time on any graph.

Lemma 1 below shows how to bypass the Nash-Williams arboricity bound (Theorem 1) and deduce the $O(m^{3/2})$ upper bound for listing all triangles in a graph from first principles.

Lemma 1. *Let G be a graph on n vertices with m edges. Then*

$$F(G) \leq 4\, m^{3/2}.$$

Proof (folklore). There are two types of edges uv:

1. $\min(\deg(u), \deg(v)) \leq 2\sqrt{m}$
2. $\min(\deg(u), \deg(v)) > 2\sqrt{m}$

There are at most m edges of type 1 and each contributes at most $2\sqrt{m}$ to $F(G)$, so the total contribution from edges of this type is at most $2m^{3/2}$.

Each edge of type 2 connects two nodes of degree $> 2\sqrt{m}$, and there are at most $2m/(2\sqrt{m}) = \sqrt{m}$ such nodes. The degree of each of them thus contributes to $F(G)$ at most $\sqrt{m}$ times and the sum of all degrees of such nodes is at most $2m$. It follows that the total contribution from edges of type 2 is at most $2m^{3/2}$.

Overall, we have $F(G) \leq 4m^{3/2}$. □

Remarks. Lemma 1 immediately gives an upper bound of $4m^{3/2}$ on the number of triangles in a graph. It is in fact known [27] that this number is at most $(2^{1/2}/3)\, m^{3/2}$ (and this bound is sharp). More generally, for every fixed $\ell \geq 3$, the number of copies of K_ℓ is $O(\alpha^{\ell-2} \cdot m) = O(m^{\ell/2})$; see [5, Thm. 3].

3 Constructions

Due to page limitations, the proofs in this section are omitted.

Lemma 2. *For every $n \geq 3$ and $3 \leq m \leq \binom{n}{2}$, there exists a graph G of order n with m edges that contains $\Theta(m^{3/2})$ triangles.*

Lemma 3. *Let $\ell \geq 3$ be a fixed integer. For every n and $b \leq n/2$, there exists a graph G of order n with $\alpha \leq b$ that has $\Omega(\alpha^{\ell-2} \cdot m)$ copies of K_ℓ, where m is the number of edges in G and α is the arboricity of G.*

4 Finding Small Complete Subgraphs Efficiently

In this section we address the problem of detecting the presence of K_ℓ for a fixed $\ell \geq 4$. We combine and refine several approaches existent in the literature of the last 40 years to obtain faster algorithms in general and for a large class of graphs with high arboricity. In particular, we will use the algorithm of Chiba and Nishizeki [5] for listing all copies of K_ℓ in $O(\alpha^{\ell-2} \cdot m)$ time. Our algorithms are formulated for the purpose of *counting* but they can be easily adapted for the purpose of *detection.*

Recall that $\omega < 2.373$ is the exponent of matrix multiplication [1,6], namely the infimum of numbers τ such that two $n \times n$ real matrices can be multiplied in $O(n^\tau)$ time (operations). Similarly, let $\omega(p, q, r)$ stand for the infimum of numbers τ such that an $n^p \times n^q$ matrix can be multiplied by an $n^q \times n^r$ matrix in $O(n^\tau)$ time (operations). For simplicity and as customary (see, e.g., [2,24]), we write that two $n \times n$ matrices can be multiplied in $O(n^\omega)$ time, since this does not affect our results; and similarly when multiplying rectangular matrices.

The Extension Method. Let $T(n, m, \ell)$ denote the running time of the algorithm running on a graph with n vertices and m edges. Write $\ell = \ell_1 + \ell_2$ for some $\ell_1, \ell_2 \geq 2$. At the beginning, we run the algorithm of Chiba and Nishizeki to form a list of subgraphs isomorphic to K_{ℓ_1}, and then for each subgraph $G_1 = (V_1, E_1)$ on the list, (i) we construct the subgraph G_2 of G induced by vertices in $V \setminus V_1$ that are adjacent to all vertices in V_1. (ii) we count (or find a) the subgraphs isomorphic to K_{ℓ_2} in G_2 (this is a *recursive* call on a smaller instance); and so the algorithm can be viewed as recursive. In other words, we count the number of extensions of the subgraph isomorphic to K_{ℓ_1} to a K_ℓ. Another formulation of this method can be found in [19]. There are $O(\alpha^{\ell_1-2} \cdot m)$ copies of K_{ℓ_1} that can be found in $O(\alpha^{\ell_1-2} \cdot m)$ time. For each fixed copy of K_{ℓ_1}, the time spent in G_2 is at most $T(n, m, \ell_2)$, and so the overall time satisfies the recurrence

$$T(n, m, \ell) = O(\alpha^{\ell_1-2} \cdot m \cdot T(n, m, \ell_2)).$$

Each copy of K_ℓ is generated exactly $\binom{\ell}{\ell_1}$ times and so the total count needs to be divided by this number in the end.

The Triangle Method. Nešetřil and Poljak [24] showed an efficient reduction of detecting and counting copies of any complete subgraph to the aforementioned method of Itai and Rodeh [14] for triangle detection and counting. To start with, consider the detection of complete subgraphs of size $\ell = 3j$. For a given graph G with n vertices, construct an auxiliary graph H with $O(n^j)$ vertices, where each vertex of H is a complete subgraph of order j in G. Two vertices V_1, V_2 in H are connected by an edge if $V_1 \cap V_2 = \emptyset$ and all edges in $V_1 \times V_2$ are present in G. The detection (or counting) of triangles in H yields an algorithm for the detection (or counting) of K_ℓ's in G, running in $O(n^{j\omega})$ time. For detecting complete subgraphs of size $\ell = 3j + i$, where $i \in \{1, 2\}$, the algorithm can be adapted so that it runs in $O(n^{j\omega+i})$ time.

For convenience, define the following integer functions

$$\beta(\ell) = \omega(\lfloor \ell/3 \rfloor, \lceil (\ell-1)/3 \rceil, \lceil \ell/3 \rceil), \quad (3)$$

$$\gamma(\ell) = \lfloor \ell/3 \rfloor \omega + \ell \pmod 3. \quad (4)$$

With this notation, the algorithm runs in $O(n^{\gamma(\ell)})$ time. Two decades later, Eisenbrand and Grandoni [10] refined the triangle method by using fast algorithms for rectangular matrix multiplication instead of those for square matrix multiplication. It partitions the graph intro three parts roughly the same size: $\ell_1 = \lfloor \ell/3 \rfloor$, $\ell_2 = \lceil (\ell-1)/3 \rceil$, $\ell_3 = \lceil \ell/3 \rceil$). The refined algorithm runs in time $O(n^{\beta(\ell)})$ time. If the rectangular matrix multiplication is carried out via the straightforward partition into square blocks and fast square matrix multiplication (see, e.g., [7, Exercise 4.2-6]), one recovers the time complexity of the algorithm of Nešetřil and Poljak; that is, $\beta(\ell) \leq \gamma(\ell)$, see [10] for details. Eisenbrand and Grandoni [10] showed that the above inequality is strict for a certain range: if $\ell \pmod 3 = 1$ and $\ell \leq 16$, or $\ell \pmod 3 = 2$. In summary, their algorithm is faster than that of Nešetřil and Poljak in these cases. Another refinement of the triangle method is considered in [19].

A Problem of Kloks, Kratsch, and Müller. The authors asked [16] whether there is an $O(m^{\ell\omega/6})$ algorithm for recognizing whether a graph contains a K_ℓ, if ℓ is a multiple of 3. Eisenbrand and Grandoni showed that this true for every multiple of 3 at least 6. Indeed, by their Theorem 2 [10], this task can be accomplished in time $O(m^{\beta(\ell)/2})$ for every $\ell \geq 6$, with $\beta(\ell)$ as in (3). If $\ell = 3j$, where $j \geq 2$, then

$$\beta(\ell) = \omega(\lfloor \ell/3 \rfloor, \lceil (\ell-1)/3 \rceil, \lceil \ell/3 \rceil) = \omega(j,j,j) = j \cdot \omega(1,1,1) = j\omega.$$

It follows that the running time is $O(m^{\beta(\ell)/2}) = O(m^{j\omega/2}) = O(m^{\ell\omega/6})$. The proof of the theorem is rather involved. Here we provide an alternative simpler argument that also yields an arboricity-sensitive bound (item (i) below).

General Case Derivations. We first consider the general cases: $\ell = 3j$, $j \geq 3$, and $\ell = 3j+1$, $j \geq 2$; and $\ell = 3j+2$, $j \geq 2$. It will be evident that our algorithms provide improved bounds for every $\ell \geq 7$. For a given $j \geq 2$, consider the interval

$$I_j = \left(\frac{\omega - 1}{j(3-\omega) + 2(\omega - 1)}, \frac{1}{2} \right).$$

(i) $\ell = 3j$, $j \geq 3$. We use the triangle method with a refined calculation. The vertices of the auxiliary graph H are subgraphs isomorphic to K_j. By the result of Chiba and Nishizeki [5], H has $O(\alpha^{j-2} \cdot m)$ vertices. The algorithm counts triangles in H in time proportional to

$$\left(\alpha^{j-2} \cdot m\right)^\omega = \alpha^{(j-2)\omega} \cdot m^\omega.$$

Since $\alpha = O(m^{1/2})$, the above expression is bounded from above as follows.

$$\alpha^{(j-2)\omega} \cdot m^\omega = O\left(\left(m^{(j-2)/2} \cdot m\right)^\omega\right) = O\left(m^{j\omega/2}\right) = O\left(m^{\ell\omega/6}\right).$$

Next, we show that for a certain high-range of α, the new bound $\alpha^{(j-2)\omega} \cdot m^\omega$ beats all previous bounds, namely $m^{j\omega/2}$ and $\alpha^{3j-2} \cdot m$, for $j \geq 3$ (or $\ell = 9, 12, 15, \ldots$). Let $\alpha = \Theta(m^x)$, where $x \in I_j$. We first verify that

$$\alpha^{(j-2)\omega} \cdot m^\omega \ll m^{j\omega/2}, \text{ or } ((j-2)x + 1)\omega < j\omega/2,$$

which holds for $x < 1/2$. Second, we verify that

$$\alpha^{(j-2)\omega} \cdot m^{\omega} \ll \alpha^{3j-2} \cdot m, \text{ or } ((j-2)x+1)\omega < (3j-2)x+1, \text{ or}$$
$$x > \frac{\omega - 1}{j(3-\omega)+2(\omega-1)},$$

which holds by the choice of the interval I_j. In particular, for $\ell = 9$, this occurs for $x \in \left(\frac{\omega-1}{7-\omega}, \frac{1}{2}\right) \supset (0.297, 0.5)$; for $\ell = 12$, this occurs for $x \in \left(\frac{\omega-1}{10-2\omega}, \frac{1}{2}\right) \supset (0.262, 0.5)$; for $\ell = 15$, this occurs for $x \in \left(\frac{\omega-1}{13-3\omega}, \frac{1}{2}\right) \supset (0.234, 0.5)$. Moreover, if $\omega = 2$, as conjectured, these intervals extend to $(1/5, 1/2)$, $(1/6, 1/2)$, and $(1/7, 1/2)$, respectively.

(ii) $\ell = 3j+1$, $j \geq 2$. The refined triangle method with $\ell_1 = j$, $\ell_2 = j+1$, $\ell_3 = j$, leads to rectangular matrix multiplication $[O(\alpha^{j-2}m) \times O(\alpha^{j-1}m)] \cdot [O(\alpha^{j-1}m) \times O(\alpha^{j-2}m)]$. Its complexity is at most $O(\alpha)$ times that of the square matrix multiplication with dimension $\alpha^{j-2}m$, the latter of which is $O(\alpha^{(j-2)\omega} \cdot m^{\omega})$. It follows that

$$T(n,m,\ell) = O(\alpha \cdot \alpha^{(j-2)\omega} \cdot m^{\omega}) = O(\alpha^{(j-2)\omega+1} \cdot m^{\omega}).$$

As before, we show that for a certain high-range of α, the new bound $\alpha^{(j-2)\omega+1} \cdot m^{\omega}$ beats all previous bounds, namely $m^{(j\omega+1)/2}$ and $\alpha^{3j-1} \cdot m$, for $j \geq 2$ (or $\ell = 7, 10, 13, \ldots$). Let $\alpha = \Theta(m^x)$, where $x \in I_j$. We first verify that

$$\alpha^{(j-2)\omega+1} \cdot m^{\omega} \ll m^{(j\omega+1)/2}, \text{ or } ((j-2)x+1)\omega < j\omega/2,$$

which holds for $x < 1/2$. Second, we verify that

$$\alpha^{(j-2)\omega+1} \cdot m^{\omega} \ll \alpha^{3j-1} \cdot m, \text{ or } ((j-2)x+1)\omega + x < (3j-1)x+1, \text{ or}$$
$$x > \frac{\omega - 1}{j(3-\omega)+2(\omega-1)},$$

which holds by the choice of the interval I_j. In particular, for $\ell = 7$, this occurs for $x \in \left(\frac{\omega-1}{4}, \frac{1}{2}\right) \supset (0.344, 0.5)$; for $\ell = 10$, this occurs for $x \in \left(\frac{\omega-1}{7-\omega}, \frac{1}{2}\right) \supset (0.297, 0.5)$; for $\ell = 13$, this occurs for $x \in \left(\frac{\omega-1}{10-2\omega}, \frac{1}{2}\right) \supset (0.262, 0.5)$. Moreover, if $\omega = 2$, as conjectured, these intervals extend to $(1/4, 1/2)$, $(1/5, 1/2)$, and $(1/6, 1/2)$, respectively.

(iii) $\ell = 3j+2$, $j \geq 2$. The refined triangle method with $\ell_1 = j+1$, $\ell_2 = j$, $\ell_3 = j+1$, leads to rectangular matrix multiplication $[O(\alpha^{j-1}m) \times O(\alpha^{j-2}m)] \cdot [O(\alpha^{j-2}m) \times O(\alpha^{j-1}m)]$. Its complexity is at most $O(\alpha^2)$ times that of the square matrix multiplication with dimension $\alpha^{j-2}m$, the latter of which is $O(\alpha^{(j-2)\omega} \cdot m^{\omega})$. It follows that

$$T(n,m,\ell) = O(\alpha^2 \cdot \alpha^{(j-2)\omega} \cdot m^{\omega}) = O(\alpha^{(j-2)\omega+2} \cdot m^{\omega}).$$

Again, we show that for a certain high-range of α, this bound beats all previous bounds, namely $m^{(j\omega+2)/2}$ and $\alpha^{3j} \cdot m$, for $j \geq 2$ (or $\ell = 8, 11, 14, \ldots$).

Let $\alpha = \Theta(m^x)$, where $x \in I_j$. We first verify that

$$\alpha^{(j-2)\omega+2} \cdot m^\omega \ll m^{(j\omega+2)/2}, \text{ or } ((j-2)x+1)\omega < j\omega/2,$$

which holds for $x < 1/2$. Second, we verify that

$$\alpha^{(j-2)\omega+2} \cdot m^\omega \ll \alpha^{3j} \cdot m, \text{ or } ((j-2)x+1)\omega + 2x < 3jx + 1, \text{ or}$$
$$x > \frac{\omega - 1}{j(3-\omega) + 2(\omega - 1)},$$

which holds by the choice of the interval I_j. In particular, for $\ell = 8$, this occurs for $x \in \left(\frac{\omega-1}{4}, \frac{1}{2}\right) \supset (0.344, 0.5)$; for $\ell = 11$, this occurs for $x \in \left(\frac{\omega-1}{7-\omega}, \frac{1}{2}\right) \supset (0.297, 0.5)$; for $\ell = 14$, this occurs for $x \in \left(\frac{\omega-1}{10-2\omega}, \frac{1}{2}\right) \supset (0.262, 0.5)$. Moreover, if $\omega = 2$, as conjectured, these intervals extend to $(1/4, 1/2)$, $(1/5, 1/2)$, and $(1/6, 1/2)$, respectively.

Running Time Derivations for Small ℓ. The previous general case derivations together with the instantiations below yield the running times listed in Table 1.

Table 1. Running time comparison for finding/counting complete subgraphs. The column on new results includes bounds in terms of m, based on up-to-date matrix multiplication times, and new arboricity-sensitive bounds in terms of m and α.

ℓ	Previous best	This paper
3	$O(\alpha \cdot m)$ [5], $O(n^{2.373})$, $O(m^{1.408})$ [2,14]	
4	$O(\alpha^2 \cdot m)$ [5], $O(n^{3.334})$, $O(m^{1.687})$ [10,16]	$O(n^{3.252})$
5	$O(\alpha^3 \cdot m)$ [5], $O(n^{4.220})$, $O(m^{2.147})$ [10,16]	$O(n^{4.090})$
6	$O(\alpha^4 \cdot m)$ [5], $O(n^{4.751})$, $O(m^{2.373})$ [10,16]	
7	$O(\alpha^5 \cdot m)$ [5], $O(n^{5.714})$, $O(m^{2.857})$ [10,16]	$O(m^{2.797})$, $O(\alpha \cdot m^{2.373})$
8	$O(\alpha^6 \cdot m)$ [5], $O(m^{3.373})$ [10,16]	$O(m^{3.252})$, $O(\alpha^2 \cdot m^{2.373})$
9	$O(\alpha^7 \cdot m)$ [5], $O(m^{3.560})$ [10,16]	$O(\alpha^{2.373} \cdot m^{2.373})$
10	$O(\alpha^8 \cdot m)$ [5], $O(m^{4.060})$ [10,16]	$O(m^4)$, $O(\alpha^{3.373} \cdot m^{2.373})$
11	$O(\alpha^9 \cdot m)$ [5], $O(m^{4.560})$ [10,16]	$O(m^{4.376})$, $O(\alpha^{4.373} \cdot m^{2.373})$
12	$O(\alpha^{10} \cdot m)$ [5], $O(m^{4.746})$ [10,16]	$O(\alpha^{4.746} \cdot m^{2.373})$
13	$O(\alpha^{11} \cdot m)$ [5], $O(m^{5.16})$ [10,16]	$O(\alpha^{5.746} \cdot m^{2.373})$
14	$O(\alpha^{11} \cdot m)$ [5], $O(m^{5.556})$ [10,16]	$O(\alpha^{6.746} \cdot m^{2.373})$
15	$O(\alpha^{13} \cdot m)$ [5], $O(m^{5.933})$ [10,16]	$O(\alpha^{7.119} \cdot m^{2.373})$

(i) $\ell = 4$. The refined triangle method with $\ell_1 = 1$, $\ell_2 = 1$, $\ell_3 = 2$, leads to rectangular matrix multiplication $[n \times n] \cdot [n \times m]$ or $[n \times n] \cdot [n \times n^2]$ in the worst case. Since according to [20, Table 3], $\omega(1,1,2) = \omega(1,2,1) \leq 3.252$, it follows that $T(n,m,4) = O(n^{3.252})$. By [10, Table 1], $T(n,m,4) = O(m^{1.682})$, but this entry appears unjustified.

(ii) $\ell = 5$. The refined triangle method with $\ell_1 = 2$, $\ell_2 = 1$, $\ell_3 = 2$, leads to rectangular matrix $[n^2 \times n] \cdot [n \times n^2]$ in the worst case. Since according to [20, Table 3], $\omega(2,1,2) = 2\omega(1,0.5,1) \leq 2 \cdot 2.045 = 4.090$, it follows that $T(n,m,5) = O(n^{4.090})$.

(iii) $\ell = 7$. The refined triangle method with $\ell_1 = 2$, $\ell_2 = 3$, $\ell_3 = 2$, leads to rectangular matrix multiplication $[m \times \alpha m] \cdot [\alpha m \times m]$ or $[m \times m^{3/2}] \cdot [m^{3/2} \times m]$ in the worst case. Since $\omega(1, 1.5, 1) \le 2.797$, it follows that $T(n, m, 7) = O(m^{2.797})$.

(iv) $\ell = 8$. The refined triangle method with $\ell_1 = \ell_2 = \ell_3 = \ell_4 = 2$, leads to counting K_4's in a graph with m vertices. Since $\omega(1, 1, 2) \le 3.252$, we have $T(n, m, 8) = O(m^{3.252})$.

(v) $\ell = 10$. By [10, Thm. 2], it takes $O(m^{\beta(10)/2})$, where $\beta(10) = \omega(3, 3, 4)$. Since according to [20, Table 3], $\omega(1, 4/3, 1) \le 8/3$, we have $\omega(3, 3, 4) = 3\omega(1, 4/3, 1) \le 3 \times 8/3 = 8$, thus $T(n, m, 10) = O(m^{\beta(10)/2}) = O(m^4)$.

(vi) $\ell = 11$. By [10, Thm. 2], it takes $O(m^{\beta(11)/2})$, where $\beta(11) = \omega(3, 4, 4)$. Since according to [20, Table 3], $\omega(1, 0.75, 1) \le 2.188$, we have $\omega(3, 4, 4) = 4\omega(1, 0.75, 1) \le 4 \times 2.188 = 8.752$, thus $T(n, m, 11) = O(m^{\beta(11)/2}) = O(m^{4.376})$.

(vii) $\ell = 13$. By [10, Thm. 2], it takes $O(m^{\beta(13)/2})$, where $\beta(13) = \omega(4, 4, 5)$. Since according to [20, Table 3], $\omega(1, 5/4, 1) \le 2.58$, we have $\omega(4, 4, 5) = 4\omega(1, 5/4, 1) \le 4 \times 2.58 = 10.32$, thus $T(n, m, 13) = O(m^{\beta(13)/2}) = O(m^{5.16})$.

(viii) $\ell = 14$. By [10, Thm. 2], it takes $O(m^{\beta(14)/2})$, where $\beta(14) = \omega(5, 4, 5)$. Since according to [20, Table 3], $\omega(1, 0.8, 1) \le 2.2223$, we have $\omega(4, 4, 5) = 5\omega(1, 0.8, 1) \le 5 \times 2.2223 = 11.1115$, thus $T(n, m, 14) = O(m^{\beta(14)/2}) = O(m^{5.556})$.

References

1. Alman, J., Williams, V.V.: A refined laser method and faster matrix multiplication. In: Proceedings of 2021 ACM-SIAM Symposium on Discrete Algorithms SODA 2021, Virtual Conference, pp. 522–539. SIAM (2021)
2. Alon, N., Yuster, R., Zwick, U.: Finding and counting given length cycles. Algorithmica **17**(3), 209–223 (1997)
3. Bar-Yehuda, R., Even, S.: On approximating a vertex cover for planar graphs. In: Proceedings of 14th ACM Symposium on Theory of Computing (STOC), pp. 303–309 (1982)
4. Björklund, A., Pagh, R., Williams, V.V., Zwick, U.: Listing triangles. In: Esparza, J., Fraigniaud, P., Husfeldt, T., Koutsoupias, E. (eds.) ICALP 2014. LNCS, vol. 8572, pp. 223–234. Springer, Heidelberg (2014). https://doi.org/10.1007/978-3-662-43948-7_19
5. Chiba, N., Nishizeki, T.: Arboricity and subgraph listing algorithms. SIAM J. Comput. **14**(1), 210–223 (1985)
6. Coppersmith, D., Winograd, S.: Matrix multiplication via arithmetic progressions. J. Symb. Comput. **9**(3), 251–280 (1990)
7. Cormen, T.H., Leiserson, C.E., Rivest, R.L., Stein, C.: Introduction to Algorithms, 3rd edn. MIT Press, Cambridge (2009)
8. Diestel, R.: Graph Theory. Springer, New York (1997)
9. Eden, T., Levi, A., Ron, D., Seshadhri, C.: Approximately counting triangles in sublinear time. SIAM J. Comput. **46**(5), 1603–1646 (2017)
10. Eisenbrand, F., Grandoni, F.: On the complexity of fixed parameter clique and dominating set. Theoret. Comput. Sci. **326**(1–3), 57–67 (2004)

11. Eppstein, D.: Arboricity and bipartite subgraph listing algorithms. Inf. Process. Lett. **51**(4), 207–211 (1994)
12. Eppstein, D., Goodrich, M.T., Mitzenmacher, M., Torres, M.R.: 2–3 cuckoo filters for faster triangle listing and set intersection. In: Proceedings of 36th SIGMOD-SIGACT-SIGAI Symposium on Principles of Database Systems, PODS, pp. 247–260 (2017)
13. Harary, F.: Graph Theory, revised. Addison-Wesley, Reading (1972)
14. Itai, A., Rodeh, M.: Finding a minimum circuit in a graph. SIAM J. Comput. **7**(4), 413–423 (1978)
15. Garey, M.R., Johnson, D.S.: Computers and Intractability: A Guide to the Theory of NP-Completeness. W.H. Freeman and Co., New York (1979)
16. Kloks, T., Kratsch, D., Müller, H.: Finding and counting small induced subgraphs efficiently. Inf. Process. Lett. **74**(3–4), 115–121 (2000)
17. Kolountzakis, M., Miller, G., Peng, R., Tsourakakis, C.: Efficient triangle counting in large graphs via degree-based vertex partitioning. Internet Math. **8**(1–2), 161–185 (2012)
18. Kopelowitz, T., Pettie, S., Porat, E.: Dynamic set intersection. In: Dehne, F., Sack, J.-R., Stege, U. (eds.) WADS 2015. LNCS, vol. 9214, pp. 470–481. Springer, Cham (2015). https://doi.org/10.1007/978-3-319-21840-3_39
19. Kowaluk, M., Lingas, A.: A multi-dimensional matrix product–a natural tool for parameterized graph algorithms. Algorithms **15**(12), 448 (2022)
20. Le Gall, F., Urrutia, F.: Improved rectangular matrix multiplication using powers of the Coppersmith-Winograd tensor. In: Proceedings of 29th Annual ACM-SIAM Symposium on Discrete Algorithms, SODA, New Orleans, pp. 1029–1046. SIAM (2018)
21. Lick, D.R., White, A.T.: k-degenerate graphs. Can. J. Math. **22**(5), 1082–1096 (1970)
22. Nash-Williams, C.S.J.A.: Edge-disjoint spanning trees of finite graphs. J. London Math. Soc. **36**, 445–450 (1961)
23. Nash-Williams, C.S.J.A.: Decomposition of finite graphs into forests. J. London Math. Soc. **1**(1), 12–12 (1964)
24. Nešetřil, J., Poljak, S.: On the complexity of the subgraph problem. Comment. Math. Univ. Carol. **26**(2), 415–419 (1985)
25. Ortmann, M., Brandes, U.: Triangle listing algorithms: back from the diversion. In: Proceedings of 16th Workshop on Algorithm Engineering and Experiments, ALENEX 2014, Portland, Oregon, pp. 1–8 (2014)
26. Papadimitriou, C.H., Yannakakis, M.: The clique problem for planar graphs. Inf. Process. Lett. **13**(4/5), 131–133 (1981)
27. Rivin, I.: Counting cycles and finite dimensional L^p norms. Adv. Appl. Math. **29**(4), 647–662 (2002)
28. Schank, T.: Algorithmic aspects of triangle-based networks analysis. Ph.D. thesis, Universität Fridericiana zu Karlsruhe (TH) (2007)
29. Tutte, W.: On the problem of decomposing a graph into n connected factors. J. Lond. Math. Soc. **36**, 221–230 (1961)
30. Zechner, N., Lingas, A.: Efficient algorithms for subgraph listing. Algorithms **7**(2), 243–252 (2014)
31. Zhou, X., Nishizeki, T.: Edge-coloring and f-coloring for various classes of graphs. J. Graph Algorithms Appl. **3**(1), 1–18 (1999)

Maximal Distortion of Geodesic Diameters in Polygonal Domains

Adrian Dumitrescu[1] and Csaba D. Tóth[2,3](✉)

[1] Algoresearch L.L.C., Milwaukee, WI, USA
ad.dumitrescu@algoresearch.org
[2] California State University Northridge, Los Angeles, CA, USA
[3] Tufts University, Medford, MA, USA
csaba.toth@csun.edu

Abstract. For a polygon P with holes in the plane, we denote by $\varrho(P)$ the ratio between the geodesic and the Euclidean diameters of P. It is shown that over all convex polygons with h convex holes, the supremum of $\varrho(P)$ is between $\Omega(h^{1/3})$ and $O(h^{1/2})$. The upper bound improves to $O(1+\min\{h^{3/4}\Delta, h^{1/2}\Delta^{1/2}\})$ if every hole has diameter at most $\Delta \cdot \mathrm{diam}_2(P)$; and to $O(1)$ if every hole is a *fat* convex polygon. Furthermore, we show that the function $g(h) = \sup_P \varrho(P)$ over convex polygons with h convex holes has the same growth rate as an analogous quantity over geometric triangulations with h vertices when $h \to \infty$.

1 Introduction

Determining the maximum distortion between two metrics on the same ground set is a fundamental problem in metric geometry. Here we study the maximum ratio between the geodesic (i.e., shortest path) diameter and the Euclidean diameter over polygons with holes. A *polygon P with h holes* (also known as a *polygonal domain*) is defined as follows. Let P_0 be a simple polygon, and let $P_1, \ldots, P_h$ be pairwise disjoint simple polygons in the interior of P_0. Then $P = P_0 \setminus \left(\bigcup_{i=1}^h P_i\right)$.

The Euclidean distance between two points $s, t \in P$ is $|st| = \|s-t\|_2$, and the shortest path distance $\mathrm{geod}(s,t)$ is the minimum arclength of a polygonal path between s and t contained in P. The triangle inequality implies that $|st| \leq \mathrm{geod}(s,t)$ for all $s, t \in P$. The *geometric dilation* (also known as the *stretch factor*) between the two distances is $\sup_{s,t\in P} \mathrm{geod}(s,t)/|st|$. The geometric dilation of P can be arbitrarily large, even if P is a (nonconvex) quadrilateral.

The *Euclidean diameter* of P is $\mathrm{diam}_2(P) = \sup_{s,t\in P} |st|$ and its *geodesic diameter* is $\mathrm{diam}_g(P) = \sup_{s,t\in P} \mathrm{geod}(s,t)$. It is clear that $\mathrm{diam}_2(P) \leq \mathrm{diam}_g(P)$. We are interested in the *distortion*

$$\varrho(P) = \frac{\mathrm{diam}_g(P)}{\mathrm{diam}_2(P)}. \tag{1}$$

Note that $\varrho(P)$ is unbounded, even for simple polygons. Indeed, if P is a zig-zag polygon with n vertices, contained in a disk of unit diameter, then $\mathrm{diam}_2(P) \leq 1$

S.-Y. Hsieh et al. (Eds.): IWOCA 2023, LNCS 13889, pp. 197–208, 2023.
https://doi.org/10.1007/978-3-031-34347-6_17

and $\text{diam}_g(P) = \Omega(n)$, hence $\varrho(P) \geq \Omega(n)$. It is not difficult to see that this bound is the best possible, that is, $\varrho(P) \leq O(n)$.

In this paper, we consider convex polygons with convex holes. Specifically, let $\mathcal{C}(h)$ denote the family of polygonal domains $P = P_0 \setminus \left(\bigcup_{i=1}^{h} P_i\right)$, where $P_0, P_1, \ldots, P_h$ are convex polygons; and let

$$g(h) = \sup_{P \in \mathcal{C}(h)} \varrho(P). \tag{2}$$

It is clear that if $h = 0$, then $\text{geod}(s,t) = |st|$ for all $s, t \in P$, which implies $g(0) = 1$. Or main result is the following.

Theorem 1. *For every $h \in \mathbb{N}$, we have $\Omega(h^{1/3}) \leq g(h) \leq O(h^{1/2})$.*

The lower bound construction is a polygonal domain in which all h holes have about the same diameter $\Theta(h^{-1/3}) \cdot \text{diam}_2(P)$. We prove a matching upper bound for all polygons P with holes of diameter $\Theta(h^{-1/3}) \cdot \text{diam}_2(P)$. In general, if the diameter of every hole is $o(1) \cdot \text{diam}_2(P)$, we can improve upon the bound $g(h) \leq O(h^{1/2})$ in Theorem 1.

Theorem 2. *If $P \in \mathcal{C}(h)$ and the diameter of every hole is at most $\Delta \cdot \text{diam}_2(P)$, then $\varrho(P) \leq O(1 + \min\{h^{3/4}\Delta, h^{1/2}\Delta^{1/2}\})$. In particular for $\Delta = O(h^{-1/3})$, we have $\varrho(P) \leq O(h^{1/3})$.*

However, if we further restrict the holes to be *fat* convex polygons, we can show that $\varrho(P) = O(1)$ for all $h \in \mathbb{N}$. In fact for every $s, t \in P$, the distortion $\text{geod}(s,t)/|st|$ is also bounded by a constant.

Informally, a convex body is *fat* if its width is comparable with its diameter. The *width* of a convex body C is the minimum width of a parallel slab enclosing C. For $0 \leq \lambda \leq 1$, a convex body C is *λ-fat* if the ratio of its width to its diameter is at least λ, that is, $\text{width}(C)/\text{diam}_2(C) \geq \lambda$; and C is *fat* if the inequality holds for a constant λ. For instance, a disk is 1-fat, a 3×4 rectangle is $\frac{3}{5}$-fat and a line segment is 0-fat. Let $\mathcal{F}_\lambda(h)$ be the family of polygonal domain $P = P_0 \setminus \left(\bigcup_{i=1}^{h} P_i\right)$, where $P_0, P_1, \ldots, P_h$ are λ-fat convex polygons.

Proposition 1. *For every $h \in \mathbb{N}$ and $P \in \mathcal{F}_\lambda(h)$, we have $\varrho(P) \leq O(\lambda^{-1})$.*

The special case when all holes are axis-aligned rectangles is also easy.

Proposition 2. *Let $P \in \mathcal{C}(h)$, $h \in \mathbb{N}$, such that all holes are axis-aligned rectangles. Then $\varrho(P) \leq O(1)$.*

Triangulations. In this paper, we focus on the diameter distortion $\varrho(P) = \text{diam}_g(P)/\text{diam}_2(P)$ for polygons $P \in \mathcal{C}(h)$ with h holes. Alternatively, we can also compare the geodesic and Euclidean diameters in n-vertex triangulations. In a *geometric graph* $G = (V, E)$, the vertices are distinct points in the plane, and the edges are straight-line segments between pairs of vertices. The *Euclidean diameter* of G, $\text{diam}_2(G) = \max_{u,v \in V} |uv|$ is the maximum distance between two vertices, and the *geodesic diameter* $\text{diam}_g(G) = \max_{u,v \in V} \text{dist}(u,v)$, where $\text{dist}(u,v)$ is the shortest path distance in G, i.e., the minimum Euclidean length of a uv-path in G. With this notation, we define $\varrho(G) = \text{diam}_g(G)/\text{diam}_2(G)$,

A Euclidean triangulation $T = (V, E)$ is a planar straight-line graph where all bounded faces are triangles, and their union is the convex hull $\text{conv}(V)$. Let

$$f(n) = \sup_{G \in \mathcal{T}(n)} \varrho(G), \tag{3}$$

where the supremum is taken over the set $\mathcal{T}(n)$ all n-vertex triangulations. Recall that $g(n)$ is the supremum of diameter distortions over polygons with n convex holes; see (2). We prove that $f(n)$ and $g(n)$ have the same growth rate.

Theorem 3. *We have* $g(n) = \Theta(f(n))$.

Alternative Problem Formulation. The following version of the question studied here may be more attractive to the escape community [9,16]. Given n pairwise disjoint convex obstacles in a convex polygon of unit diameter (e.g., a square), what is the maximum length of a (shortest) escape route from any given point in the polygon to its boundary? According to Theorem 1, it is always $O(n^{1/2})$ and sometimes $\Omega(n^{1/3})$.

Related Work. The geodesic distance in polygons with or without holes has been studied extensively from the algorithmic perspective; see [19] for a comprehensive survey. In a simple polygon P with n vertices, the geodesic distance between two given points can be computed in $O(n)$ time [17]; trade-offs are also available between time and workspace [12]. A shortest-path data structure can report the geodesic distance between any two query points in $O(\log n)$ time after $O(n)$ preprocessing time [11]. In $O(n)$ time, one can also compute the geodesic diameter [13] and radius [1].

For polygons with holes, more involved techniques are needed. Let P be a polygon with h holes, and a total of n vertices. For any $s, t \in P$, one can compute $\text{geod}(s,t)$ in $O(n + h \log h)$ time and $O(n)$ space [23], improving earlier bounds in [14,15,18,24]. A shortest-path data structure can report the geodesic distance between two query points in $O(\log n)$ query time using $O(n^{11})$ space; or in $O(h \log n)$ query time with $O(n + h^5)$ space [6]. The geodesic radius can be computed in $O(n^{11} \log n)$ time [3,22], and the geodesic diameter in $O(n^{7.73})$ or $O(n^7(\log n + h))$ time [2]. One can find an $(1+\varepsilon)$-approximation in $O((n/\varepsilon^2 + n^2/\varepsilon) \log n)$ time [2,3]. The geodesic diameter may be attained by a point pair $s, t \in P$, where both s and t lie in the interior or P; in which case it is known [2] that there are at least five different geodesic paths between s and t.

The diameter of an n-vertex triangulation with Euclidean weights can be computed in $\tilde{O}(n^{5/3})$ time [5,10]. For unweighted graphs in general, the diameter problem has been intensely studied in the fine-grained complexity community. For a graph with n vertices and m edges, breadth-first search (BFS) yields a 2-approximation in $O(m)$ time. Under the Strong Exponential Time Hypothesis (SETH), for any integer $k \geq 2$ and $\varepsilon > 0$, a $(2 - \frac{1}{k} - \varepsilon)$-approximation requires $mn^{1+1/(k-1)-o(1)}$ time [7]; see also [20].

2 Convex Polygons with Convex Holes

In this section, we prove Theorem 1. A lower bound construction is presented in Lemma 1, and the upper bound is established in Lemma 2 below.

Lower Bound. The lower bound is based on the following construction.

Lemma 1. *For every $h \in \mathbb{N}$, there exists a polygonal domain $P \in \mathcal{C}(h)$ such that $g(P) \geq \Omega(h^{1/3})$.*

Proof. We may assume w.l.o.g. that $h = k^3$ for some integer $k \geq 3$. We construct a polygon P with h holes, where the outer polygon P_0 is a regular k-gon of unit diameter, hence $\text{diam}_2(P) = \text{diam}_2(P_0) = 1$. Let $Q_0, Q_1, \ldots, Q_{k^2}$ be a sequence of $k^2 + 1$ regular k-gons with a common center such that $Q_0 = P_0$, and for every $i \in \{1, \ldots, k^2\}$, Q_i is inscribed in Q_{i-1} such that the vertices of Q_i are the midpoints of the edges of Q_{i-1}; see Fig. 1. Enumerate the k^3 edges of $Q_1, \ldots, Q_{k^2}$ as $e_1, \ldots, e_{k^3}$. For every $j = 1, \ldots, k^3$, we construct a hole as follows: Let P_j be an $(|e|-2\varepsilon) \times \frac{\varepsilon}{2}$ rectangle with symmetry axis e that contains e with the exception of the ε-neighborhoods of its endpoints. Then $P_1, \ldots, P_{k^3}$ are pairwise disjoint. Finally, let $P = P_0 \setminus \bigcup_{j=1}^{k^3} P_j$.

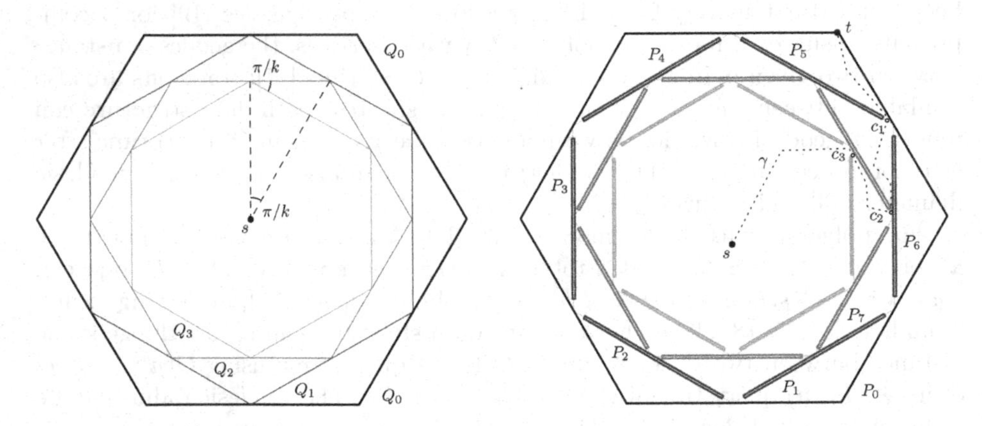

Fig. 1. Left: hexagons $Q_0, \ldots, Q_3$ for $k = 6$. Right: The 18 holes corresponding to the edges of $Q_1, \ldots, Q_3$.

Assume, w.l.o.g., that e_i is an edge of Q_i for $i \in \{0, 1, \ldots, k^2\}$. As $P_0 = Q_0$ is a regular k-gon of unit diameter, then $|e_0| \geq \Omega(1/k)$. Let us compare the edge lengths in two consecutive k-gons. Since Q_{i+1} is inscribed in Q_i, we have

$$|e_{i+1}| = |e_i| \cos\frac{\pi}{k} \geq |e_i| \left(1 - \frac{\pi^2}{2k^2}\right)$$

using the Taylor estimate $\cos x \geq 1 - x^2/2$. Consequently, for every $i \in \{0, 1, \ldots, k^2\}$,

$$|e_i| \geq |e_0| \cdot \left(1 - \frac{\pi^2}{2k^2}\right)^{k^2} \geq |e_0| \cdot \Omega(1) \geq \Omega\left(\frac{1}{k}\right).$$

It remains to show that $\mathrm{diam}_g(P) \geq \Omega(k)$. Let s be the center of P_0 and t and arbitrary vertex of P_0. Consider an st-path γ in P, and for any two points a, b along γ, let $\gamma(a, b)$ denote the subpath of γ between a and b. Let c_i be the first point where γ crosses the boundary of Q_i for $i \in \{1, \ldots, k^2\}$. By construction, c_i must be in an ε-neighborhood of a vertex of Q_i. Since the vertices of Q_{i+1} are at the midpoints of the edges of Q_i, then $|\gamma(c_i, c_{i+1})| \geq \frac{1}{2}|e_i| - 2\varepsilon \geq \Omega(|e_i|) \geq \Omega(1/k)$. Summation over $i = 0, \ldots, k^2 - 1$ yields $|\gamma| \geq \sum_{i=0}^{k^2-1} |\gamma(c_i, c_{i+1})| \geq k^2 \cdot \Omega(1/k) \geq \Omega(k) = \Omega(h^{1/3})$, as required. □

Upper Bound. Let $P \in \mathcal{C}(h)$ for some $h \in \mathbb{N}$ and let $s \in P$. For every hole P_i, let ℓ_i and r_i be points on the boundary of P_i such that $\overrightarrow{s\ell_i}$ and $\overrightarrow{sr_i}$ are tangent to P_i, and P_i lies on the left (resp., right) side of the ray $\overrightarrow{s\ell_i}$ (resp., $\overrightarrow{sr_i}$).

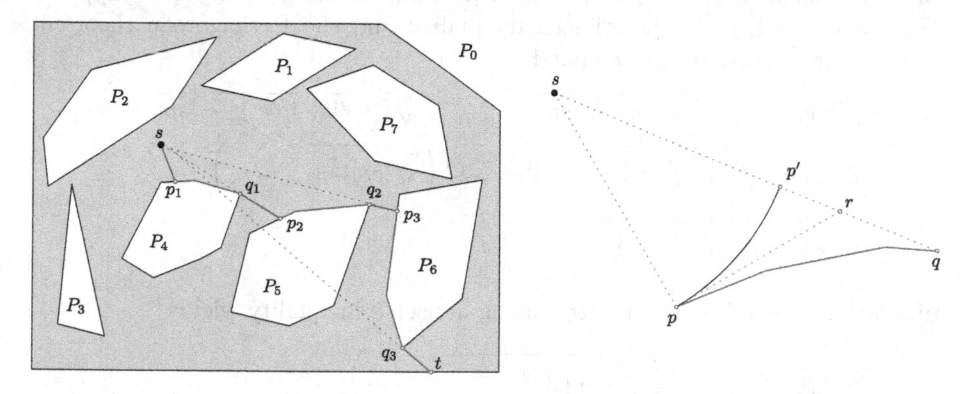

Fig. 2. Left: A polygon $P \in \mathcal{C}(7)$ with 7 convex holes, a point $s \in P$, and a path $\mathrm{greedy}_P(s, \boldsymbol{u})$ from s to a point t on the outer boundary of P. Right: A boundary arc $\widehat{pq}$, where $|\widehat{pq}| \leq |pr| + |rq|$.

We construct a path from s to some point in the outer boundary of P by the following recursive algorithm; refer to Fig. 2 (left). For a unit vector $\boldsymbol{u} \in \mathbb{S}^1$, we construct path $\mathrm{greedy}_P(s, \boldsymbol{u})$ as follows. Start from s along a ray emanating from s in direction $\boldsymbol{u}$ until reaching the boundary of P at some point p. While $p \notin \partial P_0$ do: Assume that $p \in \partial P_i$ for some $1 \leq i \leq h$. Extend the path along ∂P_i to the point ℓ_i or r_i such that the distance from s monotonically increases; and then continue along the ray $\overrightarrow{s\ell_i}$ or $\overrightarrow{sr_i}$ until reaching the boundary of P again. When $p \in \partial P_0$, the path $\mathrm{greedy}_P(s, \boldsymbol{u})$ terminates at p.

Lemma 2. *For every $P \in \mathcal{C}(h)$, every $s \in P$ and every $\boldsymbol{u} \in \mathbb{S}^1$, we have $|\mathrm{greedy}_P(s, \boldsymbol{u})| \leq O(h^{1/2}) \cdot \mathrm{diam}_2(P)$, and this bound is the best possible.*

Proof. Let P be a polygonal domain with a convex outer polygon P_0 and h convex holes. We may assume w.l.o.g. that $\mathrm{diam}_2(P) = 1$. For a point $s \in P$ and a unit vector $\boldsymbol{u}$, consider the path $\mathrm{greedy}_P(s, \boldsymbol{u})$. By construction, the distance from s monotonically increases along $\mathrm{greedy}_P(s, \boldsymbol{u})$, and so the path has no self-intersections. It is composed of *radial segments* that lie along rays emanating from s, and *boundary arcs* that lie on the boundaries of holes. By monotonicity, the total length of all radial segments is at most $\mathrm{diam}_2(P)$. Since every boundary arc ends at a point of tangency ℓ_i or r_i, for some $i \in \{1, \ldots, h\}$, the path $\mathrm{greedy}_P(s, \boldsymbol{u})$ contains at most two boundary arcs along each hole, thus the number of boundary arcs is at most $2h$. Let $\mathcal{A}$ denote the set of all boundary arcs along $\mathrm{greedy}_P(s, \boldsymbol{u})$; then $|\mathcal{A}| \leq 2h$.

Along each boundary arc $\widehat{pq} \in \mathcal{A}$, from p to q, the distance from s increases by $\Delta_{pq} = |sq| - |sp|$. By monotonicity, we have $\sum_{\widehat{pq} \in \mathcal{A}} \Delta_{pq} \leq \mathrm{diam}_2(P)$. We now give an upper bound for the length of $\widehat{pq}$. Let p' be a point in sq such that $|sp| = |sp'|$, and let r be the intersection of sq with a line orthogonal to sp passing through p; see Fig. 2 (right). Note that $|sp| < |sr|$. Since the distance from s monotonically increases along the arc $\widehat{pq}$, then q is in the closed halfplane bounded by pr that does not contain s. Combined with $|sp| < |sr|$, this implies that r lies between p' and q on the line sq, consequently $|p'r| < |p'q| = \Delta_{pq}$ and $|rq| < |p'q| = \Delta_{pq}$. By the triangle inequality and the Pythagorean theorem, these estimates give an upper bound

$$\begin{aligned}|\widehat{pq}| &\leq |pr| + |rq| = \sqrt{|sr|^2 - |sp|^2} + |rq| \leq \sqrt{(|sp'| + |p'r|)^2 - |sp|^2} + |rq| \\ &\leq \sqrt{(|sp| + \Delta_{pq})^2 - |sp|^2} + \Delta_{pq} \leq O\left(\sqrt{|sp|\Delta_{pq}} + \Delta_{pq}\right) \\ &\leq O\left(\sqrt{\mathrm{diam}_2(P) \cdot \Delta_{pq}} + \Delta_{pq}\right).\end{aligned}$$

Summation over all boundary arcs, using Jensen's inequality, yields

$$\begin{aligned}\sum_{\widehat{pq} \in \mathcal{A}} |\widehat{pq}| &\leq \sum_{\widehat{pq} \in \mathcal{A}} O\left(\sqrt{\mathrm{diam}_2(P) \cdot \Delta_{pq}} + \Delta_{pq}\right) \\ &\leq \sqrt{\mathrm{diam}_2(P)} \cdot O\left(\sum_{\widehat{pq} \in \mathcal{A}} \sqrt{\Delta_{pq}}\right) + O\left(\sum_{\widehat{pq} \in \mathcal{A}} \Delta_{pq}\right) \\ &\leq \sqrt{\mathrm{diam}_2(P)} \cdot O\left(|\mathcal{A}| \cdot \sqrt{\frac{1}{|\mathcal{A}|} \sum_{\widehat{pq} \in \mathcal{A}} \Delta_{pq}}\right) + O(\mathrm{diam}_2(P)) \\ &\leq \sqrt{\mathrm{diam}_2(P)} \cdot O\left(\sqrt{|\mathcal{A}| \cdot \mathrm{diam}_2(P)}\right) + O(\mathrm{diam}_2(P)) \\ &\leq O\left(\sqrt{|\mathcal{A}|}\right) \cdot \mathrm{diam}_2(P) \leq O\left(\sqrt{h}\right) \cdot \mathrm{diam}_2(P),\end{aligned}$$

as claimed.

We now show that the bound $|\mathrm{greedy}_P(s, \boldsymbol{u})| \leq O(h^{1/2}) \cdot \mathrm{diam}_2(P)$ is the best possible. For every $h \in \mathbb{N}$, we construct a polygon $P \in \mathcal{C}(h)$ and a point s

such that for every $\boldsymbol{u} \in \mathbb{S}^1$, we have $|\text{greedy}_P(s, \boldsymbol{u})| \geq \Omega(h^{1/2})$. Without loss of generality, we may assume $\text{diam}_2(P) = 1$ and $h = 3(k^2 + 1)$ for some $k \in \mathbb{N}$.

We start with the construction in Lemma 1 with k^3 rectangular holes in a regular k-gon P_0, where s is the center of P_0. We modify the construction in three steps: (1) Let T be a small equilateral triangle centered at s, and construct three rectangular holes around the edges of T; to obtain a total of $k^3 + 3$ holes. (2) Rotate each hole P_j counterclockwise by a small angle, such that when the greedy path reaches P_j in an ε-neighborhood of its center, it would always turn left. (3) For any $\boldsymbol{u} \in \mathbb{S}^1$, the path $\text{greedy}_P(s, \boldsymbol{u})$ exit the triangle T at a small neighborhood of a corner of T. From each corner of T, $\text{greedy}_P(s, \boldsymbol{u})$ continues to the outer boundary along the same k^2 holes. We delete all holes $\text{greedy}_P(s, \boldsymbol{u})$ does not touch for any $\boldsymbol{u} \in \mathbb{S}^1$, thus we retain $h = 3k^2 + 3$ holes. For every $\boldsymbol{u} \in \mathbb{S}^1$, we have $|\text{greedy}_P(s, \boldsymbol{u})| \geq \Omega(k)$ according to the analysis in Lemma 1, hence $|\text{greedy}_P(s, \boldsymbol{u})| \geq \Omega(h^{1/2})$, as required. □

Corollary 1. *For every $h \in \mathbb{N}$ and every polygon $P \in \mathcal{C}(h)$, we have* $\text{diam}_g(P) \leq O(h^{1/2}) \cdot \text{diam}_2(P)$.

Proof. Let $P \in \mathcal{C}(h)$ and $s_1, s_2 \in P$. By Lemma 2, there exist points $t_1, t_2 \in \partial P_0$ such that $\text{geod}(s_1, t_1) \leq O(h^{1/2}) \cdot \text{diam}_2(P)$ and $\text{geod}(s_2, t_2) \leq O(h^{1/2}) \cdot \text{diam}_2(P)$. There is a path between t_1 and t_2 along the perimeter of P_0. It is well known [21,25] that $|\partial P_0| \leq \pi \cdot \text{diam}_2(P_0)$ for every convex body P_0, hence $\text{geod}(t_1, t_2) \leq O(\text{diam}_2(P))$. The concatenation of these three paths yields a path in P connecting s_1 and s_2, of length $\text{geod}(s_1, s_2) \leq O(h^{1/2}) \cdot \text{diam}_2(P)$. □

3 Improved Upper Bound for Holes of Bounded Diameter

In this section we prove Theorem 2. Similar to the proof of Theorem 1, it is enough to bound the geodesic distance from an arbitrary point in P to the outer boundary. We give three such bounds in Lemmas 3, 4 and 7.

Lemma 3. *Let $P \in \mathcal{C}(h)$ such that $\text{diam}_2(P_i) \leq \Delta \cdot \text{diam}_2(P)$ for every hole P_i. If $\Delta \leq O(h^{-1})$, then there exists a path of length $O(\text{diam}_2(P))$ in P from any point $s \in P$ to the outer boundary ∂P_0.*

Proof. Let $s \in P$ and $t \in \partial P_0$. Construct an st-path γ as follows: Start with the straight line segment st, and whenever st intersects the interior of a hole P_i, then the segment $st \cap P_i$ is replaced by an arc along ∂P_i. Since $|\partial P_i| \leq \pi \cdot \text{diam}_2(P_i)$ for every convex hole P_i [21,25], then $|\gamma| \leq |st| + \sum_{i=1}^{h} |\partial P_i| \leq \text{diam}_2(P) + \sum_{i=1}^{h} O(\text{diam}_2(P_i)) \leq O(1 + h\Delta) \cdot \text{diam}_2(P) \leq O(\text{diam}_2(P))$, as claimed. □

Lemma 4. *Let $P \in \mathcal{C}(h)$ such that $\text{diam}_2(P_i) \leq \Delta \cdot \text{diam}_2(P)$ for every hole P_i. Then there exists a path of length $O(1 + h^{3/4}\Delta) \cdot \text{diam}_2(P)$ in P from any point $s \in P$ to the outer boundary ∂P_0.*

Proof. Assume without loss of generality that $\text{diam}_2(P) = 1$, and s is the origin. Let $\ell \in \mathbb{N}$ be a parameter to be specified later. For $i \in \{-\ell, -\ell + 1, \ldots, \ell\}$, let $H_i : y = i \cdot \Delta$ be a horizontal line, and $V_i : x = i \cdot \Delta$ a vertical line. Since any two consecutive horizontal (resp., vertical) lines are distance Δ apart, and the diameter of each hole is at most Δ, then the interior of each hole intersects at most one horizontal and at most one vertical line. By the pigeonhole principle, there are integers $a, b, c, d \in \{1, \ldots, \ell\}$ such that H_{-a}, H_b, V_{-c}, and V_d each intersects the interior of at most h/ℓ holes; see Fig. 3.

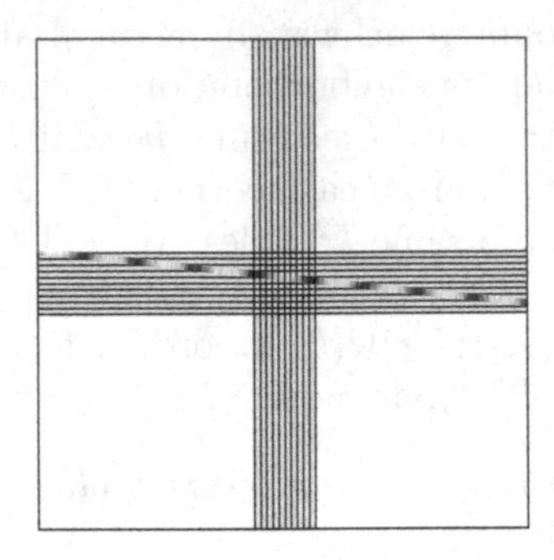

Fig. 3. Illustration for $\ell = 5$ (assuming that P is a unit square centered at s).

Let B be the axis-aligned rectangle bounded by the lines H_{-a}, H_b, V_{-c}, and V_d. Due to the spacing of the lines, we have $\text{diam}_2(B) \leq 2 \cdot \sqrt{2} \cdot \ell\Delta = O(\ell\Delta)$.

We construct a path from s to ∂P_0 as a concatenation of two paths $\gamma = \gamma_1 \oplus \gamma_2$. Let γ_1 be the initial part of $\text{greedy}_P(s, \boldsymbol{u})$ from s until reaching the boundary of $B \cap P_0$ at some point p. If $p \in \partial P_0$, then $\gamma_2 = (p)$ is a trivial one-point path. Otherwise p lies on a line $L \in \{H_{-a}, H_b, V_{-c}, V_d\}$ that intersects the interior of at most h/ℓ holes. Let γ_2 follow L from p to the boundary of P_0 such that when it encounters a hole P_i, it makes a detour along ∂P_i.

It remains to analyze the length of γ. By Lemma 2, we have $|\gamma_1| \leq O(\sqrt{h}) \cdot \text{diam}_2(B) \leq O(h^{1/2}\ell\Delta)$. The path γ_2 has edges along the line L and along the boundaries of holes whose interior intersect L. The total length of all edges along L is at most $\text{diam}_2(P) = 1$. It is well known that $\text{per}(C) \leq \pi \cdot \text{diam}_2(C)$ for every convex body [21, 25], and so the length of each detour is $O(\text{diam}_2(P_i)) \leq O(\Delta)$, and the total length of $O(h/\ell)$ detours is $O(h\Delta/\ell)$. Consequently,

$$|\gamma| \leq O(h^{1/2}\ell\Delta + h\Delta/\ell + 1). \tag{4}$$

Finally, we set $\ell = \lceil h^{1/4} \rceil$ to balance the first two terms in (4), and obtain $|\gamma| \leq O(h^{3/4}\Delta + 1)$, as claimed. □

When all holes are line segment, we construct a monotone path from s to the outer boundary. A polygonal path $\gamma = (p_0, p_1, \ldots, p_m)$ is *$\boldsymbol{u}$-monotone* for a unit vector $\boldsymbol{u} \in \mathbb{S}^1$ if $\boldsymbol{u} \cdot \overrightarrow{v_{i-1}v_i} \geq 0$ for all $i \in \{1, \ldots, m\}$; and γ is *monotone* if it is $\boldsymbol{u}$-monotone for some $\boldsymbol{u} \in \mathbb{S}^1$.

Lemma 5. *Let $P \in \mathcal{C}(h)$ such that every hole is a line segments of length at most $\Delta \cdot \text{diam}_2(P)$. If $\Delta \geq h^{-1}$, then there exists a monotone path of length $O(h^{1/2}\Delta^{1/2}) \cdot \text{diam}_2(P)$ in P from any point $s \in P$ to the outer boundary ∂P_0.*

Proof. We may assume w.l.o.g. that $\mathrm{diam}_2(P) = 1$. Denote the line segments by a_ib_i, for $i = 1, \ldots, h$, such that $x(a_i) \le x(b_i)$. Let $\ell = \lceil h^{1/2}\Delta^{1/2} \rceil$, and note that $\ell = \Theta(h^{1/2}\Delta^{1/2})$ when $\Delta \ge h^{-1}$. Partition the right halfplane (i.e., right of the y-axis) into ℓ wedges with aperture π/ℓ and apex at the origin, denoted $W_1, \ldots, W_\ell$. For each wedge W_i, let $\boldsymbol{w}_i \in \mathbb{S}$ be the direction vector of its axis of symmetry.

Partition the h segments as follows: For $j = 1, \ldots, \ell$, let $\mathcal{H}_j$ be the set of segments a_ib_i such that $\overrightarrow{a_ib_i}$ is in W_j. Finally, let $\mathcal{H}_{j^*}$ be a set with minimal cardinality, that is, $|\mathcal{H}_{j^*}| \le h/\ell = O(h^{1/2}/\Delta^{1/2})$. Let $\boldsymbol{v} = \boldsymbol{w}_{j^*}^{\perp}$. We construct a $\boldsymbol{v}$-monotone path γ from s to the outer boundary ∂P_0 as follows. Start in direction $\boldsymbol{v}$ until reaching a hole a_ib_i at some point p. While $p \notin \partial P_0$, continue along a_ib_i to one of the endpoints: to a_i if $\boldsymbol{v} \cdot \overrightarrow{a_ib_i} \ge 0$, and to b_i otherwise; then continue in direction $\boldsymbol{v}$. By monotonicity, γ visits every edge at most once.

It remains to analyze the length of γ. We distinguish between two types of edges: let E_1 be the set of edges of γ contained in $\mathcal{H}_{j^*}$, and E_2 be the set of all other edges of γ. The total length of edges in E_1 is at most the total length of all segments in $\mathcal{H}_{j^*}$, that is,

$$\sum_{e\in E_1} |e| \le |\mathcal{H}_{j^*}| \cdot \Delta \le O(h^{1/2}/\Delta^{1/2}) \cdot \Delta = O(h^{1/2}\Delta^{1/2}).$$

Every edge $e \in E_2$ makes an angle at least $\pi/(2\ell)$ with vector $\boldsymbol{v}$. Let $\mathrm{proj}(e)$ denote the orthogonal projection of e to a line of direction $\boldsymbol{v}$. Then $|\mathrm{proj}(e)| \ge |e| \sin(\pi/(2\ell))$. By monotonicity, the projections of distinct edges have disjoint interiors. Consequently, $\sum_{e\in E_2} |\mathrm{proj}(e)| \le \mathrm{diam}_2(P) = 1$. This yields

$$\begin{aligned}\sum_{e\in E_2} |e| &\le \sum_{e\in E_2} \frac{|\mathrm{proj}(e)|}{\sin(\pi/(2\ell))} = \frac{1}{\sin(\pi/(2\ell))} \sum_{e\in E_2} |\mathrm{proj}(e)| \\ &= O(\ell) = O(h^{1/2}\Delta^{1/2}).\end{aligned}$$

Overall, $|\gamma| = \sum_{e\in E_1} |e| + \sum_{e\in E_2} |e| = O(h^{1/2}\Delta^{1/2})$, as claimed. □

For extending Lemma 5 to arbitrary convex holes, we need the following technical lemma. (All omitted proofs are available in the full paper [8].)

Lemma 6. *Let P be a convex polygon with a diametral pair $a, b \in \partial P$, where $|ab| = \mathrm{diam}_2(P)$. Suppose that a line L intersects the interior of P, but does not cross the line segment ab. Let $p, q \in \partial P$ such that $pq = L \cap P$, and points a, p, q, and b appear in this counterclockwise order in ∂P; and let $\widehat{pq}$ be the counterclockwise pq-arc of ∂P. Then $|\widehat{pq}| \le \frac{4\pi\sqrt{3}}{9}|pq| < 2.42|pq|$.*

The final result is as follows.

Lemma 7. *Let $P \in \mathcal{C}(h)$ such that $\mathrm{diam}_2(P_i) \le \Delta \cdot \mathrm{diam}_2(P)$ for every hole P_i. If $\Delta \ge h^{-1}$, then there exists a path of length $O(h^{1/2}\Delta^{1/2}) \cdot \mathrm{diam}_2(P)$ in P from any point $s \in P$ to the outer boundary ∂P_0.*

4 Polygons with Fat or Axis-Aligned Convex Holes

In this section, we show that in a polygonal domain P with fat convex holes, the distortion $\mathrm{geod}(s,t)/|st|$ is bounded by a constant for all $s,t \in P$. Let C be a convex body in the plane. The *geometric dilation* of C is $\delta(C) = \sup_{s,t\in\partial C} \frac{\mathrm{geod}(s,t)}{|st|}$, where $\mathrm{geod}(s,t)$ is the shortest st-path along the boundary of C.

Lemma 8. *Let C be a λ-fat convex body. Then $\delta(C) \leq \min\{\pi\lambda^{-1}, 2(\lambda^{-1}+1)\} = O(\lambda^{-1})$.*

Corollary 2. *Let $P = P_0 \setminus \left(\bigcup_{i=1}^{h} P_i\right)$ be a polygonal domain, where $P_0, P_1, \ldots, P_h$ are λ-fat convex polygons. Then for any $s,t \in P$, we have $\mathrm{geod}(s,t) \leq O(\lambda^{-1}|st|)$.*

Proof. If the line segment st is contained in P, then $\mathrm{geod}(s,t) = |st|$, and the proof is complete. Otherwise, segment st is the concatenation of line segments contained in P and line segments $p_iq_i \subset P_i$ with $p_i, q_i \in \partial P_i$, for some indices $i \in \{1,\ldots,h\}$. By replacing each segment p_iq_i with the shortest path on the boundary of the hole P_i, we obtain an st-path γ in P. Since each hole is λ-fat, we replaced each line segment p_iq_i with a path of length $O(|p_iq_i|/\lambda)$ by Lemma 8. Overall, we have $|\gamma| \leq O(|st|/\lambda)$, as required. □

Corollary 3. *If $P = P_0\setminus\left(\bigcup_{i=1}^{h} P_i\right)$ be a polygonal domain, where $P_0, P_1, \ldots, P_h$ are λ-fat convex polygons for some $0 < \lambda \leq 1$, then $\mathrm{diam}_g(P) \leq O(\lambda^{-1}\mathrm{diam}_2(P))$, hence $\varrho(P) \leq O(\lambda^{-1})$.*

Proposition 3. *Let $P \in \mathcal{C}(h)$, $h \in \mathbb{N}$, such that every hole is an axis-aligned rectangle. Then from any point $s \in P$, there exists a path of length at most $\mathrm{diam}_2(P)$ in P to the outer boundary ∂P_0.*

Proof. Let $B = [0,a] \times [0,b]$ be a minimal axis-parallel bounding box containing P. We may assume w.l.o.g. that $x(s) \geq a/2$, $y(s) \geq b/2$, and $b \leq a$. We construct a staircase path γ as follows. Start from s in horizontal direction $\boldsymbol{d}_1 = (1,0)$ until reaching the boundary ∂P at some point p. While $p \notin \partial P_0$, make a $90°$ turn from $\boldsymbol{d}_1 = (1,0)$ to $\boldsymbol{d}_2 = (0,1)$ or vice versa, and continue. We have $|\gamma| \leq \frac{a+b}{2} \leq a \leq \mathrm{diam}_2(P)$, as claimed. □

5 Polygons with Holes Versus Triangulations

The proof of Theorem 3 is the combination of Lemmas 9 and 10 below (the proof of Lemma 9 is deferred to the full version of this paper [8]).

Lemma 9. *For every triangulation $T \in \mathcal{T}(n)$, there exists a polygonal domain $P \in \mathcal{C}(h)$ with $h = \Theta(n)$ holes such that $\varrho(P) = \Theta(\varrho(T))$.*

Every planar straight-line graph $G = (V, E)$ can be augmented to a triangulation $T = (V, E')$, with $E \subseteq E'$. A notable triangulation is the *Constrained Delaunay Triangulation*, for short, CDT(G). Bose and Keil [4] proved that CDT(G) has bounded stretch for so-called *visibility* edges: if $u, v \in V$ and uv does not cross any edge of G, then CDT(G) contains a uv-path of length $O(|uv|)$.

Lemma 10. *For every polygonal domain $P \in \mathcal{C}(h)$, there exists a triangulation $T \in \mathcal{T}(n)$ with $n = \Theta(h)$ vertices such that $\varrho(T) = \Theta(\varrho(P))$.*

Proof. Assume that $P = P_0 \setminus \bigcup_{i=1}^{h} P_i$. For all $j = 1, \ldots, h$, let $a_i, b_i \in \partial P_i$ be a diametral pair, that is, $|a_i b_i| = \mathrm{diam}_2(P_i)$. The line segments $\{a_i b_i : i = 1, \ldots, h\}$, together with the four vertices of a minimum axis-aligned bounding box of P, form a planar straight-line graph G with $2h + 4$ vertices. Let $T =$ CDT(G) be the constrained Delaunay triangulation of G.

We claim that $\varrho(T) = \Theta(\varrho(P))$. We prove this claim in two steps. For an intermediate step, we define a polygon with h line segment holes: $P' = P_0 \setminus \bigcup_{i=1}^{h} \{a_i b_i\}$. For any point pair $s, t \in P$, denote by $\mathrm{dist}(s, t)$ and $\mathrm{dist}'(s, t)$, resp., the shortest distance in P and P'. Since $P \subseteq P'$, we have $\mathrm{dist}'(s, t) \leq \mathrm{dist}(s, t)$. By Lemma 6, $\mathrm{dist}(s, t) < 2.42 \cdot \mathrm{dist}'(s, t)$ so $\mathrm{dist}'(s, t) = \Theta(\mathrm{dist}(s, t))$, $\forall s, t \in P$.

Every point $s \in P$ lies in one or more triangles in T; let s' denote a closest vertex of a triangle in T that contains s. For $s, t \in P$, let $\mathrm{dist}''(s, t)$ be the length of the st-path γ composed of the segment ss', a shortest $s't'$-path in the triangulation T, and the segment $t't$.

Since γ does not cross any of the line segments $a_j b_j$, we have $\mathrm{dist}'(s, t) \leq \mathrm{dist}''(s, t)$ for any pair of points $s, t \in P$. Conversely, every vertex in the shortest $s't'$-path in P' is an endpoint of an obstacle $a_j b_j$. Consequently, every edge is either an obstacle segment $a_j b_j$, or a visibility edge between the endpoints of two distinct obstacles. By the result of Bose and Keil [4], for every such edge pq, T contains a pq-path τ_{pq} of length $|\tau_{pq}| \leq O(|pq|)$. The concatenation of these paths is an $s't'$-path τ of length $|\tau| \leq O(\mathrm{dist}'(s', t'))$. Finally, note that the diameter of each triangle in T is at most $\mathrm{diam}_2(P')$. Consequently, if $s, t \in P$ maximizes $\mathrm{dist}(s, t)$, then $\mathrm{dist}''(s, t) = |ss'| + |\gamma| + |t't| \leq 2 \cdot \mathrm{diam}_2(P) + |\tau| \leq O(\mathrm{dist}'(s't'))$. Consequently, $\mathrm{diam}_g(T) = \Theta(\mathrm{diam}_g(P))$, which yields $\varrho(T) = \Theta(\varrho(P))$. □

Acknowledgments. Research on this paper was partially supported by the NSF awards DMS 1800734 and DMS 2154347.

References

1. Ahn, H.K., Barba, L., Bose, P., De Carufel, J.L., Korman, M., Oh, E.: A linear-time algorithm for the geodesic center of a simple polygon. Discrete Comput. Geom. **56**, 836–859 (2016)
2. Bae, S.W., Korman, M., Okamoto, Y.: The geodesic diameter of polygonal domains. Discrete Comput. Geom. **50**(2), 306–329 (2013)
3. Bae, S.W., Korman, M., Okamoto, Y.: Computing the geodesic centers of a polygonal domain. Comput. Geom. **77**, 3–9 (2019)

4. Bose, P., Keil, J.M.: On the stretch factor of the constrained Delaunay triangulation. In: Proc. 3rd IEEE Symposium on Voronoi Diagrams in Science and Engineering (ISVD), pp. 25–31 (2006)
5. Cabello, S.: Subquadratic algorithms for the diameter and the sum of pairwise distances in planar graphs. ACM Trans. Algorithms **15**(2), 1–38 (2019)
6. Chiang, Y.J., Mitchell, J.S.B.: Two-point Euclidean shortest path queries in the plane. In: Proc. 10th ACM-SIAM Symposium on Discrete Algorithms (SODA), pp. 215–224 (1999). https://doi.org/10.5555/314500.314560
7. Dalirrooyfard, M., Li, R., Williams, V.V.: Hardness of approximate diameter: Now for undirected graphs. In: Proc. 62nd IEEE Symposium on Foundations of Computer Science (FOCS), pp. 1021–1032. IEEE (2021)
8. Dumitrescu, A., Tóth, C.D.: Maximal distortion of geodesic diameters in polygonal domains. arXiv preprint arXiv:2304.03484 (2023)
9. Finch, S.R., Wetzel, J.E.: Lost in a forest. Am. Math. Mon. **111**(8), 645–654 (2004)
10. Gawrychowski, P., Kaplan, H., Mozes, S., Sharir, M., Weimann, O.: Voronoi diagrams on planar graphs, and computing the diameter in deterministic $\tilde{O}(n^{5/3})$ time. SIAM J. Comput. **50**(2), 509–554 (2021)
11. Guibas, L.J., Hershberger, J.: Optimal shortest path queries in a simple polygon. J. Comput. Syst. Sci. **39**, 126–152 (1989)
12. Har-Peled, S.: Shortest path in a polygon using sublinear space. J. Comput. Geom. **7**, 19–45 (2015)
13. Hershberger, J., Suri, S.: Matrix searching with the shortest-path metric. SIAM J. Comput. **26**(6), 1612–1634 (1997)
14. Hershberger, J., Suri, S.: An optimal algorithm for Euclidean shortest paths in the plane. SIAM J. Comput. **28**(6), 2215–2256 (1999)
15. Kapoor, S., Maheshwari, S.N., Mitchell, J.S.B.: An efficient algorithm for Euclidean shortest paths among polygonal obstacles in the plane. Discrete Comput. Geom. **18**, 377–383 (1997)
16. Kübel, D., Langetepe, E.: On the approximation of shortest escape paths. Comput. Geom. **93**, 101709 (2021)
17. Lee, D.T., Preparata, F.P.: Euclidean shortest paths in the presence of rectilinear barriers. Networks **14**, 393–410 (1984)
18. Mitchell, J.S.B.: Shortest paths among obstacles in the plane. Int. J. Comput. Geom. Appl. **6**(3), 309–332 (1996)
19. Mitchell, J.S.: Shortest paths and networks. In: Handbook of Discrete and Computational Geometry, 3 edn. CRC Press, Boca Raton, FL (2017). Chap. 31
20. Roditty, L., Williams, V.V.: Fast approximation algorithms for the diameter and radius of sparse graphs. In: Proc. 45th Symposium on Theory of Computing Conference (STOC), pp. 515–524. ACM (2013)
21. Scott, P.R., Awyong, P.W.: Inequalities for convex sets. J. Inequal. Pure Appl. Math. **1**(1), 6 (2000)
22. Wang, H.: On the geodesic centers of polygonal domains. J. Comput. Geom. **9**(1), 131–190 (2018)
23. Wang, H.: A new algorithm for Euclidean shortest paths in the plane. In: Proc. 53rd ACM Symposium on Theory of Computing (STOC), pp. 975–988 (2021)
24. Wang, H.: Shortest paths among obstacles in the plane revisited. In: Proc. 32nd ACM-SIAM Symposium on Discrete Algorithms (SODA), pp. 810–821 (2021)
25. Yaglom, I.M., Boltyanskii, V.G.: Convex Figures. Library of the Mathematical Circle (1951). Translated by Kelly, P.J., Walton, L.F., Holt, R.: Winston, New York (1961)

On 2-Strong Connectivity Orientations of Mixed Graphs and Related Problems

Loukas Georgiadis, Dionysios Kefallinos(✉), and Evangelos Kosinas

Department of Computer Science and Engineering,
University of Ioannina, Ioannina, Greece
{loukas,dkefallinos,ekosinas}@cse.uoi.gr

Abstract. Given a mixed graph G, we consider the problem of computing the maximal sets of vertices $C_1, C_2, \ldots, C_k$ with the property that by removing any edge e from G, there is an orientation R_i of $G \setminus e$ such that all vertices in C_i are strongly connected in R_i. We study properties of those sets, and show how to compute them in linear time via a reduction to the computation of the 2-edge twinless strongly connected components of a directed graph. A directed graph G is twinless strongly connected if it contains a strongly connected spanning subgraph without any pair of antiparallel (or *twin*) edges. The twinless strongly connected components (TSCCs) of G are its maximal twinless strongly connected subgraphs. A 2-*edge twinless strongly connected component (2eTSCC) of* G is a maximal subset of vertices C such that any two vertices $u, v \in C$ are in the same twinless strongly connected component TSCC of $G \setminus e$, for any edge e. These concepts are motivated by several diverse applications, such as the design of road and telecommunication networks, and the structural stability of buildings.

Keywords: Connectivity · Orientations · Mixed Graphs

1 Introduction

We investigate some connectivity problems in mixed graphs and in directed graphs (digraphs). A mixed graph G contains both undirected edges and directed edges. We denote an edge with endpoints u and v by $\{u, v\}$ if it is undirected, and by (u, v) if it is directed from u to v. An *orientation* R of G is formed by orienting all the undirected edges of G, i.e., converting each undirected edge $\{u, v\}$ into a directed edge that is either (u, v) or (v, u). Several (undirected or mixed) graph orientation problems have been studied in the literature, depending on the properties that we wish an orientation R of G to have. See, e.g., [3,15,34,36]. An orientation R of G such that R is strongly connected is called a *strong orientation of* G. More generally, an orientation R of G such that R is k-edge strongly

Research supported by the Hellenic Foundation for Research and Innovation (H.F.R.I.), under the "First Call for H.F.R.I. Research Projects to support Faculty members and Researchers and the procurement of high-cost research equipment grant", Project FANTA (eFficient Algorithms for NeTwork Analysis), number HFRI-FM17-431.

S.-Y. Hsieh et al. (Eds.): IWOCA 2023, LNCS 13889, pp. 209–220, 2023.
https://doi.org/10.1007/978-3-031-34347-6_18

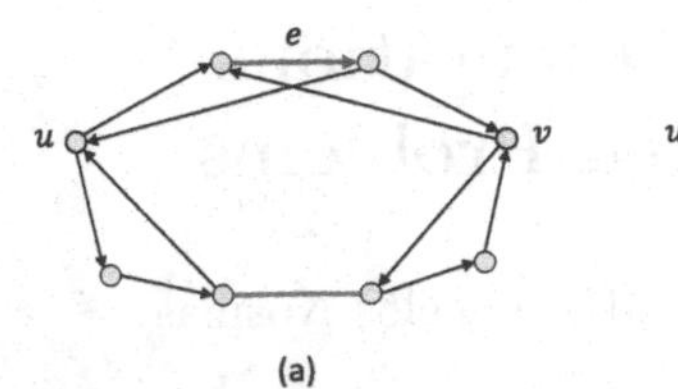

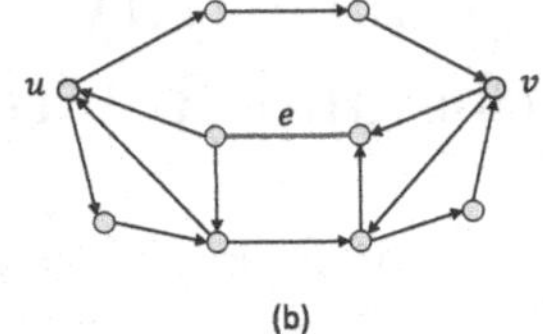

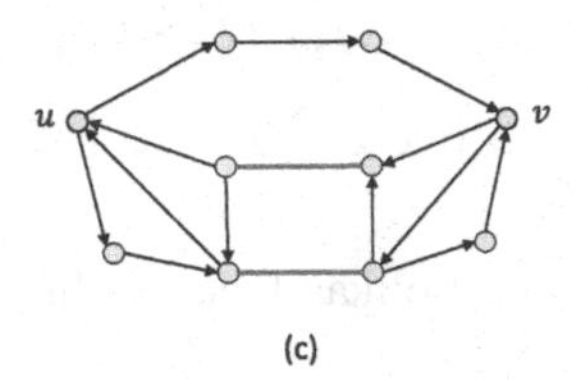

Fig. 1. (a)–(b) Vertices u and v are not in the same edge-resilient strongly orientable block of G. After the deletion of edge e (which is directed in (a) and undirected in (b)), there is no orientation of $G \setminus e$ such that u and v are strongly connected. (c) Here u and v are in the same edge-resilient strongly orientable block of G, since for any edge e, there is an orientation of $G \setminus e$ such that u and v are strongly connected.

connected is called a *k-edge strong orientation of* G. Motivated by recent work in 2-edge strong connectivity in digraphs [19,22,26], we introduce the following strong connectivity orientation problem in mixed graphs. Given a mixed graph G, we wish to compute its maximal sets of vertices $C_1, C_2, \ldots, C_k$ with the property that for every $i \in \{1, \ldots, k\}$, and every edge e of G (directed or undirected), there is an orientation R of $G \setminus e$ such that all vertices of C_i are strongly connected in R. We refer to these maximal vertex sets as the *edge-resilient strongly orientable blocks* of G. See Fig. 1. Note that when G contains only directed edges, then this definition coincides with the usual notion of 2-edge strong connectivity, i.e., each C_i is a 2-edge strongly connected component of G. We show how to solve this problem in linear time, by providing a linear-time algorithm for computing the 2-edge twinless strongly connected components [30], that we define next.

We recall some concepts in directed graphs. A digraph $G = (V, E)$ is *strongly connected* if there is a directed path from each vertex to every other vertex. The *strongly connected components* (SCCs) of G are its maximal strongly connected subgraphs. We refer to a pair of antiparallel edges, (x, y) and (y, x), of G as *twin edges*. A digraph $G = (V, E)$ is *twinless strongly connected* if it contains a strongly connected spanning subgraph (V, E') without any pair of twin edges. The *twinless strongly connected components* (TSCCs) of G are its maximal twinless strongly connected subgraphs. Two vertices $u, v \in V$ are *twinless strongly connected* if they belong to the same twinless strongly connected component of G. Raghavan [35] provided a characterization of twinless strongly connected digraphs, and, based on this characterization, presented a linear-time algorithm for computing the TSCCs. An edge (resp., a vertex) of a digraph G is a *strong bridge* (resp., a *strong articulation point*) if its removal increases the number of strongly connected components. A strongly connected digraph G is *2-edge strongly connected* if it has no strong bridges, and it is *2-vertex strongly connected* if it has at least three vertices and no strong articulation points. Two vertices $u, v \in V$ are said to be *2-edge strongly connected* (resp., *2-vertex strongly connected*) if there are two edge-disjoint (resp., two internally vertex-disjoint) directed paths from u to v and two edge-disjoint (resp., two internally vertex-disjoint) directed paths from v to u. By Menger's theorem [32] we have that u and v are 2-edge strongly connected if they remain in the same SCC after the deletion of any edge. A *2-edge strongly connected component* (resp., *2-vertex strongly connected component*) of a digraph $G = (V, E)$ is defined as a maximal subset $C \subseteq V$ such that every two vertices $u, v \in C$ are 2-edge strongly connected (resp., 2-vertex strongly connected).

An edge $e \in E$ is a *twinless strong bridge* of G if the deletion of e increases the number of TSCCs of G. Similarly, a vertex $v \in V$ is a *twinless strong articulation point* of G if the deletion of v increases the number of TSCCs of G. All twinless strong bridges and twinless strong articulation points can be computed in linear time [21]. A *2-edge twinless strongly connected component (2eTSCC) of* G is a maximal subset of vertices C such that any two vertices $u, v \in C$ are in the same TSCC of $G \setminus e$, for any edge e. Two vertices u and v are *2-edge twinless strongly connected* if they belong to the same 2eTSCC. Jaberi [30] studied some properties of 2-edge twinless strongly connected components, and presented an $O(mn)$-time algorithm for a digraph with m edges and n vertices. We provide a linear-time algorithm that is based on two notions: (i) a collection of auxiliary graphs $\mathcal{H}$ that preserve the 2-edge twinless strongly connected components of G and, for any $H \in \mathcal{H}$, the SCCs of H after the deletion of any edge have a very simple structure, and (ii) a reduction to the problem of computing the connected components of an undirected graph after the deletion of certain vertex-edge cuts.

The notions of twinless strong connectivity and mixed graph orientations are motivated by several applications, such as the design of road and telecommunication networks, the structural stability of buildings [1,7,10,35], and the analysis of biological networks [12]. The computation of edge-resilient strongly orientable blocks is related to 2-edge strong orientations of mixed graphs in the following sense. A mixed graph G has a 2-edge strong orientation only if it consists of a single edge-resilient strongly orientable block. While finding a strong orientation of a mixed graph is well understood and can be solved in linear time [7,9], computing a k-edge strong orientation for $k > 1$ seems much harder. Frank [14] gave a polynomial-time algorithm for this problem based on the concept of submodular flows. Faster algorithms were presented in [16,29]. Also, more efficient algorithms exist for undirected graphs [6,18,33].

Due to space constraints, we omit some technical details and proofs. They can be found in the full version of the paper [23].

2 Preliminaries

Let G be a (directed or undirected) graph. In general, we allow G to have multiple edges, unless otherwise specified. We denote by $V(G)$ and $E(G)$, respectively, the vertex set and the edge set of G. For a set of edges (resp., vertices) S, we let $G \setminus S$ denote the graph that results from G after deleting the edges in S (resp., the vertices in S and their incident edges). We extend this notation for mixed sets S, that may contain both vertices and edges of G, in the obvious way. Also, if S has only one element x, we abbreviate $G \setminus S$ by $G \setminus x$. Let $C \subseteq V(G)$. The induced subgraph of C, denoted by $G[C]$, is the subgraph of G with vertex set C and edge set $\{e \in E \mid \text{both endpoints of } e \text{ are in } C\}$.

For any two vertices x and y of a directed graph G, the notation $x \overset{G}{\leftrightarrow} y$ means that x and y are strongly connected in G, and the notation $x \overset{G}{\leftrightarrow}_t y$ means that x and y are twinless strongly connected in G. We omit the reference graph G from the $\overset{G}{\leftrightarrow}$ notation when it is clear from the context. Thus we may simply write $x \leftrightarrow y$ and $x \leftrightarrow_t y$. Similarly, we let $x \overset{G}{\leftrightarrow}_{2e} y$ and $x \overset{G}{\leftrightarrow}_{2et} y$ denote, respectively, that the vertices x and y are 2-edge strongly connected and 2-edge twinless strongly connected in G. Let $G = (V, E)$ be a strongly connected digraph. The *reverse digraph* of G, denoted by $G^R = (V, E^R)$, is the digraph that results from G by reversing the direction of all edges. In a digraph G, we say that a vertex x *reaches* y if there is a path in G from x to y. We say that an edge e of a strongly connected digraph G *separates* two vertices x and y if x and y belong to different strongly connected components of $G \setminus e$.

For any digraph G, the associated undirected graph G^u is the *simple* undirected graph with vertices $V(G^u) = V(G)$ and edges $E(G^u) = \{\{u, v\} \mid (u, v) \in E(G) \vee (v, u) \in E(G)\}$. Let H be an undirected graph. An edge $e \in E(H)$ is a *bridge* if its removal increases the number of connected components of H. A connected graph H is 2-edge-connected if it contains no bridges. Raghavan [35] showed that a strongly connected digraph G is twinless strongly connected if and only if its underlying undirected graph G^u is 2-edge-connected.

We introduce the concept of *marked vertex-edge blocks* of an undirected graph, which will be needed in our algorithm for computing the 2-edge twinless strongly connected components. (We note that Heinrich et al. [25] introduced the related concept of the 2.5-*connected components* of a biconnected graph.) Let G be an undirected graph where some vertices of G are marked. Let V' be the set of the marked vertices of G. Then, a marked vertex-edge block of G is a maximal subset B of $V(G) \setminus V'$ with the property that all vertices of B remain connected in $G \setminus \{v, e\}$, for every marked vertex v and any edge e. In Sect. 5 we provide a linear-time algorithm for computing the marked-vertex edge blocks of a biconnected undirected graph G, by exploiting properties of the SPQR-tree of the triconnected components of G [4,5].

Let G be a mixed graph. By *splitting* a directed edge (x, y) of a graph G, we mean that we remove (x, y) from G, and we introduce a new auxiliary vertex z and two edges $(x, z), (z, y)$. By *replacing with a gadget* an undirected edge $\{x, y\}$ of a graph G, we mean that we remove $\{x, y\}$ from G, and we introduce three new auxiliary vertices z, u, v and the edges $(x, z), (z, x), (z, u), (u, v), (v, y), (y, u), (v, z)$. We show that after splitting every directed edge and replacing every undirected edge of G with a gadget, we reduce the computation of the edge-resilient strongly orientable blocks of a mixed graph to the computation of the 2eTSCC of a digraph.

3 Connectivity-Preserving Auxiliary Graphs

In this section we describe how to construct a set of auxiliary graphs that preserve the 2-edge twinless strongly connected components of a twinless strongly connected digraph, and moreover have the property that their strongly connected components after the deletion of any edge have a very simple structure. We base our construction on the auxiliary graphs defined in [19] for computing the 2-edge strongly connected components of a digraph, and perform additional operations in order to achieve the desired properties. We note that a similar construction was given in [20] to derive auxiliary graphs (referred to as 2-connectivity-light graphs) that enable the fast computation of the 3-edge strongly connected components of a digraph. Still, we cannot apply directly the construction of [20], since we also need to maintain twinless strong connectivity.

Flow Graphs and Dominator Trees. A *flow graph* is a directed graph with a distinguished *start vertex* s such that every vertex is reachable from s. For a digraph G, we use the notation G_s in order to emphasize the fact that we consider G as a flow graph with source s. Let $G = (V, E)$ be a strongly connected graph. We will let s be a fixed but arbitrary start vertex of G. Since G is strongly connected, all vertices are reachable from s and reach s, so we can refer to the flow graphs G_s and G_s^R.

Let G_s be a flow graph with start vertex s. A vertex u is a *dominator* of a vertex v (u *dominates* v) if every path from s to v in G_s contains u; u is a *proper dominator* of v if u dominates v and $u \neq v$. The dominator relation is reflexive and transitive. Its transitive reduction is a rooted tree, the *dominator tree* $D(G_s)$: u dominates v if and only if u is an ancestor of v in $D(G_s)$. For every vertex $x \neq s$ of G_s, $d(x)$ is the immediate dominator of x in G_s (i.e., the parent of x in $D(G_s)$). For every vertex r of G_s, we let $D(r)$ denote the subtree of $D(G_s)$ rooted at r. The dominator tree can be computed in almost-linear time [31] or even in truly-linear time [2,8,13,17]. An edge (u, v) is a *bridge* of a flow graph G_s if all paths from s to v include (u, v).[1]

Property 1. ([28]) Let s be an arbitrary start vertex of G. An edge $e = (u, v)$ is strong bridge of G if and only if it is a bridge of G_s, in which case $u = d(v)$, or a bridge of G_s^R, in which case $v = d^R(u)$, or both.

Let G_s be a strongly connected digraph. For every bridge (x, y) of G_s, we say that y is a *marked* vertex. (Notice that s cannot be marked.) Property 1 implies that the bridges of G_s induce a decomposition of $D(G_s)$ into rooted subtrees. More precisely, for every bridge (x, y) of G_s, we remove the edge (x, y) from $D(G_s)$. (By Property 1, this is indeed an edge of $D(G_s)$.) Thus we have partitioned $D(G_s)$ into subtrees. Every tree T in this decomposition inherits the parent relation from $D(G_s)$, and thus it is rooted at a vertex r. We denote T as $T(r)$ to emphasize the fact that the root of T is r. Observe that the root r of a tree $T(r)$ is either a marked vertex or s. Conversely, for every vertex r that is either marked or s, there is a tree $T(r)$.

---

[1] Throughout the paper, to avoid confusion we use consistently the term *bridge* to refer to a bridge of a flow graph and the term *strong bridge* to refer to a strong bridge in the original graph.

Construction of Auxiliary Graphs. Now let G_s be a strongly connected digraph, and let r be either a marked vertex of G_s or s. We define the *auxiliary* graph $H(G_s, r)$ as follows. In G_s we shrink every $D(z)$, where z is a marked vertex such that $d(z) \in T(r)$, into z. Also, if $r \neq s$, we shrink $D(s) \setminus D(r)$ into $d(r)$. During those shrinkings we maintain all edges, except for self-loops. Also, in [19] multiple edges are converted into single edges. Here, multiple edges are converted into double edges, in order to avoid introducing new strong bridges in the auxiliary graphs. The resulting graph is $H(G_s, r)$. We consider $H(G_s, r)$ as a flow graph with start vertex r. Notice that it consists of the subgraph of G_s induced by $T(r)$, plus some extra vertices and edges. To be specific, the vertex set of $H(G_s, r)$ consists of the vertices of $T(r)$, plus all marked vertices z of G_s such that $d(z) \in T(r)$, plus $d(r)$ if $r \neq s$. The vertices of $T(r)$ are called *ordinary* in $H(G_s, r)$. The vertices of $H(G_s, r) \setminus T(r)$ are called *auxiliary* in $H(G_s, r)$. In particular, if $r \neq s$, $d(r)$ is called the *critical* vertex of $H(G_s, r)$, and $(d(r), r)$ is called the *critical* edge of $H(G_s, r)$. (Thus, $H(G_s, s)$ is the only auxiliary graph of G_s that has no critical vertex and no critical edge.)

The above construction guarantees that each path in G_s whose endpoints lie in some auxiliary graph $H(G_s, r)$ has a corresponding path in $H(G_s, r)$ with the same endpoints and vice versa. In particular, this implies that each $H(G_s, r)$ is strongly connected. Moreover, we have the following results:

Theorem 1. ([19]) *Let G_s be a strongly connected digraph, and let $r_1, \ldots, r_k$ be the marked vertices of G_s.*

(i) For any two vertices x and y of G_s, $x \overset{G_s}{\leftrightarrow}_{2e} y$ if and only if there is a vertex r (a marked vertex of G_s or s), such that x and y are both ordinary vertices of $H(G_s, r)$ and $x \overset{H(G_s,r)}{\leftrightarrow}_{2e} y$.
(ii) The collection $H(G_s, s), H(G_s, r_1), \ldots, H(G_s, r_k)$ of all the auxiliary graphs of G_s can be computed in linear time.

We provide the analogous result for 2-edge twinless strong connectivity.

Proposition 1. *Let x, y be two vertices of a strongly connected digraph G_s. Then $x \overset{G_s}{\leftrightarrow}_{2et} y$ if and only if there is a vertex r (a marked vertex of G_s or s), such that x and y are both ordinary vertices of $H(G_s, r)$ and $x \overset{H(G_s,r)}{\leftrightarrow}_{2et} y$.*

Now let G be a strongly connected digraph and let (x, y) be a strong bridge of G. We will define the *S-operation* on G and (x, y), which produces a set of digraphs as follows. Let $C_1, \ldots, C_k$ be the strongly connected components of $G \setminus (x, y)$. Now let $C \in \{C_1, \ldots, C_k\}$. We will construct a graph C' as follows. First, notice that either $x \notin C$ and $y \in C$, or $y \notin C$ and $x \in C$, or $\{x, y\} \cap C = \emptyset$. Then we set $V(C') = V(C) \cup \{x\}$, or $V(C') = V(C) \cup \{y\}$, or $V(C') = V(C) \cup \{x, y\}$, respectively. Every edge of G with both endpoints in C is included in C'. Furthermore, for every edge (u, v) of G such that $u \in C$ and $v \notin C$, we add the edge (u, x) to C'. Also, for every edge (u, v) of G such that $u \notin C$ and $v \in C$, we add the edge (y, v) to C'. Finally, we also add the edge (x, y) to C'. Now we define $S(G,(x,y)) := \{C'_1,\ldots,C'_k\}$. Note that for a strongly connected digraph G and a strong bridge e of G, every graph of $S(G,e)$ is strongly connected. Furthermore, the next proposition shows that the S-operation maintains the relation of 2-edge twinless strong connectivity.

Proposition 2. *Let G be a strongly connected digraph and let (x,y) be a strong bridge of G. Then, for any two vertices $u,v \in G$, we have $u \overset{G}{\leftrightarrow}_{2et} v$ if and only if u and v belong to the same graph C of $S(G,(x,y))$ and $u \overset{C}{\leftrightarrow}_{2et} v$.*

We can combine Propositions 1 and 2 in order to derive some auxiliary graphs that maintain the relation of 2-edge twinless strong connectivity of the original graph. Then we can exploit properties of those graphs in order to provide a linear-time algorithm for computing the 2-edge twinless strongly connected components. First we introduce some notation. Let G_s be a strongly connected digraph, and let r be either a marked vertex of G_s or s. Then we denote $H(G_s,r)$ as H_r. Furthermore, if r' is either a marked vertex of H_r or r, we denote $H(H_r^R,r')$ as $H_{rr'}$. A vertex that is ordinary in both H_r and $H_{rr'}$ is called an ordinary vertex of $H_{rr'}$; otherwise, it is called auxiliary.

Corollary 1. *Let G_s be a strongly connected digraph, and let x,y be two vertices of G_s. Then $x \overset{G_s}{\leftrightarrow}_{2et} y$ if and only if x and y are both ordinary vertices in H and $x \overset{H}{\leftrightarrow}_{2et} y$, where H is either (1) H_{ss}, or (2) H_{rr}, or (3) a graph in $S(H_{sr},(d(r),r))$, or (4) a graph in $S(H_{rr'},(d(r'),r'))$ (where r and r' are marked vertices).*

Now we can describe the structure of the strongly connected components of the graphs that appear in Corollary 1 when we remove a strong bridge from them.

Proposition 3. *Let H be one of the auxiliary graphs that appear in Corollary 1, and let $e=(x,y)$ be a strong bridge of H. Then the strongly connected components of $H \setminus e$ are given by one of the following:*

(i) $\{x\}$ *and* $H \setminus \{x\}$, *where x is an auxiliary vertex*
(ii) $\{y\}$ *and* $H \setminus \{y\}$, *where y is an auxiliary vertex*
(iii) $\{x\}$, $\{y\}$, *and* $H \setminus \{x,y\}$, *where x,y are both auxiliary vertices*

4 Computing 2eTSCCs

We assume that G is a twinless strongly connected digraph, since otherwise we can compute the twinless strongly connected components in linear time and process each one separately [35]. We let E_t denote the set of twinless strong bridges of G, and let E_s denote the set of strong bridges of G. (Note that $E_s \subseteq E_t$.) Jaberi [30] gave an $O(mn)$-time algorithm for computing the 2eTSCCs of G. We provide a faster algorithm that processes separately the edges in $E_t \setminus E_s$ and the edges in E_s, and partitions the vertices of G accordingly.

Let e be an edge in $E_t \setminus E_s$. Then the TSCCs of $G \setminus e$ are given by the 2-edge-connected components of $G^u \setminus \{e^u\}$, where e^u is the undirected counterpart of e [35]. Thus, we can simply remove the bridges of $G^u \setminus \{e^u\}$, in order to get the partition into the TSCCs that is due to e. To compute the partition that is due to all edges in $E_t \setminus E_s$ at once, we may use the cactus graph Q which is given by contracting the 3-edge-connected components of G^u into single nodes [11]. Q comes together with a function $\phi : V(G^u) \rightarrow V(Q)$ (the quotient map) that maps every vertex of G^u to the node of Q that contains it, and induces a natural correspondence between edges of G^u and edges of Q. The cactus graph of the 3-edge-connected components provides a clear representation of the 2-edge cuts of an undirected graph; by definition, it has the property that every edge of it belongs to exactly one cycle. Thus, Algorithm 1 shows how we can compute in linear time the partition of 2eTSCCs that is due to the edges in $E_t \setminus E_s$.

Algorithm 1: Compute the partition of 2eTSCCs of G that is due to the twinless strong bridges that are not strong bridges.

1 compute the cactus Q of the 3-edge-connected components of G^u, and let $\phi : V(G^u) \rightarrow V(Q)$ be the quotient map
2 **foreach** *edge e of Q* **do**
3 | **if** *e corresponds to a single edge of G that has no twin and is not a strong bridge* **then** remove from Q the edges of the cycle that contains e
4 **end**
5 let Q' be the graph that remains after all the removals in the previous step
6 let $C_1, \ldots, C_k$ be the connected components of Q'
7 **return** $\phi^{-1}(C_1), \ldots, \phi^{-1}(C_k)$

Now we consider the problem of computing the partition of the 2eTSCCs of G due to the strong bridges. Here we reduce the problem to the auxiliary graphs that appear in Corollary 1, and we apply the information provided by Proposition 3 as follows. Let H be one of those auxiliary graphs. For every strong bridge e of H, we define the subset X_e of $V(H)$ as $X_e = \{x\}$, or $X_e = \{y\}$, or $X_e = \{x, y\}$, depending on whether e satisfies (*i*), (*ii*), or (*iii*), respectively, of Proposition 3. Then, X_e satisfies (1) $H[V \setminus X_e]$ is a strongly connected component of $H \setminus e$, and (2) X_e contains only auxiliary vertices.

Now we can apply the following procedure to compute the partition of 2-edge twinless strongly connected components of the ordinary vertices of H due to the strong bridges. Initially, we let $\mathcal{P}$ be the trivial partition of V (i.e., $\mathcal{P} = \{V\}$). Then, for every strong bridge e of H, we compute the TSCCs of $H \setminus X_e$, and we refine $\mathcal{P}$ according to those TSCCs. By [35], the computation of the TSCCs of $H \setminus X_e$ is equivalent to determining the 2-edge-connected components of $H^u \setminus X_e$. Observe that this procedure does not run in linear time in total, since it has to be performed for every strong bridge e of H.

Thus our goal is to perform the above procedure for all strong bridges e of H at once. We can do this by first taking H^u, and then shrinking every X_e in

H^u into a single marked vertex, for every strong bridge e of H. Let H' be the resulting graph. Then we simply compute the marked vertex-edge blocks of H'. The whole procedure is shown in Algorithm 2. We note that, given an auxiliary graph H as above, we can compute all sets X_e in linear time by first computing all strong bridges of H [28], and then checking which case of Proposition 3 applies for each strong bridge.

Algorithm 2: A linear-time algorithm for computing the partition of 2-edge twinless strongly connected components of an auxiliary graph H due to the strong bridges

input : An auxiliary graph H equipped with the following information: for every strong bridge e of H, the set X_e defined as above
output: The partition of 2-edge twinless strongly connected components of the ordinary vertices of H due to the strong bridges

```
1  begin
2      compute the underlying undirected graph H^u
3      foreach strong bridge e of H do
4          contract X_e into a single vertex in H^u, and mark it
5      end
6      let H' be the graph with the marked contracted vertices derived from H^u
7      compute the partition B_ve of the marked vertex-edge blocks of H'
8      let O be the partition of V consisting of the set of the ordinary vertices of
       H and the set of the auxiliary vertices of H
9      return B_ve refined by O
10 end
```

The final 2eTSCCs of (the subset of the ordinary vertices of) an auxiliary graph are given by the mutual refinement of the partitions computed by Algorithms 1 and 2. (The mutual refinement of two partitions can be computed in linear time using bucket sort.) Hence, by Corollary 1, we obtain the 2eTSCCs of the original strongly connected digraph.

It remains to establish that Algorithm 2 runs in linear time. For this we provide a linear-time procedure for Step 7. Observe that the marked vertices of H' have the property that their removal from H leaves the graph strongly connected, and thus they are not articulation points of the underlying graph H^u. This allows us to reduce the computation of the marked vertex-edge blocks of H' to the computation of marked vertex-edge blocks in biconnected graphs. Specifically, we first partition H' into its biconnected components, which can be done in linear time [37]. Then we process each biconnected component separately, and we compute the marked vertex-edge blocks that are contained in it. Finally, we "glue" the marked vertex-edge blocks of all biconnected components, guided by their common vertices that are articulation points of the graph. In the next section we provide a linear-time algorithm for computing the marked vertex-edge blocks of a biconnected graph.

5 Computing Marked Vertex-Edge Blocks

Here we consider the problem of computing the marked vertex-edge blocks of a biconnected undirected graph G. Let V' be the set of the marked vertices of G. We provide a linear-time algorithm for computing all marked-vertex edge blocks of G, by exploiting properties of the SPQR-tree [4,5] of the triconnected components of G [27].

An SPQR tree $\mathcal{T}$ for G represents the triconnected components of G [4,5]. It has several applications in dynamic graph algorithms and graph drawing, and can be constructed in linear time [24]. Each node $\alpha \in \mathcal{T}$ is associated with an undirected graph G_α. Each vertex of G_α corresponds to a vertex of the original graph G. An edge of G_α is either a *virtual edge* that corresponds to a separation pair of G, or a *real edge* that corresponds to an edge of the original graph G. Let $e = \{x, y\}$ be an edge of G such that $\{v, e\}$ is a vertex-edge cut-pair of G. Then, $\mathcal{T}$ must contain an S-node α such that v, x and y are vertices of G_α and $\{x, y\}$ is not a virtual edge. The above observation implies that we can use $\mathcal{T}$ to identify all vertex-edge cut-pairs of G as follows. A vertex-edge cut-pair (v, e) is such that $v \in V(G_\alpha)$ and e is a real edge of G_α that is not adjacent to v, where α is an S-node [21,25]. Now we define the *split operation* of v as follows. Let e_1 and e_2 be the edges incident to v in G_α. We split v into two vertices v_1 and v_2, where v_1 is incident only to e_1 and v_2 is incident only to e_2. (In effect, this makes S a path with endpoints v_1 and v_2.) To find the connected components of $G \setminus \{v, e\}$, we execute a split operation on v and delete e from the resulting path. Note that $e \neq e_1, e_2$, and e does not have a copy in any other node of the SPQR tree since it is a real edge. Then, the connected components of $G \setminus \{v, e\}$ are represented by the resulting subtrees of $\mathcal{T}$.

Here, we need to partition the ordinary vertices of G according to the vertex-edge cut-pairs (v, e), where v is a marked auxiliary vertex. To do this efficiently, we can process all vertices simultaneously as follows. First, we note that we only need to consider the marked vertices that are in S-nodes that contain at least one real edge. Let α be such an S-node. We perform the split operation on each marked (auxiliary) vertex v, and then delete all the real edges of α. This breaks $\mathcal{T}$ into subtrees, and the desired partition of the ordinary vertices is formed by the ordinary vertices of each subtree. Hence, we obtain:

Theorem 2. *The marked vertex-edge blocks of an undirected graph can be computed in linear time.*

By Sect. 4 we now have:

Theorem 3. *The 2-edge twinless strongly connected components of a directed graph can be computed in linear time.*

Finally, by the reduction of Sect. 2, we have:

Theorem 4. *The edge-resilient strongly orientable blocks of a mixed graph can be computed in linear time.*

References

1. Aamand, A., Hjuler, N., Holm, J., Rotenberg, E.: One-way trail orientations. In: 45th International Colloquium on Automata, Languages, and Programming (ICALP 2018). LIPIcs, vol. 107, pp. 6:1–6:13 (2018)
2. Alstrup, S., Harel, D., Lauridsen, P.W., Thorup, M.: Dominators in linear time. SIAM J. Comput. **28**(6), 2117–32 (1999)
3. Bang-Jensen, J., Gutin, G.: Digraphs: Theory, Algorithms and Applications (Springer Monographs in Mathematics). Springer, London (2002). https://doi.org/10.1007/978-1-84800-998-1
4. Battista, G.D., Tamassia, R.: On-line maintenance of triconnected components with SPQR-trees. Algorithmica **15**(4), 302–318 (1996)
5. Battista, G.D., Tamassia, R.: On-line planarity testing. SIAM J. Comput. **25**(5), 956–997 (1996)
6. Bhalgat, A., Hariharan, R.: Fast edge orientation for unweighted graphs. In: Proceedings of the Twentieth Annual ACM-SIAM Symposium on Discrete Algorithms (SODA 2009), pp. 265–272 (2009)
7. Boesch, F., Tindell, R.: Robbins's theorem for mixed multigraphs. Am. Math. Mon. **87**(9), 716–719 (1980)
8. Buchsbaum, A.L., Georgiadis, L., Kaplan, H., Rogers, A., Tarjan, R.E., Westbrook, J.R.: Linear-time algorithms for dominators and other path-evaluation problems. SIAM J. Comput. **38**(4), 1533–1573 (2008)
9. Chung, F.R.K., Garey, M.R., Tarjan, R.E.: Strongly connected orientations of mixed multigraphs. Networks **15**(4), 477–484 (1985)
10. Conte, A., Grossi, R., Marino, A., Rizzi, R., Versari, L.: Directing road networks by listing strong orientations. In: Mäkinen, V., Puglisi, S.J., Salmela, L. (eds.) IWOCA 2016. LNCS, vol. 9843, pp. 83–95. Springer, Cham (2016). https://doi.org/10.1007/978-3-319-44543-4_7
11. Dinitz, E.: The 3-edge-components and a structural description of all 3-edge-cuts in a graph. In: Mayr, E.W. (ed.) WG 1992. LNCS, vol. 657, pp. 145–157. Springer, Heidelberg (1993). https://doi.org/10.1007/3-540-56402-0_44
12. Elberfeld, M., Segev, D., Davidson, C.R., Silverbush, D., Sharan, R.: Approximation algorithms for orienting mixed graphs. Theor. Comput. Sci. **483**, 96–103 (2013). Special Issue Combinatorial Pattern Matching 2011
13. Fraczak, W., Georgiadis, L., Miller, A., Tarjan, R.E.: Finding dominators via disjoint set union. J. Discrete Algorithms **23**, 2–20 (2013)
14. Frank, A.: An algorithm for submodular functions on graphs. In: Bonn Workshop on Combinatorial Optimization, North-Holland Mathematics Studies, vol. 66, pp. 97–120. North-Holland (1982)
15. Frank, A.: Connections in Combinatorial Optimization, 1st edn. Oxford University Press, Oxford (2011)
16. Gabow, H.N.: A framework for cost-scaling algorithms for submodular flow problems. In: Proceedings of 1993 IEEE 34th Annual Foundations of Computer Science, pp. 449–458 (1993). https://doi.org/10.1109/SFCS.1993.366842
17. Gabow, H.N.: The minset-poset approach to representations of graph connectivity. ACM Trans. Algorithms **12**(2), 1–73 (2016)
18. Gabow, H.N.: Efficient splitting off algorithms for graphs. In: Proceedings of the Twenty-Sixth Annual ACM Symposium on Theory of Computing (STOC 1994), pp. 696–705 (1994)

19. Georgiadis, L., Italiano, G.F., Laura, L., Parotsidis, N.: 2-edge connectivity in directed graphs. ACM Trans. Algorithms **13**(1), 1–24 (2016)
20. Georgiadis, L., Kipouridis, E., Papadopoulos, C., Parotsidis, N.: Faster computation of 3-edge-connected components in digraphs. In: Proceedings of 34th ACM-SIAM Symposium on Discrete Algorithms (SODA 2023), pp. 2489–2531 (2023)
21. Georgiadis, L., Kosinas, E.: Linear-time algorithms for computing twinless strong articulation points and related problems. In: 31st International Symposium on Algorithms and Computation (ISAAC 2020), vol. 181, pp. 38:1–38:16 (2020)
22. Georgiadis, L., Italiano, G.F., Parotsidis, N.: Strong connectivity in directed graphs under failures, with applications. SIAM J. Comput. **49**(5), 865–926 (2020)
23. Georgiadis, L., Kefallinos, D., Kosinas, E.: On 2-strong connectivity orientations of mixed graphs and related problems. arXiv preprint arXiv:2302.02215 (2023)
24. Gutwenger, C., Mutzel, P.: A linear time implementation of SPQR-trees. In: Marks, J. (ed.) GD 2000. LNCS, vol. 1984, pp. 77–90. Springer, Heidelberg (2001). https://doi.org/10.1007/3-540-44541-2_8
25. Heinrich, I., Heller, T., Schmidt, E., Streicher, M.: 2.5-connectivity: unique components, critical graphs, and applications. In: Adler, I., Müller, H. (eds.) WG 2020. LNCS, vol. 12301, pp. 352–363. Springer, Cham (2020). https://doi.org/10.1007/978-3-030-60440-0_28
26. Henzinger, M., Krinninger, S., Loitzenbauer, V.: Finding 2-edge and 2-vertex strongly connected components in quadratic time. In: Halldórsson, M.M., Iwama, K., Kobayashi, N., Speckmann, B. (eds.) ICALP 2015. LNCS, vol. 9134, pp. 713–724. Springer, Heidelberg (2015). https://doi.org/10.1007/978-3-662-47672-7_58
27. Hopcroft, J.E., Tarjan, R.E.: Dividing a graph into triconnected components. SIAM J. Comput. **2**(3), 135–158 (1973)
28. Italiano, G.F., Laura, L., Santaroni, F.: Finding strong bridges and strong articulation points in linear time. Theoret. Comput. Sci. **447**, 74–84 (2012)
29. Iwata, S., Kobayashi, Y.: An algorithm for minimum cost arc-connectivity orientations. Algorithmica **56**, 437–447 (2010)
30. Jaberi, R.: 2-edge-twinless blocks. Bulletin des Sciences Mathématiques **168**, 102969 (2021)
31. Lengauer, T., Tarjan, R.E.: A fast algorithm for finding dominators in a flowgraph. ACM Trans. Program. Lang. Syst. **1**(1), 121–41 (1979)
32. Menger, K.: Zur allgemeinen kurventheorie. Fundam. Math. **10**(1), 96–115 (1927)
33. Nagamochi, H., Ibaraki, T.: Deterministic $\tilde{O}$(nm) time edge-splitting in undirected graphs. In: Proceedings of the Twenty-Eighth Annual ACM Symposium on Theory of Computing (STOC 1996), pp. 64–73 (1996)
34. Nash-Williams, C.S.J.A.: On orientations, connectivity and odd-vertex-pairings in finite graphs. Can. J. Math. **12**, 555–567 (1960)
35. Raghavan, S.: Twinless strongly connected components. In: Alt, F.B., Fu, M.C., Golden, B.L. (eds.) Perspectives in Operations Research. Operations Research/Computer Science Interfaces Series, vol. 36, pp. 285–304. Springer, Boston (2006). https://doi.org/10.1007/978-0-387-39934-8_17
36. Schrijver, A.: Combinatorial Optimization - Polyhedra and Efficiency. Springer, Heidelberg (2003)
37. Tarjan, R.E.: Depth-first search and linear graph algorithms. SIAM J. Comput. **1**(2), 146–160 (1972)

Make a Graph Singly Connected by Edge Orientations

Tim A. Hartmann and Komal Muluk(✉)

Department of Computer Science, RWTH Aachen University, Aachen, Germany
{hartmann,muluk}@algo.rwth-aachen.de

Abstract. A *directed* graph D is *singly connected* if for every ordered pair of vertices (s, t), there is at most one path from s to t in D. Graph orientation problems ask, given an *undirected* graph G, to find an orientation of the edges such that the resultant *directed* graph D has a certain property. In this work, we study the graph orientation problem where the desired property is that D is singly connected. Our main result concerns graphs of a fixed girth g and coloring number c. For every $g, c \geq 3$, the problem restricted to instances of girth g and coloring number c, is either NP-complete or in P. As further algorithmic results, we show that the problem is NP-hard on planar graphs and polynomial time solvable distance-hereditary graphs.

Keywords: directed graphs · chromatic number · girth

1 Introduction

Graph orientation problems are well-known class of problems in which given an undirected graph, the task is to decide whether we can assign a direction to each edge such that a particular desired property is satisfied on the resulting directed graph. One can ask questions like, for an undirected graph, can we assign directions to all edges such that the digraph thus formed is acyclic, or it is strongly connected, or it contains a directed Euler circuit, and many more such interesting properties. These problems are important because of the emerging field of networks designing; networks are modeled by directed graphs. Consider the problem of converting an undirected graph to a directed acyclic graph by means of such an orientation of edges. All simple graphs can be converted into directed acyclic graphs by simply considering a DFS tree on the graph and directing all edges from the ancestor to the descendant. This trivially solves the decision version of the graph orientation problem for all simple graphs.

In this work we consider 'singly-connected' as the desired property of the graph orientation property. A directed graph D is *singly connected* if for every ordered pair of vertices (s, t), there is at most one path from s to t in D. For an undirected graph G, an $E(G)$-orientation is a mapping σ that maps each edge

This research has been supported by the DFG RTG 2236 "UnRAVeL" - Uncertainty and Randomness in Algorithms, Verification and Logic.

S.-Y. Hsieh et al. (Eds.): IWOCA 2023, LNCS 13889, pp. 221–232, 2023.
https://doi.org/10.1007/978-3-031-34347-6_19

$\{u, v\} \in E(G)$ to either of its endpoints u or v, which results into a directed graph G_σ. An $E(G)$-orientation σ is an *sc-orientation* if G_σ is singly connected. A graph G is said to be *sc-orientable* if there exists an sc-orientation of G. Hence we study the following problem.

Problem: SC-ORIENTATION (SCO)
Input: A simple undirected graph $G = (V, E)$.
Question: Is G sc-orientable?

There has been several studies around the singly connected graphs in the literature. The problem of testing whether a graph is singly connected or not was introduced in an exercise in Cormen et al. in [4]. Buchsbaum and Carlisle, in 1993, gave an algorithm running in time $O(n^2)$ [3]. There has been other attempts to improve this result, for example [8,12]. For our SC-ORIENTATION problem, this means that the edge orientation σ that makes G_σ singly connected serves as an NP-certificate. Thus SCO is in NP.

Das et al. [6] studied the vertex deletion version of the problem. They asked, given a *directed* graph D, does there exists a set of k vertices whose deletion renders a singly connected graph. Alongside, they pointed a close link of the SINGLY CONNECTED VERTEX DELETION (SCVD) problem to the classical FEEDBACK VERTEX SET (FVS) problem. They argued how SCVD is *a* directed counterpart of FVS, which makes it theoretically important. Our problem is a natural follow-up question in this direction which looks at the graph orientation version.

In the next section we prove that if a graph has an sc-orientation, then it also has an acyclic sc-orientation. So, in addition for a graph to orient to an acyclic digraph, SC-ORIENTATION also demands the orientation to a singly connected digraph. As discussed earlier, it is trivial to find an acyclic orientation of the graph. However, in contrast to that, we prove, the additional singly connected condition makes it harder to even decide whether such an orientation exits or not.

1.1 Our Results

First, we prove that SC-ORIENTATION is NP-complete even for planar graphs. Thus the hardness also holds for all graphs restricted to girth at most 4. Complementing this result, we show that planar graphs of girth at least 5 are always sc-orientable, hence, the decision problem is trivial.

We aim to show analogous dichotomy for general graphs. As an upper bound, we show, for any graph with girth at least twice the chromatic number, the graph is sc-orientable. Hence the problem is trivially solvable for such inputs. To account for this, we study SC-ORIENTATION restricted to the graph class $\mathcal{G}_{g,c}$, the class of graphs that have girth g and chromatic number c, for some positive integers g, c. Our main result is that, for each pair $g, c \geq 3$, SC-ORIENTATION restricted to $\mathcal{G}_{g,c}$ is either NP-complete or trivially solvable. We do not pinpoint the exact boundary for the values of g and c where the transformation occurs. The approach is to compile an NP-hardness proof that solely relies upon a single no-instance of the given girth g and chromatic number c.

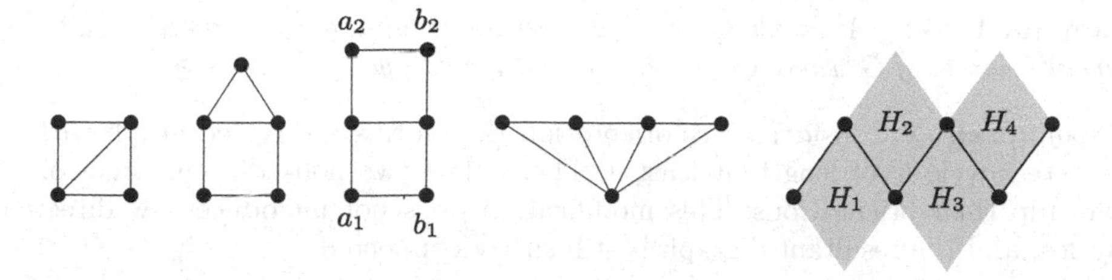

Fig. 1. From left to right: a diamond, a house, a domino (a $G_{2,3}$), a gem, and a gadget for the construction of Lemma 5. Non-sc-orientable are the diamond, the house and hence also the gem. The domino is sc-orientable.

As further results, we provide polynomial time algorithms of SCO for a special graph class such as distance-hereditary graphs. Here, we give a concise definition of the yes-instances by forbidden subgraphs.

From the technical sides, we provide two simplification of the problem. First, one can restrict the search for an sc-orientation to the orientations that also avoid directed cycles. Second, we show how we can assume that the input is triangle free.

In Sect. 2, we begin with the Preliminaries and the structural results. Section 3 shows NP-completeness of SC-ORIENTATION on planar graphs. Section 4 contains our dichotomy result. Then, Sect. 5 considers perfect and distance-hereditary graphs. We conclude with Sect. 6.

2 Notation and Technical Preliminaries

For a graph G, let $V(G)$ be its vertex set, and $E(G)$ be its set of edges. Elements in $V(G)$ are called nodes or vertices. A k-coloring of a graph G, for $k \in \mathbb{N}$, is a mapping $f\colon V(G) \to \{1, \dots, k\}$ such that $f(u) \neq f(v)$ if $\{u, v\} \in E(G)$. The coloring number, $\chi(G)$, of graph G is the smallest integer k such that G has a k-coloring. A graph has girth g if the length of the shortest cycle in the graph is g. By definition, a graph without any cycles has girth ∞. Given a path $P = v_1, v_2, \dots, v_n$, we denote the length of a path to be the total number of vertices in the path. Throughout the paper we consider different small graphs, typically as forbidden induced subgraphs, drawn in Fig. 1. We denote the grid graph of width x and height y as $G_{x,y}$. For further reading related to the concepts in graph theory we refer reader to the book by Diestel [7].

The following technical observations are useful for our later results. The upcoming lemma shows when looking for an sc-orientation of a graph, it suffices to look for an sc-orientation that additionally avoids cycles of length greater than 3.

Lemma 1. (⋆)[1] *A graph G is sc-orientable if and only if there is an sc-orientation σ of G where G_σ contains no directed cycle of length ≥ 4.*

Proof (Sketch). Consider an sc-orientation that forms a directed graph with a directed cycle C of length at least 4. Then select two non-adjacent edges of C and flip their orientations. This modification does not introduce new directed cycles, and the resultant digraph is still singly connected. □

Further, we can remove triangles from the input graph unless it contains a diamond or a house as an induced subgraph. It is easy to see, if a graph G contains a diamond or a house graph as an induced subgraph, then G is not sc-orientable.

Lemma 2. (⋆) *Let a (diamond, house)-free graph G contain a triangle uvw. Then G has an sc-orientation, if and only if G', resulting from G by contracting the edges in uvw, has an sc-orientation.*

Proof (Sketch). Note that an sc-orientation of G must form a directed cycle of uvw. Then an sc-orientation of G is essentially also an sc-orientation of G'. □

3 NP-Hardness on Planar Graphs

This section derives NP-hardness of SC-ORIENTATION on planar graphs. As observed in the Sect. 1, SC-ORIENTATION is in NP. To show hardness, we give a reduction from a variant (to be shortly defined) of the PLANAR 3-SAT problem. Given a 3-SAT formula ϕ, a variable-clause graph is a graph obtained by taking variables and clauses representatives as the vertex set, and the edge set contains an edge between a variable-clause pair if the variable appears in the clause. We use the variant of 3-SAT where there is a planar embedding of the variable-clause graph, in which there are some additional edges (only connecting clause nodes) which form a cycle traversing through all clause nodes. We denote this cycle as *clause-cycle*, and the problem as CLAUSE-LINKED PLANAR 3-SAT. Kratochvíl et al. proved the NP-hardness of CLAUSE-LINKED PLANAR 3-SAT [13].

Problem: CLAUSE-LINKED PLANAR 3-SAT
Input: A boolean formula ϕ with clauses $C = \{c_1, \ldots, c_m\}$ and variables $X = \{v_1, \ldots, v_n\}$, and a planar embedding of the variable-clause graph together with a clause-cycle.
Question: Is ϕ satisfiable?

Theorem 1. (⋆) SC-ORIENTATION *is* NP*-complete even for planar graphs.*

[1] The full proofs of the statements marked with (⋆) have been omitted due to limited space availability.

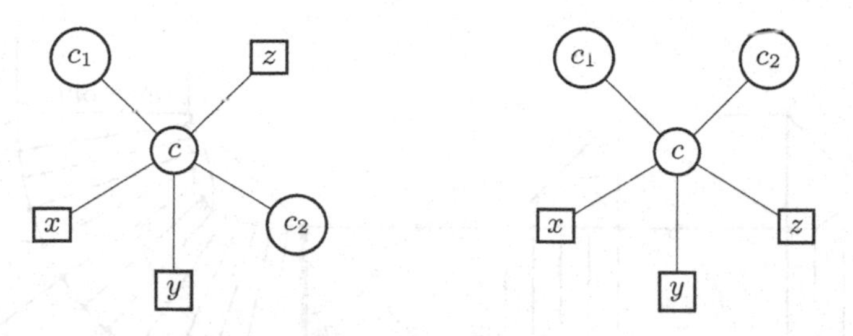

Fig. 2. Type 1 (left) and type 2 (right) of clause nodes according to their neighbourhoods in the planar embedding Π of $G_\eta(\phi)$.

Proof. Our idea is to take the given planar embedding, and use this as a backbone for our construction. Let η be an ordering of clause nodes such that the clause-cycle traverses through C as per the ordering η. Let the given planar embedding be Π, and the underlying graph be $G_\eta(\phi)$. For the reduction, we define a clause gadget and a variable gadget to replace the clause nodes and variable nodes in Π. To begin with, let us start with the following interesting graph structure:

Ladder. The domino (the 2×3-grid graph, see Fig. 1), has exactly two possible sc-orientations. If we fix the orientation of one edge, all orientations of the domino are fixed up to singly-connectedness (every vertex becomes either a sink or a source for the domino). Specifically, we are interested in the following properties regarding the bottom edge $\{a_1, b_1\}$ and the top edge $\{a_2, b_2\}$:

(1) $(a_1, b_1), (a_2, b_2)$ are *coupled*, that is, for every sc-orientation σ, either $\sigma(\{a_x, b_x\}) = b_x$ for $x \in \{1, 2\}$, or $\sigma(\{a_x, b_x\}) = a_x$ for $x \in \{1, 2\}$,
(2) there is an sc-orientation without any directed path from $\{a_1, b_1\}$ to $\{a_2, b_2\}$.

Since the directions of coupling edges matter, we refer to them as ordered pairs, for example, edge $\{a, b\} \in V(G)$ becomes (a, b) or (b, a). The coupling property is transitive, in the sense that, when we take 2 dominos and identify the top edge of one to the bottom edge of another (forming a 2×5-grid), the very top and the very bottom edges are also coupled. Further extrapolation can be made to form a *ladder* (a $2 \times n$-grid). Additionally, the bottom and the middle edges of a domino are in the reverse direction in every sc-orientation. Hence they are the *reverse-coupled* edges. The coupling property of dominos helps us attach as many 4-cycles as we want to a ladder without affecting the two possible sc-orientations of the structure. Let us call the ladder with the additional 4-cycles an *extended ladder* ($\mathcal{L}$). For our refernce, let us call the bottom of the ladder the 0th-step. Let us call the set of edges which are coupled with the 0th-step by *even edges*, and the set of reverse-coupled edges by *odd edges*. All in all, if we fix the orientation of the 0th-step, the orientation of the whole ladder is fixed up to an sc-orientation. Let us construct a large enough extended ladder $\mathcal{L}_U$, we call this the universal ladder.

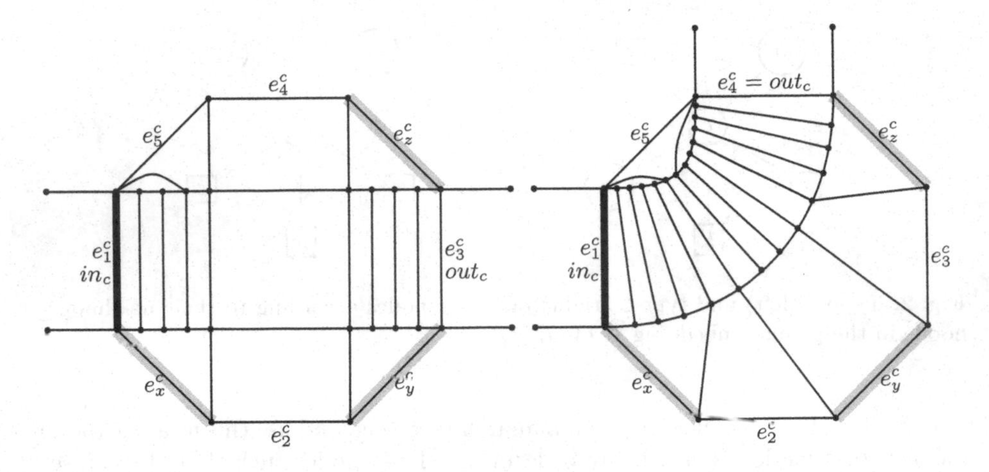

Fig. 3. Clause gadget for type 1 (left) and type 2 (right)

Clause Gadget. In the planar embedding Π, each clause node c is adjacent to three variable nodes, say x, y, z, and two more clause nodes, say c_1 and c_2. There are two types in which c can be surrounded by nodes x, y, z, c_1, c_2 in Π: only two variables appear together on one side of the clause-cycle, or all three variables appear consecutively on one side of the clause cycle, see Fig. 2. For a replacement to these two types of configurations, we define two different clause gadgets (see Fig. 3). For $c \in C$, the basic structure of the clause gadget G_c includes an 8-cycle which contains 3 edges e_x^c, e_y^c, and e_z^c (corresponding to x, y, and z), and five generic edges $e_1^c, e_2^c, e_3^c, e_4^c, e_5^c$. In addition, the universal ladder $\mathcal{L}_U$ passes through the gadget as shown in Fig. 3. For each of the two types of clause gadgets, we distinguish two edges of the clause gadget from where the universal ladder enters and exits by the notations in_c and out_c respectively.

Variable Gadget. Each variable node x in Π is replaced by a sufficiently large ladder-cycle $\mathcal{LC}_x$, see Fig. 4. The bold edges in Fig. 4 correspond to the 0th-steps of $\mathcal{LC}_x$. The orientation of all 0th-steps in $\mathcal{LC}_x$ can either be clockwise or anticlockwise in an sc-orientation of the graph. This is because of the cyclic ladder formed in the gadget. The two types of orientations are in one-to-one correspondence with the two boolean assignments of the variable in the 3-SAT formula.

Construction. We make the following replacements while constructing our input graph G of SCO from the planar embedding Π given in the input of the Clause-linked Planar 3-SAT problem:

- Replace each clause node c in Π by a clause gadget G_c in G.
- Replace each variable node x in Π by a variable gadget $\mathcal{LC}_x$ in G.

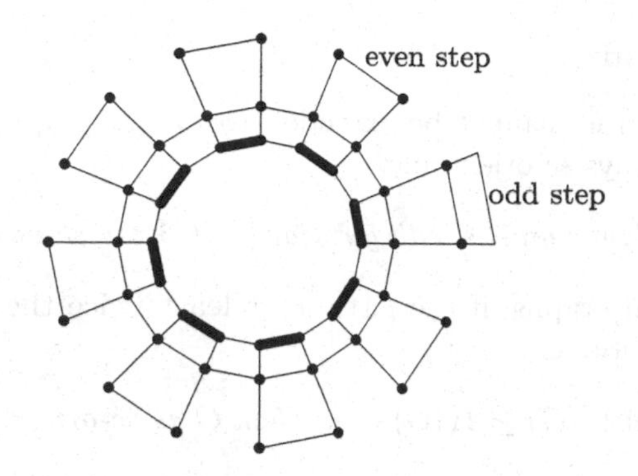

Fig. 4. Variable gadget (the laddercycle $\mathcal{LC}_x$)

- Replace each edge $\{x, c\}$ in Π by extending a ladder from the variable gadget $\mathcal{LC}_x$ to the edge e_x^c of the clause gadget G_c. If $\bar{x}$ appears in C, then we identify an odd step of $\mathcal{LC}_x$ to e_x^c. If x appears in c, then we identify an even step of $\mathcal{LC}_x$ to the corresponding edge e_x^c.
- For a $\{c_i, c_j\}$ edge in Π, where c_i appears before c_j in η, we identify the dangling edges next to out_{c_i} to that of the dangling edges next to in_{c_j} while maintaining the planarity. Moreover, let us fix the 0th-step of the universal ladder $\mathcal{L}_U$ to be the in_{c_1} edge of the clause gadget G_{c_1}.

Note that all clause and variable gadgets are planar, and the edge replacements by ladder also maintains the planarity of the resulting graph. As per our construction of the clause gadgets, for each structure G_c, the in_c and out_c edges together are either oriented clockwise or anti-clockwise in every sc-orientation of $G_\eta(\phi)$. Moreover, all the in_c edges are oriented in one direction for all $c \in C$, and all the out_c edges are oriented in the other direction. If we fix the orientation of the 0th-step of $\mathcal{L}_U$, it fixes the orientation of all edges associated with $\mathcal{L}_U$ and consequently fixes the orientation of the in_c and out_c edges for all $c \in C$. It also fixes the orientation of all generic edges in G_c for all $c \in C$. We remain to prove that the CLAUSE-LINKED PLANAR 3-SAT instance is a yes-instance if and only if the corresponding constructed instance of the SC-ORIENTATION problem is a yes-instance. The proof of correctness of the above statement can be found in the appendix. □

4 Girth and Coloring Number

Notably the NP-hardness reduction so far heavily relied on a cycle of length 4, as part of a 'ladder'. Hence the constructed graphs have girth 4. We recall the *girth* of a graph is the length of its shortest cycle. Can we extend this result to girth 5 graphs or for graphs of even higher girth?

4.1 Upper Bounds

The NP-hardness proof cannot be extended to planar graphs of girth 5, since such graphs are always sc-orientable.

Lemma 3. ($\star$) *Planar graphs with girth at least 5 are sc-orientable.*

Also, for general graphs, if the girth is at least twice the coloring number, the graph is sc-orientable.

Theorem 2. *Let* $\text{girth}(G) \geq 2\chi(G) - 1$. *Then* G *is sc-orientable.*

Proof. Consider a proper vertex-coloring $c : V(G) \to \{1, \dots, |\chi(G)|\}$. Let σ be an orientation which orients each edge $\{u, v\}$ to the vertex which corresponds to the higher color class, that is, to $\text{argmax}_{w \in \{u,v\}} c(w)$. We claim that G_σ is singly connected. Assuming the contrary, there are $s, t \in V(G)$ with disjoint s, t-paths P_1, P_2. Then P_1, P_2 form a cycle of length at least $\text{girth}(G) \geq 2|\chi(G)| - 1$. At least one of the paths, say P_1, consists of $> \chi(G)$ vertices. Then there are vertices u, v on path P_1 with the same color. Without loss of generality say u appears before v in P_1. To form an s, t-path the successor of u on path P_1 must have a higher color than u and so on. Thus a later vertex v with the same color contradicts that P_1 is an s, t-path. □

As a remark, by Grötzsch's Theorem every planar graph of girth at least 4 is 3-colorable [10]. Hence, Lemma 3 also follows from Theorem 2.

4.2 The Dichotomy

We do not know whether there even exists a graph of girth at least 5 that is not sc-orientable. Intriguingly, however, we do not need to answer this question to follow a dichotomy between the P-time and NP-completeness. That is, for every $g, c \geq 3$, SC-ORIENTATION when restricted to graphs of girth g and coloring number c is either NP-complete or trivially solvable. Let the class of graphs of girth g and chromatic number c be $\mathcal{G}_{g,c}$. Our proof does not pinpoint the exact boundary between P-time and NP-completeness. We only know, for a fixed c, there is some upper bound g' for the values of g such that the hardness is only unclear on $\mathcal{G}_{g,c}$, where $g < g'$. See Theorem 2. Moreover, we showed in Sect. 3, the NP-hardness on $\mathcal{G}_{g,c}$ for $g = 3, 4$. The key ingredient for our dichotomy result is to compile an NP-hardness proof that only relies upon a single no-instance of a certain girth g and coloring number c.

The main building block of our construction is a *coupling gadget*. It is a graph that couples the orientation of two special edges but that does *not* enforce a coloring depending on the orientation. An sc-orientable graph H with special edges $\{a_1, b_1\}, \{a_2, b_2\} \in E(H)$ is a *coupling gadget* on $(a_1, b_1), (a_2, b_2)$ if

(1) $(a_1, b_1), (a_2, b_2)$ are *coupled*, that is, for every sc-orientation σ, either $\sigma(\{a_x, b_x\}) = b_x$ for $x \in \{1, 2\}$, or $\sigma(\{a_x, b_x\}) = a_x$ for $x \in \{1, 2\}$,

(2) there is an sc-orientation without any directed path from $\{a_1, b_1\}$ to $\{a_2, b_2\}$, and
(3) there are two $\chi(H)$-colorings f, f' of H such that $f(a_1) = f(a_2)$, $f(b_1) = f(b_2)$ and $f'(a_1) = f'(b_2)$, $f'(b_1) = f'(a_2)$.

For example, $G_{2,3}$ ladder (which we use for Theorem 1) satisfies only properties (1) and (2). In turn, $G_{2,4} \cup K_3$ is a coupling gadget.

Lemma 4. *For every $g, c \geq 3$, there is an sc-orientable graph $G \in \mathcal{G}_{g,c}$.*

Proof. We know that there is at least one graph $G_{g',c}$ of girth g' and chromatic number c, see [9]. When $g = g' \geq 2c - 1$, Theorem 2 yields that $G_{g,c}$ is sc-orientable. For $g = g' < 2c - 1$, the graph, $G_{2c-1,c} \cup C_g$, is sc-orientable; it has girth g and chromatic number c. □

Lemma 5. (⋆) *For $g \geq 3$, $c \geq 4$, there is a non-sc-orientable graph $G \in \mathcal{G}_{g,c}$, if and only if there is a coupling gadget $H \in \mathcal{G}_{g,c}$.*

Proof (Sketch). ($\Rightarrow$) Consider a non-sc-orientable graph $G \in G_{g,c}$. We assume that G contains at least one edge $\{u, v\}$ such that subdividing $\{u, v\}$ yields an sc-orientable graph. If that is not the case, pick an edge $\{u, v\}$ and consider the graph G' resulting from G where $\{u, v\}$ is subdivided into a path uwv. To assure termination, initially assume all edges unmarked, and consider subdividing an unmarked edge $\{u, v\}$ before considering the marked edges. Whenever an edge $\{u, v\}$ is subdivided, mark the new edges $\{u, w\}$ and $\{w, v\}$. The procedure terminates at the latest when all edges are marked since then the resulting graph is bipartite and hence trivially sc-orientable.

Now, let H' consists of a copy of G where edge $\{u, v\}$ is subdivided resulting in a path uwv. Then H' has an sc-orientation σ. Crucially, σ couples edges (u, w), (w, v), meaning it orientates $\{u, w\}, \{w, v\}$ either both away from w or both towards w. To see this, assume an sc-orientation σ' that, up to symmetry, has $\sigma'(\{u, w\}) = w$ and $\sigma'(\{w, v\}) = v$. Then σ' extended with $\sigma'(\{u, v\}) = v$ forms an sc-orientation of G, despite it being non-sc-orientable. Thus indeed edges $(u, w), (w, v)$ are coupled.

To construct H (which also satisfies properties (2), (3)), we introduce four copies H_1, H_2, H_3, H_4 of H' naming their vertices with subscript $1, 2, 3, 4$ respectively. We identify the vertices w_1, u_2 and vertices v_1, w_2, u_3 and vertices v_2, w_3, u_4 and vertices v_3, w_4; see the gadget in Fig. 1. Then H forms a coupling gadget on edges $(u_1, w_1), (w_4, v_4)$. So far, the construction does not decrease the girth and does not increase the coloring number. By adding suitable sc-orientable components to H, we can ensure the girth and chromatic number remains the same as that of G. □

Theorem 3. (⋆) *For $g, c \geq 3$,* SC-ORIENTATION *restricted to the graphs in $\mathcal{G}_{g,c}$ is either* NP*-complete or trivially solvable.*

Proof (Sketch). Consider fixed $g, c \geq 3$. If all graphs $G \in \mathcal{G}_{g,c}$ are sc-orientable, then an algorithm may simply always answer 'yes'. Otherwise, Lemma 5 provides a coupling gadget $H \in \mathcal{G}_{g,c}$. Due to Property 2 of the coupling gadgets, all coupled edges are disjoint; this allows us to couple an arbitrary pair of disjoint edges by introducing a coupling gadget between them. To follow NP-hardness for graphs in $\mathcal{G}_{g,c}$, we adapt the NP-hardness construction of Theorem 1 to output graphs in $\mathcal{G}_{g,c}$. □

5 Perfect Graphs

This section considers perfect graphs. First we show that the NP-hardness proof from Theorem 1 can be adapted to perfect graphs. Then later we study the distance-hereditary graphs as a special case.

We modify our NP-hardness reduction from Theorem 1 as follows: In the reduced instance, for each edge $e = \{u, v\}$, introduce two new vertices w_e and w'_e. Delete edge e, and instead add edges such that there is a path uw_ev, and a triangle $uw_ew'_e$.

Then according to Lemma 2, the modified instance is equivalent to the original instance in the sense that the modified graph is sc-orientable if and only if the originally constructed graph is sc-orientable. Further, the modified graph is 3-colorable and has a maximum clique of size 3; hence it is perfect. From this we conclude the following result.

Theorem 4. SC-ORIENTATION *is* NP-*hard even for perfect graphs.*

5.1 Distance-Hereditary Graphs

Now we derive two classifications of the distance hereditary graphs that are sc-orientable. One way is to simply restrict the recursive definition to (locally) avoid a diamond subgraph. The second is a concise classification by forbidden subgraphs.

A graph is *distance hereditary* if it can be constructed starting from a single isolated vertex with the following operations [1]: a *pendant* vertex to $u \in V(G)$, that is, add a vertex v and edge $\{u, v\}$; *true twin* on a vertex $u \in V(G)$, that is, add a vertex v with neighborhood $N(v) = N(u)$ and edge $\{u, v\}$; *false twin* on a vertex $u \in V(G)$, that is, add a vertex v with neighborhood $N(v) = N(u)$.

We denote a graph as *strongly distance hereditary* if it can be constructed from a single isolated vertex with the following restricted operations:

- a pendant vertex to $u \in V(G)$;
- true twin on a vertex $u \in V(G)$, restricted to u where $|N(u)| = 1$; and
- false twin on a vertex $u \in V(G)$, restricted to u where $N(u)^2 \cap E(G) = \emptyset$.

Observe that the forbidden operations would immediately imply a diamond as a subgraph. Thus every distance hereditary graph that has an sc-orientation must be strongly distance hereditary. Now, we show also the converse is true.

Theorem 5. ($\star$) *The sc-orientable distance hereditary graphs are exactly the strongly distance hereditary graphs. They can be recognized in polynomial time.*

Note that distance hereditary graphs are exactly the (house, hole, domino, gem)-free graphs [1], where a hole is any cycle of length ≥ 5. When we replace 'gem' with 'diamond' we exactly end up with the strongly distance hereditary graphs.

Lemma 6. ($\star$) *The strongly distance hereditary graphs are exactly the graphs with house, hole, domino and diamond as forbidden subgraph.*

6 Conclusion

Among others results, we have shown that for every $g, c \geq 3$, the restriction to graphs of girth g and chromatic number c is either NP-complete or in P. While we know NP-completeness for low girth and chromatic number and we know that for relatively large girth compared to $\chi(G)$ the problem is trivial, it yet remains to pinpoint the exact boundary of the NP-hard and P cases. As shown, to extend the NP-hardness result, one needs to simply find a no-instance of the girth and chromatic number of interest (or alternatively merely a coupling gadget), for example graphs of girth 5 and chromatic number 4. On the other hand, one might improve Theorem 2 to also hold for smaller values of girth g.

References

1. Bandelt, H., Mulder, H.M.: Distance-hereditary graphs. J. Comb. Theory Ser. B **41**(2), 182–208 (1986). https://doi.org/10.1016/0095-8956(86)90043-2
2. Borodin, O.V., Glebov, A.N.: On the partition of a planar graph of girth 5 into an empty and an acyclic subgraph (2001)
3. Buchsbaum, A.L., Carlisle, M.C.: Determining uni-connectivity in directed graphs. Inf. Process. Lett. **48**(1), 9–12 (1993). https://doi.org/10.1016/0020-0190(93)90261-7
4. Cormen, T.H., Leiserson, C.E., Rivest, R.L., Stein, C.: Introduction to Algorithms, 2nd edn. The MIT Press, Cambridge (2001)
5. Damiand, G., Habib, M., Paul, C.: A simple paradigm for graph recognition: application to cographs and distance hereditary graphs. Theor. Comput. Sci. **263**(1–2), 99–111 (2001). https://doi.org/10.1016/S0304-3975(00)00234-6
6. Das, A., Kanesh, L., Madathil, J., Muluk, K., Purohit, N., Saurabh, S.: On the complexity of singly connected vertex deletion. Theor. Comput. Sci. **934**, 47–64 (2022). https://doi.org/10.1016/j.tcs.2022.08.012
7. Diestel, R.: Graph Theory, 4th edn., vol. 173. Graduate Texts in Mathematics. Springer (2012)
8. Dietzfelbinger, M., Jaberi, R.: On testing single connectedness in directed graphs and some related problems. Inf. Process. Lett. **115**(9), 684–688 (2015)
9. Erdős, P.: Graph theory and probability. Can. J. Math. **11**, 34–38 (1959)
10. Grötzsch, H.: Ein dreifarbensatz für dreikreisfreie netze auf der kugel. Wiss. Z. Martin Luther Univ. Halle-Wittenberg Math. Nat. Reihe **8**, 109–120 (1959)

11. Hammer, P.L., Maffray, F.: Completely separable graphs. Discret. Appl. Math. **27**(1–2), 85–99 (1990). https://doi.org/10.1016/0166-218X(90)90131-U
12. Khuller, S.: An $\mathcal{O}(|V|^2)$ algorithm for single connectedness. Inf. Process. Lett. **72**(3), 105–107 (1999). https://doi.org/10.1016/S0020-0190(99)00135-0
13. Kratochvíl, J., Lubiw, A., Nesetril, J.: Noncrossing subgraphs in topological layouts. SIAM J. Discret. Math. **4**(2), 223–244 (1991). https://doi.org/10.1137/0404022

Computing the Center of Uncertain Points on Cactus Graphs

Ran Hu, Divy H. Kanani, and Jingru Zhang(✉)

Cleveland State University, Cleveland, OH 44115, USA
{r.hu,d.kanani}@vikes.csuohio.edu, j.zhang40@csuohio.edu

Abstract. In this paper, we consider the (weighted) one-center problem of uncertain points on a cactus graph. Given are a cactus graph G and a set of n uncertain points. Each uncertain point has m possible locations on G with probabilities and a non-negative weight. The (weighted) one-center problem aims to compute a point (the center) x^* on G to minimize the maximum (weighted) expected distance from x^* to all uncertain points. No previous algorithm is known for this problem. In this paper, we propose an $O(|G| + mn \log mn)$-time algorithm for solving it. Since the input is $O(|G| + mn)$, our algorithm is almost optimal.

Keywords: Algorithms · One-Center · Cactus Graph · Uncertain Points

1 Introduction

Problems on uncertain data have attracted an increasing amount of attention due to the observation that many real-world measurements are inherently accompanied with uncertainty. For example, the k-center model has been considered a lot on uncertain points in many different settings [1,3,4,9,10,12,15,18]. In this paper, we study the (weighted) one-center problem of uncertain points on a cactus graph.

Let $G = (V, E)$ be a cactus graph where any two cycles do not share edges. Every edge e on G has a positive length. A point $x = (u, v, t)$ on G is characterized by being located at a distance of t on edge (u, v) from vertex u. Given any two points p and q on G, the distance $d(p, q)$ between p and q is defined as the length of their shortest path on G.

Let $\mathcal{P}$ be a set of n uncertain points $P_1, P_2, \cdots, P_n$ on G. Each $P_i \in \mathcal{P}$ has m possible locations (points) $p_{i1}, p_{i2}, \cdots, p_{im}$ on G. Each location p_{ij} is associated with a probability $f_{ij} \geq 0$ for P_i appearing at p_{ij}. Additionally, each $P_i \in \mathcal{P}$ has a weight $w_i \geq 0$.

Assume that all given points (locations) on any edge $e \in G$ are given sorted so that when we visit e, all points on e can be traversed in order.

Consider any point x on G. For any $P_i \in \mathcal{P}$, the (weighted) expected distance $\mathsf{Ed}(P_i, x)$ from P_i to x is defined as $w_i \cdot \sum_{j=1}^{m} f_{ij} d(p_{ij}, x)$. The center of G with respect to $\mathcal{P}$ is defined to be a point x^* on G that minimizes the maximum expected distance $\max_{1 \leq i \leq n} \mathsf{Ed}(P_i, x)$. The goal is to compute center x^* on G.

S.-Y. Hsieh et al. (Eds.): IWOCA 2023, LNCS 13889, pp. 233–245, 2023.
https://doi.org/10.1007/978-3-031-34347-6_20

If G is a tree network, then center x^* can be computed in $O(mn)$ time by [17]. To the best of our knowledge, however, no previous work exists for this problem on cacti. In this paper, we propose an $O(|G| + mn \log mn)$-time algorithm for solving the problem where $|G|$ is the size of G. Note that our result matches the $O(|G| + n \log n)$ result [6] for the weighted *deterministic* case where each uncertain point has exactly one location.

1.1 Related Work

The deterministic one-center problem on graphs have been studied a lot. On a tree, the (weighted) one-center problem has been solved in linear time by Megiddo [14]. On a cactus, an $O(|G| + n \log n)$ algorithm was given by Ben-Moshe [6]. Note that the unweighted cactus version can be solved in linear time [13]. When G is a general graph, the center can be found in $O(|E| \cdot |V| \log |V|)$ time [11], provided that the distance-matrix of G is given. See [5,19,20] for variations of the general k-center problem.

When it comes to uncertain points, a few of results for the one-center problem are available. When G is a path network, the center of $\mathcal{P}$ can be found in $O(mn)$ time [16]. On tree graphs, the problem can be addressed in linear time [18] as well. See [9,12,18] for the general k-center problem on uncertain points.

1.2 Our Approach

Lemma 5 shows that the general one-center problem can be reduced in linear time to a *vertex-constrained* instance where all locations of $\mathcal{P}$ are at vertices of G and every vertex of G holds at least one location of $\mathcal{P}$. Our algorithm focuses on solving the vertex-constrained version.

As shown in [8], a cactus graph is indeed a block graph and its skeleton is a tree where each node uniquely represents a cycle block, a graft block (i.e., a maximum connected tree subgraph), or a hinge (a vertex on a cycle of degree at least 3) on G. Since center x^* lies on an edge of a circle or a graft block on G, we seek for that block containing x^* by performing a binary search on its tree representation T. Our $O(mn \log mn)$ algorithm requires to address the following problems.

We first solve the one-center problem of uncertain points on a cycle. Since each $\mathsf{Ed}(P_i, x)$ is piece-wise linear but non-convex as x moves along the cycle, our strategy is computing the local center of $\mathcal{P}$ on every edge. Based on our useful observations, we can resolve this problem in $O(mn \log mn)$ time with the help of the dynamic convex-hull data structure [2,7].

Two more problems are needed to be addressed during the search for the node containing x^*. First, given any hinge node h on T, the problem requires to determine if center x^* is on h, i.e., at hinge G_h h represents, and otherwise, which split subtree of h on T contains x^*, that is, which hanging subgraph of G_h on G contains x^*. In addition, a more general problem is the *center-detecting* problem: Given any block node u on T, the goal is to determine whether x^* is on u (i.e., on block G_u on G), and otherwise, which split tree of the H-subtree of u on T contains x^*, that is, which hanging subgraph of G_u contains x^*.

These two problems are more general problems on cacti than the tree version [17] since each $\mathsf{Ed}(P_i, x)$ is no longer a convex function in x on any path of G. We however observe that the median of any $P_i \in \mathcal{P}$ always fall in the hanging subgraph of a block whose probability sum of P_i is at least 0.5. Based on this, with the assistance of other useful observations and lemmas, we can efficiently solve each above problem in $O(mn)$ time.

Outline. In Sect. 2, we introduce some notations and observations. In Sect. 3, we present our algorithm for the one-center problem on a cycle. In Sect. 4, we discuss our algorithm for the problem on a cactus.

2 Preliminary

In the following, unless otherwise stated, we assume that our problem is the vertex-constrained case where every location of $\mathcal{P}$ is at a vertex on G and every vertex holds at least one location of $\mathcal{P}$. Note that Lemma 5 shows that any general case can be reduced in linear time into a vertex-constrained case.

Some terminologies arc borrowed from the literature [8]. A G-vertex is a vertex on G not included in any cycle, and a hinge is one on a cycle of degree greater than 2. A graft is a maximum (connected) tree subgraph on G where every leaf is either a hinge or a G-vertex, all hinges are at leaves, and no two hinges belong to the same cycle. A cactus graph is indeed a block graph consisting of graft blocks and cycle blocks so that the skeleton of G is a trec T where for each block on G, a node is joined by an edge to its hinges. See Fig. 1 for an example.

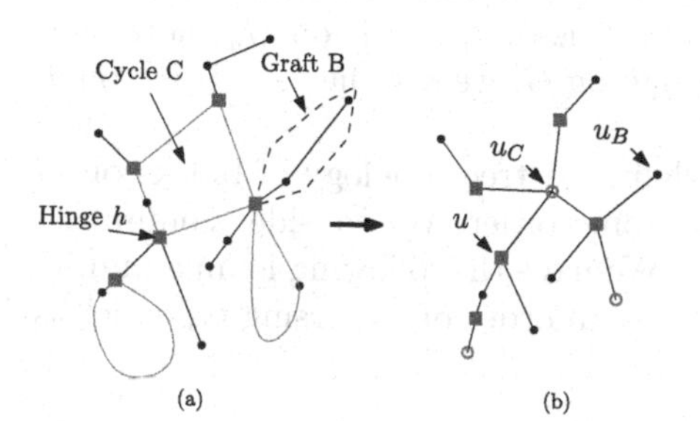

Fig. 1. (a) Illustrating a cactus G that consists of 3 cycles, 5 hinges (squares) and 6 G-vertices (disks); (b) Illustrating G's skeleton T where circular and disk nodes represent cycles and grafts of G (e.g., nodes u, u_C and u_B respectively representing hinge h, cycle C and graft B on G).

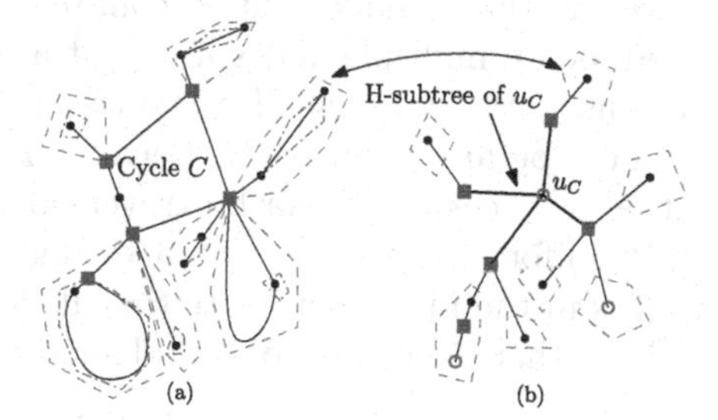

Fig. 2. (a) Cycle C on G has 7 split subgraphs (blue dash doted lines) and accordingly 7 hanging subgraphs (red dashed lines); (b) on T, the H-subtree of node u_c representing cycle C has 7 split subtrees each of which represents a distinct hanging subgraph of C on G. (Color figure online)

In fact, T represents a decomposition of G so that we can traverse nodes on T in a specific order to traverse G blocks by blocks in the according order. Our algorithm thus works on T to compute center x^*. Tree T can be computed by a depth-first-search on G [6,8] so that each node on T is attached with a block or a hinge of G. We say that a node u on T is a block (resp., hinge) node if it represents a block (resp., hinge) on G. In our preprocessing work, we construct the skeleton T with additional information maintained for nodes of T to fasten the computation.

Denote by G_u the block (resp., hinge) on G of any block (resp., hinge) node u on T. More specifically, we calculate and maintain the cycle circumstance for every cycle node on T. For any hinge node h on T, h is attached with hinge G_h on G (i.e., h represents G_h). For each adjacent node u of h, vertex G_h also exists on block G_u but with only adjacent vertices of G_u (that is, there is a copy of G_h on G_u but with adjacent vertices only on G_u). We associate each adjacent node u in the adjacent list of h with vertex G_h (the copy of G_h) on G_u, and also maintain the link from vertex G_h on G_u to node h.

Clearly, the size $|T|$ of T is $O(mn)$ due to $|G| = O(mn)$. It is not difficult to see that all preprocessing work can be done in $O(mn)$ time. As a result, the following operations can be done in constant time.

1. Given any vertex v on G, finding the node on T whose block v is on;
2. Given any hinge node h on T, finding vertex G_h on the block of every adjacent node of h on T;
3. Given any block node u on T, for any hinge on G_u, finding the hinge node on T representing it.

Consider every hinge on the block of every block node on T as an *open* vertex that does not contain any locations of $\mathcal{P}$. To be convenient, for any point x on G, we say that a node u on T contains x or x is on u if x is on G_u; note that x may be on multiple nodes if x is at a hinge on G; we say that a subtree on T contains x if x is on one of its nodes.

Let x be any point on G. Because T defines a tree topology of blocks on G so that vertices on G can be traversed in some order. We consider computing $\mathsf{Ed}(P_i, x)$ for all $1 \le i \le n$ by traversing T. We have the following lemma and its proof is in the full paper. Note that it defines an order of traversing G, which is used in other operations of our algorithm.

Lemma 1. *Given any point x on G, $\mathsf{Ed}(P_i, x)$ for all $1 \le i \le n$ can be computed in $O(mn)$ time.*

We say that a point x on G is an *articulation* point if x is on a graft block; removing x generates several connected disjoint subgraphs; each of them is called a *split* subgraph of x; the subgraph induced by x and one of its split subgraphs is called a *hanging* subgraph of x.

Similarly, any connected subgraph G' of G has several split subgraphs caused by removing G', and each split subgraph with adjacent vertice(s) on G' contributes a hanging subgraph. See Fig. 2 (a) for an example.

Consider any uncertain point $P_i \in \mathcal{P}$. There exists a point x_i^* on G so that $\mathsf{Ed}(P_i, x)$ reaches its minimum at $x = x_i^*$; point x_i^* is called the *median* of P_i on G. For any subgraph G' on G, we refer to value $\sum_{p_{ij} \in G'} f_{ij}$ as P_i's *probability sum* of G'; we refer to value $w_i \cdot \sum_{p_{ij} \in G'} f_{ij} \cdot d(p_{ij}, x)$ as P_i's (weighted) *distance sum* of G' to point x.

Notice that we say that median x_i^* of P_i (resp., center x^*) is on a hanging subgraph of a subgraph G' on G iff x_i^* (resp., x^*) is likely to be on that split subgraph of G' it contains. The below lemma holds and its proof is in the full paper.

Lemma 2. *Consider any articulation point x on G and any uncertain point $P_i \in \mathcal{P}$.*

1. *If x has a split subgraph whose probability sum of P_i is greater than 0.5, then its median x_i^* is on the hanging subgraph including that split subgraph;*
2. *The point x is x_i^* if P_i's probability sum of each split subgraph of x is less than 0.5;*
3. *The point x is x_i^* if x has a split subgraph with P_i's probability sum equal to 0.5.*

For any point $x \in G$, we say that P_i is a dominant uncertain point of x if $\mathsf{Ed}(P_i, x) \geq \mathsf{Ed}(P_j, x)$ for each $1 \leq j \leq n$. Point x may have multiple dominant uncertain points. Lemma 2 implies the following corollary.

Corollary 1. *Consider any articulation point x on G.*

1. *If x has one dominant uncertain point whose median is at x, then center x^* is at x;*
2. *If two dominant uncertain points have their medians on different hanging subgraphs of x, then x^* is at x;*
3. *Otherwise, x^* is on the hanging subgraph that contains all their medians.*

Let u be any block node on T; denote by T_u^H the subtree on T induced by u and its adjacent (hinge) nodes; we refer to T_u^H as the *H-subtree* of u on T. Each hanging subgraph of block G_u on G is represented by a split subtree of T_u^H on T. See Fig. 2 (b) for an example. Lemma 2 also implies the following corollary.

Corollary 2. *Consider any block node u on T and any uncertain point P_i of $\mathcal{P}$.*

1. *If the H-subtree T_u^H of u has a split subtree whose probability sum of P_i is greater than 0.5, then x_i^* is on the split subtree of T_u^H;*
2. *If the probability sum of P_i on each of T_u^H's split subtree is less than 0.5, then x_i^* is on u (i.e., block G_u of G);*
3. *If T_u^H has a split subtree whose probability sum of P_i is equal to 0.5, then x_i^* is on that hinge node on T_u^H that is adjacent to the split subtree.*

Moreover, we have the following lemma. The proof is in the full paper.

Lemma 3. *Given any articulation point x on G, we can determine in $O(mn)$ time whether x is x^*, and otherwise, which hanging subgraph of x contains x^*.*

Consider any hinge node u on T. Lemma 3 implies the following corollary.

Corollary 3. *Given any hinge node u on T, we can determine in $O(mn)$ time whether x^* is on u (i.e., at hinge G_u on G), and otherwise, which split subtree of u contains x^*.*

3 The One-Center Problem on a Cycle

In this section, we consider the one-center problem for the case of G being a cycle. A general expected distance is considered: each $P_i \in \mathcal{P}$ is associated with a constant c_i so that the (weighted) distance of P_i to x is equal to their (weighted) expected distance plus c_i. With a little abuse of notations, we refer to it as the expected distance $\mathsf{Ed}(P_i, x)$ from P_i to x.

Our algorithm focuses on the vertex-constrained version where every location is at a vertex on G and every vertex holds at least one location. Since G is a cycle, it is easy to see that any general instance can be reduced in linear time to a vertex-constrained instance.

Let $u_1, u_2, \cdots, u_M$ be the clockwise enumeration of all vertices on G, and $M \leq mn$. Let $l(G)$ be G's circumstance. Every u_i has a *semicircular* point $x_{i'}$ with $d(u_i, x_{i'}) = l(G)/2$ on G. Because sequence $x_{1'}, \cdots, x_{M'}$ is in the clockwise order. $x_{1'}, \cdots, x_{M'}$ can be computed in order in $O(mn)$ time by traversing G clockwise.

Join these semicircular points $x_{1'}, \cdots, x_{M'}$ to G by merging them and $u_1, \cdots, u_M$ in clockwise order; simultaneously, reindex all vertices on G clockwise; hence, a clockwise enumeration of all vertices on G is generated in $O(mn)$ time. Clearly, the size N of G is now at most $2mn$. Given any vertex u_i on G, there exists another vertex u_{i^c} so that $d(u_i, u_{i^c}) = l(G)/2$. Importantly, $i^c = [(i-1)^c + 1]\%N$ for $2 \leq i \leq N$ and $1^c = (N^c + 1)$.

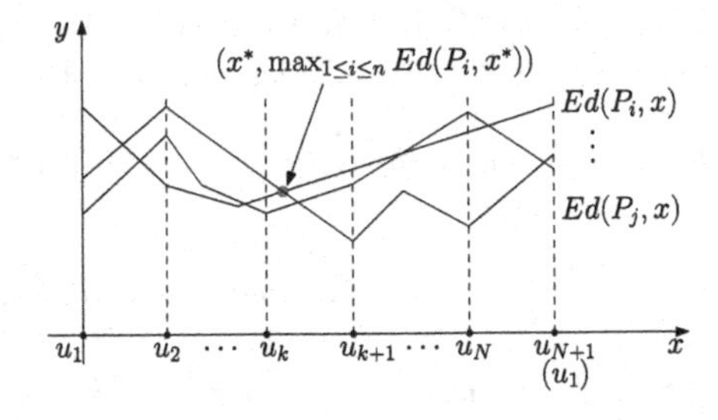

Fig. 3. Consider $y = \mathsf{Ed}(P_i, x)$ in x, y-coordinate system by projecting cycle G onto x-axis; $\mathsf{Ed}(P_i, x)$ of each $P_i \in \mathcal{P}$ is linear in x on any edge of G; center x^* is decided by the projection on x-axis of the lowest point on the upper envelope of all $y = \mathsf{Ed}(P_i, x)$'s.

Let x be any point on G. Consider the expected distance $y = \mathsf{Ed}(P_i, x)$ in the x, y-coordinate system. We set u_1 at the origin and project vertices $u_1, u_2, \cdots, u_N\ u_{N+1}, \cdots, u_{2N}$ on x-axis in order so that $u_{N+i} = u_i$. Denote by x_i the x-coordinate of u_i on x-axis. For $1 \leq i < j \leq N$, the clockwise distance between u_i and u_j on G is exactly $(x_j - x_i)$ and their counterclockwise distance is equal to $(x_{i+N} - x_j)$.

As shall be analyzed below, each $\mathsf{Ed}(P_i, x)$ is linear in $x \in [x_s, x_{s+1}]$ for each $1 \leq s \leq N$ but may be neither convex nor concave for $x \in [x_1, x_{N+1}]$, which is different to the deterministic case [6]. See Fig. 3. Center x^* is determined by the lowest point of the upper envelope of all $\mathsf{Ed}(P_i, x)$ for $x \in [x_1, x_{N+1}]$. Our strategy is computing the lowest point of the upper envelope on interval $[x_s, x_{s+1}]$, i.e., computing the local center $x^*_{s,s+1}$ of $\mathcal{P}$ on $[x_s, x_{s+1}]$, for each $1 \leq s \leq N$; center x^* is obviously decided by the lowest one among all of them.

For each $1 \leq s \leq N+1$, vertex u_s has a *semicircular* point x' on x-axis with $x_s - x' = l(G)/2$; x' must be at a vertex on x-axis in that u_s on G has its semicircular point at vertex u_{s^c}; we still let u_{s^c} be u_s's semicircular vertex on x-axis. Clearly, for each $1 \leq s \leq N$, $(s+1)^c = s^c + 1$, and the semicircular point of any point in $[x_s, x_{s+1}]$ lies in $[x_{s^c}, x_{(s+1)^c}]$. Indices $1^c, 2^c, \cdots, (N+1)^c$ can be easily determined in order in $O(mn)$ time and so we omit the details.

Consider any uncertain point P_i of $\mathcal{P}$. Because for any $1 \leq s \leq N$, interval $[x_{s+1}, x_{s+N}]$ contains all locations of $\mathcal{P}$ uniquely. We denote by x_{ij} the x-coordinate of location p_{ij} in $[x_{s+1}, x_{s+N}]$; denote by $F_i(x_s, x_{s^c})$ the probability sum of P_i's locations in $[x_s, x_{s^c}]$; let $D_i(x_{s+1}, x_{s^c})$ be value $w_i \cdot \sum_{p_{ij} \in [x_{s+1}, x_{s^c}]} f_{ij} x_{ij}$ and $D^c_i(x_{s^c+1}, x_{s+N})$ be value $w_i \cdot \sum_{p_{ij} \in [x_{s^c+1}, x_{s+N}]} f_{ij}(l(G) - x_{ij})$. Due to $F_i(x_{s+1}, x_{s^c}) + F_i(x_{s^c+1}, x_{s+N}) = 1$, we have that $\mathsf{Ed}(P_i, x)$ for $x \in [x_s, x_{s+1}]$ can be formulated as follows.

$$
\begin{aligned}
\mathsf{Ed}(P_i, x) &= c_i + w_i \sum_{p_{ij} \in [x_{s+1}, x_{s^c}]} f_{ij} \cdot (x_{ij} - x) + w_i \sum_{p_{ij} \in [x_{s^c+1}, x_{s+N}]} f_{ij} \cdot [l(G) - (x_{ij} - x)] \\
&= c_i + w_i \Big(\sum_{p_{ij} \in [x_{s^c+1}, x_{s+N}]} f_{ij} - \sum_{p_{ij} \in [x_{s+1}, x_{s^c}]} f_{ij} \Big) \cdot x + w_i \sum_{p_{ij} \in [x_{s+1}, x_{s^c}]} f_{ij} x_{ij} \\
&\quad - w_i \sum_{p_{ij} \in [x_{s^c+1}, x_{s+N}]} f_{ij}(l(G) - x_{ij}) \\
&= w_i [1 - 2F_i(x_{s+1}, x_{s^c})] \cdot x + c_i + D_i(x_{s+1}, x_{s^c}) - D^c_i(x_{s^c+1}, x_{s+N})
\end{aligned}
$$

It turns out that each $\mathsf{Ed}(P_i, x)$ is linear in $x \in [x_s, x_{s+1}]$ for each $1 \leq s \leq N$, and it turns at $x = x_s$ if P_i has locations at points u_s, u_{s^c}, or u_{s+N}. Note that $\mathsf{Ed}(P_i, x)$ may be neither convex nor concave. Hence, each $\mathsf{Ed}(P_i, x)$ is a piecewise linear function of complexity at most m for $x \in [x_1, x_{N+1}]$. It follows that the local center $x^*_{s,s+1}$ of $\mathcal{P}$ on $[x_s, x_{s+1}]$ is decided by the x-coordinate of the lowest point of the upper envelope on $[x_s, x_{s+1}]$ of functions $\mathsf{Ed}(P_i, x)$'s for all $1 \leq i \leq n$.

Consider the problem of computing the lowest points on the upper envelope of all $\mathsf{Ed}(P_i, x)$'s on interval $[x_s, x_{s+1}]$ for all $1 \leq s \leq N$ from left to right.

Let L be the set of lines by extending all line segments on $\mathsf{Ed}(P_i, x)$ for all $1 \le i \le n$, and $|L| \le mn$. Since the upper envelope of lines is geometric dual to the convex (lower) hull of points, the dynamic convex-hull maintenance data structure of Brodal and Jacob [7] can be applied to L so that with $O(|L| \log |L|)$-time preprocessing and $O(|L|)$-space, our problem can be solved as follows.

Suppose that we are about to process interval $[x_s, x_{s+1}]$. The dynamic convex-hull maintenance data structure Φ currently maintains the information of only n lines caused by extending the line segment of each $\mathsf{Ed}(P_i, x)$'s on $[x_{s-1}, x_s]$. Let $\mathcal{P}_s$ be the subset of uncertain points of $\mathcal{P}$ whose expected distance functions turn at $x = x_s$. For each $P_i \in \mathcal{P}_s$, we delete from Φ function $\mathsf{Ed}(P_i, x)$ for $x \in [x_{s-1}, x_s]$ and then insert the line function of $\mathsf{Ed}(P_i, x)$ on $[x_s, x_{s+1}]$ into Φ. After these $2|\mathcal{P}_s|$ updates, we compute the local center $x^*_{s,s+1}$ of $\mathcal{P}$ on $[x_s, x_{s+1}]$ as follows.

Perform an extreme-point query on Φ in the vertical direction to compute the lowest point of the upper envelope of the n lines. If the obtained point falls in $[x_s, x_{s+1}]$, $x^*_{s,s+1}$ is of same x-coordinate as this point and its y-coordinate is the objective value at $x^*_{s,s+1}$; otherwise, it is to left of line $x = x_s$ (resp., to right of $x = x_{s+1}$) and thereby $x^*_{s,s+1}$ is of x-coordinate equal to x_s (resp., x_{s+1}); accordingly, we then compute the objective value at $x = x_s$ (resp., $x = x_{s+1}$) by performing another extreme-point query in direction $y = -x_s \cdot x$ (resp., $y = -x_{s+1} \cdot x$).

Note that $\mathcal{P}_1 = \mathcal{P}$ for interval $[x_1, x_2]$ and $\sum_{s=1}^{N} |P_s| = |L| \le mn$. Since updates and queries each takes $O(\log |L|)$ amortized time, for each interval $[x_s, x_{s+1}]$, we spend totally $O(|\mathcal{P}_s| \cdot \log |L|)$ amortized time on computing $x^*_{s,s+1}$. It implies that the time complexity for all updates and queries on Φ is $O(mn \log mn)$ time. Therefore, the total running time of computing the local centers of $\mathcal{P}$ on $[x_s, x_{s+1}]$ for all $1 \le s \le N$ is $O(mn \log mn)$ plus the time on determining functions $\mathsf{Ed}(P_i, x)$ of each $P_i \in \mathcal{P}_s$ on $[x_s, x_{s+1}]$ for all $1 \le s \le N$.

We now present how to determine $\mathsf{Ed}(P_i, x)$ of each $P_i \in \mathcal{P}_s$ in $x \in [x_s, x_{s+1}]$ for all $1 \le s \le N$. Recall that $\mathsf{Ed}(P_i, x) = w_i \cdot [1 - 2F_i(x_{s+1}, x_{s^c})] \cdot x + D_i(x_{s+1}, x_{s^c}) - D_i^c(x_{s^c+1}, x_{s+N}) + c_i$ for $x \in [x_s, x_{s+1}]$. It suffices to compute the three coefficients $F_i(x_{s+1}, x_{s^c})$, $D_i(x_{s+1}, x_{s^c})$ and $D_i^c(x_{s^c+1}, x_{s+N})$ for each $1 \le i \le n$ and $1 \le s \le N$.

We create auxiliary arrays $X[1 \cdots n]$, $Y[1 \cdots n]$ and $Z[1 \cdots n]$ to maintain the three coefficients of $\mathsf{Ed}(P_i, x)$ for x in the current interval $[x_s, x_{s+1}]$, respectively; another array $I[1 \cdots n]$ is also created so that $I[i] = 1$ indicates that $P_i \in \mathcal{P}_s$ for the current interval $[x_s, x_{s+1}]$; we associate with u_s for each $1 \le s \le N$ an empty list $\mathcal{A}_s$ that will store the coefficients of $\mathsf{Ed}(P_i, x)$ on $[x_s, x_{s+1}]$ for each $P_i \in \mathcal{P}_s$. Initially, $X[1 \cdots n]$, $Y[1 \cdots n]$, and $Z[1 \cdots n]$ are all set as zero; $I[1 \cdots n]$ is set as one due to $\mathcal{P}_1 = \mathcal{P}$.

For interval $[x_1, x_2]$, we compute $F_i(x_2, x_{1^c})$, $D_i(x_2, x_{1^c})$ and $D_i^c(x_{1^c+1}, x_{N+1})$ for each $1 \le i \le n$: for every location p_{ij} in $[x_2, x_{1^c}]$, we set $X[i] = X[i] + f_{ij}$ and $Y[i] = Y[i] + w_i \cdot f_{ij} \cdot x_{ij}$; for every location p_{ij} in $[x_{1^c+1}, x_{N+1}]$, we set $Z[i] = Z[i] + w_i \cdot (l(G) - x_{ij})$. Since x_{1^c} is known in $O(1)$ time, it is easy to see that for all $P_i \in \mathcal{P}_1$, functions $\mathsf{Ed}(P_i, x)$ for $x \in [x_1, x_2]$ can be determined in $O(mn)$ time.

Next, we store in list $\mathcal{A}_1$ the coefficients of $\mathsf{Ed}(P_i, x)$ of all $P_i \in \mathcal{P}_1$ on $[x_1, x_2]$: for each $I[i] = 1$, we add tuples $(i, w_i \cdot X[i], c_i + Y[i] - Z[i])$ to $\mathcal{A}_1$ and then set $I[i] = 0$. Clearly, list $\mathcal{A}_1$ for u_1 can be computed in $O(mn)$ time.

Suppose we are about to determine the line function of $\mathsf{Ed}(P_i, x)$ on $[x_s, x_{s+1}]$, i.e., coefficients $F_i(x_{s+1}, x_{s^c})$, $D_i(x_{s+1}, x_{s^c})$ and $D_i^c(x_{s^c+1}, x_{s+N})$, for each $P_i \in \mathcal{P}_s$. Note that if P_i has no locations at u_s, u_{s^c} and u_{s+N}, then P_i is not in $\mathcal{P}_s$; otherwise, $\mathsf{Ed}(P_i, x)$ turns at $x = x_s$ and we need to determine $\mathsf{Ed}(P_i, x)$ for $x \in [x_s, x_{s+1}]$.

Recall that for $x \in [x_{s-1}, x_s]$, $\mathsf{Ed}(P_i, x) = c_i + w_i \cdot [1 - 2F_i(x_s, x_{(s-1)^c})] \cdot x + D_i(x_s, x_{(s-1)^c}) - D_i^c(x_{(s-1)^c+1}, x_{s-1+N})$. On account of $s^c = (s-1)^c + 1$, for $x \in [x_s, x_{s+1}]$, we have $F_i(x_{s+1}, x_{s^c}) = F_i(x_s, x_{(s-1)^c}) - F_i(x_s, x_s) + F_i(x_{s^c}, x_{s^c})$, $D_i(x_{s+1}, x_{s^c}) = D_i(x_s, x_{(s-1)^c}) - D_i(x_s, x_s) + D_i(x_{s^c}, x_{s^c})$, and $D_i^c(x_{s^c+1}, x_{s+N}) = D_i^c(x_{(s-1)^c+1}, x_{s-1+N}) - D_i^c(x_{s^c}, x_{s^c}) + D_i^c(x_{s+N}, x_{s+N})$. Additionally, for each $1 \leq i \leq n$, $\mathsf{Ed}(P_i, x)$ on $[x_{s-1}, x_s]$ is known, and its three coefficients are respectively in entries $X[i]$, $Y[i]$, and $Z[i]$. We can determine $\mathsf{Ed}(P_i, x)$ of each $P_i \in \mathcal{P}_s$ on $[x_s, x_{s+1}]$ as follows.

For each location p_{ij} at u_s, we set $X[i] = X[i] - f_{ij}$, $Y[i] = Y[i] - w_i f_{ij} x_{ij}$ and $I[i] = 1$; for each location p_{ij} at u_{s^c}, we set $X[i] = X[i] + f_{ij}$, $Y[i] = Y[i] + w_i f_{ij} x_{ij}$, $Z[i] = Z[i] - w_i f_{ij}(l(G) - x_{ij})$ and $I[i] = 1$; further, for each location p_{ij} at u_{s+N}, we set $Z[i] = Z[i] + w_i f_{ij}(l(G) - x_{ij})$ and $I[i] = 1$. Subsequently, we revisit locations at u_s, u_{s^c} and u_{s+N}: for each location p_{ij}, if $I[i] = 1$ then we add a tuple $(i, w_i \cdot X[i], c_i + Y[i] - Z[i])$ to $\mathcal{A}_s$ and set $I[i] = 0$, and otherwise, we continue our visit.

For each $2 \leq s \leq N$, clearly, functions $\mathsf{Ed}(P_i, x)$ on $[x_s, x_{s+1}]$ of all $P_i \in \mathcal{P}_s$ can be determined in the time linear to the number of locations at the three vertices u_s, u_{s^c} and u_{s+N}. It follows that the time complexity for determining $\mathsf{Ed}(P_i, x)$ of each $P_i \in \mathcal{P}_s$ for all $1 \leq s \leq N$, i.e., computing the set L, is $O(mn)$; that is, the time complexity for determining $\mathsf{Ed}(P_i, x)$ for each $P_i \in \mathcal{P}$ on $[x_1, x_{N+1}]$ is $O(mn)$.

Combining all above efforts, we have the following theorem.

Theorem 1. *The one-center problem of $\mathcal{P}$ on a cycle can be solved in $O(|G| + mn \log mn)$ time.*

4 The Algorithm

In this section, we shall present our algorithm for computing the center x^* of $\mathcal{P}$ on cactus G. We first give the lemma for solving the base case where a node of T, i.e., a block of G, is known to contain center x^*. The proof is in the full paper.

Lemma 4. *If a node u on T is known to contain center x^*, then x^* can be computed in $O(mn \log mn)$ time.*

Now we are ready to present our algorithm that performs a recursive search on T to locate the node, i.e., the block on G, that contains center x^*. Once the node is found, Lemma 4 is then applied to find center x^* in $O(mn \log mn)$ time.

On the tree, a node is called a *centroid* if every split subtree of this node has no more than half nodes, and the centroid can be found in $O(|T|)$ time by a traversal on the tree [11,14].

We first compute the centroid c of T in $O(|T|)$ time. If c is a hinge node, then we apply Corollary 3 to c, which takes $O(mn)$ time; if x^* is on c, we then immediately return its hinge G_c on G as x^*; otherwise, we obtain a split subtree of c on T representing the hanging subgraph of hinge G_c on G that contains x^*.

On the other hand, c is a block node. We then solve the *center-detecting* problem for c that is to decide which split subtree of c's H-subtree T_c^H on T contains x^*, that is, determine which hanging subgraph of block G_c contains x^*. As we shall present in Sect. 4.1, the center-detecting problem can be solved in $O(mn)$ time. It follows that x^* is either on one of T_c^H's split subtrees or T_c^H. In the later case, since G_c is represented by T_c^H, we can apply Lemma 4 to c so that the center x^* can be obtained in $O(mn \log mn)$ time.

In general, we obtain a subtree T' that contains center x^*. The size of T' is no more than half of T. Further, we continue to perform the above procedure recursively on the obtained T'. Similarly, we compute the centroid c of T' in $O(|T'|)$ time; we then determine in $O(mn)$ time whether x^* is on node c, and otherwise, find the subtree of T' containing x^* but of size at most $|T'|/2$.

As analyzed above, each recursive step takes $O(mn)$ time. After $O(\log mn)$ recursive steps, we obtain one node on T that is known to contain center x^*. At this moment, we apply Lemma 4 to this node to compute x^* in $O(mn \log mn)$ time. Therefore, the vertex-constrained one-center problem can be solved in $O(mn \log mn)$ time.

Recall that in the general case, locations of $\mathcal{P}$ could be anywhere on the given cactus graph rather than only at vertices. To solve the general one-center problem, we first reduce the given general instance to a vertex-constrained instance by Lemma 5, and then apply our above algorithm to compute the center. The proof for Lemma 5 is in the full paper.

Lemma 5. *The general case of the one-center problem can be reduced to a vertex-constrained case in* $O(|G| + mn)$ *time.*

Theorem 2. *The one-center problem of n uncertain points on cactus graphs can be solved in* $O(|G| + mn \log mn)$ *time.*

4.1 The Center-Detecting Problem

Given any block node u on T, the center-detecting problem is to determine which split subtree of u's H-subtree T_u^H on T contains x^*, i.e., which hanging subgraph of block G_u contains x^*. If G is a tree, this problem can be solved in $O(mn)$ time [17]. Our problem is on cacti and a new approach is proposed below.

Let $G_1(u), \cdots, G_s(u)$ be all hanging subgraphs of block G_u on G; for each $G_k(u)$, let v_k be the hinge on $G_k(u)$ that connects its vertices with $G/G_k(u)$. $G_1(u), \cdots, G_s(u)$ are represented by split subtrees $T_1(u), \cdots, T_s(u)$ of T_u^H on T, respectively.

Let u be the root of T. For each $1 \leq k \leq s$, $T_k(u)$ is rooted at a block node u_k; hinge v_k is an (open) vertex on block G_{u_k}; the parent node of u_k on T is the hinge node h_k on T_u^H that represents v_k; note that h_k might be h_t for some $1 \leq t \neq k \leq s$. For all $1 \leq k \leq s$, $T_k(u)$, h_k, and v_k on block G_{u_k} can be obtained in $O(mn)$ time via traversing subtrees rooted at $h_1, \cdots, h_s$.

For each $1 \leq k \leq s$, there is a subset $\mathcal{P}_k$ of uncertain points so that each $P_i \in \mathcal{P}_k$ has its probability sum of $G_k(u)/\{u_k\}$, i.e., $T_k(u)$, greater than 0.5. Clearly, $\mathcal{P}_i \cap \mathcal{P}_j = \emptyset$ holds for any $1 \leq i \neq j \leq s$.

Define $\tau(G_k(u)) = \max_{P_i \in \mathcal{P}_k} \mathsf{Ed}(P_i, v_k)$. Let γ be the largest value of $\tau(G_k(u))$'s for all $1 \leq k \leq s$. We have the below observation whose proof is in the full paper.

Observation 1. *If $\tau(G_k(u)) < \gamma$, then center x^* cannot be on $G_k(u)/\{v_k\}$; if $\tau(G_r(u)) = \tau(G_t(u)) = \gamma$ for some $1 \leq r \neq t \leq s$, then center x^* is on block G_u.*

Below, we first describe the approach for solving the center-detecting problem and then present how to compute values $\tau(G_k(u))$ for all $1 \leq k \leq s$.

First, we compute $\gamma = \max_{k=1}^{s} \tau(G_k(u))$ in $O(s)$ time. We then determine in $O(s)$ time if there exists only one subgraph $G_r(u)$ with $\tau(G_r(u)) = \gamma$. If yes, then center x^* is on either $G_r(u)$ or G_u. Their only common vertex is v_r, and v_r and its corresponding hinge h_r on T are known in $O(1)$ time. For this case, we further apply Corollary 3 to h_r on T; if x^* is at v_r then we immediately return hinge v_r on G as the center; otherwise, we obtain the subtree on T that represents the one containing x^* among $G_r(u)$ and G_u, and return it.

On the other hand, there exist at least two subgraphs, e.g., $G_r(u)$ and $G_t(u)$, so that $\tau(G_r(u)) = \tau(G_t(u)) = \gamma$ for $1 \leq r \neq t \leq s$. By Observation 1, x^* is on G_u and thereby node u on T is returned. Due to $s \leq mn$, we can see that all the above operations can be carried out in $O(mn)$ time.

To solve the center-detecting problem, it remains to compute $\tau(G_k(u))$ for all $1 \leq k \leq s$. We first consider the problem of computing the distance $d(v_k, x)$ for any given point x and any given v_k on G. We have the following result and its proof is in the full paper.

Lemma 6. *With $O(mn)$-time preprocessing work, given any hinge v_k and any point x on G, the distance $d(v_k, x)$ can be known in constant time.*

We now consider the problem of computing $\tau(G_k(u))$ for each $1 \leq k \leq s$, which is solved as follows.

First, we determine the subset $\mathcal{P}_k$ for each $1 \leq k \leq s$: Create auxiliary arrays $A[1 \cdots n]$ initialized as zero and $B[1 \cdots n]$ initialized as null. We do a pre-order traversal on $T_k(u)$ from node u_k to compute the probability sum of each P_i on $G_k(u)/v_k$; during the traversal, for each location p_{ij}, we add f_{ij} to $A[i]$ and continue to check if $A[i] > 0.5$; if yes, we set $B[i]$ as u_k, and otherwise, we continue our traversal on $T_k(u)$; once we are done, we traverse $T_k(u)$ again to reset $A[i] = 0$ for every location p_{ij} on $T_k(u)$. Clearly, $B[i] = u_k$ iff $P_i \in \mathcal{P}_k$. After traversing $T_1(u), \cdots, T_s(u)$ as the above, given any $1 \leq i \leq n$, we can know to which subset P_i belongs by accessing $B[i]$.

To compute $\tau(G_k(u))$ for each $1 \leq k \leq s$, it suffices to compute $\mathsf{Ed}(P_i, v_k)$ for each $P_i \in \mathcal{P}_k$. In details, we first create an array $L[1 \cdots n]$ to maintain values $\mathsf{Ed}(P_i, v_k)$ of each $P_i \in \mathcal{P}_k$ for all $1 \leq k \leq s$. We then traverse G directly to compute values $\mathsf{Ed}(P_i, v_k)$. During the traversal on G, for each location p_{ij}, if $B[i]$ is u_k, then P_i is in $\mathcal{P}_k$; we continue to compute in constant time the distance $d(p_{ij}, v_k)$ by Lemma 6; we then add value $w_i \cdot f_{ij} \cdot d(p_{ij}, v_k)$ to $L[i]$. It follows that in $O(mn)$ time we can compute values $\mathsf{Ed}(P_i, v_k)$ of each $P_i \in \mathcal{P}_k$ for all $1 \leq k \leq s$.

With the above efforts, $\tau(G_k(u))$ for all $1 \leq k \leq s$ can be computed by scanning $L[1 \cdots n]$: Initialize each $\tau(G_k(u))$ as zero; for each $L[i]$, supposing $B[i]$ is u_k, we set $\tau(G_k(u))$ as the larger of $\tau(G_k(u))$ and $L[i]$; otherwise, either $L[i] = 0$ or $B[i]$ is null, and hence we continue our scan. These can be carried out in $O(n)$ time.

In a summary, with $O(mn)$-preprocessing work, values $\tau(G_k(u))$ for all $1 \leq k \leq s$ can be computed in $O(mn)$ time. Once values $\tau(G_k(u))$ are known, as the above stated, the center-detecting problem for any given block node u on T can be solved in $O(mn)$ time. The following lemma is thus proved.

Lemma 7. *Given any block node u on T, the center-detecting problem can be solved in $O(mn)$ time.*

References

1. Abam, M., de Berg, M., Farahzad, S., Mirsadeghi, M., Saghafian, M.: Preclustering algorithms for imprecise points. Algorithmica **84**, 1467–1489 (2022)
2. Agarwal, P., Matoušek, J.: Dynamic half-space range reporting and its applications. Algorithmica **13**(4), 325–345 (1995)
3. Averbakh, I., Bereg, S.: Facility location problems with uncertainty on the plane. Discret. Optim. **2**, 3–34 (2005)
4. Averbakh, I., Berman, O.: Minimax regret p-center location on a network with demand uncertainty. Locat. Sci. **5**, 247–254 (1997)
5. Bai, C., Kang, L., Shan, E.: The connected p-center problem on cactus graphs. Theor. Comput. Sci. **749**, 59–65 (2017)
6. Ben-Moshe, B., Bhattacharya, B., Shi, Q., Tamir, A.: Efficient algorithms for center problems in cactus networks. Theor. Comput. Sci. **378**(3), 237–252 (2007)
7. Brodal, G., Jacob, R.: Dynamic planar convex hull. In: Proceedings of the 43rd IEEE Symposium on Foundations of Computer Science (FOCS), pp. 617–626 (2002)
8. Burkard, R., Krarup, J.: A linear algorithm for the pos/neg-weighted 1-median problem on cactus. Computing **60**(3), 498–509 (1998)
9. Huang, L., Li, J.: Stochastic k-center and j-flat-center problems. In: Proceedings of the 28th ACM-SIAM Symposium on Discrete Algorithms (SODA), pp. 110–129 (2017)
10. Kachooei, H.A., Davoodi, M., Tayebi, D.: The p-center problem under uncertainty. In: Proceedings of the 2nd Iranian Conference on Computational Geometry, pp. 9–12 (2019)
11. Kariv, O., Hakimi, S.: An algorithmic approach to network location problems. I: the p-centers. SIAM J. Appl. Math. **37**(3), 513–538 (1979)

12. Keikha, V., Aghamolaei, S., Mohades, A., Ghodsi, M.: Clustering geometrically-modeled points in the aggregated uncertainty model. CoRR abs/2111.13989 (2021). https://arxiv.org/abs/2111.13989
13. Lan, Y., Wang, Y., Suzuki, H.: A linear-time algorithm for solving the center problem on weighted cactus graphs. Inf. Process. Lett. **71**(5), 205–212 (1999)
14. Megiddo, N.: Linear-time algorithms for linear programming in R^3 and related problems. SIAM J. Comput. **12**(4), 759–776 (1983)
15. Nguyen, Q., Zhang, J.: Line-constrained l_∞ one-center problem on uncertain points. In: Proceedings of the 3rd International Conference on Advanced Information Science and System, vol. 71, pp. 1–5 (2021)
16. Wang, H., Zhang, J.: A note on computing the center of uncertain data on the real line. Oper. Res. Lett. **44**, 370–373 (2016)
17. Wang, H., Zhang, J.: Computing the center of uncertain points on tree networks. Algorithmica **78**(1), 232–254 (2017)
18. Wang, H., Zhang, J.: Covering uncertain points on a tree. Algorithmica **81**, 2346–2376 (2019)
19. Yen, W.: The connected p-center problem on block graphs with forbidden vertices. Theor. Comput. Sci. **426–427**, 13–24 (2012)
20. Zmazek, B., Žerovnik, J.: The obnoxious center problem on weighted cactus graphs. Discret. Appl. Math. **136**(2), 377–386 (2004)

Cosecure Domination: Hardness Results and Algorithms

Kusum[(✉)] and Arti Pandey

Department of Mathematics, Indian Institute of Technology Ropar,
Rupnagar 140001, Punjab, India
{2018maz0011,arti}@iitrpr.ac.in

Abstract. For a simple graph $G = (V, E)$ without any isolated vertex, a cosecure dominating set S of G satisfies two properties, (i) S is a dominating set of G, (ii) for every vertex $v \in S$, there exists a vertex $u \in V \setminus S$ such that $uv \in E$ and $(S \setminus \{v\}) \cup \{u\}$ is a dominating set of G. The minimum cardinality of a cosecure dominating set of G is called the cosecure domination number of G and is denoted by $\gamma_{cs}(G)$. The Minimum Cosecure Domination problem is to find a cosecure dominating set of a graph G of cardinality $\gamma_{cs}(G)$. The decision version of the problem is known to be NP-complete for bipartite, planar, and split graphs. Also, it is known that the Minimum Cosecure Domination problem is efficiently solvable for proper interval graphs and cographs.

In this paper, we work on various important graph classes in an effort to reduce the complexity gap of the Minimum Cosecure Domination problem. We show that the decision version of the problem remains NP-complete for doubly chordal graphs, chordal bipartite graphs, star-convex bipartite graphs and comb-convex bipartite graphs. On the positive side, we give an efficient algorithm to compute the cosecure domination number of chain graphs, which is an important subclass of bipartite graphs. In addition, we show that the problem is linear-time solvable for bounded tree-width graphs. Further, we prove that the computational complexity of this problem varies from the classical domination problem.

Keywords: Cosecure Domination · Bipartite Graphs · Doubly Chordal Graphs · Bounded Tree-width Graphs · NP-completeness

1 Introduction

In this paper, $G = (V, E)$ denotes a graph, where V is the set of vertices and E is the set of edges in G. The graphs considered in this article are assumed to be finite, simple, and undirected. A set $D \subseteq V$ is a dominating set of a graph G, if the closed neighbourhood of D is the vertex set V, that is, $N[D] = V$.

Kusum—Research supported by University Grants Commission(UGC), India, under File No.: 1047/(CSIR-UGC NET DEC. 2017)

A. Pandey—Research supported by CRG project, Grant Number-CRG/2022/008333, Science and Engineering Research Board (SERB), India.

S.-Y. Hsieh et al. (Eds.): IWOCA 2023, LNCS 13889, pp. 246–258, 2023.
https://doi.org/10.1007/978-3-031-34347-6_21

The domination number of G, denoted by $\gamma(G)$, is the minimum cardinality of a dominating set of G. Given a graph G, the MINIMUM DOMINATION (MDS) problem is to compute a dominating set of G of cardinality $\gamma(G)$. The decision version of the MDS problem is DOMINATION DECISION problem, notated as DM problem; takes a graph G and a positive integer k as an input and asks whether there exists a dominating set of cardinality at most k. The MINIMUM DOMINATION problem and many of its variations has been vastly studied in the literature and interested readers may refer to [8,9].

One of the important variations of domination is secure domination and this concept was first introduced by Cockayne et al. [6] in 2005. A set $S \subseteq V$ is a *secure dominating set* of G, if S is dominating set of G and for every $u \in V \setminus S$, there exists $v \in S$ such that $uv \in E$ and $(S \setminus \{v\}) \cup \{u\}$ forms a dominating set of G. The minimum cardinality of a secure dominating set of G is called the secure domination number of G and is denoted by $\gamma_s(G)$. The SECURE DOMINATION problem is to compute a secure dominating set of G of cardinality $\gamma_s(G)$. Several researchers have contributed to the study of this problem and its many variants [1,4,6,14,15,21]. For a detailed survey of this problem, one can refer to [8].

Consider a situation in which the goal is to protect the graph by using a subset of guards and simultaneously provide a backup or substitute (non-guard) for each guard such that the resultant arrangement after one substitution still protects the graph. Motivated by the above situtaion, another interesting variation of domination known as the cosecure domination was introduced in 2014 by Arumugam et al. [2], which was then further studied in [12,16,18,23]. This variation is partly related to secure domination and it is, in a way, a complement to secure domination. A set $S \subseteq V$ is said to be a *cosecure dominating set*, abbreviated as CSDS of G, if S is a dominating set of G and for every $u \in S$, there exists a vertex $v \in V \setminus S$ (replacement of u) such that $uv \in E$ and $(S \setminus \{u\}) \cup \{v\}$ is a dominating set of G. In this definition, we simply can say that v S*-replaces* u or v is replacement for u. A simple observation is that V can never be a cosecure dominating set of G. It should be noted that no cosecure dominating set exists, if the graph have isolated vertices. Also, we remark that the cosecure domination number of a disconnected graph G is simply sum of the cosecure domination number of the connected components of G. So, in this paper, we will just consider only the connected graphs without any isolated vertics.

Given a graph G without isolated vertex, the MINIMUM COSECURE DOMINATION problem (MCSD problem) is an optimization problem in which we need to compute a cosecure dominating set of G of cardinality $\gamma_{cs}(G)$. Given a graph G without isolated vertex and a positive integer k, the COSECURE DOMINATION DECISION problem, abbreviated as CSDD problem, is to determine whether there exists a cosecure dominating set of G of cardinality at most k. Clearly, $\gamma(G) \leq \gamma_{cs}(G)$.

The CSDD problem is known to be NP-complete for bipartite, chordal or planar graphs [2]. The bound related study on the cosecure domination number is done for some families of the graph classes [2,12]. The Mycielski graphs having the cosecure domination number 2 or 3 are characterized and a sharp upper bound was given for $\gamma_{cs}(\mu(G))$, where $\mu(G)$ is the Mycielski of a graph G. In 2021, Zou et al. proved that $\gamma_{cs}(G)$ of a proper interval graph G can be computed in linear-time [23]. Recently in [16], Kusum et al. augmented the complexity results and proved that the cosecure domination number of cographs can be determined in linear-time. They also demonstrated that the CSDD problem remains NP-complete for split graphs. In addition, they proved that the problem is APX-hard for bounded degree graphs and provided an inapproximability result for the problem. Further, they proved that the problem APX-complete for perfect graphs with bounded degree.

In this paper, we build on the existing research by examining the complexity status of the Minimum Cosecure Domination problem in many graph classes of significant importance, namely, doubly chordal graphs, bounded tree-width graphs, chain graphs, chordal bipartite graphs, star-convex bipartite graphs and comb-convex bipartite graphs. The content and structure of the paper is as follows. In Sect. 2, we give some preliminaries. In Sect. 3, we reduce the gap regarding the complexity status of the problem by showing that the CSDD problem is NP-complete for chordal bipartite graphs, star-convex bipartite graphs, and comb-convex bipartite graphs, which are all subclasses of bipartite graphs. In Sect. 4, we prove that the complexity of the cosecure domination and that of domination varies for some graph classes and we identify two of those. We prove that the CSDD problem remains NP-complete for doubly chordal graphs, for which the domination problem is easily solvable. In Sect. 5 and 6, we present some positive results, we prove that the MCSD problem is linear-time solvable for bounded tree-width graphs, and we present a polynomial-time algorithm for computing the cosecure domination number of chain graphs, respectively. Finally, in Sect. 7, we conclude the paper.

2 Preliminaries

We refer to [22] for graph theoretic definitions and notations. A graph $G = (V, E)$ is said to be a *bipartite graph*, if V can be partitioned into P and Q such that for any $uv \in E$, either $u \in P$ and $v \in Q$, or $u \in Q$ and $v \in P$. Such a partition (P, Q) of V is said to be a *bipartition* of V and the sets P, Q are called the partites of V. We denote a bipartite graph $G = (V, E)$ with bipartition (P, Q) as $G = (P, Q, E)$ with $n_1 = |P|$ and $n_2 = |Q|$. A bipartite graph $G = (P, Q, E)$ is said to be a *chordal bipartite graph*, if every cycle of length at least six has a chord.

A bipartite graph $G = (P, Q, E)$ is said to be a tree-convex (star-convex or comb-convex) bipartite graph, if we can define a tree (star or comb) $T = (P, F)$ such that for every $u \in Q$, $T[N_G(u)]$ forms a connected induced subgraph of T [11]. A bipartite graph $G = (P, Q, E)$ is said to be a *chain graph* (also known as bipartite chain graph or bi-split graph), if there exists a linearly ordering $(p_1, p_2, \ldots, p_{n_1})$ of the vertices of the partite P such that $N(p_1) \subseteq N(p_2) \subseteq \cdots \subseteq N(p_{n_1})$. If $G = (P, Q, E)$ is a chain graph, then a linear ordering, say $(q_1, q_2, \ldots, q_{n_2})$ of the vertices of the partite Q also exists such that $N(q_1) \supseteq N(q_2) \supseteq \cdots \supseteq N(q_{n_2})$. For a chain graph $G = (P, Q, E)$, a *chain ordering* is an ordering $\alpha = (p_1, p_2, \ldots, p_{n_1}, q_1, q_2, \ldots, q_{n_2})$ of $P \cup Q$ such that $N(p_1) \subseteq N(p_2) \subseteq \cdots \subseteq N(p_{n_1})$ and $N(q_1) \supseteq N(q_2) \supseteq \cdots \supseteq N(q_{n_2})$. Note that a chain ordering of a chain graph can be computed in linear-time [10].

Let $G = (V, E)$ be a graph. A vertex $x \in V$ is called a *simplicial vertex* of G, if the subgraph induced on $N[x]$ is complete. A vertex $y \in N[x]$ is said to be a *maximum neighbour* of x, if for each $z \in N[x]$, $N[z] \subseteq N[y]$. A vertex $x \in V$ is said to be a *doubly simplicial vertex*, if x is a simplicial vertex and has a maximum neighbour. A *doubly perfect elimination ordering* of the vertex set V of G, abbreviated as DPEO of G, is an ordering $(u_1, u_2, \ldots, u_n)$ of V such that for every $i \in [n]$, u_i is a doubly simplicial vertex of the subgraph induced on $\{u_i, u_{i+1}, \ldots, u_n\}$ of G. A graph is said to be *doubly chordal*, if it is chordal as well as dually chordal. A characterization of doubly chordal graph says that a graph G is doubly chordal if and only if G has a DPEO [19].

The following results regarding the cosecure domination are already known in the literature [2].

Lemma 1. [2] *For a complete bipartite graph $G = (X, Y, E)$ with $|X| \leq |Y|$,*

$$\gamma_{cs}(G) = \begin{cases} |Y| & \text{if } |X| = 1; \\ 2 & \text{if } |X| = 2; \\ 3 & \text{if } |X| = 3; \\ 4 & \text{otherwise.} \end{cases} \tag{1}$$

Lemma 2. [2] *Let L_u be the set of pendent vertices that are adjacent to a vertex u in graph G. If $|L_u| \geq 2$, then for every cosecure dominating set D of G, $L_u \subseteq D$ and $u \notin D$.*

The proofs of the results marked with $\star$ are omitted due to space constraints and are included in the full version of the paper [17].

3 NP-Completeness in Some Subclasses of Bipartite Graphs

In this section, we study the NP-completeness of the CSDD problem. The CSDD problem is known to be NP-complete for bipartite graphs and here we strengthen the complexity status of the CSDD problem by showing that it remains NP-complete for chordal bipartite graphs, star-convex bipartite graphs and comb-convex bipartite graphs, which are subclasses of bipartite graphs. For that we will be using an already known result regarding the NP-completeness of the DM problem for bipartite graphs.

Theorem 1. [3,20] *The DM problem is NP-complete for chordal bipartite graphs, and bipartite graphs.*

3.1 Chordal Bipartite Graphs

In this subsection, we prove that the decision version of the MINIMUM COSECURE DOMINATION problem is NP-complete, when restricted to chordal bipartite graphs. The proof of this follows by using a polynomial-time reduction from the DOMINATION DECISION problem to the COSECURE DOMINATION DECISION problem.

Now, we illustrate the reduction from an instance G, k, where G is a chordal bipartite graph and k is a positive integer, of the DM problem to an instance G', k' of the CSDD problem. Given a graph $G = (V, E)$ with $V = \{v_i \mid 1 \leq i \leq n\}$, we construct a graph $G' = (V', E')$ from G by attaching a path (v_i, v_{i_1}, v_{i_2}) to each vertex $v_i \in V$, where $V' = V \cup \{v_{i_1}, v_{i_2} \mid 1 < i \leq n\}$ and $E' = E \cup \{v_i v_{i_1}, v_{i_1} v_{i_2} \mid 1 \leq i \leq n\}$. It is easy to see that the above defined reduction can be done in polynomial-time. Note that if G is a chordal bipartite graph, then G' is also a chordal bipartite graph.

Lemma 3. ⋆ *G has a dominating set of cardinality at most k if and only if G' has a cosecure dominating set of cardinality at most $k' = k + |V(G)|$.*

The proof of following theorem directly follows from Theorem 1 and Lemma 3.

Theorem 2. *The CSDD problem is NP-complete for chordal bipartite graphs.*

3.2 Star-Convex Bipartite Graphs

In this subsection, we prove that the decision version of the MINIMUM COSECURE DOMINATION problem is NP-complete, when restricted to connected star-convex

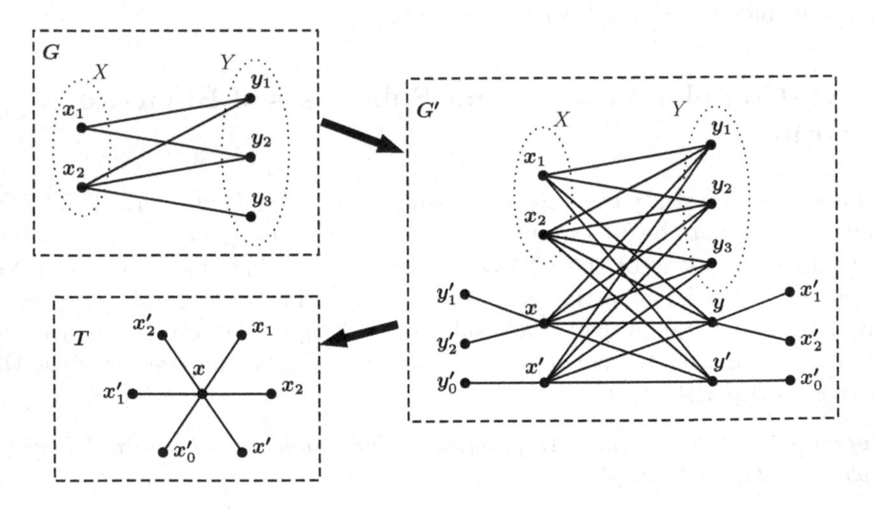

Fig. 1. Illustrating the construction of graph G' from a graph G.

bipartite graphs. The proof of this follows by using a reduction from a polynomial-time reduction from the DOMINATION DECISION problem to the COSECURE DOMINATION DECISION problem.

Theorem 3. *The CSDD problem is NP-complete for star-convex bipartite graphs.*

Proof. Clearly, the CSDD problem is in NP for star-convex bipartite graphs. In order to prove the NP-completeness, we give a polynomial-time reduction from the DM problem for bipartite graphs to the CSDD problem for star-convex bipartite graphs.

Suppose that a bipartite graph $G = (X, Y, E)$ is given, where $X = \{x_i \mid 1 \leq i \leq n_1\}$ and $Y = \{y_i \mid 1 \leq i \leq n_2\}$. We construct a star-convex bipartite graph $G' = (X', Y', E')$ from G in the following way:

- $X' = X \cup \{x, x', x'_0, x'_1, x'_2\}$,
- $Y' = Y \cup \{y, y', y'_0, y'_1, y'_2\}$, and
- $E' = E \cup \{xy_i, x'y_i \mid 1 \leq i \leq n_2\} \cup \{yx_i, y'x_i \mid 1 \leq i \leq n_1\} \cup \{xy'_i, yx'_i \mid 1 \leq i \leq 2\} \cup \{xy, xy', x'y, x'y', x'y'_0, y'x'_0\}$.

Here, $|X'| = n_1 + 5$, $|Y'| = n_2 + 5$ and $|E'| = |E| + 2n_1 + 2n_2 + 10$. It is easy to see that G' can be constructed from G in polynomial-time. Also, the newly constructed graph G' is a star-convex bipartite graph with star $T = (X', F)$, where $F = \{xx_i \mid 1 \leq i \leq n_1\} \cup \{xx', xx'_i \mid 0 \leq i \leq 2\}$ and x is the center of the star T. Figure 1 illustrates the construction of G' from G.

Claim 1 * *G has a dominating set of cardinality at most k if and only if G' has a cosecure dominating set of cardinality at most $k + 6$.*

This completes the proof of the result. □

3.3 Comb-Convex Bipartite Graphs

In this subsection, we prove that the decision version of the MINIMUM COSECURE DOMINATION problem is NP-complete for comb-convex bipartite graphs. The proof of this follows by using a polynomial-time reduction from an instance of the DM problem to an instance of the CSDD problem.

Theorem 4. * *The CSDD problem is NP-complete for comb-convex bipartite graphs.*

4 Complexity Difference Between Domination and Cosecure Domination

In this section, we demonstrate that the complexity of the MINIMUM DOMINATION problem may vary from the complexity of the MINIMUM COSECURE DOMINATION problem for some graph classes and we identify two such graph classes.

4.1 GY4-Graphs

In this subsection, we define a graph class which we call as GY4-graphs, and we prove that the MCSD problem is polynomial-time solvable for GY4-graphs, whereas the decision version of the MDS problem is NP-complete.

Let S^4 denote a star graph on 4 vertices. For $1 \leq i \leq n$, let $\{S_i^4 \mid 1 \leq i \leq n\}$ be collection of n star graphs of order 4 such that v_i^1, v_i^2, v_i^3 denote the pendent vertices and v_i^4 denote the center vertex. Now, we formally define the graph class GY4-graphs as follows:

Definition 1. *GY4-graphs: A graph $G^Y = (V^Y, E^Y)$ is said to be a GY4-graph, if it can be constructed from a graph $G = (V, E)$ with $V = \{v_1, v_2, \ldots, v_n\}$, by making pendent vertex v_i^1 of a star graph S_i^4 adjacent to vertex $v_i \in V$, for each $1 \leq i \leq n$.*

Note that $|V^Y| = 5n$ and $|E^Y| = 4n + |E|$. So, $n = |V^Y|/5$. First, we show that the cosecure domination number of GY4-graphs can be computed in linear-time.

Theorem 5. ⋆ *For a GY4-graph $G^Y = (V^Y, E^Y)$, $\gamma_{cs}(G^Y) = \frac{3}{5}|V^Y|$.*

Next, we show that the DM problem is NP-complete for GY4-graphs. In order to do this, we prove that the MINIMUM DOMINATION problem for general graph G is efficiently solvable if and only if the problem is efficiently solvable for the corresponding GY4-graph G^Y.

Lemma 4. ⋆ *Let $G^Y = (V^Y, E^Y)$ be a GY4-graph corresponding to a graph $G = (V, E)$ of order n and $k \leq n$. Then, G has a dominating set of cardinality at most k if and only if G^Y has a dominating set of cardinality at most $k + n$.*

As the DOMINATION DECISION problem is NP-complete for general graphs [3]. Thus, the NP-completeness of the DOMINATION DECISION problem follows directly from Lemma 4.

Theorem 6. *The DM problem is NP-complete for GY4-graphs.*

4.2 Doubly Chordal Graphs

Note that the MDS problem is already known to be linear-time solvable for doubly chordal graphs [5]. In this subsection, we show the NP-completeness of the CSDD problem for doubly chordal graphs. In order to prove this, we give a reduction from an instance of the SET COVER DECISION problem to an instance of the COSECURE DOMINATION DECISION problem.

Before doing that, first, we formally define the SET COVER DECISION problem. Given a pair (A, S) and a positive integer k, where A is a set of p elements and S is a collection of q subsets of A, the SET COVER DECISION problem asks whether there exists a subset S' of C such that $\cup_{B \in S'} B = A$. The NP-completeness of the SET COVER DECISION problem is already known [13].

Theorem 7. ⋆ *The CSDD problem is NP-complete for doubly chordal graphs.*

5 Bounded Tree-Width Graphs

In this section, we prove that the MINIMUM COSECURE DOMINATION problem can be solved in linear-time. First, we formally define the parameter tree-width of a graph. For a graph $G = (V, E)$, its *tree decomposition* is a pair (T, S), where $T = (U, F)$ is a tree, and $S = \{S_u \mid u \in U\}$ is a collection of subsets of V such that

- $\cup_{u \in U} S_u = V$,
- for each $xy \in E$, there exists $u \in U$ such that $x, y \in S_u$, and
- for all $x \in V$, the vertices in the set $\{u \in U \mid x \in S_u\}$ forms a subtree of T.

The width of a tree decomposition (T, S) of a graph G is defined as $(\max\{|S_u| \mid u \in U\} - 1)$. The *tree-width* of a graph G is the minimum width of any tree decomposition of G. A graph is said to be a *bounded tree-width graph*, if its tree-width is bounded. Now, we prove that the cosecure domination problem can be formulated as CMSOL.

Theorem 8. * *For a graph $G = (V, E)$ and a positive integer k, the CSDD problem can be expressed in CMSOL.*

The famous Courcelle's Theorem [7] states that any problem which can be expressed as a CMSOL formula is solvable in linear-time for graphs having bounded tree-width. From Courcelle Theorem and above theorem, the following result directly follows.

Theorem 9. *For bounded tree-width graphs, the CSDM problem is solvable in linear-time.*

6 Algorithm for Chain Graphs

In this section, we present an efficient algorithm to compute the cosecure domination number of chain graphs. Recall that a bipartite graph $G = (X, Y, E)$ is a *chain graph*, if there exists a *chain ordering* $\alpha = (x_1, x_2, \ldots, x_{n_1}, y_1, y_2, \ldots, y_{n_2})$ of $X \cup Y$ such that $N(x_1) \subseteq N(x_2) \subseteq \cdots \subseteq N(x_{n_1})$ and $N(y_1) \supseteq N(y_2) \supseteq \cdots \supseteq N(y_{n_2})$. For a chain graph, its chain ordering can be computed in linear-time [10].

Now, we define a relation R on X as follows: x_i and x_j are related if $N(x_i) = N(x_j)$. Observe that R is an equivalence relation. Assume that $X_1, X_2, \ldots, X_k$ is the partition of X based on the relation R. Define $Y_1 = N(X_1)$ and $Y_i = N(X_i) \setminus \cup_{j=1}^{i-1} N(X_j)$ for $i = 2, 3, \ldots k$. Then, $Y_1, Y_2, \ldots, Y_k$ forms a partition of Y. Such partition $X_1, X_2, \ldots, X_k, Y_1, Y_2, \ldots, Y_k$ of $X \cup Y$ is called a *proper ordered chain partition* of $X \cup Y$. Note that the number of sets in the partition of X (or Y) are k. Next, we remark that the set of pendent vertices of G is contained in $X_1 \cup Y_k$.

Throughout this section, we consider a chain graph G with a proper ordered chain partition $X_1, X_2, \ldots, X_k$ and $Y_1, Y_2, \ldots, Y_k$ of X and Y, respectively. For $i \in [k]$, let $X_i = \{x_{i1}, x_{i2}, \ldots, x_{ir}\}$ and $Y_i = \{y_{i1}, y_{i2}, \ldots, y_{ir}\}$. Note that $k = 1$

if and only if G is a complete bipartite graph. From now onwards, we assume that G is a connected chain graph with $k \geq 2$.

The following lemma directly follows from Lemma 2.

Lemma 5. *If there are more than one pendent from X, then every cosecure dominating set S contains X_1 and does not contain y_{11}.*

Now, we assume that there are more than one pendent from X in the chain graph G. In Lemma 6, we prove that the cosecure domination number of $G[X_1 \cup Y_1]$ and the remaining graph can be computed independently and their sum will give the cosecure domination number of G. Similar result follows when there are more than one pendent from Y.

Lemma 6. * *Let G be a chain graph such that there are more than one pendent vertex from X. Define $G_1 = G[X_1 \cup Y_1]$, $G_2 = G[\cup_{i=2}^{k}(X_i \cup Y_i)]$. Then, $\gamma_{cs}(G) = \gamma_{cs}(G_1) + \gamma_{cs}(G_2)$.*

In a chain graph G, if there are more than one pendent vertex from both X and Y. Let $G_1 = G[X_1 \cup Y_1]$, $G_2 = G[\cup_{i=2}^{k-1}(X_i \cup Y_i)]$ and $G_3 = G[X_k \cup Y_k]$, then using Lemma 6, it follows that $\gamma_{cs}(G) = \sum_{i=1}^{3} \gamma_{cs}(G_i)$.

Now, we consider a chain graph G having $|X| \geq 4$ and $|Y| \geq 4$. In Lemma 7, we give a lower bound on the cosecure domination number of G.

Lemma 7. * *Let G be a chain graph such that $|X| \geq 4$ and $|Y| \geq 4$. Then, $\gamma_{cs}(G) \geq 4$.*

In the next lemma, we consider the case when G is a chain graph with $k = 2$ and determine the cosecure domination number in all the possible cases.

Lemma 8. * *Let G be a chain graph such that $k = 2$. Then, one of the following case occurs.*

1. *If there does not exist any pendent vertex in G and $|X| = 3$ or $|Y| = 3$, then $\gamma_{cs}(G) = 3$, otherwise, $\gamma_{cs}(G) = 4$.*
2. *If there exist more than one pendent vertex from X or Y or both. Define $G_1 = G[X_1 \cup Y_1]$ and $G_2 = G[X_2 \cup Y_2]$. Then, $\gamma_{cs}(G) = \gamma_{cs}(G_1) + \gamma_{cs}(G_2)$.*
3. *If there exist at most one pendent vertex from X and Y both. If $|X| = 2$ or $|Y| = 2$, then $\gamma_{cs}(G) = 2$. Else-if $|X| = 3$ or $|Y| = 3$, then $\gamma_{cs}(G) = 3$, otherwise, $\gamma_{cs}(G) = 4$.*

From now onwards, we assume that G is a chain graph and $k \geq 3$. In the following lemma, we will consider the case when the chain graph G has no pendent vertex and we give the exact value of the cosecure domination number of G.

Lemma 9. * *If G is a chain graph without any pendent vertices, then $\gamma_{cs}(G) = 4$.*

Now, we assume that there is at most one pendent from X and Y both in a chain graph G. In Lemma 10, we determine the value of the cosecure domination number of G.

Algorithm 1: CSDN_Chain(G, k)

Input: A connected chain graph $G = (V, E)$ with proper ordered chain partition $X_1, X_2, \ldots, X_k$ and $Y_1, Y_2, \ldots, Y_k$ of X and Y.
Output: Cosecure domination number of G, that is, $\gamma_{cs}(G)$.

if $(k = 2)$ **then**
  if $(|Y_1| > 1$ *and* $|X_2| > 1)$ **then**
    $X = X_1 \cup X_2$, $Y = Y_1 \cup Y_2$;
    if $(|X| = 3$ *or* $|Y| = 3)$ **then**
      $\gamma_{cs}(G) = 3$;
    else
      $\gamma_{cs}(G) = 4$;
  else if $((|X_1| > 1$ *and* $|Y_1| = 1)$ *or* $(|X_2| = 1$ *and* $|Y_2| > 1))$ **then**
    Let $G_1 = G[X_1 \cup Y_1]$ and $G_2 = G[X_2 \cup Y_2]$;
    Let $p_1 = \min\{|X_1|, |Y_1|\}$, $q_1 = \max\{|X_1|, |Y_1|\}$, $p_2 = \min\{|X_2|, |Y_2|\}$ and $q_2 = \max\{|X_2|, |Y_2|\}$;
    $\gamma_{cs}(G) =$ **CSDN_CB**$(G_1, p_1, q_1)+$**CSDN_CB**(G_2, p_2, q_2);
  else if $((|X_1| = |Y_1| = 1)$ *or* $(|X_2| = |Y_2| = 1))$ **then**
    if $(|X| = 2$ *or* $|Y| = 2)$ **then**
      $\gamma_{cs}(G) = 2$;
    else if $(|X| = 3$ *or* $|Y| = 3)$ **then**
      $\gamma_{cs}(G) = 3$;
    else if $(|X| \geq 4$ *and* $|Y| \geq 4)$ **then**
      $\gamma_{cs}(G) = 4$;

if $(k \geq 3)$ **then**
  if $(|Y_1| > 1$ *and* $|X_k| > 1)$ **then**
    $\gamma_{cs}(G) = 4$;
  else if $((|X_1| > 1$ *and* $|Y_1| = 1)$ *and* $(|Y_k| > 1$ *and* $|X_k| = 1))$ **then**
    Let $G' = G[\cup_{i=2}^{k-1}(X_i \cup Y_i)]$;
    $\gamma_{cs}(G) = |X_1| + |Y_k| +$**CSDN_Chain**$(G', k-2)$;
  else if $(|X_1| > 1$ *and* $|Y_1| = 1)$ **then**
    Let $G' = G[\cup_{i=2}^{k}(X_i \cup Y_i)]$;
    $\gamma_{cs}(G) = |X_1|+$**CSDN_Chain**$(G', k-1)$;
  else if $(|Y_k| > 1$ *and* $|X_k| = 1)$ **then**
    Let $G' = G[\cup_{i=1}^{k-1}(X_i \cup Y_i)]$;
    $\gamma_{cs}(G) = |Y_k|+$**CSDN_Chain**$(G', k-1)$;
  else
    if $(|X| = 3$ *or* $|Y| = 3)$ **then**
      $\gamma_{cs}(G) = 3$;
    else
      $\gamma_{cs}(G) = 4$;

return $\gamma_{cs}(G)$;

Lemma 10. $\star$ *Let G be a chain graph with at most one pendent vertex from X and Y both. If $|X| = 3$ or $|Y| = 3$ then $\gamma_{cs}(G) = 3$, otherwise, $\gamma_{cs}(G) = 4$.*

Finally, we assume that G is a chain graph such that there are at least two pendent from X or Y or both. In Lemma 11, we give an expression to determine the value of the cosecure domination number of G in every possible case.

Lemma 11. $\star$ *Let G be a chain graph with $k \geq 3$. Then,*

1. *If there exist more than one pendent vertex from X and at most one pendent from Y. Define $G' = G[\cup_{i=2}^{k}(X_i \cup Y_i)]$. Then, $\gamma_{cs}(G) = |X_1| + \gamma_{cs}(G')$.*
2. *If there exist more than one pendent vertex from Y and at most one pendent from X. Define $G' = G[\cup_{i=1}^{k-1}(X_i \cup Y_i)]$. Then, $\gamma_{cs}(G) = |Y_k| + \gamma_{cs}(G')$.*
3. *If there exist more than one pendent vertex from X and Y both. $G' = G[\cup_{i=2}^{k-1}(X_i \cup Y_i)]$. Then, $\gamma_{cs}(G) = |X_1| + |Y_k| + \gamma_{cs}(G')$.*

Assume that an algorithm, namely **CSDN_CB**(G, p, q) is designed using Lemma 1, which takes a complete bipartite graph and cardinalities of the partite sets, namely p, q (satisfying $p \leq q$) as input and returns $\gamma_{cs}(G)$ as output.

Now, on the basis of above lemmas, we design a recursive algorithm, namely, **CSDN_Chain**(G, k) to find the cosecure domination number of chain graphs.

The algorithm takes a connected chain graph $G = (V, E)$ with a proper ordered chain partition $X_1, X_2, \ldots, X_k$ and $Y_1, Y_2, \ldots, Y_k$ of X and Y as an input. While executing the algorithm, we call the algorithm **CSDN_CB**(G, p, q) whenever we encounter a complete bipartite graph.

Let G be a connected chain graph and $X_1, X_2, \ldots, X_k$ and $Y_1, Y_2, \ldots, Y_k$ be the proper ordered chain partition of X and Y, respectively. The case when $k = 2$ works as base case of our algorithm. The correctness of the base case follows from Lemma 8. Then, Lemma 11 helps us in designing the algorithm using the recursive approach and proves that the correctness of the algorithm.

Now, we state the main result of this section. The proof of the following theorem directly follows from combining Lemma 9, Lemma 10 and Lemma 11. As the running time of our algorithm **CSDN_Chain**(G, k) is polynomial, therefore, the cosecure domination number of a connected chain graph can be computed in polynomial-time.

Theorem 10. *Given a connected chain graph $G = (X, Y, E)$ with proper ordered chain partition $X_1, X_2, \ldots, X_k$ and $Y_1, Y_2, \ldots, Y_k$ of X and Y. Then, the cosecure domination number of G can be computed in polynomial-time.*

7 Conclusion

We have resolved the complexity status of the Minimum Cosecure Domination problem on various important graph classes, namely, chain graphs, chordal bipartite graphs, star-convex bipartite graphs, comb-convex bipartite graphs, doubly chordal graphs and bounded tree-width graphs. It was known that the Cosecure Domination Decision problem is NP-complete for bipartite graphs. Extending this, we showed that the problem remains NP-complete even when restricted to star-convex bipartite graphs, comb-convex bipartite graphs and chordal bipartite graphs, which are all subclasses of bipartite graphs. We have also proved that the problem is NP-complete for doubly chordal graphs. Further, we showed that the computational complexity the CSDD problem varies from that of the classical domination problem, as the domination problem is efficiently solvable for doubly chordal graphs. Additionally, we have defined a graph class for which the MCSD problem is solvable in linear-time, whereas the domination problem is NP-complete. On the positive side, we have proved that the Minimum Cosecure Domination problem is efficiently solvable for chain graphs and bounded tree-width graphs. Naturally, it would be interesting to do the complexity study of the Minimum Cosecure Domination problem in many other important graphs classes for which the problem is still open.

References

1. Araki, T., Yamanaka, R.: Secure domination in cographs. Discret. Appl. Math. **262**, 179–184 (2019)
2. Arumugam, S., Ebadi, K., Manrique, M.: Co-secure and secure domination in graphs. Util. Math. **94**, 167–182 (2014)
3. Bertossi, A.A.: Dominating sets for split and bipartite graphs. Inform. Process. Lett. **19**(1), 37–40 (1984)
4. Boumediene Merouane, H., Chellali, M.: On secure domination in graphs. Inform. Process. Lett. **115**(10), 786–790 (2015)
5. Brandstädt, A., Chepoi, V.D., Dragan, F.F.: The algorithmic use of hypertree structure and maximum neighbourhood orderings. Discret. Appl. Math. **82**(1–3), 43–77 (1998)
6. Cockayne, E.J., Grobler, P.J.P., Gründlingh, W.R., Munganga, J., van Vuuren, J.H.: Protection of a graph. Util. Math. **67**, 19–32 (2005)
7. Courcelle, B.: The monadic second-order logic of graphs. I. Recognizable sets of finite graphs. Inform. Comput. **85**(1), 12–75 (1990)
8. Haynes, T.W., Hedetniemi, S.T., Henning, M.A. (eds.): Topics in Domination in Graphs. DM, vol. 64. Springer, Cham (2020). https://doi.org/10.1007/978-3-030-51117-3
9. Haynes, T.W., Hedetniemi, S.T., Henning, M.A. (eds.): Structures of Domination in Graphs. DM, vol. 66. Springer, Cham (2021). https://doi.org/10.1007/978-3-030-58892-2
10. Heggernes, P., Kratsch, D.: Linear-time certifying recognition algorithms and forbidden induced subgraphs. Nordic J. Comput. **14**(1–2), 87–108 (2008) (2007)
11. Jiang, W., Liu, T., Ren, T., Xu, K.: Two hardness results on feedback vertex sets. In: Atallah, M., Li, X.-Y., Zhu, B. (eds.) Frontiers in Algorithmics and Algorithmic Aspects in Information and Management. LNCS, vol. 6681, pp. 233–243. Springer, Heidelberg (2011). https://doi.org/10.1007/978-3-642-21204-8_26
12. Joseph, A., Sangeetha, V.: Bounds on co-secure domination in graphs. Int. J. Math. Trends Technol. **55**(2), 158–164 (2018)
13. Karp, R.M.: Reducibility among combinatorial problems. In: Complexity of computer computations (Proc. Sympos., IBM Thomas J. Watson Res. Center, Yorktown Heights, N.Y., 1972), pp. 85–103. Plenum, New York (1972)

14. Kišek, A., Klavžar, S.: Correcting the algorithm for the secure domination number of cographs by Jha, Pradhan, and Banerjee. Inform. Process. Lett. **172**, Paper No. 106155, 4 (2021)
15. Klostermeyer, W.F., Mynhardt, C.M.: Secure domination and secure total domination in graphs. Discuss. Math. Graph Theory **28**(2), 267–284 (2008)
16. Kusum, Pandey, A.: Complexity results on cosecure domination in graphs. In: Bagchi, A., Muthu, R. (eds.) Algorithms and Discrete Applied Mathematics. CALDAM 2023. LNCS, vol. 13947, pp. 335–347. Springer, Cham (2023). https://doi.org/10.1007/978-3-031-25211-2_26
17. Kusum, Pandey, A.: Cosecure domination: Hardness results and algorithm (2023). https://doi.org/10.48550/arXiv.2302.13031
18. Manjusha, P., Chithra, M.R.: Co-secure domination in Mycielski graphs. J. Comb. Math. Comb. Comput. **113**, 289–297 (2020)
19. Moscarini, M.: Doubly chordal graphs, Steiner trees, and connected domination. Networks **23**(1), 59–69 (1993)
20. Müller, H., Brandstädt, A.: The NP-completeness of Steiner tree and dominating set for chordal bipartite graphs. Theor. Comput. Sci. **53**(2–3), 257–265 (1987)
21. Wang, H., Zhao, Y., Deng, Y.: The complexity of secure domination problem in graphs. Discuss. Math. Graph Theory **38**(2), 385–396 (2018)
22. West, D.B., et al.: Introduction to Graph Theory, vol. 2. Prentice hall, Upper Saddle River (2001)
23. Zou, Y.H., Liu, J.J., Chang, S.C., Hsu, C.C.: The co-secure domination in proper interval graphs. Discret. Appl. Math. **311**, 68–71 (2022)

Optimal Cost-Based Allocations Under Two-Sided Preferences

Girija Limaye and Meghana Nasre(✉)

Indian Institute of Technology Madras, Chennai, India
{girija,meghana}@cse.iitm.ac.in

Abstract. The Hospital Residents setting models important problems like school choice, assignment of undergraduate students to degree programs, among many others. In this setting, fixed quotas are associated with programs that limit the number of agents that can be assigned to them. Motivated by scenarios where *all* agents must be matched, we propose and study the cost-based allocation setting, which allows controlled flexibility with respect to quotas.

In our model, we seek to compute a matching that matches all agents and is optimal with respect to preferences, and minimizes either a local or a global objective on costs. We show that there is a sharp contrast – minimizing the local objective is polynomial-time solvable, whereas minimizing the global objective is hard to approximate within a specific constant factor unless P = NP. On the positive side, we present approximation algorithms for the global objective in the general case and a particular hard case. We achieve the approximation guarantee for the particular case via a linear programming based algorithm.

Keywords: Matchings under two-sided preferences · Envy-freeness · Cost-based allocation · Approximation algorithms · Linear Programming

1 Introduction

The problem of computing optimal many-to-one matchings under two-sided preferences is extensively investigated in the literature [2,4,10,12,17]. This setting is commonly known as the Hospital Residents (HR) setting. It captures important applications like assigning medical interns to hospitals [17], allocating undergraduate students to degree programs [4] and assigning children to schools [2].

This setting can be modelled as a bipartite graph $G = (\mathcal{A} \cup \mathcal{P}, E)$ where $\mathcal{A}$ and $\mathcal{P}$ denote a set of agents and a set of programs respectively. An edge $(a, p) \in E$ indicates that agent a and program p are mutually acceptable. For a vertex v, let $\mathcal{N}(v)$ denote the vertices adjacent to v. Each agent and every program has a preference ordering over its mutually acceptable partners. For a vertex $v \in \mathcal{A} \cup \mathcal{P}$, if v prefers u over w then we write it as $u \succ_v w$. A program

This work was partially supported by the grant CRG/2019/004757.

S.-Y. Hsieh et al. (Eds.): IWOCA 2023, LNCS 13889, pp. 259–270, 2023.
https://doi.org/10.1007/978-3-031-34347-6_22

p has an upper-quota $q(p)$ denoting the maximum number of agents that can be assigned to it. The goal is to compute a matching, that is, an assignment between agents and programs such that an agent is matched to at most one program and a program is matched to at most its upper-quota many agents.

In certain applications of the HR setting, *every* agent must be matched. For instance, in school choice [2] every child must find a school; while matching sailors to billets in the US Navy [16,19], every sailor must be assigned to some billet. In the HR setting, the *rigid* upper-quotas limit the number of agents that can be matched in any matching. Thus, a matching that matches every agent (an $\mathcal{A}$-perfect matching) cannot be guaranteed.

Motivated by the need to match a large number of agents, the problem of *capacity expansion* is investigated very recently in [3,5–7]. In the capacity expansion problem, the quotas of programs are *augmented* to improve the *welfare* of the agents. In another work, Gajulapalli et al. [9] study a two-round mechanism for school admissions wherein the goal of the second round is to accommodate more students by suggesting quota increments to schools. However, in none of the works mentioned above except [7], $\mathcal{A}$-perfect matchings are guaranteed. Furthermore, the above models assume that the upper-quotas of programs are flexible. Allowing unbounded flexibility can lead to skewed assignments since certain programs are *popular*. For instance, to ensure $\mathcal{A}$-perfectness in one of the problems investigated in [7], the capacity of a single program may have to be increased by a huge amount. Thus, there is a need to control flexibility.

Cost-based allocations are recently studied in [1] under *one-sided* preferences, wherein only agents have preferences. Costs allow *controlled* flexibility w.r.t. quotas. We propose the cost-based setting under two-sided preferences to achieve $\mathcal{A}$-perfectness and denote it as the cost-controlled quota (CCQ) setting. An instance in the CCQ setting is similar to the HR instance, except that instead of an upper-quota, every program p specifies a non-negative integral, finite cost $c(p)$ that denotes the cost of matching a single agent to p. Our goal is to compute an $\mathcal{A}$-perfect matching that is *optimal* with respect to preferences as well as costs.

Optimality w.r.t. Preferences. Let M be a matching. Then, $M(a)$ denotes the program matched to agent a (if unmatched, $M(a) = \perp$) and $M(p)$ denote the set of agents matched to program p. Program p is *under-subscribed* in M if $|M(p)| < q(p)$. Stability [10] is a de-facto notion of optimality in the HR setting.

Definition 1 (Stable matchings in HR setting). *A pair $(a,p) \in E \setminus M$ is a blocking pair w.r.t. the matching M if $p \succ_a M(a)$ and p is either under-subscribed in M or there exists at least one agent $a' \in M(p)$ such that $a \succ_p a'$. A matching M is* stable *if there is no blocking pair w.r.t. M.*

It is well known that every HR instance admits a stable matching that can be computed efficiently [10]. The notion of stability inherently assumes the existence of input quotas. In the CCQ setting, quotas are not a part of the input, and we let the costs control the extent to which a program is matched. Envy-freeness, a relaxation of stability is defined *independent* of input quotas.

Definition 2 (Envy-free matchings). *Given a matching M, an agent a has a* justified envy *(here onwards called envy) towards a matched agent a', where*

$M(a') = p$ and $(a, p) \in E$ if $p \succ_a M(a)$ and $a \succ_p a'$. The pair (a, a') is an envy-pair w.r.t. M. A matching M is envy-free if there is no envy-pair w.r.t. M.

Let M be an envy-free matching in a CCQ instance H. Let G denote the HR instance wherein the preferences are borrowed from H and for every program p, $q(p)$ is set to $|M(p)|$, then M is stable in G. Thus, envy-freeness is a natural substitute for stability in the CCQ setting.

While we use envy-freeness in the CCQ setting, prior to this, structural properties of envy-free matchings in HR setting are studied in [18]. Envy-free matchings are also studied in HR setting with *lower-quotas* [8,14,20].

Optimality with Respect to Costs. Recall that the cost of a program p denotes the cost of matching a single agent to p which implicitly captures the logistic constraints. Thus, minimizing a cost-based criterion results in an optimal way of assigning agents. Let M be a matching in a CCQ instance. Then the maximum cost spent at any program in M is defined as $\max_{p \in \mathcal{P}}\{|M(p)| \cdot c(p)\}$ and the total cost of M is defined as $\sum_{p \in \mathcal{P}}(|M(p)| \cdot c(p))$ We consider the following two natural criteria w.r.t. costs: **(i)** a local criterion, that is, to minimize the maximum cost spent at any program and **(ii)** a global criterion, that is, to minimize the total cost spent.

Problem Definition. Based on the two optimality notions defined, we propose and investigate the following two optimization problems: let H be a CCQ instance. Let MINMAX denote the problem of computing an $\mathcal{A}$-perfect envy-free matching in H that minimizes the maximum cost spent at any program. Let MINSUM denote the problem of computing an $\mathcal{A}$-perfect envy-free matching in H that minimizes the total cost.

Example. Consider a CCQ instance with three agents and three programs. The preference lists of agents and programs are as follows: $p_1 \succ_{a_1} p_0$, $p_2 \succ_{a_2} p_1$ and a_3 ranks only p_1; p_0 ranks only a_1, p_2 ranks only a_2 and $a_1 \succ_{p_1} a_2 \succ_{p_1} a_3$. Let $c(p_0) = 0, c(p_1) = 1$ and $c(p_2) = 2$. The matchings $M_1 = \{(a_1, p_1), (a_2, p_1), (a_3, p_1)\}$ and $M_2 = \{(a_1, p_1), (a_2, p_2), (a_3, p_1)\}$ are both envy-free and M_1 with total cost of 3 is optimal for MINSUM whereas M_2 with a max-cost of 2 is optimal for MINMAX.

1.1 Our Contributions

This is the first work that investigates the cost-controlled quotas under two-sided preferences. We show that there exists an efficient algorithm for the MINMAX problem whereas in a sharp contrast, the MINSUM turns out to be NP-hard. The proofs of the results marked with ($\star$) can be found in the full version [15].

Theorem 1. MINMAX *problem is solvable in $O(m \log m)$ time where $m = |E|$.*

Theorem 2 ($\star$). MINSUM *problem is strongly* NP*-hard even when the instance has two distinct costs.* MINSUM *problem cannot be approximated within a factor $\frac{7}{6} - \epsilon, \epsilon > 0$, unless* P = NP.

The inapproximability result above does not imply the NP-hardness for two distinct costs – a special hard case that we work with later in the paper. We complement our hardness results with the following approximation algorithms for general instances.

Theorem 3. MINSUM *problem admits an* ℓ_p*-approximation algorithm and a* $|\mathcal{P}|$*-approximation algorithm where* ℓ_p *denotes the length of the longest preference list of a program.*

The analysis of our ℓ_p-approximation algorithm uses a natural lower bound on the MINSUM problem. We show that ℓ_p is the best approximation guarantee one can achieve using this lower bound. We establish that the optimal cost of the MINMAX problem also serves as a lower bound for the MINSUM problem on the same instance and this gives us the $|\mathcal{P}|$-approximation algorithm.

A Special Hard Case of MINSUM. Let $\mathsf{CCQ}_{c1,c2}$ denote the CCQ instance with two distinct costs c_1 and c_2 such that $0 \leq c_1 < c_2$. This particular case occurs when programs can be partitioned as – one set having all low-cost programs of cost c_1 and another set having all high-cost programs of cost c_2. As stated in Theorem 2, MINSUM problem is NP-hard even for $\mathsf{CCQ}_{c1,c2}$. We show an improved approximation algorithm for this special case.

Let ℓ_a denote the length of the longest preference list of an agent. In practice, ℓ_a is typically small compared to ℓ_p and $|\mathcal{P}|$, many times a constant [11,13]. We show that the MINSUM problem on $\mathsf{CCQ}_{c1,c2}$ instances is ℓ_a-approximable. We achieve our approximation guarantee via a technically involved linear programming (LP) based algorithm. Although our approximation guarantee is for a particular hard case, our linear program is for general CCQ instances.

Theorem 4. MINSUM *problem admits an* ℓ_a*-approximation algorithm on* $\mathsf{CCQ}_{c1,c2}$ *instances.*

1.2 Other Models and Relation to CCQ Setting

Firstly, we remark that there are alternate ways to formulate an optimization problem in the CCQ setting: (i) given a total budget $\mathcal{B}$, compute a largest envy-free matching with cost at most $\mathcal{B}$. (ii) given an HR instance and a cost for every program, *augment* the input quotas to compute an $\mathcal{A}$-perfect envy-free matching with minimum total cost. The NP-hardness for both these problems can be proved by easily modifying the NP-hardness reduction for MINSUM.

Next, we discuss the approaches proposed in the literature to circumvent rigid upper quotas. As mentioned earlier, the capacity planning problem with similar motivation as ours is studied in [3,5–7,9]. In the two-round school choice problem studied by Gajulapalli et al. [9], their goal in round-2 is to match *all* agents in a particular set. This set is derived from the matching in round-1 and they need to match the agents in an envy-free manner (called stability preserving in their work). This can still leave certain agents unassigned. It can be shown that the CCQ setting generalizes the matching problems in round-2. We remark that in [9] the authors state that a variant of MINSUM problem (Problem 33, Sect. 7) is NP-hard. However, they do not investigate the problem in detail.

In the very recent works [3,5,6] the authors consider the problem of distributing extra seats (beyond the input quotas) limited by a *budget* that leads to the best outcome for agents. Their setting does not involve costs, and importantly, $\mathcal{A}$-perfectness is not guaranteed. Bobbio et al. [6] show the NP-hardness of their problem. Bobbio et al. [5] and Abe et al. [3] propose a set of approaches which include heuristics along with empirical evaluations. In our work, we present algorithms with theoretical guarantees. Chen and Csáji [7] investigate a variant of the capacity augmentation problem mentioned earlier and present hardness, approximation algorithms and parameterized complexity results.

Finally, Santhini et al. [1] consider cost-based quotas under one-sided preferences. The *signature* of a matching allows encoding requirements about the number of agents matched to a particular rank. They consider the problem of computing a min-cost matching with a desired signature and show that it is efficiently solvable. This result is in contrast to the hardness and inapproximability results we show for a similar optimization problem under two-sided preferences.

2 Algorithmic Results: General Case

In this section we present an approximation algorithm for MINSUM with guarantee of ℓ_p. Next, we present a polynomial time algorithm for the MINMAX problem and prove that the output is a $|\mathcal{P}|$-approximation to the MINSUM problem.

2.1 ℓ_p-approximation for MINSUM

Let H be a CCQ instance. For an agent a, let p_a^* denote the least-cost program in its preference list. If multiple programs have the same least cost, we let p_a^* be the most-preferred such program. It is easy to observe that any $\mathcal{A}$-perfect matching in H has cost at least $\sum_{a \in \mathcal{A}} c(p_a^*)$. Let OPT denote an optimal solution, and $c(\mathsf{OPT})$ be its cost. Since OPT is $\mathcal{A}$-perfect, $c(\mathsf{OPT}) \geq \sum_{a \in \mathcal{A}} c(p_a^*)$. We use this lower bound, denoted as lb_1, to prove our approximation guarantee.[1]

Our algorithm (Algorithm 1) starts by matching every agent a to p_a^*. Note that this matching is $\mathcal{A}$-perfect and has a minimum cost but is not necessarily envy-free. Now the algorithm considers programs in an arbitrary order. For a program p, we consider agents in the reverse preference list ordering of p. If there exists agent $a \notin M(p)$ such that $p \succ_a M(a)$ and there exists $a' \in M(p)$ such that $a \succ_p a'$, then (a, a') form an envy-pair. We resolve this by *promoting* a to p. The algorithm stops after considering every program.

Note that in Algorithm 1 an agent may get promoted (line 5) but never gets demoted. Further, program p is assigned agents in the **for** loop (line 2) only when at least one agent is matched to p in line 1. Therefore, if program p is assigned at

[1] ℓ_p is the best guarantee achievable using lb_1. See [15] for the details.

Algorithm 1. An ℓ_p-approximation algorithm for MINSUM

1: let $M = \{(a,p) \mid a \in \mathcal{A} \text{ and } p = p_a^*\}$
2: **for** every program p **do**
3: **for** a in reverse preference list ordering of p **do**
4: **if** $\exists a' \in M(p)$ s.t. $a \succ_p a'$ and $p \succ_a M(a)$ **then**
5: $M = M \setminus \{(a, M(a))\} \cup \{(a,p)\}$
6: return M

least one agent in the final output matching, then $p = p_a^*$ for some agent $a \in \mathcal{A}$. It is clear that the computed matching M is $\mathcal{A}$-perfect. In Lemma 1, we show envy-freeness and an approximation guarantee of M.

Lemma 1 (⋆). *The matching M is envy-free and an ℓ_p-approximation to* MINSUM.

2.2 MINMAX and Its Relation to MINSUM

In this section, we present an efficient algorithm for MINMAX. Let H be a CCQ instance. Let t^* denote the cost of an optimal solution for the MINMAX problem in H. Note that t^* must be an integral multiple of $c(p)$ for some program p. Therefore, considering a specific set of cost values as described below suffices. We show that there exists $O(m)$ distinct possibilities for t^*, among which the optimal one can be found using a binary search. We further establish that t^* is a lower bound on the optimal cost for MINSUM in H and use this to show the approximation guarantee of $|\mathcal{P}|$ for MINSUM.

Algorithm for MINMAX. Let M^* be an optimal solution for the MINMAX problem on H. Then $t^* = \max_{p\in\mathcal{P}} \{c(p) \cdot |M^*(p)|\}$. For any integer t, we define an HR instance G_t where the preference lists are borrowed from H and for each $p \in \mathcal{P}$, $q(p) = \left\lfloor \frac{t}{c(p)} \right\rfloor$. Our algorithm is based on the following observations:

- For any $t < t^*$, the HR instance G_t does not admit an $\mathcal{A}$-perfect stable matching. Otherwise, this contradicts the optimality of M^* since stability implies envy-freeness.
- The optimal value t^* lies in the range $\min_{p\in\mathcal{P}} \{c(p)\}$ to $\max_{p\in\mathcal{P}} \{(len(p) \cdot c(p))\}$, where $len(p)$ denotes the length of the preference list of program p.
- For any $t \geq t^*$, G_t admits an $\mathcal{A}$-perfect stable matching.

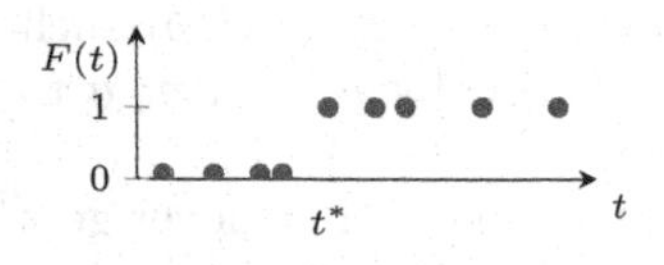

The adjoining plot illustrates these observations. For an integer t, let $F(t)$ be 1 if G_t admits an $\mathcal{A}$-perfect stable matching, 0 otherwise. Our algorithm constructs a sorted array $\hat{c}_p$ for each program $p \in \mathcal{P}$ such that for $1 \leq i \leq len(p)$, we have $\hat{c}_p[i] = i \cdot c(p)$. Therefore, these arrays together contain $\sum_{p\in\mathcal{P}} len(p) = |E| = m$ many values. We merge these arrays to construct a sorted array $\hat{c}$ of distinct costs. Then we perform a binary search for the optimal value of t^* in the sorted array $\hat{c}$: for a particular value $t = \hat{c}[k]$ we construct G_t by setting appropriate upper-quotas. If a stable matching in G_t is not $\mathcal{A}$-perfect, then we search in the *upper-range*; otherwise, we check if $G_{t'}$ admits an $\mathcal{A}$-perfect stable matching for $t' = \hat{c}[k-1]$. If not, we return t; otherwise, we search for the optimal in the *lower-range*.

Construction of array $\hat{c}$ takes $O(m \log m)$ time. We perform binary search over $O(m)$ distinct values, and in each iteration, we compute at most two stable matchings in $O(m)$ time [10]. Therefore the algorithm runs in time $O(m \log m)$. This establishes Theorem 1.

Relation to MINSUM. We prove that an optimal solution for the MINMAX problem is a $|\mathcal{P}|$-approximation for the MINSUM problem.

Lemma 2 (⋆). *The optimal solution for the* MINMAX *problem is a $|\mathcal{P}|$-approximation for the* MINSUM *problem on the same instance.*

Proof (sketch). Let H be a CCQ instance. Let M^* and N^* be an optimal solution respectively for the MINMAX problem and for the MINSUM problem on H. Let t^* and y^* denote the cost of M^* and N^* respectively. We show that $y^* \geq t^*$. Further, we note that since t^* denotes the maximum cost incurred at any program in M^*, the total cost of M^* is upper bounded by $|\mathcal{P}| \cdot t^* \leq |\mathcal{P}| \cdot y^*$.

This establishes Theorem 3.

3 MINSUM: A Special Case

Recall that $\text{CCQ}_{c1,c2}$ denotes the CCQ instance with two distinct costs c_1, c_2 such that $0 \leq c_1 < c_2$. By Theorem 2 we know that MINSUM is NP-hard even under this restriction. Recall that ℓ_a denotes the length of the longest preference list of any agent, and in practice, it can be much smaller than ℓ_p and $|\mathcal{P}|$.

Linear Program for General Instances. We present an LP relaxation (primal and dual) for general instances of MINSUM. We use this LP relaxation to design an ℓ_a-approximation algorithm for $\text{CCQ}_{c1,c2}$.

Let $H = (\mathcal{A} \cup \mathcal{P}, E)$ be a CCQ instance. Figure 1(a) shows primal for the MINSUM problem. Let $x_{a,p}$ be a variable for the edge $(a,p) \in E$: $x_{a,p}$ is 1 if a is matched to p, 0 otherwise. The objective of the primal LP (Eq. 1) is to minimize the total cost of all matched edges. Equation 2 encodes the envy-freeness constraint: if agent a is matched to p, then every agent $a' \succ_p a$ must be matched to either p or a higher-preferred program than p, otherwise, a' envies a. In the primal LP, the envy-freeness constraint is present for a triplet (a', p, a) where $a' \succ_p a$. We call such a triplet a *valid* triplet. Equation 3 encodes $\mathcal{A}$-perfectness constraint.

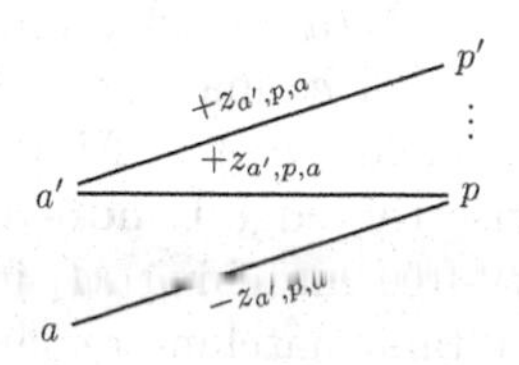

Fig. 2. The edges shown in the figure are those whose dual constraint contains the variable $z_{a',p,a}$ in either positive or negative form for a valid triplet (a', p, a) and any $p' \succ_{a'} p$.

minimize

$$\sum_{p\in\mathcal{P}} c(p) \cdot \sum_{(a,p)\in E} x_{a,p} \quad (1)$$

subject to

$$\sum_{\substack{p':\\ p'=p \text{ or}\\ p'\succ_{a'}p}} x_{a',p'} \geq x_{a,p}, \; \forall (a',p) \in E, a \prec_p a' \quad (2)$$

$$\sum_{(a,p)\in E} x_{a,p} = 1, \quad \forall a \in \mathcal{A} \quad (3)$$

$$x_{a,p} \geq 0, \quad \forall (a,p) \in E \quad (4)$$

(a) Primal

maximize

$$\sum_{a\in\mathcal{A}} y_a \quad (5)$$

subject to

$$y_a + \sum_{\substack{p':\\ p'=p \text{ or}\\ p'\prec_a p}} \sum_{\substack{a':\\ a'\prec_{p'} a}} z_{a,p',a'} - \sum_{\substack{a':\\ a\prec_p a'}} z_{a',p,a} \leq c(p), \quad \forall (a,p) \in E \quad (6)$$

$$z_{a',p,a} \geq 0, \quad \forall (a',p) \in E, a \prec_p a' \quad (7)$$

(b) Dual

Fig. 1. Linear Program relaxation for MINSUM problem

In the dual LP (Fig. 1(b)), we have two kinds of variables, the y variables, which correspond to every agent, and the z variables that correspond to every valid triplet in the primal program. The dual constraint (Eq. 6) is for every edge (a, p) in E. The y_a variable corresponding to an agent a appears in the dual constraint corresponding to every edge incident on a. For a valid triplet (a', p, a), the corresponding dual variable $z_{a',p,a}$ appears in negative form in exactly one constraint, and it is for the edge (a, p). The same variable $z_{a',p,a}$ appears in positive form in the constraint for every edge (a', p') such that $p' = p$ or $p' \succ_{a'} p$ (refer Fig. 2).

3.1 ℓ_a-Approximation Algorithm for $\mathbf{CCQ_{c1,c2}}$

For a given dual setting, an edge is ***tight*** if Eq. 6 is satisfied with equality; otherwise, it is ***slack***. Let M be a matching in H. For every program p, the ***threshold agent*** of p, denoted as $thresh(p)$, is the most-preferred agent a such that $p \succ_a M(a)$, if such an agent exists, otherwise we let $thresh(p)$ to be $\perp$. For an envy-free matching M, and an agent a (matched or unmatched), we say that an edge $(a, p) \notin M$ is ***matchable*** if (a, p) is tight and $a = thresh(p)$, otherwise the edge is **non-matchable**. It is straightforward to verify that for an envy-free matching M, if we match agent a along a matchable edge, then the resultant matching remains envy-free. Note that an update in dual variables can make edges tight but not necessarily matchable. However, a subset of these tight edges may become matchable as the algorithm proceeds. Our algorithm uses *free-promotions* routine that takes care of matching such edges (see [15]). Recall that for an unmatched agent a, we let $M(a) = \perp$.

Description of the Algorithm. Algorithm 2 gives the pseudo-code. We begin with an empty matching M and by setting all y variables to c_1 and all z variables

Algorithm 2. compute an ℓ_a-approximation of MINSUM on $\mathsf{CCQ}_{c1,c2}$

1: let $M = \emptyset$, all y variables are set to c_1 and all z variables are set to 0
2: **for** every agent $a \in \mathcal{A}$ s.t. $\exists p \in \mathcal{N}(a)$ such that $c(p) = c_1$ **do**
3: let p be the most-preferred program in $\mathcal{N}(a)$ s.t. $c(p) = c_1$ and let $M = M \cup \{(a,p)\}$
4: compute $thresh(p)$ for every program $p \in \mathcal{P}$
5: **while** M is not $\mathcal{A}$-perfect **do**
6: let a be an unmatched agent
7: **while** a is unmatched **do**
8: set $y_a = y_a + c_2 - c_1$
9: **if** there exists a matchable edge incident on a **then**
10: $M = M \cup \{(a,p) \mid (a,p)$ is the most-preferred matchable edge for $a\}$
11: perform free-promotions routine and re-compute thresholds
12: **else**
13: $\mathcal{P}(a) = \{p \in \mathcal{N}(a) \mid p \succ_a M(a), (a,p)$ is tight and $thresh(p) \neq a\}$
14: **while** $\mathcal{P}(a) \neq \emptyset$ **do**
15: let a' be the threshold agent of some program in $\mathcal{P}(a)$
16: let $\mathcal{P}(a,a')$ denote the set of programs in $\mathcal{P}(a)$ whose threshold agent is a'
17: let p be the least-preferred program for a' in $\mathcal{P}(a,a')$
18: set $z_{a',p,a} = c_2 - c_1$
19: let (a',p') be the most-preferred matchable edge incident on a'. Unmatch a' if matched and let $M = M \cup \{(a',p')\}$
20: execute free-promotions routine, re-compute thresholds and the set $\mathcal{P}(a)$
21: return M

to 0. For this dual setting, all edges incident on programs with cost c_1 are tight. For every agent a which has a program with cost c_1 in its list we match a with the most-preferred tight edge (**for** loop, line 2). We observe that after this step, if an agent is unmatched, then all programs in its preference list have cost c_2.

Our goal is to match these unmatched agents *without* introducing envies. We compute threshold agents for all programs w.r.t. M (line 4). As long as M is not $\mathcal{A}$-perfect, we perform the following steps: we pick an arbitrary unmatched agent a. We maintain an invariant that allows us to update the dual variable y_a (line 8) such that the dual setting is feasible and at least one edge incident on a becomes tight (but not necessarily matchable). If at least one edge incident on a also becomes matchable, then we match a along the most-preferred such edge (line 10). Since M is modified, some non-matchable tight edges may become matchable now. We execute the free-promotions routine to match such edges and re-compute the threshold agents.

Otherwise, every tight edge incident on a is non-matchable. It implies that for every program p such that (a,p) is tight, we have $thresh(p) \succ_p a$. We pick a threshold agent a', which could be a threshold agent at multiple programs. We *carefully select* the program p (line 17) and update $z_{a',p,a}$ variable. The choice of p and an invariant maintained by our algorithm together ensure that this update is dual feasible and at least one edge incident on a', which is higher-preferred over $M(a')$ becomes matchable.

We unmatch a' if matched. If (a', p') is the most-preferred matchable edge incident on a', we match a' to p'. Since M is updated, like before, we execute the free-promotions routine and re-compute the threshold agents. We repeat this process till either a is matched along the most-preferred edge or all higher-preferred edges incident on a become slack (line 14).

3.2 Proof of Correctness of Algorithm 2

We present the important observations and the proof sketch. Refer [15] for the detailed proofs. We observe the following properties.

- **(P1)** At line 4, no agent is assigned to any program with cost c_2 and for every agent a (matched or unmatched), every program $p \succ_a M(a)$ has cost c_2.
- **(P2)** A matched agent never gets demoted.
- **(P3)** A tight edge incident on a matched agent remains tight.
- **(P4)** All matched edges are tight at the end of the algorithm.

Property **(P1)** is a simple observation about the matching at line 4. Whenever a matched agent a changes its partner from $M(a)$ to p', we have $thresh(p') = a$. By the definition of the threshold agent, $p' \succ_a M(a)$, which implies **(P2)**. Note that the only edge that can become slack during the execution is the edge (a, p) which is incident on an unmatched agent a (line 18). This implies **(P3)**. We observe that when the edge is matched, it is tight. By **(P3)**, a matched edge (being incident on a matched agent) always remains tight, implying **(P4)**.

Proving Envy-Freeness. We prove envy-freeness using induction on the iterations of the **for** loop in line 5. By the induction hypothesis, the matching is envy-free before an iteration. Only matchable edge (a, p) is matched during an iteration. By definition, $thresh(p) = a$; therefore, no other agent envies a. Since a is promoted, a doesn't envy other agents. This observation and the induction hypothesis ensure that the matching is envy-free before the next iteration.

Dual Feasibility, $\mathcal{A}$-Perfectness and Termination. For the edge (a, p), let $slack(a, p)$ denote its slack. We categorize agents as follows (see Fig. 3): an agent a is called a ***type-1*** agent if for every program $p \succ_a M(a)$, $slack(a, p) = c_2 - c_1$. An agent a is called a ***type-2*** agent if a is matched and for every program $p \succ_a M(a)$, $slack(a, p) = 0$ and $thresh(p) \neq a$, that is, (a, p) is a non-matchable tight edge. A type-1 agent could be matched or unmatched, but a type-2 agent is always matched. We show that our algorithm maintains the following invariant.

Lemma 3 ($\star$). *Before every iteration of the loop starting at line 7, an agent is either a type-1 or a type-2 agent.*

The slack value of the edges incident on type-1 or type-2 agents, Lemma 3 and the dual updates performed in line 8 and line 18 together guarantee that the dual setting is feasible. Using Lemma 3 and the dual updates, we claim that

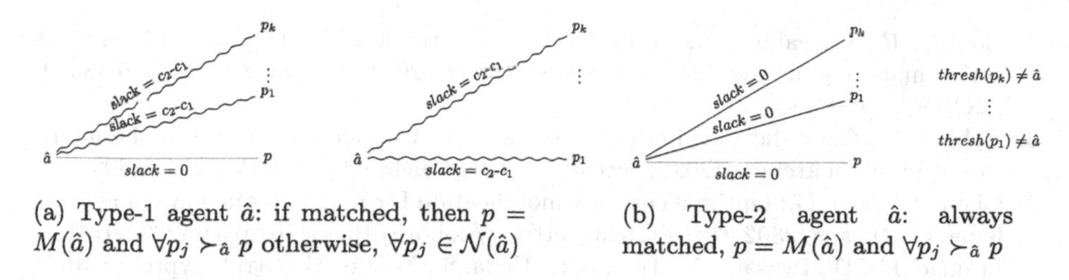

(a) Type-1 agent $\hat{a}$: if matched, then $p = M(\hat{a})$ and $\forall p_j \succ_{\hat{a}} p$ otherwise, $\forall p_j \in \mathcal{N}(\hat{a})$

(b) Type-2 agent $\hat{a}$: always matched, $p = M(\hat{a})$ and $\forall p_j \succ_{\hat{a}} p$

Fig. 3. Types of agents defined in the proof of correctness of Algorithm 2

the algorithm makes progress. Then by observing the condition of the **while** loop (line 5), we show that our algorithm terminates by computing an $\mathcal{A}$-perfect matching.

ℓ_a-Approximation Guarantee. The cost of the matching is the summation of the left-hand side of Eq. 6 of every matched edge. We *charge* a unit update in $z_{a',p,a}$ variable to a unit update in y_a variable such that at most ℓ_a many z variables are charged to a fixed y_a variable. This charging argument and the update in y variables give us an approximation guarantee of $\ell_a + 1$. To improve the guarantee to ℓ_a, we carefully inspect the cases wherein a z variable appearing as a positive term in Eq. 6 for a matched edge is canceled with that appearing as a negative term in Eq. 6 of another matched edge (refer Fig. 2). This enables us to charge at most $\ell_a - 1$ (instead of ℓ_a) many z variables to a fixed y_a variable, thereby improving the guarantee to ℓ_a.

This establishes Theorem 4.

Discussion: In this work, we propose and investigate the cost-controlled quota setting for two-sided preferences. A specific open direction is to bridge the gap between the upper bound and lower bound for MINSUM. It is also interesting to extend the LP-based algorithm for general instances.

References

1. Santhini, K.A., Sankar, G.S., Nasre, M. : Optimal matchings with one-sided preferences: fixed and cost-based quotas. In: 21st International Conference on Autonomous Agents and Multiagent Systems, AAMAS 2022, Auckland, New Zealand, 9–13 May 2022, pp. 696–704. International Foundation for Autonomous Agents and Multiagent Systems (IFAAMAS) (2022). https://doi.org/10.5555/3535850.3535929
2. Abdulkadiroğlu, A., Sönmez, T.: School choice: a mechanism design approach. Am. Econ. Rev. **93**(3), 729–747 (2003). https://doi.org/10.1257/000282803322157061
3. Abe, K., Komiyama, J., Iwasaki, A.: Anytime capacity expansion in medical residency match by monte carlo tree search. In: Proceedings of the Thirty-First International Joint Conference on Artificial Intelligence, IJCAI 2022, Vienna, Austria, 23–29 July 2022, pp. 3–9. ijcai.org (2022). https://doi.org/10.24963/ijcai.2022/1
4. Baswana, S., Chakrabarti, P.P., Chandran, S., Kanoria, Y., Patange, U.: Centralized admissions for engineering colleges in India. Interfaces **49**(5), 338–354 (2019). https://doi.org/10.1287/inte.2019.1007

5. Bobbio, F., Carvalho, M., Lodi, A., Rios, I., Torrico, A.: Capacity planning in stable matching: an application to school choice (2021). https://doi.org/10.48550/ARXIV.2110.00734
6. Bobbio, F., Carvalho, M., Lodi, A., Torrico, A.: Capacity variation in the many-to-one stable matching (2022). https://doi.org/10.48550/ARXIV.2205.01302
7. Chen, J., Csáji, G.: Optimal capacity modification for many-to-one matching problems. CoRR abs/2302.01815 (2023). https://doi.org/10.48550/arXiv.2302.01815
8. Fragiadakis, D., Iwasaki, A., Troyan, P., Ueda, S., Yokoo, M.: Strategyproof matching with minimum quotas. ACM Trans. Econ. Comput. **4**(1), 6:1–6:40 (2015). https://doi.org/10.1145/2841226, http://doi.acm.org/10.1145/2841226
9. Gajulapalli, K., Liu, J.A., Mai, T., Vazirani, V.V.: Stability-preserving, time-efficient mechanisms for school choice in two rounds. In: 40th IARCS Annual Conference on Foundations of Software Technology and Theoretical Computer Science, FSTTCS 2020). LIPIcs, vol. 182 (2020). https://doi.org/10.4230/LIPIcs.FSTTCS.2020.21
10. Gale, D., Shapley, L.S.: College admissions and the stability of marriage. Am. Math. Mon. **69**(1), 9–15 (1962). http://www.jstor.org/stable/2312726
11. Irving, R.W.: Matching medical students to Pairs of hospitals: a new variation on a well-known theme. In: Bilardi, G., Italiano, G.F., Pietracaprina, A., Pucci, G. (eds.) ESA 1998. LNCS, vol. 1461, pp. 381–392. Springer, Heidelberg (1998). https://doi.org/10.1007/3-540-68530-8_32
12. Irving, R.W., Leather, P., Gusfield, D.: An efficient algorithm for the "optimal" stable marriage. J. ACM **34**(3), 532–543 (1987). https://doi.org/10.1145/28869.28871
13. Irving, R.W., Manlove, D.F., O'Malley, G.: Stable marriage with ties and bounded length preference lists. J. Discret. Algorithms **7**(2), 213–219 (2009). https://doi.org/10.1016/j.jda.2008.09.003, selected papers from the 2nd Algorithms and Complexity in Durham Workshop ACiD 2006
14. Krishnaa, P., Limaye, G., Nasre, M., Nimbhorkar, P.: Envy-freeness and relaxed stability: Hardness and approximation algorithms. In: Harks, T., Klimm, M. (eds.) SAGT 2020. LNCS, vol. 12283, pp. 193–208. Springer, Cham (2020). https://doi.org/10.1007/978-3-030-57980-7_13
15. Limaye, G., Nasre, M.: Envy-free matchings with cost-controlled quotas. CoRR abs/2101.04425 (2021). https://arxiv.org/abs/2101.04425
16. Robards, P.A.: Applying two-sided matching processes to the united states navy enlisted assignment process. Technical report, NAVAL POSTGRADUATE SCHOOL MONTEREY CA (2001). https://hdl.handle.net/10945/10845
17. Roth, A.E.: On the allocation of residents to rural hospitals: a general property of two-sided matching markets. Econometrica **54**(2), 425–427 (1986). http://www.jstor.org/stable/1913160
18. Wu, Q., Roth, A.E.: The lattice of envy-free matchings. Games Econ. Behav. **109**, 201–211 (2018). https://doi.org/10.1016/j.geb.2017.12.016
19. Yang, W., Giampapa, J., Sycara, K.: Two-sided matching for the us navy detailing process with market complication. Technical report, CMU-RI-TR-03-49, Robotics Institute, Carnegie-Mellon University (2003). https://www.ri.cmu.edu/publications/two-sided-matching-for-the-u-s-navy-detailing-process-with-market-complication/
20. Yokoi, Yu.: Envy-free matchings with lower quotas. Algorithmica **82**(2), 188–211 (2018). https://doi.org/10.1007/s00453-018-0493-7

Generating Cyclic Rotation Gray Codes for Stamp Foldings and Semi-meanders

Bowie Liu and Dennis Wong(✉)

Macao Polytechnic University, Macao, China
p1709065@mpu.edu.mo, cwong@uoguelph.ca

Abstract. We present a simple algorithm that generates cyclic rotation Gray codes for stamp foldings and semi-meanders, where consecutive strings differ by a stamp rotation. These are the first known Gray codes for stamp foldings and semi-meanders, and we thus solve an open problem posted by Sawada and Li in [Electron. J. Comb. 19(2), 2012]. The algorithm generates each stamp folding and semi-meander in constant amortized time and $O(n)$-amortized time per string respectively, using a linear amount of memory.

Keywords: Stamp foldings · Meanders · Semi-meanders · Reflectable language · Binary reflected Gray code · Gray code · CAT algorithm

1 Introduction

A *stamp folding* is a way to fold a linear strip of n stamps into a single pile, with the assumption that the perforations between the stamps are infinitely elastic. As an example, Fig. 1 illustrates the sixteen stamp foldings for $n = 4$. We always orient a pile of stamps horizontally, with the stamps facing left and right and the perforations facing up and down. Additionally, the perforation between stamp 1 and stamp 2 is located at the bottom. The sixteen piles of stamps for $n = 4$ can thus be obtained by contracting the horizontal lines in Fig. 1. Each stamp folding can be represented by a unique permutation $(\pi(1)\pi(2)\cdots\pi(n))$, where $\pi(i)$ is the stamp at position i in the pile when considering it from left to right. The permutations that correspond to the sixteen stamp foldings in Fig. 1 for $n = 4$ are as follows:

$$1234, 1243, 1342, 1432, 2134, 2143, 2341, 2431,$$
$$3124, 3214, 3412, 3421, 4123, 4213, 4312, 4321.$$

In the rest of this paper, we use this permutation representation to represent stamp foldings. An alternative permutation representation can be found in [4]. Note that not all permutations correspond to valid stamp foldings. For example, the permutation 1423 requires the strip of stamps to intersect itself and is not a valid stamp folding.

A *semi-meander* is a stamp folding with the restriction that stamp n can be seen from above when n is even, and stamp n can be seen from below when n is odd (i.e. stamp n is not blocked by perforations between other stamps and can be seen from either above or below depending on its parity). For example, 1234 is a semi-meander, while 2143 is not a semi-meander since stamp 4 is blocked by the perforation between

S.-Y. Hsieh et al. (Eds.): IWOCA 2023, LNCS 13889, pp. 271–281, 2023.
https://doi.org/10.1007/978-3-031-34347-6_23

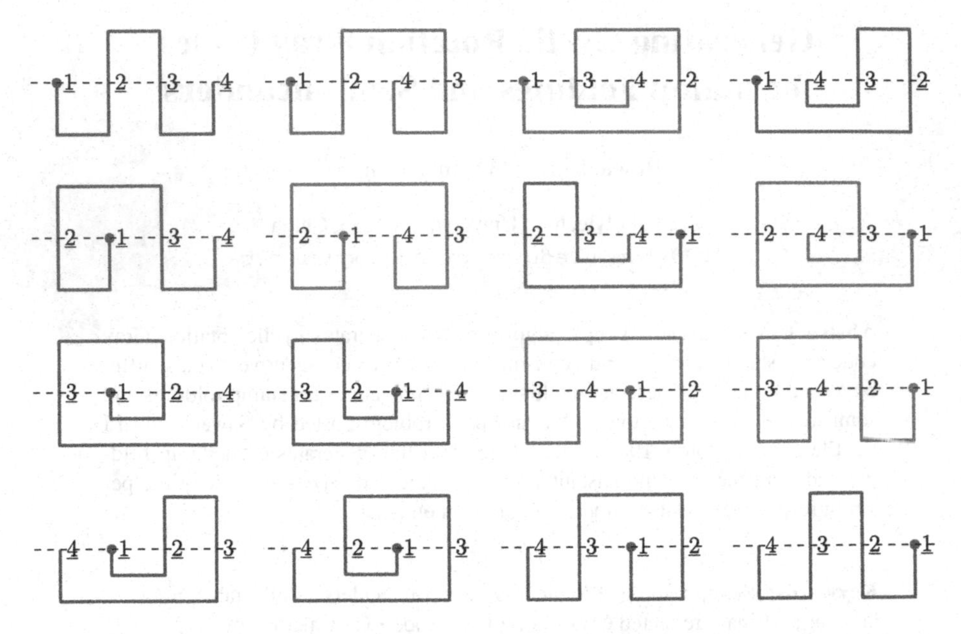

Fig. 1. Stamp foldings of order four. The stamp with a dot is stamp 1. Stamp foldings that are underlined (in red color) are semi-meanders. (Color figure online)

stamp 2 and stamp 3 and cannot be seen from above. The permutations that correspond to the ten semi-meanders in Fig. 1 for $n = 4$ are given as follows:

$$1234, 1432, 2134, 2341, 3124, 3214, 4123, 4213, 4312, 4321.$$

The study of stamp foldings and relatives has a long history and has traditionally attracted considerable interest from mathematicians [1–3, 6–8, 10, 11, 14, 15, 19, 20]. For example, the five foldings of four blank stamps appear at the front cover of the book *A Handbook of Integer Sequences* by Sloane [19], which is the ancestor of the well-known *Online Encyclopedia of Integer Sequences* [13]. Lucas [10] first stated the enumeration problem for stamp foldings in 1891 by asking in how many ways a strip of n stamps can be folded. See [8] for a brief history of the development of the enumeration problem of stamp foldings. Stamp foldings and relatives have lots of applications, ranging from robot coverage path planning [21] and conditioned random walks [5], to protein folding [18].

The enumeration sequences for stamp foldings and semi-meanders are A000136 and A000682 in the Online Encyclopedia of Integer Sequences, respectively [13]. The first ten terms for the enumeration sequences of stamp foldings and semi-meanders are as follows:

- Stamp foldings: 1, 2, 6, 16, 50, 144, 462, 1392, 4536, and 14060;
- Semi-meanders: 1, 2, 4, 10, 24, 66, 174, 504, 1406, and 4210.

Although a large number of terms of the folding sequences have been computed, no closed formula has been found for both enumeration sequences.

One of the most important aspects of combinatorial generation is to list the instances of a combinatorial object so that consecutive instances differ by a specified *closeness condition* involving a constant amount of change. Lists of this type are called *Gray codes*. This terminology is due to the eponymous *binary reflected Gray code* (BRGC) by Frank Gray, which orders the 2^n binary strings of length n so that consecutive strings differ by one bit. For example, when $n = 4$ the order is

$$0000, 1000, 1100, 0100, 0110, 1110, 1010, 0010,$$
$$0011, 1011, 1111, 0111, 0101, 1101, 1001, 0001.$$

We note that the order above is *cyclic* because the last and first strings also differ by the closeness condition, and this property holds for all n. The BRGC listing is a 1-Gray code in which consecutive strings differ by one symbol change. In this paper, we focus on rotation Gray code in which consecutive strings differ by a stamp rotation.

An interesting related problem is thus to discover Gray codes for stamp foldings and semi-meanders. There are several algorithms to generate stamp foldings and semi-meanders. Lunnon [11] in 1968 provided a backtracking algorithm that exhaustively generates stamp foldings after considering rotational equivalence and equivalence of the content between the first crease and the second crease. More recently, Sawada and Li [15] provided an efficient algorithm that exhaustively generates stamp foldings and semi-meanders in constant amortized time per string. However, the listings produced by these algorithms are not Gray codes. The problem of finding Gray codes for stamp foldings and semi-meanders is listed as an open problem in the paper by Sawada and Li [15], and also in the Gray code survey by Mütze [12]. In this paper, we solve this open problem by providing the first known cyclic rotation Gray codes for stamp foldings and semi-meanders, where consecutive strings differ by a stamp rotation. The formal definition of stamp rotation is provided in Sect. 2. The algorithm generates each stamp folding in constant amortized time per string, and each semi-meander in $O(n)$-amortized time per string respectively, using a linear amount of memory.

The rest of the paper is outlined as follows. In Sect. 2, we describe a simple algorithm to generate cyclic rotation Gray codes for stamp foldings and semi-meanders. Then in Sect. 3, we prove the Gray code property and show that the algorithm generates each stamp folding and semi-meander in constant amortized time and $O(n)$-amortized time per string respectively, using a linear amount of memory.

2 Gray Codes for Semi-meanders and Stamp Foldings

In this section, we first describe a simple recursive algorithm to generate a cyclic rotation Gray code for semi-meanders. We then describe simple modifications to the algorithm to generate a cyclic rotation Gray code for stamp foldings.

Consider a stamp folding $\alpha = p_1p_2 \cdots p_n$. A *stamp rotation* of the i-th to j-th symbols of a stamp folding α into its k-th position with $k < i \leq j \leq n$, denoted by $\text{rotate}_\alpha(i, j, k)$, is the string $p_1p_2 \cdots p_{k-1}p_ip_{i+1} \cdots p_jp_kp_{k+1} \cdots p_{i-1}p_{j+1}p_{j+2} \cdots p_n$. As an example, given a stamp folding (semi-meander) $\alpha = 6345127$, Fig. 2 illustrates the stamp folding (semi-meander) 6512347 obtained by applying $\text{rotate}_\alpha(4, 6, 2)$ on α

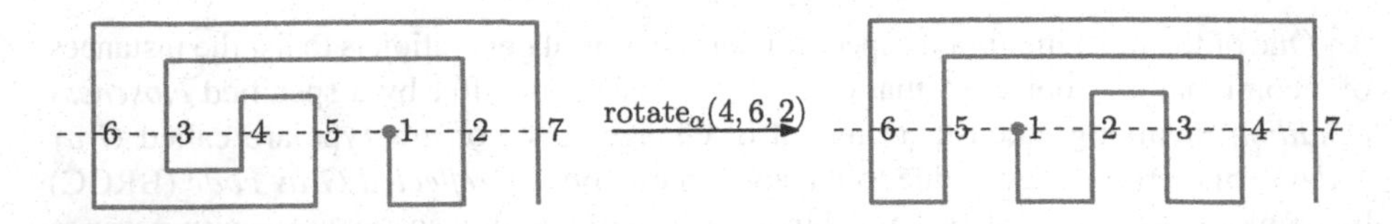

Fig. 2. Stamp rotation. The stamp folding 6512347 can be obtained by applying a stamp rotation $\text{rotate}_\alpha(4, 6, 2)$ on $\alpha = 6345127$.

(illustrated as $6\overleftarrow{34\underline{512}}7 \rightarrow 6512347$). Note that not all stamp rotations with $k < i \leq j \leq n$ can generate a valid stamp folding. To simplify the notation, we also define two special stamp rotations. A *left rotation* of the suffix starting at position i of a stamp folding α, denoted by $\overleftarrow{\text{rotate}}_\alpha(i)$, is the string obtained by rotating the suffix $p_i p_{i+1} \cdots p_n$ to the front of α, that is $\overleftarrow{\text{rotate}}_\alpha(i) = \text{rotate}_\alpha(i, n, 1) = p_i p_{i+1} \cdots p_n p_1 p_2 \cdots p_{i-1}$. Similarly, a *right rotation* of the prefix ending at position j of a stamp folding α, denoted by $\overrightarrow{\text{rotate}}_\alpha(j)$, is the string obtained by rotating the prefix $p_1 p_2 \cdots p_j$ to the end of α, that is $\overrightarrow{\text{rotate}}_\alpha(j) = \text{rotate}_\alpha(j+1, n, 1) = p_{j+1} p_{j+2} \cdots p_n p_1 p_2 \cdots p_j$. Observe that $\overleftarrow{\text{rotate}}_\alpha(t+1) = \overrightarrow{\text{rotate}}_\alpha(t)$. A *string rotation* of a string $\alpha = p_1 p_2 \cdots p_n$ is the string $p_2 p_3 \cdots p_n p_1$ which is obtained by taking the first character p_1 of α and placing it in the last position. The set of stamp foldings that are equivalent to a stamp folding α under string rotation is denoted by $\text{Rots}(\alpha)$. For example, $\text{Rots}(1243) = \{1243, 2431, 4312, 3124\}$. The strings in $\text{Rots}(\alpha)$ can be obtained by applying left rotation $\overleftarrow{\text{rotate}}_\alpha(i)$ on α for all integers $1 \leq i \leq n$, or applying right rotation $\overrightarrow{\text{rotate}}_\alpha(j)$ on α for all integers $1 \leq j \leq n$. We also define $I(e, \alpha)$ as the index of an element e within a string α. For example, $I(p_2, \alpha) = 2$ and $I(5, 6512347) = 2$.

Lemma 1. [14] *If $\alpha = p_1 p_2 \cdots p_n$ is a stamp folding, then $\beta \in \text{Rots}(\alpha)$ is also a stamp folding.*

Lemma 1 implies that the set of stamp foldings is partitioned into equivalence classes under string rotation. Also, note that this property does not hold for semi-meanders. For example, the string 3124 is a semi-meander but $1243 \in \text{Rots}(3124)$ is not a semi-meander.

The following lemmas follow from the definition of semi-meander.

Lemma 2. *The string $\alpha = p_1 p_2 \cdots p_n$ with $p_1 = n$ is a stamp folding of order n if and only if $p_2 p_3 \cdots p_n$ is a semi-meander of order $n - 1$.*

Proof. The backward direction is trivial. For the forward direction, assume by contrapositive that $p_2 p_3 \cdots p_n$ is not a semi-meander. If $p_2 p_3 \cdots p_n$ is not even a stamp folding of order $n-1$, then clearly α is not a stamp folding. Otherwise if $p_2 p_3 \cdots p_n$ is a stamp folding but not a semi-meander, now suppose $p_t = n - 1$ and $2 \leq t \leq n$. Observe that by the definition of semi-meander, p_t is blocked by a perforation between other stamps and thus it cannot connect to $p_1 = n$ without crossing any stamp. Thus α is also not a stamp folding.

Corollary 1. *If $\alpha = p_1 p_2 \cdots p_n$ is a semi-meander with $p_1 = n$, then $p_2 p_3 \cdots p_n p_1$ is also a semi-meander.*

Proof. By Lemma 2, α is a stamp folding and thus $p_2p_3\cdots p_n$ is a semi-meander of order $n-1$. Thus stamp $n-1$ of $p_2p_3\cdots p_n$ is not blocked by any perforation between other stamps and can connect to a stamp at position 1 or at position n to produce the string α or $p_2p_3\cdots p_np_1$. Lastly, stamps at position 1 and position n are at the boundary and cannot be blocked by any perforation between other stamps, thus $p_2p_3\cdots p_np_1$ is a semi-meander.

Lemma 3. *Suppose* $\alpha = p_1p_2\cdots p_n$ *is a semi-meander where* $p_n \neq n$ *and* $\beta = \overleftarrow{rotate}_\alpha(n-k+1)$ *with* $k \geq 1$ *being the smallest possible integer such that* β *is a semi-meander, then*

- $k = 1$ *if* $p_n = 1$ *and* n *is even;*
- $k = n - I(p_n+1, \alpha) + 1$ *if* p_n *and* n *have the same parity;*
- $k = n - I(p_n-1, \alpha) + 1$ *if* p_n *and* n *have different parities.*

Proof. If $p_n = 1$ and n is even, then clearly $k = 1$ since p_n only connects to $p_i = 2$ for some $i < n$ and the perforation between p_n and p_i is at the bottom while stamp n is extending in the upward direction. Otherwise, if p_n and n have the same parity, then assume W.L.O.G. that the perforation between p_n and p_n+1 is at the bottom. Clearly, $p_t = n$ is also extending in the downward direction. Since α is a semi-meander, $p_t = n$ can be seen from below and thus $t < I(p_n+1, \alpha) < n$. Now consider the string $\beta = \overleftarrow{\text{rotate}}_\alpha(n-k+1)$. When $1 < k < I(p_n+1, \alpha)$, the perforation between p_n+1 and p_n of β is at the bottom and would block stamp n making β not a semi-meander, and thus $k \geq I(p_n+1, \alpha)$. Furthermore, observe that there is no perforation between p_i and p_j at the bottom with $I(p_n+1, \alpha) < i$ and $j < I(p_n+1, \alpha)$ as otherwise the perforation intersects with the perforation between p_n+1 and p_n. Thus, $\beta = \overleftarrow{\text{rotate}}_\alpha(n-k+1)$ is a semi-meander when $k = I(p_n+1, \alpha)$. The proof is similar when p_n and n have different parities.

Lemma 4. *Suppose* $\alpha = p_1p_2\cdots p_n$ *is a semi-meander where* $p_1 \neq n$ *and* $\beta = \overrightarrow{rotate}_\alpha(k)$ *with* $k \geq 1$ *being the smallest possible integer such that* β *is a semi-meander, then*

- $k = 1$ *if* $p_1 = 1$ *and* n *is even;*
- $k = I(p_1+1, \alpha)$ *if* p_1 *and* n *have the same parity;*
- $k = I(p_1-1, \alpha)$ *if* p_1 *and* n *have different parities.*

Proof. The proof is similar to the one for Lemma 3.

In [9], Li and Sawada developed a framework to generate Gray codes for reflectable languages. A language L over an alphabet set σ is said to be reflectable if for every $i > 1$ there exist two symbols x_i and y_i in σ such that if $w_1w_2\cdots w_{i-1}$ is a prefix of a word in L, then both $w_1w_2\cdots w_{i-1}x_i$ and $w_1w_2\cdots w_{i-1}y_i$ are also prefixes of words in L. By reflecting the order of the children and using the special symbols x_i and y_i as the first and last children of each node at level $i-1$, Li and Sawada devised a generic recursive algorithm to generate Gray codes for reflectable languages which include lots of combinatorial objects such as k-ary strings, restricted growth functions and k-ary

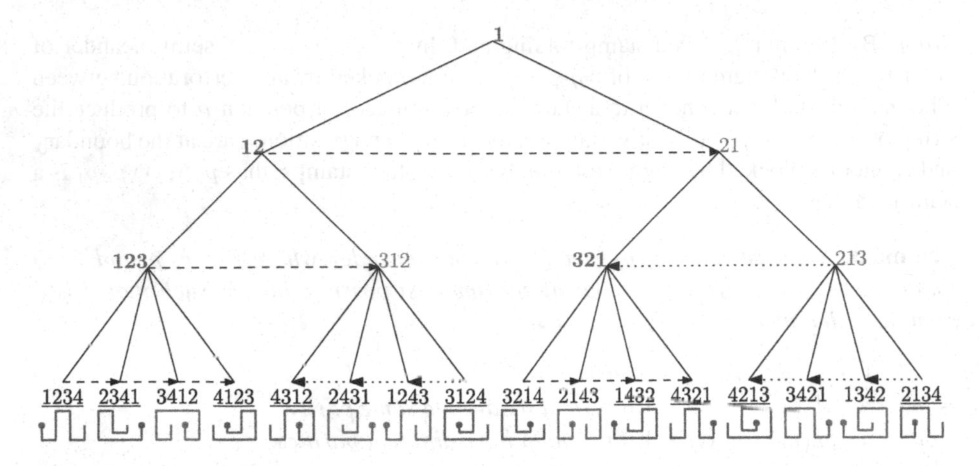

Fig. 3. Recursive computation tree constructed by our algorithm that outputs stamp foldings and semi-meanders for $n = 4$ in cyclic rotation Gray code order. The nodes at the last level are generated by the PRINT function and carry no sign. For nodes from level 1 to level $n - 1$, the nodes in bold (in blue color) carry a positive sign, and the rest of the nodes (in red color) carry a negative sign. A dashed arrow indicates applying right rotations to generate its neighbors, while a dotted arrow indicates applying left rotations to generate its neighbors. The underlined strings are semi-meanders. (Color figure online)

trees. For example, the algorithm generates the binary reflected Gray code in Sect. 1 when we set $x_i = 0$ and $y_i = 1$ for every $i > 1$.

Here we use a similar idea to recursively generate a Gray code for semi-meanders. The algorithm can be easily modified to generate stamp foldings. Depending on the level of a node in the recursive computation tree, there are two possibilities for its children at level t:

- The root of the recursive computation tree is the stamp 1 of order 1;
- If a node $p_1p_2 \cdots p_{t-1}$ is at level $t - 1$ where $t - 1 \geq 1$, then its children are the semi-meanders in $\text{Rots}(tp_1p_2 \cdots p_{t-1})$, which includes the strings $tp_1p_2 \cdots p_{t-1}$ and $p_1p_2 \cdots p_{t-1}t$ (by Corollary 1).

To generate the Gray code for semi-meanders, we maintain an array $q_1q_2 \cdots q_n$ which determines the sign of a new node in the recursive computation tree. The current level of the recursive computation tree is given by the parameter t. The sign of a new node at level t is given by the parameter q_t, where $q_t = 1$ denotes the new node has a positive sign, and $q_t = 0$ denotes the new node has a negative sign. We also initialize the sign q_t at each level as positive ($q_t = 1$) at the beginning of the algorithm. The word being generated is stored in a doubly linked list $p = p_1p_2 \cdots p_n$. Now if the current node has a positive sign, we generate the child $p_1p_2 \cdots p_{t-1}t$ and then keep applying right rotation to generate all semi-meanders in $\text{Rots}(p_1p_2 \cdots p_{t-1}t)$ until it reaches the string $tp_1p_2 \cdots p_{t-1}$. Then if the current node has a negative sign, we generate the child $tp_1p_2 \cdots p_{t-1}$ and then keep applying left rotation to generate all semi-meanders in $\text{Rots}(tp_1p_2 \cdots p_{t-1})$ until it reaches the string $p_1p_2 \cdots p_{t-1}t$. Finally when we reach

Algorithm 1. The algorithm that finds the number of left rotations or right rotations required to reach the next semi-meander.

```
1: function NEXTSEMIMEANDER(p, t, d)
2:     if d = 1 then j ← p_1
3:     else j ← p_t
4:     if j = 1 and t is even then return 1
5:     else if j and t have the same parity then
6:         if d = 1 then return I(j + 1, p)
7:         else return t − I(j + 1, p) + 1
8:     else
9:         if d = 1 then return I(j − 1, p)
10:        else return t − I(j − 1, p) + 1
```

Algorithm 2. The recursive algorithm that generates rotation Gray codes for stamp foldings and semi-meanders.

```
1: procedure GEN(p, t)
2:     i ← 1
3:     j ← 0
4:     while i ≤ t do
5:         if t ≥ n then PRINT(p)
6:         else
7:             if q_{t+1} = 1 then GEN(p · (t + 1), t + 1)
8:             else GEN((t + 1) · p, t + 1)
9:             q_{m+1} ← ¬q_{m+1}
10:        if t ≥ n and generating stamp foldings then j ← 1
11:        else j ← NEXTSEMIMEANDER(p, t, q_m)
12:        if q_m = 1 then p ← rotate→(j)        ▷ right rotation
13:        else p ← rotate←(n − j + 1)           ▷ left rotation
14:        i ← i + j
```

level $t = n$, we print out the semi-meanders that are in $\text{Rots}(p_1 p_2 \cdots p_{n-1} n)$. The next semi-meander to be obtained by applying left rotation or right rotation is determined by the function NextSemiMeander(p, t, d), which is a direct implementation of Lemma 3 and Lemma 4 (Algorithm 1). We also complement q_t every time a node is generated at level t. This way, at the start of each recursive call we can be sure that the previous word generated had stamp t at the same position. Finally to generate stamp foldings, notice that removing stamp n from a stamp folding always creates a semi-meander of order $n-1$ (Lemma 2). The recursive computation tree for stamp foldings is thus the same as the one generating semi-meanders, except at level n we print all stamp foldings that are in $\text{Rots}(p_1 p_2 \cdots p_{n-1} n)$, that is all strings in $\text{Rots}(p_1 p_2 \cdots p_{n-1} n)$. Pseudocode of the algorithm is shown in Algorithm 2. To run the algorithm, we make the initial call Gen$(1, 1)$ which sets $p = 1$ and $t = 1$.

As an example, the algorithm generates the following cyclic rotation Gray codes for stamp foldings and semi-meanders of length five respectively:

- 12345, 23451, 34512, 45123, 51234, 52341, 15234, 41523, 34152, 23415, 41235, 12354, 23541, 35412, 54123, 54312, 25431, 12543, 31254, 43125, 31245, 12453, 24531, 45312, 53124, 53214, 45321, 14532, 21453, 32145, 14325, 43251, 32514, 25143, 51432, 54321, 15432, 21543, 32154, 43215, 42135, 21354, 13542, 35421, 54213, 52134, 45213, 34521, 13452, 21345;
- 12345, 34512, 51234, 52341, 23415, 41235, 54123, 54312, 12543, 43125, 31245, 53124, 53214, 32145, 14325, 51432, 54321, 21543, 43215, 42135, 54213, 52134, 34521, 21345.

The Gray code listing of semi-meanders can be obtained by *filtering* the Gray code listing of stamp foldings. For more about filtering Gray codes, see [16,17]. Figure 3 illustrates the recursive computation tree when $n = 4$.

3 Analyzing the Algorithm

In this section, we prove that our algorithm generates cyclic rotation Gray codes for stamp foldings and semi-meanders in constant amortized time and $O(n)$-amortized time per string respectively.

We first prove the Gray code property for semi-meanders. The proof for the Gray code property for stamp foldings follows from the one for semi-meanders.

Lemma 5. *Each consecutive semi-meanders in the listing generated by the algorithm* GEN *differ by a stamp rotation rotate*$_\alpha(i, j, k)$ *for some* $k < i \leq j \leq n$.

Proof. The proof is by induction on n. In the base case when $n = 2$, the generated semi-meanders are 12 and 21 and clearly they differ by a stamp rotation. Inductively, assume consecutive semi-meanders generated by the algorithm GEN differ by a stamp rotation when $n = k - 1$. Consider the case when $n = k$, clearly the first $k - 1$ levels of the recursive computation tree for $n = k$ are exactly the same as the recursive computation tree for $n = k - 1$. If two consecutive nodes at level k have the same parent at level $k - 1$, then clearly the semi-meanders differ by a left rotation or a right rotation. Otherwise if two consecutive nodes at level k have different parents α and β at level $k - 1$. Observe that by the algorithm GEN, α and β are consecutive nodes at level $k - 1$ and W.L.O.G. assume α comes before β. Then by the inductive hypothesis the corresponding semi-meanders for α and β differ by a stamp rotation. Also, since α and β are consecutive nodes at level $k - 1$, their signs are different by the algorithm and thus the two children can only be the last child of α and the first child of β. If α carries a positive sign, then the two children are $n \cdot \alpha$ and $n \cdot \beta$. Otherwise, the two children are $\alpha \cdot n$ and $\beta \cdot n$. In both cases, the two children differ by a stamp rotation.

Lemma 6. *The first and last strings generated by the algorithm* GEN *are* $123 \cdots n$ *and* $2134 \cdots n$ *respectively.*

Proof. The leftmost branch of the recursive computation tree corresponds to the first string generated by the algorithm. Observe that all nodes of the leftmost branch carry positive signs and are the first child of their parents. Thus the first string generated by the algorithm is $123 \cdots n$. Similarly, the rightmost branch of the recursive computation tree corresponds to the last string generated by the algorithm. The rightmost node at level two of the recursive computation tree corresponds to the string 21. Furthermore, observe that the number of semi-meanders and the number of stamp foldings are even numbers when $n > 1$ (for each stamp folding (semi-meander) $p_1p_2 \cdots p_n$, its reversal $p_np_{n-1} \cdots p_1$ is also a stamp folding (semi-meander)). Thus the rightmost nodes at each level $t > 1$ carry negative signs and are the last child of their parents. Therefore, the last string generated by the algorithm is $2134 \cdots n$.

Lemma 7. *The algorithm* GEN *exhaustively generates all semi-meanders of length* n.

Proof. The proof is by induction on n. In the base case when $n = 2$, the generated semi-meanders are 12 and 21 which are all the possible semi-meanders for $n = 2$. Inductively, assume the algorithm generates all semi-meanders for $n = k - 1$. Consider the case when $n = k$, by Lemma 2 clearly the algorithm generates either one of the semi-meanders of the form $\{kp_1p_2 \cdots p_{k-1}, p_1p_2 \cdots p_{k-1}k\}$ for all possible $p_1p_2 \cdots p_{k-1}$ when considering the first child produced by each node at level $k - 1$. Since each semi-meander has a symbol $p_t = k$, generating all valid string rotations of the first child exhaustively lists out all semi-meanders for $n = k$.

Together, Lemma 5, Lemma 6 and Lemma 7 prove the following theorem.

Theorem 1. *The algorithm* GEN *produces a list of all semi-meanders of order* n *in a cyclic rotation Gray code order.*

We then prove the Gray code property for stamp foldings.

Lemma 8. *Each consecutive stamp foldings in the listing generated by the algorithm* GEN *differ by a stamp rotation* $rotate_\alpha(i, j, k)$ *for some* $k < i \leq j \leq n$.

Proof. Since the first $n - 1$ levels of the recursive computation tree for stamp foldings of length n are exactly the same as the recursive computation tree for semi-meanders of length $n - 1$, by Lemma 5 the strings correspond to consecutive nodes at level $n - 1$ differ by a stamp rotation. Therefore by the same argument as the inductive step of Lemma 5, consecutive stamp foldings at level n also differ by a stamp rotation.

Lemma 9. *The algorithm* GEN *exhaustively generates all stamp foldings of length* n.

Proof. Since the first $n - 1$ levels of the recursive computation tree for stamp foldings of length n are exactly the same as the recursive computation tree for semi-meanders of length $n - 1$, by Lemma 7 the algorithm exhaustively generates all nodes correspond to semi-meanders of length $n - 1$. Thus by Lemma 2, the algorithm generates either one of the stamp foldings of the form $\{kp_1p_2 \cdots p_{k-1}, p_1p_2 \cdots p_{k-1}k\}$ for all possible $p_1p_2 \cdots p_{k-1}$ when considering the first child produced by each node at level $k - 1$. Since the set of stamp foldings is partitioned into equivalence classes under string rotation, the algorithm generates all string rotations of the first child and thus exhaustively lists out all stamp foldings.

Similarly, Lemma 8, Lemma 6 and Lemma 9 prove the following theorem.

Theorem 2. *The algorithm* GEN *produces a list of all stamp foldings of order* n *in a cyclic rotation Gray code order.*

Finally, we prove the time complexity of our algorithm.

Theorem 3. *Semi-meanders and stamp foldings of length* n *can be generated in cyclic rotation Gray order in* $O(n)$*-amortized time and constant amortized time per string respectively, using* $O(n)$ *space.*

Proof. Clearly for each node at level $t < n$, each recursive call of the algorithm GEN only requires $O(n)$ amount of work and a linear amount of space to generate all rotations of a semi-meander. By Corollary 1, since each call to GEN makes at least two recursive calls and there is no dead ends in the computation tree, the algorithm generates each node at level $n-1$ of the computation tree in $O(n)$-amortized time per node using a linear amount of space. If we are generating semi-meanders, by the same argument each node at level n requires $O(n)$ amount of work and thus each string can be generated in $O(n)$-amortized time per string using a linear amount of space. Otherwise if we are generating stamp foldings, by Lemma 1 each node at level $n-1$ of the computation tree has exactly n children, while as discussed above each node at level $n-1$ can be generated in $O(n)$-amortized time per node. Therefore, the algorithm generates stamp foldings in constant amortized time per string using a linear amount of space.

Acknowledgements. This research is supported by the Macao Polytechnic University research grant (Project code: RP/FCA-02/2022) and the National Research Foundation of Korea (NRF) grant funded by the Ministry of Science and ICT (MSIT), Korea (No. 2020R1F1A1A01070666).

References

1. Bobier, B., Sawada, J.: A fast algorithm to generate open meandric systems and meanders. ACM Trans. Algorithms **6**(2), 1–12 (2010)
2. COS++. The combinatorial object server (2023): Generate meanders and stamp foldings (2023)
3. France, M., van der Poorten, A.: Arithmetic and analytic properties of paper folding sequences. Bull. Aust. Math. Soc. **24**(1), 123–131 (1981)
4. Hoffmann, K., Mehlhorn, K., Rosenstiehl, P., Tarjan, R.: Sorting Jordan sequences in linear time using level-linked search trees. Inf. Control **68**(1), 170–184 (1986)
5. Iordache, O.: Implementing Polytope Projects for Smart Systems, pp. 65–80. Springer, Cham (2017). Conditioned walks
6. Jensen, I.: A transfer matrix approach to the enumeration of plane meanders. J. Phys. A: Math. Gen. **33**(34), 5953 (2000)
7. Koehler, J.: Folding a strip of stamps. J. Comb. Theory **5**(2), 135–152 (1968)
8. Legendre, S.: Foldings and meanders. Australas. J Comb. **58**, 275–291 (2014)
9. Li, Y., Sawada, J.: Gray codes for reflectable languages. Inf. Process. Lett. **109**(5), 296–300 (2009)
10. Lucas, E.: Théorie des Nombres, vol. 1, p. 120. Gauthier-Villars, Paris (1891)
11. Lunnon, W.: A map-folding problem. Math. Comput. **22**, 193–199 (1968)
12. Mütze, T.: Combinatorial Gray codes - an updated survey. arXiv preprint arXiv:2202.01280 (2022)
13. OEIS Foundation Inc., The on-line encyclopedia of integer sequences, published electronically at (2023). http://oeis.org
14. Sainte-Lagüe, M.: Avec des nombres et des lignes, pp. 147–162. Vuibert, Paris (1937)

15. Sawada, J., Li, R.: Stamp foldings, semi-meanders, and open meanders: fast generation algorithms. Electron. J. Comb. **19**(2), 43 (2012)
16. Sawada, J., Williams, A., Wong, D.: Inside the binary reflected Gray code: Flip-swap languages in 2-Gray code order. In: Lecroq, T., Puzynina, S. (eds.) WORDS 2021. LNCS, vol. 12847, pp. 172–184. Springer, Cham (2021). https://doi.org/10.1007/978-3-030-85088-3_15
17. Sawada, J., Williams, A., Wong, D.: Flip-swap languages in binary reflected Gray code order. Theor. Comput. Sci. **933**, 138–148 (2022)
18. Schweitzer-Stenner, R., Uversky, V.: Protein and peptide folding, misfolding, and non-folding. Wiley Series in Protein and Peptide Science. Wiley, Hoboken (2012)
19. Sloane, N.: A Handbook of Integer Sequences. MIT Press, Cambridge (1973)
20. Touchard, J.: Contribution a létude du probleme des timbres poste. Can. J. Math. **2**, 385–398 (1950)
21. Zhu, L., Yao, S., Li, B., Song, A., Jia, Y., Mitani, J.: A geometric folding pattern for robot coverage path planning. In: 2021 IEEE International Conference on Robotics and Automation (ICRA), pp. 8509–8515 (2021)

On Computing Large Temporal (Unilateral) Connected Components

Isnard Lopes Costa[2], Raul Lopes[3,4], Andrea Marino[1], and Ana Silva[1,2(✉)]

[1] Dipartimento di Statistica, Informatica, Applicazioni, Università degli Studi di Firenze, Firenze, Italy
andrea.marino@unifi.it

[2] Departamento de Matemática, Universidade Federal do Ceará, Fortaleza, CE, Brazil
isnard.lopes@alu.ufc.br, anasilva@mat.ufc.br

[3] Université Paris-Dauphine, PSL University, CNRS UMR7243, LAMSADE, Paris, France

[4] DIENS, Ecole normale supérieure de Paris, CNRS, Paris, France
raul.lopes@ens.psl.eu

Abstract. A temporal (directed) graph is a graph whose edges are available only at specific times during its lifetime, τ. Paths are sequences of adjacent edges whose appearing times are either strictly increasing or non-strictly increasing (i.e., non-decreasing) depending on the scenario. Then, the classical concept of connected components and also of unilateral connected components in static graphs naturally extends to temporal graphs. In this paper, we answer the following fundamental questions in temporal graphs. (i) What is the complexity of deciding the existence of a component of size k, parameterized by τ, by k, and by $k + \tau$? We show that this question has a different answer depending on the considered definition of component and whether the temporal graph is directed or undirected. (ii) What is the minimum running time required to check whether a subset of vertices are pairwise reachable? A quadratic algorithm is known but, contrary to the static case, we show that a better running time is unlikely unless SETH fails. (iii) Is it possible to verify whether a subset of vertices is a component in polynomial time? We show that depending on the definition of component this test is NP-complete.

1 Introduction

A *(directed) temporal graph* (G, λ) *with lifetime* τ consists of a (directed) graph G together with a *time-function* $\lambda : E(G) \to 2^{[\tau]}$ which tells when each edge $e \in E(G)$ is available along the discrete time interval $[\tau]$. Given $i \in [\tau]$, the

(Partially) supported by: FUNCAP MLC-0191-00056.01.00/22 and PNE-0112-00061.01.00/16, CNPq 303803/2020-7, Italian PNRR CN4 Centro Nazionale per la Mobilità Sostenibile, and the group Casino/ENS Chair on Algorithmics and Machine Learning. Thanks to Giulia Punzi and Mamadou Kanté for interesting discussions. A full version of this paper is available at: https://arxiv.org/abs/2302.12068.

S.-Y. Hsieh et al. (Eds.): IWOCA 2023, LNCS 13889, pp. 282–293, 2023.
https://doi.org/10.1007/978-3-031-34347-6_24

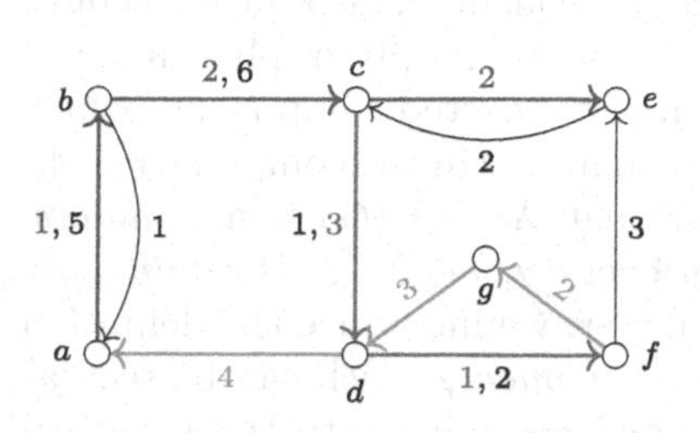

$A' = \{a, b\}$ is a closed connected set, as a and b reach each other without using external vertices.

$A = \{a, b, c, d\}$ is a maximal closed connected set, i.e. a CLOSED TCC.

$B = \{a, b, c, d, e\}$ is a CLOSED TUCC but not a CLOSED TCC as, using only vertices in B, a, b, c, d reach each other, e reaches all the vertices in B and vice versa, except for d, which does not reach e. B is also a TCC, as d can reach e using the external vertex f.

$C = \{a, b, c, d, e, f\}$ is a TUCC as B forms a CLOSED TUCC, f is able to reach every other vertex directly or via the external vertex g. However, C is not a TCC as a, b, c, e cannot reach f.

Fig. 1. On the left a temporal graph, where on each edge e we depict $\lambda(e)$. Some of its components according to the non-strict model are reported on the right.

snapshot G_i refers to the subgraph of G containing exactly the edges available in time i. Temporal graphs, also appearing in the literature under different names, have attracted a lot of attention in the past decade, as many works have extended classical notions of Graph Theory to temporal graphs (we refer the reader to the survey [11] and the seminal paper [10]).

A crucial characteristic of temporal graphs is that a u, v-walk/path in G is valid only if it traverses a sequence of adjacent edges $e_1, \ldots, e_k$ at non-decreasing times $t_1 \leq \ldots \leq t_k$, respectively, with $t_i \in \lambda(e_i)$ for every $i \in [k]$. Similarly, one can consider strictly increasing sequences, i.e. with $t_1 < \ldots < t_k$. The former model is referred to as *non-strict* model, while the latter as *strict*. In both settings, we call such sequence a *temporal u, v-walk/path*, and we say that *u reaches v*. For instance, in Fig. 1, both blue and green paths are valid in the non-strict model, but only the green one is valid in the strict model, as the blue one traverses two edges with label 2. The red path is not valid in either model.

The non-strict model is more appropriate in situations where the time granularity is relatively big. This is the case in a disease-spreading scenario [19], where the spreading speed might be unclear or in "time-varying graphs", as in [14], where a single snapshot corresponds to all the edges available in a time interval, e.g. the set of all the streets available in a day. As for the strict model, it can represent the connections of the public transportation network of a city which are available only at precise scheduled times. All in all, there is a rich literature on both models and this is why we explore both settings.

Connected Sets and Components. Given a temporal graph $\mathcal{G} = (G, \lambda)$, we say that $X \subseteq V(G)$ is a *temporal connected set* if u reaches v *and* v reaches u, for every $u, v \in X$. Extending the classical notion of connected components in static graphs, in [2] the authors define a *temporal connected component* (TCC for short) as a maximal connected set of $\mathcal{G}$. Such constraint can be strengthened to the existence of such paths using only vertices of X. Formally, X is a *closed temporal connected component* (CLOSED TCC for short) if, for every $u, v \in X$, we have that u reaches v *and* v reaches u through temporal paths contained in X. See Fig. 1 for an example of TCC and CLOSED TCC.

Unilateral Connected Components. In the same fashion, also the concept of *unilateral connected components* can be extended to temporal graphs. In static graph theory, they are a well-studied relaxation of connected components which asks for a path from u to v or vice versa, for every pair u, v in the component [1,4]. More formally, in a directed graph G, we say that $X \subseteq V(G)$ is a *unilateral connected set* if u reaches v or v reaches u, for every $u, v \in X$. X is a *unilateral connected component* if it is maximal. In this paper, we introduce the definition of a *(closed) unilateral temporal connected set/component*, which can be seen as the immediate translation of unilateral connected component to the temporal context. Formally, $X \subseteq V(G)$ is a *temporal unilateral connected set* if u reaches v *or* v reaches u, for every $u, v \in X$, and it is a *closed unilateral connected set* if this holds using paths contained in X. Finally, a *(closed) temporal unilateral connected component* ((CLOSED) TUCC for short) is a maximal (closed) temporal unilateral connected set. See again Fig. 1 for an example.

Problems. In this paper, we deal with four different definitions of temporal connected components, depending on whether they are unilateral or not, and whether they are closed or not. In what follows, we pose three questions, and we comment on partial knowledge about each of them. Later on, we discuss our results, which close almost all the gaps found in the literature. We start by asking the following.

Question 1 (Parameterized complexity). What is the complexity of deciding the existence of temporal components of size at least k parameterized by *(i)* τ, i.e. the lifetime, *(ii)* k, and *(iii)* $k + \tau$?

In order to answer Question 1 for the strict model, there is a very simple parameterized reduction from k-clique, known to be W[1]-hard when parameterized by k [7], to deciding the existence of connected components (both closed or not and both unilateral or not) of size at least k in undirected temporal graphs. This reduction has appeared in [5]. Given an undirected graph G, we can simply consider the temporal graph $\mathcal{G} = (G, \lambda)$ where $\lambda(uv) = \{1\}$ for all $uv \in E(G)$ (i.e., $\mathcal{G}$ is equal to G itself). As u temporally reaches v if and only if $uv \in E(G)$, one can see that all those problems are now equivalent to deciding the existence of a k-clique in G. Observe that we get W[1]-hardness when parameterized by k or $k + \tau$, and para-NP-completeness when parameterized by τ, both in the undirected and the directed case.[1] However, this reduction does not work in the case of the *non-strict* model, leaving Question 1 open. Indeed the reductions in [2] and in [6] for (CLOSED) TCCs, which work indistinctly for both the strict or the non-strict models, are not parameterized reductions. We also observe that the aforementioned reductions work on the non-strict model only for $\tau \geq 4$.

Another question of interest is the following. Letting n be the number of vertices in $\mathcal{G}$ and M be the number of *temporal edges*,[2] it is known that, in

[1] In the directed case, it suffices to replace each edge of the input graph with two opposite directed edges between the same endpoints.

[2] $M = \sum_{e \in E(\mathcal{G})} |\lambda(e)|$.

Table 1. A summary of our results for the parameterized complexity of computing components of size at least k of a temporal graph $\mathcal{G}$ having lifetime τ in the *non-strict* model. "W[1]-h" stands for W[1]-hardness and "p-NP" stands for para-NP-completeness. For the strict model the entries are W[1]-h in the third and fourth columns and p-NP in the second one already for $\tau = 1$, both for the directed and the undirected case.

	PAR. τ	PAR. k	PAR. $k + \tau$
TCC	p-NP $\tau \geq 2$ (Theorem 1)	W[1]-h Dir. $\tau \geq 2$ (Theorem 3) and Undir. (Theorem 2)	W[1]-h Dir. (Theorem 3) FPT Undir. (Theorem 5)
TUCC		W[1]-h Dir. $\tau \geq 2$ (Theorem 3) FPT Undir. (Theorem 5)	
CLOSED TCC		W[1]-h Dir. $\tau \geq 3$ (Theorem 4)	W[1]-h Dir. (Theorem 4) FPT Undir. (Theorem 5)
CLOSED TUCC		W[1]-h Dir. $\tau \geq 3$ (Theorem 4) FPT Undir. (Theorem 5)	

order to verify whether $X \subseteq V(G)$ is a connected set in $\mathcal{G}$, we can simply apply $O(n)$ single source "best" path computations (see e.g. [17]), resulting in a time complexity of $O(n \cdot M)$. This is $O(M^2)$ if $\mathcal{G}$ has no isolated vertices, a natural assumption when dealing with connectivity problems. As in static graphs testing connectivity can be done in linear time [8], we ask whether the described algorithm can be improved.

Question 2 (Lower bound on checking connectivity). Given a temporal graph $\mathcal{G}$ and a subset $X \subseteq V(\mathcal{G})$, what is the minimum running time required to check whether X is a (unilateral) connected set?

Finally we focus on one last question.

Question 3 (Checking maximality). Given a temporal graph $\mathcal{G}$ and a subset $X \subseteq V(\mathcal{G})$, is it possible to verify, in polynomial time, whether X is a component, i.e. a maximal (closed) (unilateral) connected set?

For Question 3, we first observe that the property of being a temporal (unilateral) connected set is hereditary (forming an independence system [12]), meaning that every subset of a (unilateral) connected set is still a (unilateral) connected set. For instance, in Fig. 1, every subset of the connected set $B = \{a, b, c, d, e\}$ is a connected set. Also, checking whether $X' \subseteq V(G)$ is a temporal (unilateral) connected set can be done in time $O(n \cdot M)$, as discussed above. We can then check whether X is a maximal such set in time $O(n^2 \cdot M)$: it suffices to test, for every $v \in V(G) \backslash X$, whether by adding v to X we still get a temporal (unilateral) connected set. On the other hand, *closed* connected (unilateral) sets are not hereditary, because by removing vertices from the set we could destroy the paths between other members of the set. This is the case for the closed connected set $A = \{a, b, c, d\}$ in Fig. 1, since by removing d there are no temporal paths from c to a nor b using only vertices in the remainder of the set. This implies that the same approach as before does not work, i.e., we cannot check whether X is maximal by adding to X a *single* vertex at a time, then checking for connectivity.

For instance, the closed connected set $A' = \{a, b\}$ in Fig. 1 cannot be grown into the closed connected set A by adding one vertex at a time, since both $A' \cup \{c\}$ and $A' \cup \{d\}$ are not closed connected sets. Hence, the answer to Question 3 for closed sets does not seem easy, and until now was still open.

Our Results. Our results concerning Question 1 are reported in Table 1 for the non-strict model, since for the strict model all the entries would be W[1]-hard or para-NP-complete already for $\tau = 1$, as we argued before. In the non-strict model, we observe instead that the situation is much more granulated. If $\tau = 1$, then all the problems become the corresponding traditional ones in static graphs, which are all polynomial (see Paragraph "Related works"). As for bigger values of τ, the complexity depends on the definition of component being considered, and whether the temporal graph is directed or not. Table 1 considers $\tau > 1$, reporting on negative results, "$\tau \geq x$" for some x meaning that the negative result starts to hold for temporal graphs of lifetime at least x.

The second column of Table 1 addresses Question 1 *(i)*, i.e., parameterization by τ. We prove that, for all the definitions of components being considered, the related problem becomes immediately para-NP-complete as soon as τ increases from 1 to 2; this is done in Theorem 1. This reduction improves upon the reduction of [2], which holds only for $\tau \geq 4$.

Question 1 *(ii)* (parameterization by k) is addressed in the third column of Table 1. Considering first directed temporal graphs, we prove that all the problems are W[1]-hard. In particular, deciding the existence of a TCC or TUCC of size at least k is W[1]-hard already for $\tau \geq 2$ (Theorem 3). As for the existence of closed components, W[1]-hardness also holds as long as $\tau \geq 3$ (Theorem 4). Observe that, since τ is constant in both results, these also imply the W[1]-hardness results presented in the last column, thus answering also Question 1 *(iii)* (parameterization by $k + \tau$) for directed graphs. On the other hand, if the temporal graph is undirected, then the situation is even more granulated. Deciding the existence of a TCC of size at least k remains W[1]-hard, but only if τ is unbounded. This is complemented by the answer to Question 1 *(iii)*, presented in the last column of Table 1: TCC and (even) CLOSED TCC are FPT on undirected graphs when parameterized by $k + \tau$ (Theorem 5). We also give FPT algorithms when parameterized by k for unilateral components, namely TUCC and CLOSED TUCC. Observe how this differs from TCC, whose corresponding problem is W[1]-hard, meaning that unilateral and traditional components behave very differently when parameterized by k.

In summary, Table 1 answers Question 1 for almost all the definitions of components, both for directed and undirected temporal graphs, leaving open only the problems of, given an undirected temporal graph, deciding the existence of a CLOSED TCC of size at least k when parameterized by k, and solving the same problem for CLOSED TCC and CLOSED TUCC in directed temporal graphs where $\tau = 2$.

Concerning Questions 2 and 3, our results are summarized in Table 2. All these results hold both for the strict and the non-strict models. For Question 2, we prove that the trivial $O(M^2)$ algorithm to test whether S is a (closed)

Table 2. Our results for Question 2 and Question 3, holding for both the *strict* and the *non-strict* models. Recall that a component is a (inclusion-wise) maximal connected set. The $O(\cdot)$ result is easy and explained in the introduction. M (resp. n) denotes the number of temporal edges (resp. nodes) in $\mathcal{G}$.

	CHECK WHETHER $X \subseteq V$ IS A CONNECTED SET	CHECK WHETHER $X \subseteq V$ IS A COMPONENT
TCC	$\Theta(M^2)$ (Theorem 6)	$O(n^2 \cdot M)$
TUCC		
CLOSED TCC		NP-c (Theorem 7)
CLOSED TUCC		

(unilateral) connected set is best possible, unless the Strong Exponential Time Hypothesis (SETH) fails [9]. For Question 3, in the case of TCC and TUCC, we have already seen that checking whether a set $X \subseteq V$ is a component can be done in $O(n^2 \cdot M)$. Interestingly, for CLOSED TCC and CLOSED TUCC, we answer negatively (unless P = NP) to Question 3.

Related Work. The known reductions for temporal connected components in the literature [2,6] which considers the non-strict setting are not parameterized and leave open the case when $\tau = 2$ or 3. The reductions we give here are parameterized (Theorems 3 and 4) and Theorem 1 closes also the cases $\tau = 2$ and 3. Furthermore, in [6] they show a series of interesting transformations but none of them allows us to apply known negative results for the strict model to the non-strict one. There are many other papers about temporal connected components in the literature, including [5], where they give an example where there can be an exponential number of temporal connected components in the strict model. In [14], the authors show that the problem of computing TCCs is a particular case of finding cliques in the so-called *affine graph*. This does not imply that the problem is NP-complete as claimed, since in order to prove hardness one needs to do a reduction on the opposite direction, i.e., a reduction from a hard problem to TCC instead. Finally, we remark that there are many results in the literature concerning unilateral components in static graphs [1], also with applications to community detection [13]. Even though the number of unilateral components in a graph is exponential [1], deciding whether there is one of size at least k is polynomial.

2 Parameterized Complexity Results

Parameterization by τ. We start by proving the result in the first column of Table 1, about para-NP-completeness wrt to the lifetime τ, which applies to all the definitions of temporal components. For (CLOSED) TCC we present a reduction from the NP-complete problem MAXIMUM EDGE BICLIQUE (MEBP for short) [16]. A *biclique* in a graph G is a complete bipartite subgraph of G. The MEBP problem consists of, given a bipartite graph G and an integer k, deciding whether G has a biclique with at least k edges. Using the same construction, we prove hardness of (CLOSED) TUCC reducing from the NP-complete problem $2K_2$-FREE EDGE SUBGRAPH [18]. In this problem we are given a bipartite graph

G and an integer k, and are trying to decide whether G has a $2K_2$-free subgraph with at least k edges.

The main idea of the reductions is to generate a temporal graph $\mathcal{G}$ whose underlying graph is the line graph L of a bipartite graph H with parts X, Y. Recall that, for each $u \in X \cup Y$, there is a clique in L formed by all the edges incident to u; denote such clique by C_u. We make active in timestep 1 the edges within C_u for every $u \in X$, and in timestep 2 the edges within C_u for every $u \in Y$. Doing so, we ensure that any pair of vertices of $\mathcal{G}$ associated with a biclique in H reach one another in $\mathcal{G}$. We prove that there exists a biclique in H with at least k edges if and only if there exists a CLOSED TCC in $\mathcal{G}$ of size at least k. The result extends to TCCs, as every TCC is also a CLOSED TCC. For the unilateral case, we can relax the biclique to a $2K_2$-free graph since only one reachability relation is needed. As a result, we get the following.

Theorem 1. *For every fixed $\tau \geq 2$ and given a temporal graph $\mathcal{G} = (G, \lambda)$ of lifetime τ and an integer k, it is NP-complete to decide if $\mathcal{G}$ has a (CLOSED) TCC or a (CLOSED) TUCC of size at least k, even if G is the line graph of a bipartite graph.*

W[1]-hardness by k. We now focus on proving the W[1]-hardness results in the second column of Table 1 concerning parameterization by k, which also imply some of the results of the third column. The following W[1]-hardness results (Theorem 2, 4, and 3) are parameterized reductions from k-CLIQUE. The general objective is constructing a temporal graph $\mathcal{G}$ in a way that vertices in $\mathcal{G}$ are in the same component if and only if the corresponding nodes in the original graph are adjacent. Notice that we have to do this while: (i) ensuring that the size of the desired component is $f(k)$ for some computable function k (i.e., this is a parameterized reduction); and (ii) avoiding that the closed neighborhood of a vertex forms a component, so as to not a have a false "yes" answer to k-CLIQUE. To address these tasks, we rely on different techniques. The first reduction concerns TCC in undirected graphs and requires τ to be unbounded, as for τ bounded we show that the problem is FPT by $k + \tau$ (Theorem 5). The technique used is a parameterized evolution of the so-called *semaphore* technique used in [2,6], which in general replaces edges by labeled diamonds to control paths of the original graph. However, while the original reduction gives labels in order to ensure that paths longer than one are broken, the following one allows the existence of paths longer than one. But if a temporal path from u to v exists for $uv \notin E(G)$, then the construction ensures the non-existence of temporal paths from v to u. Because of this property, the reduction does not extend to TUCCs, which we prove to be FPT when parameterized by k instead (Theorem 5).

Theorem 2. *Given a temporal graph $\mathcal{G}$ and an integer k, deciding if $\mathcal{G}$ has a TCC of size at least k is W[1]-hard with parameter k.*

Proof. We make a parameterized reduction from k-CLIQUE. Let G be graph and $k \geq 3$ be an integer. We construct the temporal graph $\mathcal{G} = (G', \lambda)$ as follows.

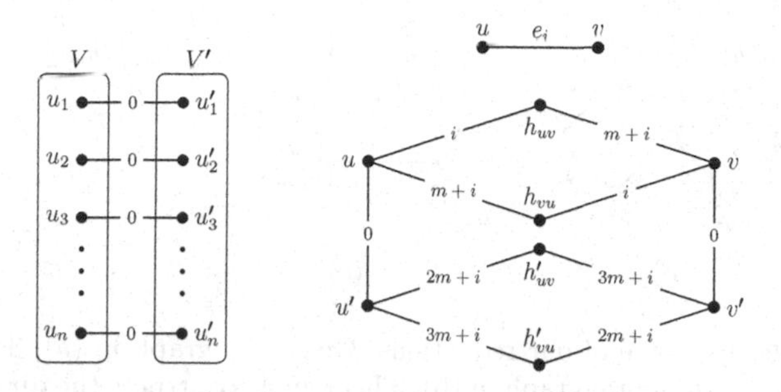

Fig. 2. Construction used in the proof of Theorem 2. On the left, the two copies of $V(G)$ and the edges between them, active in timestep 0. On the right, the edge $e_i \in E(G)$ and the associated gadget in $\mathcal{G}$.

See Fig. 2 to follow the construction. First, add to G' every vertex in $V(G)$ and make $V = V(G)$. Second, add to G' a copy u' of every vertex $u \in V$ and define $V' = \{u' \mid u \in V\}$. Third, for every pair u, u' with $u \in V$ and $u' \in V'$ add the edge uu' to G' and make all such edges active at timestep 0. Fourth, consider an arbitrary ordering $e_1, \ldots, e_m$ of the edges of G and, for each edge $e_i = uv$, create four new vertices $\{h_{uv}, h_{vu}, h'_{uv}, h'_{vu} \mid uv \in E(G)\}$, adding edges:

- uh_{uv} and vh_{vu}, active at time i;
- $u'h'_{uv}$ and $v'h'_{vu}$, active at time $2m + i$;
- $h_{vu}u$ and $h_{uv}v$, active at time $m + i$; and
- $h'_{vu}u'$ and $h'_{uv}v'$, active at time $3m + i$.

Denote the set $\{h_{uv}, h_{vu} \mid uv \in E(G)\}$ by H, and the set $\{h'_{uv}, h'_{vu} \mid uv \in E(G)\}$ by H'. We now prove that G has a clique of size at least k if and only if $\mathcal{G}$ has a TCC of size at least $2k$. Given a clique C in G, it is easy to check that $C \cup \{u \in V' \mid u \in C\}$ is a TCC.

Now, let $S \subseteq V(G')$ be a TCC of $\mathcal{G}$ of size at least $2k$. We want to show that either $C = \{u \in V(G) \mid u \in S \cap V\}$ or $C' = \{u \in V(G) \mid u' \in S \cap V'\}$ is a clique of G of size at least k. This part of the proof combines a series of useful facts, which we cannot include here due to space constraints. In what follows we present a sketch of it.

First, we argue that both C and C' are cliques in G. Then, by observing that the only edges between $V \cup H$ and $V' \cup H'$ are those incident to V and V' at timestep 0, we conclude that either $S \subseteq V \cup H$ or $S \subseteq V' \cup H'$. Since the cases are similar, we assume the former. If $|S \cap V| \geq k$, then C contains a clique of size at least k and the result follows. Otherwise, we define $E_S = \{uv \in E(G) \mid \{h_{uv}, h_{vu}\} \cap S \neq \emptyset\}$. That is, E_S is the set of edges of G related to vertices in $S \cap H$. We then prove the following claim.

Claim. Let $a, b \in S \cap H$ be associated with distinct edges g, g' of G sharing an endpoint v. If u and w are the other endpoints of g and g', respectively, then u and w are also adjacent in G. Additionally, either $|S \cap \{h_{xy}, h_{yx}\}| \leq 1$ for every $xy \in E(G)$, or $|S \cap H| \leq 2$.

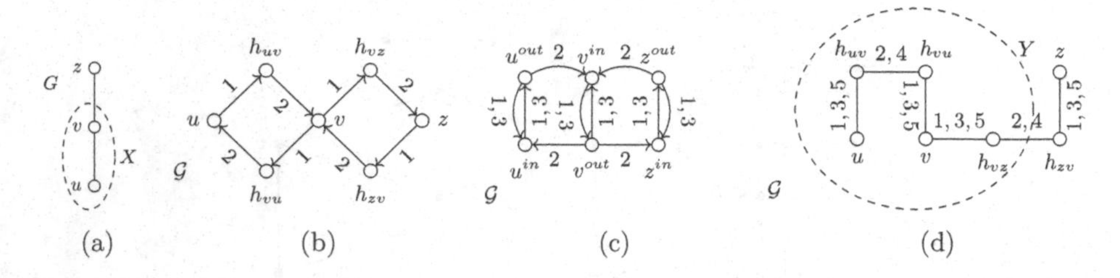

Fig. 3. Examples for some of our reductions. Given the graph in (a), Theorem 3 constructs the directed temporal graph in (b), Theorem 4 constructs the directed temporal graph in (c), and, given additionally set X in (a), Theorem 7 contructs the temporal graph $\mathcal{G}$ and set Y in (d).

To finish the proof, we first recall that we are in the case $|S \cap H| \geq k+1$. By our assumption that $k \geq 3$, note that the above claim gives us that $|S \cap \{h_{xy}, h_{yx}\}| \leq 1$ for every $xy \in E(G)$, which in turn implies that $|E_S| = |S \cap H|$. Additionally, observe that, since $|S \cap H| \geq 4$, the same claim also gives us that there must exist $w \in V$ such that e is incident to w for every $e \in E_S$. Indeed, the only way that 3 distinct edges can be mutually adjacent without being all incident to the same vertex is if they form a triangle. Supposing that 3 edges in E_S form a triangle $T = (a, b, c)$, since $|E_S| \geq 4$, there exists an edge $e \in E_S \setminus E(T)$. But now, since G is a simple graph, e is incident to at most one between a, b and c, say a. We get a contradiction wrt. the aforementioned claim as in this case e is not incident to edge $bc \in E_S$. Finally, by letting $C'' = \{v_1, \dots, v_k\}$ be any choice of k distinct vertices such that $\{wv_1, \dots, wv_k\} \subseteq E_S$, our claim gives us that v_i and v_j are adjacent in G, for every $i, j \in [k]$; i.e., C'' is a k-clique in G. □

The following result concerns TCC and TUCC in directed temporal graphs. It is important to remark that for TCC and τ unbounded, we already know that the problem is W[1]-hard because of Theorem 2 which holds for undirected graphs and extends to directed ones. However, the following reduction applies specifically for directed ones already for $\tau = 2$. The technique used here is the previously mentioned semaphore technique, made parameterized by exploiting the direction of the edges. Namely, we reduce from k-CLIQUE by replacing every edge uv of G by two vertices w_{uv} and w_{vu} and the directed temporal paths $(u, 1, w_{uv}, 2, v)$ and $(v, 1, w_{vw}, 2, u)$. See Fig. 3(b) to see the temporal graph obtained from the graph in Fig. 3(a). One can check that G has a clique of size at least k if and only if $\mathcal{G}$ has a TCC of size at least k. For TUCC, we only need to add one of w_{uv} or w_{vu}.

Theorem 3. *Given a directed temporal graph $\mathcal{G}$ and an integer k, deciding if $\mathcal{G}$ has a* TCC *of size at least k is* W[1]*-hard with parameter k, even if $\mathcal{G}$ has lifetime 2. The same holds for* TUCC.

The next result concerns closed TCCs and TUCCs. In this case, we also reduce from k-CLIQUE, but we cannot apply the semaphore technique as before. Indeed, as we are dealing with closed components, nodes must be reachable using vertices inside the components, while the semaphore technique would make them reachable via additional nodes, which do not necessarily reach each other. For this reason, in the following we introduce a new technique subdividing nodes, instead of edges, in order to break paths of the original graph of length longer than one, being careful to allow that these additional nodes reach each other. The construction is shown in Fig. 3, which shows how to construct temporal graph $\mathcal{G}$ in Fig. 3(c), given graph G in Fig. 3(a) in a way that graph G has a clique of size k if and only if $\mathcal{G}$ has a CLOSED TCC (TUCC) of size at least $2k$.

Theorem 4. *Given a directed temporal graph $\mathcal{G}$ and an integer k, deciding if $\mathcal{G}$ has a* CLOSED TCC *of size at least k is* W[1]*-hard with parameter k, even if $\mathcal{G}$ has lifetime 3. The same holds for* CLOSED TUCC.

FPT Algorithms. We now show our FPT algorithms to find (CLOSED) TCCs and (CLOSED) TUCCs in undirected temporal graphs, as for directed temporal graphs we have proved W[1]-hardness. In particular, we prove the following result.

Theorem 5. *Given a temporal graph $\mathcal{G} = (G, \lambda)$ on n vertices and with lifetime τ, and a positive integer k, there are algorithms running in time*

1. *$O(k^{k\cdot\tau} \cdot n)$ that decides whether there is a* TCC *of size at least k;*
2. *$O(2^{k^\tau} \cdot n)$ that decides whether there is a* CLOSED TCC *of size at least k;*
3. *$O(k^{k^2} \cdot n)$ that decides whether there is a* TUCC *of size at least k; and*
4. *$O(2^{k^k} \cdot n)$ that decides whether there is a* CLOSED TUCC *of size at least k.*

Proof. The *reachability digraph* R associated to $\mathcal{G}$ is a directed graph with the same vertex set as $\mathcal{G}$, and such that uv is an edge in R if and only u reaches v in $\mathcal{G}$ and $u \neq v$. This is related to the *affine* graph in [14]. Observe that finding a TCC (resp. TUCC) in $\mathcal{G}$ of size at least k is equivalent to finding a set $S \subseteq V(\mathcal{G})$ in R of size *exactly* k such that $uv \in E(R)$ and (resp. or) $vu \in E(R)$ for every pair $u, v \in V(R)$. As for finding a CLOSED TCC (resp. CLOSED TUCC), we need to have the same property, except that all subsets of size *at least* k must be tested (recall that being a closed connected (unilateral) set is not hereditary). Therefore, if Δ is the maximum degree of R, then testing connectivity takes time $O(k^{\Delta} \cdot n)$ (it suffices to test all subsets of size $k-1$ in $N(u)$, for all $u \in V(R)$), while testing closed connectivity takes time $O(2^{\Delta} \cdot n)$ (it suffices to test all subsets of size *at least* $k-1$ in $N(u)$, for all $u \in V(R)$). The proofs then consist in bounding the value Δ in each case. □

It is important to observe that, for unilateral components, these bounds depend only on k, while for TCCs and CLOSED TCCs they depend on both k and τ. This is consistent with the fact that we have proved that for TCC the problem is W[1]-hard when parameterized just by k (Theorem 2).

3 Checking Connectivity and Maximality

This section is focused on Questions 2 and 3. The former is open for all definitions of components for both the strict and the non-strict models. We answer to question providing the following conditional lower bound, which holds for both models, where the notation $\tilde{O}(\cdot)$ ignores polylog factors. We apply the technique used for instance in [3] to prove lower bounds for polynomial problems, which intuitively consists of an exponential reduction from SAT to our problem.

Theorem 6. *Consider a temporal graph $\mathcal{G}$ on M temporal edges. There is no algorithm running in time $\tilde{O}(M^{2-\epsilon})$, for some ϵ, that decides whether G is temporally (unilaterally) connected, unless* SETH *fails.*

We now focus on Question 3. We prove the results in the second column of Table 2, about the problem of deciding whether a subset of vertices Y of a temporal graph is a component, i.e. a maximal connected set. The question is open both for the strict and the non-strict model. We argued already in the introduction that this is polynomial for TCC and TUCC for both models. In the following we prove NP-completeness for CLOSED TCC and CLOSED TUCC on undirected graphs. The results extend to directed graphs as well.

Theorem 7. *Let $\mathcal{G}$ be a (directed) temporal graph, and $Y \subseteq V(\mathcal{G})$. Deciding whether Y is a* CLOSED TCC *is* NP*-complete. The same holds for* CLOSED TUCC.

Proof. We reduce from the problem of deciding whether a subset of vertices X of a given a graph G is a maximal 2-club, where a 2-club is a set of vertices C such that $G[C]$ has diameter at most 2. This problem has been shown to be NP-complete in [15]. Let us first focus on the strict model. In this case, given G we can build a temporal graph $\mathcal{G}$ with only two snapshots, both equal to G. Observe that X is a 2-club in G if and only if X is a CLOSED TCC in $\mathcal{G}$. Indeed, because we can take only one edge in each snapshot and $\tau = 2$, we get that temporal paths will always have length at most 2. This also extends to CLOSED TUCC by noting that all paths in $\mathcal{G}$ can be temporally traversed in both directions.

In the case of the non-strict model, the situation is more complicated as in each snapshot we can take an arbitrary number of edges resulting in paths arbitrarily long. We show the construction for CLOSED TCC in what follows. Let $\mathcal{G}$ be obtained from G by subdividing each edge uv of G twice, creating vertices h_{uv} and h_{vu}, with $\lambda(uh_{uv}) = \lambda(vh_{vu}) = \{1,3,5\}$, and $\lambda(h_{uv}h_{vu}) = \{2,4\}$. See Fig. 3 (d) for an illustration.

Given (G, X), the instance of maximal 2-club, we prove that X is a maximal 2-club in G iff $Y = X \cup N_H(X)$ is a CLOSED TCC in $\mathcal{G}$. For this, it suffices to prove that, given $X' \subseteq V(G)$ and defining Y' similarly as before w.r.t. X', we have that $G[X']$ has diameter at most 2 iff Y' is a closed temporal connected set. The proof extends to CLOSED TUCC by proving that every CLOSED TCC is also a CLOSED TUCC and vice-versa. □

References

1. Arjomandi, E.: On finding all unilaterally connected components of a digraph. Inf. Process. Lett. **5**(1), 8–10 (1976)
2. Bhadra, S., Ferreira, A.: Complexity of connected components in evolving graphs and the computation of multicast trees in dynamic networks. In: Pierre, S., Barbeau, M., Kranakis, E. (eds.) ADHOC-NOW 2003. LNCS, vol. 2865, pp. 259–270. Springer, Heidelberg (2003). https://doi.org/10.1007/978-3-540-39611-6_23
3. Borassi, M., Crescenzi, P., Habib, M.: Into the square: on the complexity of some quadratic-time solvable problems. In: ICTCS. Electronic Notes in Theoretical Computer Science, vol. 322, pp. 51–67. Elsevier (2015)
4. Borodin, A.B., Munro, I.: Notes on efficient and optimal algorithms. Technical report, U. of Toronto and U. of Waterloo, Canada (1972)
5. Casteigts, A.: Finding structure in dynamic networks. arXiv preprint arXiv:1807.07801 (2018)
6. Casteigts, A., Corsini, T., Sarkar, W.: Simple, strict, proper, happy: a study of reachability in temporal graphs. arXiv preprint arXiv:2208.01720 (2022)
7. Downey, R.G., Fellows, M.R.: Fixed-parameter tractability and completeness ii: on completeness for w [1]. Theor. Comput. Sci. **141**(1–2), 109–131 (1995)
8. Hopcroft, J., Tarjan, R.: Algorithm 447: efficient algorithms for graph manipulation. Commun. ACM **16**(6), 372–378 (1973)
9. Impagliazzo, R., Paturi, R.: On the complexity of k-sat. J. Comput. Syst. Sci. **62**(2), 367–375 (2001)
10. Kempe, D., Kleinberg, J., Kumar, A.: Connectivity and inference problems for temporal networks. In: Yao, F.F., Luks, E.M., eds, Proceedings of the Thirty-Second Annual ACM Symposium on Theory of Computing, 21–23 May 2000, Portland, pp. 504–513. ACM (2000)
11. Latapy, M., Viard, T., Magnien, C.: Stream graphs and link streams for the modeling of interactions over time. Soc. Netw. Anal. Min. **8**(1), 1–29 (2018). https://doi.org/10.1007/s13278-018-0537-7
12. Lawler, E.L., Lenstra, J.K., Rinnooy Kan, A.H.G.: Generating all maximal independent sets: Np-hardness and polynomial-time algorithms. SIAM J. Comput. **9**(3), 558–565 (1980)
13. Levorato, V., Petermann, C.: Detection of communities in directed networks based on strongly p-connected components. In: 2011 International Conference on Computational Aspects of Social Networks (CASoN), pp. 211–216. IEEE (2011)
14. Nicosia, V., Tang, J., Musolesi, M., Russo, G., Mascolo, C., Latora, V.: Components in time-varying graphs. Chaos: Interdisc. J. Nonlinear Sci. **22**(2), 023101 (2012)
15. Foad Mahdavi Pajouh and Balabhaskar Balasundaram: On inclusionwise maximal and maximum cardinality k-clubs in graphs. Discret. Optim. **9**(2), 84–97 (2012)
16. Peeters, R.: The maximum edge biclique problem is np-complete. Discret. Appl. Math. **131**(3), 651–654 (2003)
17. Huanhuan, W., Cheng, J., Huang, S., Ke, Y., Yi, L., Yanyan, X.: Path problems in temporal graphs. Proc. VLDB Endowment **7**(9), 721–732 (2014)
18. Yannakakis, M.: Computing the minimum fill-in is np-complete. SIAM J. Algebraic Discrete Methods **2**(1), 77–79 (1981)
19. Zschoche, P., Fluschnik, T., Molter, H., Niedermeier, R.: The complexity of finding small separators in temporal graphs. J. Comput. Syst. Sci. **107**, 72–92 (2020)

On Integer Linear Programs for Treewidth Based on Perfect Elimination Orderings

Sven Mallach(✉)

High Performance Computing & Analytics Lab, University of Bonn, Friedrich-Hirzebruch-Allee 8, 53115 Bonn, Germany
sven.mallach@cs.uni-bonn.de

Abstract. We analyze two well-known integer programming formulations for determining the treewidth of a graph that are based on perfect elimination orderings. For the first time, we prove structural properties that explain their limitations in providing convenient lower bounds and how the latter are constituted. Moreover, we investigate a flow metric approach that proved promising to achieve approximation guarantees for the pathwidth of a graph, and we show why these techniques cannot be carried over to improve the addressed formulations for the treewidth. Via computational experiments, we provide an impression on the quality and proportionality of the lower bounds on the treewidth obtained with different relaxations of perfect ordering formulations.

Keywords: treewidth · linear programming · integer programming

1 Introduction

The treewidth of a graph and several related width parameters are of theoretical as well as of practical interest. In particular, while deciding the treewidth of an arbitrary graph is itself $\mathcal{NP}$-complete [1,21], many $\mathcal{NP}$-hard problems on graphs can be solved efficiently on instances of bounded treewidth, see e.g. [4,9].

While structural relations between width parameters and algorithms exploiting them are vital fields of research receiving increasing attention, recent progress regarding the exact computation of the treewidth of general graphs $G = (V, E)$ has been mainly achieved by developments related to the PACE challenges in 2016 and 2017 [12,13]. Here, Tamaki's approach based on dynamic programming over maximal cliques of minimal triangulations of G (to obtain upper bounds) and lower bound computations via minors of G obtained using contraction-based algorithms [25] proved particularly successful. Indeed, dynamic programming algorithms have a long tradition for the computation of treewidth, see e.g. [6,7].

Interestingly, another successful algorithm that emerged from PACE, developed by Bannach, Bernds, and Ehlers [2], and called Jdrasil, relies on a SAT-approach based on perfect elimination orderings (PEO) which we define formally in Sect. 2. PEO-based models for treewidth have been considered early as well. For SAT formulations, they were used e.g. in [3], and also the first and perhaps

S.-Y. Hsieh et al. (Eds.): IWOCA 2023, LNCS 13889, pp. 294–306, 2023.
https://doi.org/10.1007/978-3-031-34347-6_25

most intuitive integer linear programming (ILP) formulations for treewidth from 2004 were based on PEOs, see Grigoriev et al. [20]. However, over the last decade the latter have not been brought to success to compute the treewidth routinely even for moderately sized graphs as recently pointed out by Grigoriev [19].

While this may be partly explained by the lack of an according algorithmic framework applying preprocessing techniques as aggressively as e.g. in [2], it is folklore that another major obstacle consists in the weak lower bounds provided by the linear programming relaxations of PEO-based ILP formulations [19]. Yet, little is known about the reasons for this weakness whose identification also serves as a first orientation regarding the potential improvement of these formulations.

In order to bridge this gap, we provide first structural evidence and explanations for the major limitations of PEO-based ILP formulations for computing the treewidth. In particular, we prove that the central class of constraints responsible to ensure the acyclicness of the ordering relation does not contribute to the lower bounds obtained from a relaxation at first hand, and analyze the role and impact of triangulation constraints standalone and when combined with the former. Along the way, we provide insights on how certain treewidth bounds are constituted respectively induced by the graph structure. Moreover, we investigate the flow metric approach by Bornstein and Vempala [10] that proved promising to derive approximation guarantees for the pathwidth and other related problems, and reveal why it cannot be combined with the existing PEO-based formulations in order to compute better lower bounds for the treewidth. Finally, we demonstrate by computational experiments how the lower bounds obtained by forming different relaxations of these formulations relate to each other.

The outline of this paper is as follows. Section 2 introduces the basic concepts of treewidth and the PEO-based ILP formulations to compute it, thereby addressing related work. In Sect. 3, we present our structural analysis of the PEO-based formulations, and in Sect. 4, we investigate the flow metrics approach in the context of treewidth. The results of our experimental study are the subject of Sect. 5. The paper closes with a conclusion and outlook in Sect. 6.

2 Treewidth and Perfect Elimination Orderings

2.1 Basic Definitions

A tree decomposition of an undirected graph $G = (V, E)$ is a collection of sets $X_i \subseteq V$, $i \in I$, along with a tree $T = (I, F)$ such that (a) $\bigcup_{i \in I} X_i = V$, (b) there is a set X_i, $i \in I$, with $\{v, w\} \subseteq X_i$ for each $\{v, w\} \in E$, and (c) for each $j \in V$, the tree-edges F connect all tree-vertices $i \in I$ where $j \in X_i$ [24]. The width of a tree decomposition is defined as $\max_{i \in I} |X_i| - 1$ and the treewidth $tw(G)$ of G is the minimum width among all its tree decompositions.

A well-known alternative definition of the treewidth is based on linear orderings and triangulations [5]. A linear ordering of a graph $G = (V, E)$ is a bijection $\pi : V \to \{1, \ldots, n\}$. A perfect elimination ordering (or scheme) is a linear ordering π such that for each vertex $v \in V$, the set of vertices $R_v^\pi := \{w \in V : \pi(w) > \pi(v) \text{ and } \{v, w\} \in E\}$ forms a clique. A PEO exists if and only if G is chordal [18]. On the other hand, each linear ordering π defines a triangulation (chordalization) G' of G by means of augmenting edges between higher ranked but non-adjacent neighbors of the vertices in the order given by π, see also [5]. Denote by R'^{π}_{v} this (potential) extension of R^{π}_{v} for $v \in V$, then the treewidth of G can be derived as

$$tw(G) = \min_{\pi \text{PEO}} \max_{v \in V} |R'^{\pi}_{v}|,$$

i.e., by finding a triangulation (chordalization) of G with minimum clique size [5].

2.2 PEO-based Integer Programming Formulations and Relaxations

From a coarse perspective, two principal approaches to formulate PEO-based ILPs have been investigated. Both have in common that the edges of (a chordalization of) an undirected graph $G = (V, E)$ are interpreted in an oriented fashion, i.e., we consider (arcs augmenting) the bidirected pendant $D = (V, A)$ of G where $\{i, j\} \in E \Leftrightarrow \{(i, j), (j, i)\} \subseteq A$. Moreover, we denote by $A^* := \{(i, j) : i, j \in V,\ i \neq j\}$ the arc set of a (conceptually) completed digraph D.

We refer to the first and more common model as TW_{LO} as it relies on the determination of a (total) linear ordering of the vertices of the graph under consideration, i.e., it models the relative order of each vertex pair. It originates from Bodlaender and Koster, see also [19,20], and can be written as follows:

$$
\begin{array}{llll}
\min w & & & \\
\text{s.t.} \sum\limits_{j \in V : \{i,j\} \in E} x_{ij} + \sum\limits_{j \in V : \{i,j\} \notin E} y_{ij} & \leq w & \text{for all } i \in V & (1) \\
x_{ij} + x_{ji} & = 1 & \text{for all } i, j \in V, i < j & (2) \\
x_{ij} + x_{jk} + x_{ki} & \leq 2 & \text{for all } i, j, k \in V, i \neq j \neq k \neq i & (3) \\
y_{ij} + y_{ik} - y_{jk} - y_{kj} & \leq 1 & \text{for all } \{j, k\} \notin E, i \in V \setminus \{j, k\} & (4) \\
y_{ij} - x_{ij} & = 0 & \text{for all } i, j \in V, \{i, j\} \in E & (5) \\
y_{ij} - x_{ij} & \leq 0 & \text{for all } i, j \in V, i \neq j, \{i, j\} \notin E & (6) \\
x_{ij} & \geq 0 & \text{for all } i, j \in V, i \neq j & \\
y_{ij} & \geq 0 & \text{for all } i, j \in V, i \neq j & \\
w & \in \mathbb{R} & & \\
x_{ij} & \in \{0, 1\} & \text{for all } i, j \in V, i \neq j &
\end{array}
$$

The interpretation associated with TW_{LO} is that for all $i, j \in V$, $i \neq j$, one has $\pi(i) < \pi(j)$ if and only if $x_{ij} = 1$, and that $\{i, j\}$ is an edge of the triangulation if and only if $y_{ij} = 1$ or $y_{ji} = 1$. Equations (2) enforce that each vertex pair is ordered while the three-di-cycle inequalities (3) ensure consistency of the ordering expressed in the x-variables, i.e., transitivity of the precedence relation, see also [23]. Inequalities (4) impose the necessary augmentations of (so-called "fill-in") edges $\{j, k\} \notin E$ to obtain a chordalization where each R'^{π}_{i} is a clique. Moreover, Eq. (5) ensure that each original edge has a correctly oriented counterpart in the triangulation, and inequalities (6) let the augmented edges be oriented consistently with the ordering as well. In combination with constraints (4), they also establish the integrality of y if x is integral. Based on that, the objective function and the constraints (1) finally ensure that the variable w captures the treewidth as derived in the previous subsection.

While our results are shown to apply also for $\mathrm{TW_{LO}}$, we will mainly work with a second PEO formulation used also in [16,27] that we refer to as $\mathrm{TW_{AD}}$. $\mathrm{TW_{AD}}$ enforces a partial ordering only for the endpoints of the edges that constitute the actual triangulation, and can be written slightly more compactly as follows:

$$
\begin{aligned}
\min\ & w & & & \\
\text{s.t.}\ & \sum_{j \in V} x_{ij} & \leq w & \quad \text{for all } i \in V & (7)\\
& x_{ij} + x_{ji} & = 1 & \quad \text{for all } \{i,j\} \in E & (8)\\
& \sum_{(i,j) \in C} x_{ij} & \leq |C| - 1 & \quad \text{for all di-cycles } C \subseteq A^* & (9)\\
& x_{ij} + x_{ik} - x_{jk} - x_{kj} & \leq 1 & \quad \text{for all } \{j,k\} \notin E, i \in V \setminus \{j,k\} & (10)\\
& x_{ij} & \geq 0 & \quad \text{for all } i,j \in V, i \neq j & (11)\\
& w & \in \mathbb{R} & & \\
& x_{ij} & \in \{0,1\} & \quad \text{for all } i,j \in V, i \neq j & (12)
\end{aligned}
$$

The formulation encodes both the partial ordering and the triangulation into the x-variables. To this end, constraints (8) now only enforce that each original edge $\{i,j\} \in E$ is oriented while inequalities (10) impose the same if $\{j,k\} \notin E$ is added to the triangulation (but otherwise $x_{jk} + x_{kj} = 0$ is possible in $\mathrm{TW_{AD}}$). Since they have the same effect as (4) in $\mathrm{TW_{LO}}$, we call the constraints (10) *simpliciality inequalities* like in [27]. The di-cycle inequalities (9) ensure that the digraph $F = (V, \bar{A})$ where $\bar{A} = \{(i,j) \in A : x_{ij} = 1\}$ is acyclic. Finally, constraints (7) determine the treewidth of G based on the triangulation $T := \{\{i,j\} \in V \times V, i \neq j : x_{ij} + x_{ji} = 1\}$.

Since both presented formulations model the treewidth of an undirected graph $G = (V, E)$ exactly, a lower bound on the treewidth of G is obtained when removing any of their constraints (compare also Sect. 5). This is true in particular for the integrality restrictions on the ordering variables whose removal gives the linear programming (LP) relaxation of the respective ILP.

3 Analysis of Perfect Elimination Ordering ILPs

3.1 Preliminary Considerations

For $\mathrm{TW_{LO}}$ and $\mathrm{TW_{AD}}$ it is known that their LP relaxations are "weak" in the sense that the lower bound obtained can be much smaller than the treewidth. In fact, the corresponding gap may even be made infinitely large, e.g. for n-by-n grids whose treewidth is n but where the lower bound obtained by the LP relaxations of $\mathrm{TW_{LO}}$ and $\mathrm{TW_{AD}}$ is at most two [19]. This follows directly from the fact that $x_{ij} = \frac{1}{2}$ ($y_{ij} = \frac{1}{2}$) for all $i, j \in V$, $i \neq j$, is always feasible for the respective LP relaxations, which also implies for general graphs $G = (V, E)$ that the lower bound obtained is always less or equal to $\max_{v \in V} \frac{\deg(v)}{2}$ where $\deg(v)$ denotes the degree of $v \in V$. Indeed, this bound is attained for regular graphs.

Another known undesirable property of PEO formulations (applying beyond ILPs) is their inherent symmetry, given e.g. by the fact that usually several linear orderings will admit an optimal triangulation in terms of the treewidth. Moreover, every chordalization imposes symmetry e.g. in terms of the classic result that each chordal (but not complete) (sub)graph has two non-adjacent simplicial vertices [14] that could be placed at the end of the ordering without loss of generality [8]. The same applies to cliques [6].

In the next subsections, we shed light on the structural reasons for the weakness of TW_{LO}, TW_{AD} and their relaxations in terms of the constraint structure.

3.2 Weakness of Di-Cycle Relaxations

We first consider the following basic relaxation of TW_{AD}, referred to as $\text{TW}_{\text{AD}}^{\text{B}}$, that results from removing (9), (10), and (12), and rearranging terms slightly.

$$
\begin{array}{llll}
\min w & & & \\
\text{s.t. } w - \sum_{j \in V} x_{ij} & \geq 0 & \text{for all } i \in V & (7) \\
x_{ij} + x_{ji} & = 1 & \text{for all } \{i,j\} \in E & (8) \\
x_{ij} & \geq 0 & \text{for all } i,j \in V, i \neq j & \\
w & \in \mathbb{R} & &
\end{array}
$$

The following first observation is immediate and will be of utility.

Observation 1. *There is always an optimum solution $x^* \in \mathbb{R}^{A^*}$ to TW_{AD}^B where $x^*_{ij} = x^*_{ji} = 0$ if $\{i,j\} \notin E$.*

We now direct our attention to a step-wise proof of the following major theorem which implies that the lower bound obtained by $\text{TW}_{\text{AD}}^{\text{B}}$ cannot be improved by the addition of the (possibly exponentially many) di-cycle inequalities (9).

Theorem 2. *There is always an optimum solution $x^* \in \mathbb{R}^{A^*}$ to TW_{AD}^B that satisfies any of the (left out) di-cycle inequalities (9).*

It appears that proving Theorem 2 is most appropriate via linear programming duality which will also provide further structural results on the way.

To this end, consider first the linear program, in the following referred to as $\text{TW}_{\text{AD}}^{\text{C}}$, that results from $\text{TW}_{\text{AD}}^{\text{B}}$ by reinterpreting (7) based on Observation 1 as

$$
w - \sum_{j:\{i,j\} \in E} x_{ij} \geq 0 \qquad \text{for all } i \in V
$$

and by appending the di-cycle constraints (9) rewritten (partly based on Observation 1 as well) as

$$
- \sum_{(i,j) \in C} x_{ij} \geq 1 - |C| \qquad \text{for all di-cycles } C \subseteq A.
$$

Let $C^1_{ij} := \{C \subseteq A \text{ di-cycle} : (i,j) \in C\}$ and $C^2_{ij} := \{C \subseteq A \text{ di-cycle} : (j,i) \in C\}$ for $\{i,j\} \in E$ and $i < j$. Then the dual (linear program) of $\mathrm{TW}^{\mathrm{C}}_{\mathrm{AD}}$ is:

$$
\begin{array}{llll}
\max & \sum_{\{i,j\} \in E} \mu_{ij} + \sum_{C \subseteq A \text{ di-cycle}} (1 - |C|)\, \lambda_C & & \\
\text{s.t.} & \sum_{i \in V} \pi_i & \leq 1 & \\
& \mu_{ij} - \pi_i - \sum_{C \in C^1_{ij}} \lambda_C & \leq 0 & \text{for all } \{i,j\} \in E,\ i < j \quad (13) \\
& \mu_{ij} - \pi_j - \sum_{C \in C^2_{ij}} \lambda_C & \leq 0 & \text{for all } \{i,j\} \in E,\ i < j \quad (14) \\
& \pi_i & \geq 0 & \text{for all } i \in V \\
& \lambda_C & \geq 0 & \text{for all di-cycles } C \subseteq A \\
& \mu_{ij} & \in \mathbb{R} & \text{for all } \{i,j\} \in E
\end{array}
$$

Theorem 3. *There is always an optimum solution to the dual of TW^C_{AD} such that $\lambda_C = 0$ for all di-cycles $C \subseteq A$.*

Proof. First, in absence of the variables λ_C (i.e., when considering the dual of $\mathrm{TW}^{\mathrm{B}}_{\mathrm{AD}}$), it is easy to see that $\mu_{ij} \leq \min\{\pi_i, \pi_j\}$ for all $\{i,j\} \in E$, i.e., μ and thus the dual objective is only bounded from above by this relation to π. Further, it is observed from μ_{ij}'s occurrence in both (13) and (14) that any *further* increase of μ_{ij} by some amount $\delta_{ij} > 0$ requires that we have both $\delta_{ij} \leq \sum_{C \in C^1_{ij}} \lambda_C =: \nu_1$ and $\delta_{ij} \leq \sum_{C \in C^2_{ij}} \lambda_C =: \nu_2$. By equally distributing the constant contributions associated with a variable λ_C over the arcs of the respective di-cycle C, we may rewrite the objective as $\sum_{\{i,j\} \in E,\, i<j} \left(\mu_{ij} + \sum_{C \in C^1_{ij}} \frac{\lambda_C}{|C|} + \sum_{C \in C^2_{ij}} \frac{\lambda_C}{|C|} \right) - \sum_{C \subseteq A \text{ di-cycle}} |C|\, \lambda_C.$

It is now apparent that increasing μ_{ij} by $\delta_{ij} \leq \min\{\nu_1, \nu_2\}$ imposes a direct "reward" of $\delta_{ij} \leq \frac{1}{2}(\nu_1 + \nu_2)$ in the objective, there is another per-edge "reward" of at most $\frac{1}{3}(\nu_1 + \nu_2)$ (as any cycle has length at least three), but also a "loss" of at least $\nu_1 + \nu_2$ as (i,j) and (j,i) contribute to the cardinalities of the respective di-cycles. Thus, a solution where $\lambda_C > 0$ for any di-cycle $C \subseteq A$ cannot be optimal.

Proof (of Theorem 2). Since, by Theorem 3, there is an optimum solution to the dual of $\mathrm{TW}^{\mathrm{C}}_{\mathrm{AD}}$ with $\lambda_C = 0$ for all di-cycles $C \subseteq A$, the dual objective values in presence and absence of the di-cycle inequalities coincide, and so do, by the strong duality theorem for linear programming, the objective values of $\mathrm{TW}^{\mathrm{B}}_{\mathrm{AD}}$ and $\mathrm{TW}^{\mathrm{C}}_{\mathrm{AD}}$. This in turn implies that there is a solution to $\mathrm{TW}^{\mathrm{B}}_{\mathrm{AD}}$ that satisfies all di-cycle inequalities in addition without increasing the objective value[1].

[1] In general, the di-cycle inequalities still pose a restriction of the feasible region of $\mathrm{TW}^{\mathrm{B}}_{\mathrm{AD}}$, i.e., an optimum solution to the latter may violate some of them, but adding them does not change the objective value. An impact of the di-cycle inequalities is still possible in the presence of (e.g.) the simpliciality constraints, compare Sect. 5.

Recall that $\mathrm{TW_{LO}}$ the latter formulation is obtained from $\mathrm{TW_{AD}}$ by imposing (8) to all vertex pairs (i.e., (2)), by replacing the cycle inequalities (9) by the three-di-cycle inequalities (3), and by introducing additional variables y_{ij} for all $i, j \in V$, $i \neq j$. Moreover, forming the two analogue relaxations $\mathrm{TW_{LO}^B}$ and $\mathrm{TW_{LO}^C}$, one immediately observes (as an analogue to Observation 1) that $y_{ij} = x_{ij}$ for all $\{i, j\} \in E$, and that there is always an optimum solution with $y_{ij} = y_{ji} = 0$ for all $\{i, j\} \notin E$. So we may restrict $\mathrm{TW_{LO}^C}$ to the x-variables as well, and then its according dual looks almost like the one of $\mathrm{TW_{AD}^C}$. The only differences are that constraints (13) and (14) are present for all $i, j \in V$, $i \neq j$ (but reduce to $\mu_{ij} \leq 0$ if $\{i, j\} \notin E$), and that for $\{i, j\} \in E$ they summarize over dual variables for three-di-cycle inequalities instead of di-cycle inequalities. Thus, the same argumentation as for Theorem 2 applies to prove the following.

Theorem 4. *There is an optimum solution $(x^*, y^*) \in \mathbb{R}^{A^*} \times \mathbb{R}^{A^*}$ to TW_{LO}^B with $y_{ij} = x_{ij}$ for all $\{i, j\} \in E$ and $y_{ij} = y_{ji} = 0$ for all $\{i, j\} \notin E$ that satisfies any of the (left out) three-di-cycle inequalities (3).*

Finally, the insights about how the dual objective function is bounded immediately give the following result for both formulations.

Corollary 1. *Let $G = (V, E)$ be an undirected graph. Then the lower bound on the treewidth of G provided by TW_{AD}^B, TW_{AD}^C, TW_{LO}^B and TW_{LO}^C is equal to* $\max\{|E(U)|\frac{1}{U} : U \subseteq V\}$.

Proof. As argued before, the objectives of the respective dual linear programs resolve to the maximization of $\sum_{\{i,j\} \in E} \mu_{ij}$. Since $\mu_{ij} \leq \min\{\pi_i, \pi_j\}$ for all $\{i, j\} \in E$ and $\sum_{i \in V} \pi_i = 1$, such a maximum is attained for a densest (or, equivalently, maximum average density) subgraph of G, i.e., for a set $U \subseteq V$ that maximizes $|E(U)|\frac{1}{U}$, see also [11,17].

The actual problem solved to find an optimum dual solution is thus a densest subgraph problem. A similar observation was made by Grigoriev [19] in a slightly different (dual) scenario for a pathwidth formulation.

3.3 The Role and Effect of Simpliciality Inequalities

We now consider the full LP relaxation of $\mathrm{TW_{AD}}$, respectively the effect of adding the so far neglected simpliciality inequalities (10) again.

These inequalities are formulated for each $\{j, k\} \notin E$ and all $i \in V \setminus \{j, k\}$. Assuming the integrality of x, the straightforward implication they impose is that one of the arcs (j, k) and (k, j) must be augmented if both the arcs (i, j) and (i, k) are present as well (binary conjunction). Consequently, the following two observations related to the impact of these inequalities are understood.

Observation 5. *The simpliciality inequalities (10) are (partial) linearizations of the quadratic constraints $x_{jk} + x_{kj} \geq x_{ij} \cdot x_{ik}$ for all $\{j, k\} \notin E$, $i \in V \setminus \{j, k\}$.*

Observation 5 refers to a partial linearization because (10) only enforces $x_{jk} + x_{kj}$ to be one if the product is, while the objective ensures that it will be zero if not enforced otherwise but of relevance for the objective value. On the other hand, in fractional terms when solving the LP relaxation, the sum $x_{jk} + x_{kj}$ is only enforced by (10) to be at least $x_{ij} + x_{ik} - 1$ if this is larger than zero. So typically, for some $\{j, k\} \notin E$ where $x_{ij} + x_{ik} \leq 1$ for all $i \in V$, and in pathological cases where even $x_{ij} = x_{ji} = \frac{1}{2}$ for all $\{i, j\} \in E$ gives an optimum solution to the relaxation, these inequalities do not have any effect at all.

Observation 6. *For all $\{j, k\} \notin E$, at least $|V| - 3$ of the $|V| - 2$ simpliciality inequalities (10) are redundant for any solution to TW_{AD} or its LP relaxation.*

Observation 6 subsumes the fact that if the aforementioned effect on $x_{jk}+x_{kj}$ is imposed at all, then it is imposed the strongest by a single $\hat{i} \in V$ such that $x_{\hat{i}j} + x_{\hat{i}k}$ is maximum among all $i \in V \setminus \{j, k\}$.

Our computational experiments in Sect. 5 nevertheless show that the simpliciality inequalities typically do have some impact on the obtained solutions and lower bounds though it is a weak one. In particular, only in their presence some of the variables x_{jk}, $\{j, k\} \notin E$, may be enforced to a non-zero value at all.

4 On Flow Metric Extensions for PEO-based ILPs

Connected with the hope for stronger LP relaxations, it is proposed in [26] and [19] to combine PEO formulations with flow metric variables and constraints as described by Bornstein and Vempala in [10].

As already observed in the two former articles, when applying the flow metric formulation in the context of treewidth, it suffices to restrict to the variables $g_k^{ij} \geq 0$, $i, j, k \in V$ supposed to represent the flow from i to j that goes through k. In addition[2], we rule out the variables g_k^{ii}, $i, k \in V$, and reformulate the constraints described in [10] only for pairwise different vertex pairs as follows:

$$g_j^{ij} + g_i^{ji} = 1 \quad \text{for all } i, j \in V, i \neq j \tag{15}$$

$$g_k^{ij} + g_k^{ji} + g_j^{ik} + g_j^{ki} + g_i^{jk} + g_i^{kj} = 1 \quad \text{for all } i, j, k \in V, i < j < k \tag{16}$$

$$d(i, j) - \sum_{k \in V} g_k^{ij} = 0 \quad \text{for all } i, j \in V, i \neq j \tag{17}$$

$$d(i, j) + d(j, k) - d(i, k) \geq 0 \quad \text{for all } i, j, k \in V, i \neq j \neq k \neq i \tag{18}$$

[2] Concerning the original formulation in [10], observe that if variables g_i^{ii} were defined, then their value would be implied to be $\frac{1}{2}$ by the original pendant of (15). Similarly, the pendant of (16) would render the problem infeasible if imposed for $i = j = k$. Moreover, variables g_k^{ii}, $i,k \in V$, $k \neq i$, are not sensible with regard to their informal definition, and cause another issue with (16). Namely, if (16) were stated for $i = j \neq k$ (or any symmetric selection), then an equation of the form $2g_k^{ii} + 2g_i^{ik} + 2g_i^{ki} = 1$ would result, forcing any of these variables to be at most $\frac{1}{2}$ which is clearly not intended. We conclude that the constraints are only plausible in the way displayed which is also suggested by the proof of Theorem 2.2 in [10] where the final sum would not comply with the binomial coefficients if it was not meant for $i < j < k$.

Equations (15) impose that for each pair of different vertices $i, j \in V$, the value of the flow from i to j plus the flow from j to i is one, and Eq. (16) enforce for each vertex triple that the flow from i to j that goes through k, the flow from i to k that goes through j, and the flow from j to k that goes through i sum up to one. While (17) are only auxiliary constraints to define the flow-based distances between each pair of different vertices, constraints (18) finally impose the triangle inequality on these. As Bornstein and Vempala point out in [10], flow-based distances satisfying the above constraints also satisfy the so-called spreading constraints $\sum_{i,j \in S} d(i,j) \geq \binom{|S|}{3}$ for all $S \subseteq V$, $|S| \geq 3$.

Applying the flow metric extensions to the PEO-based formulations for treewidth, the variables g_j^{ij} shall be identified with x_{ij} for $i, j \in V$, $i \neq j$, as described in [19,26] while the variables g_i^{ij} could as well be eliminated as described below.

The following observation exposes an essential difference compared to e.g. the linear ordering formulations for the pathwidth where the variables $g_v^{i,j}$ for $v \in V \backslash \{i, j\}$ all have a non-zero coefficient in the relaxation to be solved (cf. [10]).

Observation 7. *Among the additional variables $g_k^{ij} \geq 0$, $i, j, k \in V$, $k \neq i \neq j$, the variables g_j^{ij} for $i, j \in V$, $i \neq j$, are the only ones taking part in the definition of the objective function of TW_{LO} via the constraints (1).*

In other words, only if there exists a scenario where the constraints (15)–(18) must have an impact on at least one variable g_j^{ij} (x_{ij}), $i, j \in V$, $i \neq j$, a strengthening of the lower bounds obtained with the LP relaxation of $\mathrm{TW}_{\mathrm{LO}}$ is possible. Unfortunately, such a scenario does not exist as the following result states.

Proposition 1. *When extending the LP relaxation of TW_{LO} by the constraints (15)–(18) linked to its variables by the identity $g_j^{ij} = x_{ij}$ for $i, j \in V$, $i \neq j$, the lower bound obtained remains the same as with the LP relaxation of TW_{LO}.*

Proof. Under the identification of x_{ij}- and g_j^{ij}-variables, clearly constraints (15) coincide with (2). Moreover, the variables occurring in constraints (16) are entirely disjoint from those in (15). It follows that an impact on any g_j^{ij}, $i, j \in V$, $i \neq j$, could only be established indirectly via the auxiliary sum (17) and constraints (18). But in (17), g_j^{ij} is accumulated only with the variables g_k^{ij} for $k \in V \setminus \{j\}$. This leaves freedom to set (e.g.) g_i^{ij} to zero, and $g_k^{ij} = \frac{1}{6}$ for all $k \in V \setminus \{i, j\}$ which will satisfy all equations (16) and establish that $d(i,j) = g_j^{ij} + (|V| - 2)\frac{1}{6}$ for all $i, j \in V$, $i \neq j$. In particular, $d(i,j)$ and g_j^{ij} (x_{ij}) then coincide up to a constant which is equal for all vertex pairs. Finally, considering a three-di-cycle inequality with an index shift to a cycle $j \to i \to k \to j$, it can be rewritten by exploiting (2):

$$\begin{aligned} & x_{ji} + x_{kj} + x_{ik} && \leq 2 && \text{for all } i, j, k \in V, i \neq j \neq k \\ \Leftrightarrow\; & -x_{ji} - x_{kj} - x_{ik} && \geq -2 && \text{for all } i, j, k \in V, i \neq j \neq k \\ \Leftrightarrow\; & x_{ij} + x_{jk} - x_{ik} && \geq 0 && \text{for all } i, j, k \in V, i \neq j \neq k \end{aligned}$$

Thus, if x satisfies the three-di-cycle inequalities, there is always a corresponding solution to the extended formulation that satisfies the triangle inequalities (18) as well and that has the same objective (lower bound on the treewidth).

Table 1. Lower bounds obtained with different PEO-based relaxations for some selected graph instances, while $LB_S^{LP} \leq LB_R^{LP} \leq LB_B^{IP}$, $LB_B^{IP} \leq LB_C^{IP}$, and $LB_B^{IP} \leq LB_S^{IP}$.

Instance	|V|	|E|	tw	$LB_{B,C}^{LP}$	LB_S^{LP}	LB_R^{LP}	LB_B^{IP}	LB_C^{IP}	LB_S^{IP}
barley	48	126	7	3.00	3.20	3.20	3	5	5
huck	74	301	10	5.46	5.53	5.53	6	10	6
jean	80	254	9	5.39	5.57	5.57*	6	9	7
mainuk	48	198	7	4.46	4.67	4.67	5	7	6
mildew	35	80	4	2.37	2.44	2.44	3	3	3
myciel2	5	5	2	1.00	1.00	1.00	1	2	1
myciel3	11	20	5	1.81	2.08	2.08	2	3	3
myciel4	23	71	10	3.08	3.83	3.83	4	5	6
water	32	123	9	4.25	4.68	4.68*	5	6	6
bcspwr01	39	46	3	1.25	1.36	1.36	2	2	2
bcspwr02	49	59	3	1.35	1.54	1.54	2	2	2
bcsstk02	66	2145	65	32.50	32.50	32.50	33	65	33
can___24	24	68	5	2.83	2.93	2.93	3	4	4
can___62	62	78	3	1.40	1.59	1.59	2	2	3
curtis54	54	124	5	2.44	2.63	2.63	3	4	4
dwt___66	66	127	2	1.95	1.95	1.95	2	2	2
dwt___72	72	75	2	1.06	1.14	1.14	2	2	2
dwt___87	87	227	7	2.96	3.48	3.48*	3	4	5
steam3	80	424	7	5.30	5.30	5.30	6	7	6
will57	57	127	4	2.62	2.70	2.70	3	4	3
grid_5	25	40	5	1.60	1.86	1.86	2	2	3
grid_6	36	60	6	1.67	1.91	1.91	2	2	3
grid_7	49	84	7	1.71	1.95	1.95	2	2	4
p40_18_32	18	32	4	1.81	2.09	2.09	2	2	3
p50_19_25	19	25	3	1.44	1.55	1.55	2	2	2
p60_20_22	20	22	2	1.18	1.28	1.28	2	2	2
p70_21_25	21	25	3	1.36	1.60	1.60	2	2	2
p80_22_30	22	30	3	1.46	1.66	1.66	2	2	2
p90_23_35	23	35	4	1.62	1.97	1.97	2	2	3
p100_24_34	24	34	3	1.53	1.85	1.85	2	2	3

5 Computational Experiments

To provide an impression on the relation between the lower bounds obtained with different relaxations of PEO-based ILP formulations, we compiled a testbed of

30 well accessible graphs that have been used in previous experimental studies for various width parameters (see e.g. [5,15,22] for the respective repositories).

We define and compute lower bounds based on relaxations of $\mathrm{TW_{AD}}$ which all involve the constraints (7), (8), and (11). While these completely define $\mathrm{LB}^{\mathrm{LP}}_{B,C}$, (10) is considered in addition for $\mathrm{LB}^{\mathrm{LP}}_{S}$, and both (9) and (10) are added for $\mathrm{LB}^{\mathrm{LP}}_{R}$ (LP relaxation). Similarly, only the integrality restrictions (12) are employed in addition for $\mathrm{LB}^{\mathrm{IP}}_{B}$, (9) and (12) for $\mathrm{LB}^{\mathrm{IP}}_{C}$, and finally (10) and (12) for $\mathrm{LB}^{\mathrm{IP}}_{S}$.

The results are displayed in Table 1 along with the treewidth of the graphs (column tw). Generally, they confirm that the lower bounds obtained with the LP relaxations of $\mathrm{TW_{AD}}$ and $\mathrm{TW_{LO}}$, here represented by $\mathrm{LB}^{\mathrm{LP}}_{R}$, are not satisfactory. Even the simplest (and typically quickly solved) ILP-relaxation provides a lower bound $\mathrm{LB}^{\mathrm{IP}}_{B}$ that is frequently (but not always) a bit stronger (even though usually inconvenient as well). It is also apparent that the simpliciality inequalities may lead to a slight but often negligible improvement of the lower bound ($\mathrm{LB}^{\mathrm{LP}}_{S}$) compared to $\mathrm{LB}^{\mathrm{LP}}_{B,C}$. In a few cases (marked with an asterisk), $\mathrm{LB}^{\mathrm{LP}}_{R}$ improves over $\mathrm{LB}^{\mathrm{LP}}_{S}$ in insignificant digits due to the additional di-cycle inequalities. Finally, when looking at the ILP relaxations, it turns out that the impact of the di-cycle inequalities ($\mathrm{LB}^{\mathrm{IP}}_{C}$) is frequently more significant than the one of the simpliciality inequalities ($\mathrm{LB}^{\mathrm{IP}}_{S}$). However, these relaxations are expensive to solve and often still too weak to successively close the gap to the treewidth.

6 Conclusion

We have provided first structural evidence and explanations for the weakness of LP relaxations of PEO formulations for computing the treewidth. In particular, we showed that there is always an optimum solution to a basic relaxation that satisfies all di-cycle inequalities, i.e., the addition of the latter does not have an impact on the obtained lower bounds, and this impact was shown to be also insignificant when simpliciality inequalities are taken into account as well. The objective value of the basic relaxation is determined by a densest subgraph. Solving it with integrality restrictions, the obtained lower bounds are frequently (but not always) slightly better than with a complete LP relaxation. Adding di-cycle inequalities or simpliciality inequalities in this context has a more significant but still unsatisfactory effect especially since these relaxations are expensive to solve. We also revealed that combining the LP relaxation of existing PEO formulations and the LP formulation for flow metrics does not lead to improved lower bounds.

References

1. Arnborg, S., Corneil, D.G., Proskurowski, A.: Complexity of finding embeddings in a k-tree. SIAM J. Algebraic Discrete Meth. **8**(2), 277–284 (1987)
2. Bannach, M., Berndt, S., Ehlers, T.: Jdrasil: A modular library for computing tree decompositions. In: Iliopoulos, C.S., Pissis, S.P., Puglisi, S.J., Raman, R. (eds.) 16th International Symposium on Experimental Algorithms (SEA 2017), vol. 75 of Leibniz Intern. Proceedings in Informatics (LIPIcs), pp. 28:1–28:21, Dagstuhl, Germany (2017). Schloss Dagstuhl-Leibniz-Zentrum fuer Informatik

3. Berg, J., Järvisalo, M.: SAT-based approaches to treewidth computation: an evaluation. In: 2014 IEEE 26th International Conference on Tools with Artificial Intelligence, pp. 328–335. IEEE Computer Society (2014)
4. Bodlaender, H.L.: Dynamic programming on graphs with bounded treewidth. In: Lepistö, T., Salomaa, A. (eds.) ICALP 1988. LNCS, vol. 317, pp. 105–118. Springer, Heidelberg (1988). https://doi.org/10.1007/3-540-19488-6_110
5. Bodlaender, H.L., Fomin, F.V., Koster, A.M.C.A., Kratsch, D., Thilikos, D.M.: On exact algorithms for treewidth. In: Azar, Y., Erlebach, T. (eds.) ESA 2006. LNCS, vol. 4168, pp. 672–683. Springer, Heidelberg (2006). https://doi.org/10.1007/11841036_60
6. Bodlaender, H.L., Fomin, F.V., Koster, A.M.C.A., Kratsch, D., Thilikos, D.M.: A note on exact algorithms for vertex ordering problems on graphs. Theory Comput. Syst. **50**(3), 420–432 (2012)
7. Bodlaender, H.L., Kloks, T.: Efficient and constructive algorithms for the pathwidth and treewidth of graphs. J. Algorithms **21**(2), 358–402 (1996)
8. Bodlaender, H.L., Koster, A.M., Eijkhof, F.V.D.: Pre-processing rules for triangulation of probabilistic networks. Comput. Intell. **21**(3), 286–305 (2005)
9. Bodlaender, H.L., Koster, A.M.C.A.: Combinatorial optimization on graphs of bounded treewidth. Comput. J. **51**(3), 255 (2007)
10. Bornstein, C.F., Vempala, S.: Flow metrics. In: Rajsbaum, S. (ed.) LATIN 2002. LNCS, vol. 2286, pp. 516–527. Springer, Heidelberg (2002). https://doi.org/10.1007/3-540-45995-2_45
11. Charikar, M.: Greedy approximation algorithms for finding dense components in a graph. In: Jansen, K., Khuller, S. (eds.) APPROX 2000. LNCS, vol. 1913, pp. 84–95. Springer, Heidelberg (2000). https://doi.org/10.1007/3-540-44436-X_10
12. Dell, H., Husfeldt, T., Jansen, B.M., Kaski, P., Komusiewicz, C., Rosamond, F.A.: The first parameterized algorithms and computational experiments challenge. In: Guo, J., Hermelin, D. (eds.) 11th International Symposium on Parameterized and Exact Computation (IPEC 2016), vol. 63 of Leibniz Intern. Proceedings in Informatics (LIPIcs), pp. 30:1–30:9, Dagstuhl, Germany (2017). Schloss Dagstuhl-Leibniz-Zentrum fuer Informatik
13. Dell, H., Komusiewicz, C., Talmon, N., Weller, M.: The PACE 2017 parameterized algorithms and computational experiments challenge: the second iteration. In: Lokshtanov, D., Nishimura, N. (eds.) 12th international symposium on Parameterized and Exact Computation (IPEC 2017), vol. 89 of Leibniz Intern. Proceedings in Informatics (LIPIcs), pp. 30:1–30:12, Dagstuhl, Germany (2018). Schloss Dagstuhl-Leibniz-Zentrum fuer Informatik
14. Dirac, G.A.: On rigid circuit graphs. Abh. Math. Semin. Univ. Hambg. **25**(1), 71–76 (1961)
15. Duarte, A., Escudero, L.F., Martí, R., Mladenovic, N., Pantrigo, J.J., Sánchez-Oro, J.: Variable neighborhood search for the vertex separation problem. Comput. Oper. Res. **39**(12), 3247–3255 (2012)
16. Feremans, C., Oswald, M., Reinelt, G.: A Y-Formulation for the Treewidth. Heidelberg University, Germany, Preprint (2002)
17. Goldberg, A.V.: Finding a maximum density subgraph. Technical report, Berkeley, USA (1984)
18. Golumbic, M.C.: Algorithmic Graph Theory and Perfect Graphs. Academic Press, New York (1980)

19. Grigoriev, A.: Possible and impossible attempts to solve the treewidth problem via ILPs. In: Fomin, F.V., Kratsch, S., van Leeuwen, E.J. (eds.) Treewidth, Kernels, and Algorithms. LNCS, vol. 12160, pp. 78–88. Springer, Cham (2020). https://doi.org/10.1007/978-3-030-42071-0_7
20. Grigoriev, A., Ensinck, H., Usotskaya, N.: Integer linear programming formulations for treewidth. Research Memorandum 030, Maastricht University, Maastricht Research School of Economics of Technology and Organization (METEOR) (2011)
21. Lengauer, T.: Black-white pebbles and graph separation. Acta Inform. **16**(4), 465–475 (1981)
22. Martí, R., Campos, V., Piñana, E.: A branch and bound algorithm for the matrix bandwidth minimization. Europ. J. Oper. Res. **186**(2), 513–528 (2008)
23. Martí, R., Reinelt, G.: The Linear Ordering Problem. Springer, Boston (2011)
24. Robertson, N., Seymour, P.D.: Graph minors. II. algorithmic aspects of tree-width. J. Algorithms **7**(3), 309–322 (1986)
25. Tamaki, H.: Heuristic computation of exact treewidth. In: Schulz, C., Uçar, B., (eds.) 20th International Symposium on Experimental Algorithms (SEA 2022), vol. 233 of Leibniz International Proceedings in Informatics (LIPIcs), pp. 17:1–17:16, Dagstuhl, Germany (2022). Schloss Dagstuhl - Leibniz-Zentrum für Informatik
26. Usotskaya, N.: Exploiting geometric properties in combinatorial optimization = Benutten van geomwetrische eigenschappen in combinatiorische optimalisatie. PhD thesis, Maastricht University, Netherlands (2011)
27. Yüceoglu, B.: Branch-and-cut algorithms for graph problems. PhD thesis, Maastricht University (2015)

Finding Perfect Matching Cuts Faster

Neeldhara Misra(✉) and Yash More

Indian Institute of Technology, Gandhinagar, Gandhinagar, India
{neeldhara.m,yash.mh}@iitgn.ac.in

Abstract. A cut (X, Y) is a perfect matching cut if and only if each vertex in X has exactly one neighbor in Y and each vertex in Y has exactly one neighbor in X. The computational problem of determining if a graph admits a perfect matching cut is NP-complete, even when restricted to the class of bipartite graphs of maximum degree 3 and arbitrarily large girth. Assuming ETH, the problem does not admit an algorithm subexponential in n. On the other hand, the problem is known to be polynomial-time solvable when restricted to interval graphs, permutation graphs, and some other special classes of graphs. It is also known to be FPT parameterized by treewidth or cliquewidth and in XP when parameterized by mim-width (maximum induced matching-width) of a given decomposition of the graph.

The ETH-hardness of the problem has motivated the study of exact algorithms for PMC, and the best-known running time has complexity (We use the $\mathcal{O}^*$ notation which suppresses polynomial factors.) $\mathcal{O}^*(1.2721^n)$ [Le and Telle, WG 2021]. In this contribution, we design a mildly improved branching algorithm using an arguably simpler approach, whose running time is $\mathcal{O}^*(1.2599^n)$. This addresses an open problem posed by Le and Telle. We also demonstrate an $\mathcal{O}^*(1.1938^n)$ algorithm for graphs of maximum degree 3 that have girth at least six.

Keywords: Perfect Matching Cut · Branch and Bound

1 Introduction

The Perfect Matching Cut problem was introduced by Heggernes and Telle (1998) in the context of their study of generalized domination problems. Generalized dominating sets are parameterized by two sets of nonnegative integers σ and ρ as follows. A set S of vertices of a graph is said to be a (σ, ρ)-set if $\forall v \in S, |N(v) \cap S| \in \sigma$ and $V_u \notin S, | N(v) \cap S | \in \rho$. The (k, σ, ρ)-partition problem asks for the existence of a partition $V_1, V_2, \ldots, V_h$ of vertices of a given graph G such that $V_i, i = 1, 2, \ldots, k$ is a (σ, ρ)-set of G. One of the special cases of this problem that the authors demonstrated to be NP-complete was when $k = 2$, $\sigma = \mathbb{N}$ and $\rho = 1$. Note that this asks if the graph can be partitioned into two parts, say X and Y, so that every vertex in X has exactly one neighbor in Y and every vertex in Y has exactly one neighbor in X. This is indeed the definition of a perfect matching cut.

Supported by IIT Gandhinagar and the SERB ECR grant MTR/2017/001033.

S.-Y. Hsieh et al. (Eds.): IWOCA 2023, LNCS 13889, pp. 307–318, 2023.
https://doi.org/10.1007/978-3-031-34347-6_26

This formulation of the perfect matching cut makes it a specific instance of what is called a (σ, ρ) 2-partitioning problem (Telle and Proskurowski, 1997), where the problem is also phrased the following equivalent labelling task: label the vertices with two labels such that each vertex has exactly one neighbor labelled differently from itself. Recent developments by Édouard Bonnet et al. (2023) show that the problem is NP-complete even on 3-connected cubic bipartite planar graphs.

We note that the PERFECT MATCHING CUT problem can be thought of as a more demanding variation of the MATCHING CUT problem (Chvátal, 1984), which asks if the graph can be partitioned into two parts, say X and Y as before, so that every vertex in X has *at most* one neighbor in Y and every vertex in Y has *at most* one neighbor in X. This way the edges in the cut induced by (X, Y) form a matching, but not necessarily one that is perfect. Neither a matching cut nor a perfect matching cut is guaranteed to exist. The computational question is to determine if they do. One could also ask for a matching cut that cuts at least k edges, with the perfect matching cut question being the special case when $k = n/2$. For a representative sample of recent developments on the algorithmic aspects of the MATCHING CUT problem, we refer the reader to the works of Komusiewicz et al. (2018); Golovach et al. (2021); Chen et al. (2021).

From results of Bui-Xuan et al. (2013) and Telle and Proskurowski (1997), it turns out that the PERFECT MATCHING CUT problem is FPT in the treewidth or cliquewidth of the input graph and XP in the maximum induced matching-width (*mim-width*) of a given decomposition of the graph. In a more recent development, Le and Telle (2022) show that the problem can be solved in polynomial time on two other graph classes as well: the first one includes claw-free graphs and graphs without an induced path on five vertices, while the second one properly contains all chordal graphs. Finally, Feghali et al. (2023) obtain a complete complexity classification of PERFECT MATCHING CUT for $\mathcal{H}$-subgraph-free graphs where $\mathcal{H}$ is any finite set of graphs.

In contrast with MATCHING CUT, which is known to be polynomial-time solvable when restricted to graphs of maximum degree 3, Le and Telle (2022) show that PERFECT MATCHING CUT is NP-complete in the class of bipartite graphs of maximum degree 3 and arbitrarily large girth. They also show that the problem cannot be solved in $\mathcal{O}^*\left(2^{o(n)}\right)$ time for n-vertex bipartite graphs and cannot be solved in $\mathcal{O}^*\left(2^{o(\sqrt{n})}\right)$ time for bipartite graphs with maximum degree 3 and arbitrarily large girth, assuming the Exponential Time Hypothesis. Finally, they demonstrate the first exact algorithm to solve PERFECT MATCHING CUT on n-vertex graphs, with a runtime of $\mathcal{O}^*(1.2721^n)$ using a branch-and-bound technique. In this work, we continue this line work and propose two algorithms with faster running times for PERFECT MATCHING CUT — one of them works on general graphs while the other takes advantage of the structure of graphs with maximum degree 3 and girth at least six.

Specifically, we show that PERFECT MATCHING CUT can be solved in $\mathcal{O}^*(1.2599^n)$ time on n-vertex graphs and in $\mathcal{O}^*(1.1938^n)$ time on n-vertex graphs with maximum degree 3 and girth at least six. Our overall approach in both cases follows the framework introduced by Le and Telle (2022). Given a graph G, we will maintain and develop a partial matching cut (A, B), where A and B are disjoint subsets of $V(G)$ with the property that every vertex in A has at most one neighbor in B and every vertex in B has at most one neighbor in A. We say that a perfect matching cut (X, Y) *extends* a partial matching cut (A, B) if $A \subseteq X$ and $B \subseteq Y$.

We make progress by committing more and more vertices in $V(G) \setminus (A \cup B)$ to either A or B in a manner that is either forced[1] or by exhaustive exploration[2]. We do this until we either have evidence that there is no perfect matching cut that extends the current partial matching cut[3] or until we fully develop (A, B) into a partition of $V(G)$. We report YES if any of the fully developed partitions induce a perfect matching cut, and NO otherwise. The correctness of this approach follows from the exhaustiveness of the choices we make for branching.

Both algorithms have the following overall structure. Assuming that the input is a YES-instance, guess an edge (uv) in the perfect matching cut. This gives us an initial partial matching cut $(A = \{u\}, B = \{v\})$ at a cost of $\mathcal{O}(m)$ overhead in the running time. We preprocess instances using some polynomial-time reduction rules to ensure that the partial matching cut satisfies a few key invariants. If none of the reduction rules apply to an instance, it is said to be *reduced*. Given a reduced instance that is not fully resolved (i.e: $V \setminus (A \cup B)$ is non-empty), we solve the instance according to a collection of specific branching rules.

As is standard for branch-and-bound algorithms (Fomin and Kratsch, 2010), we have a measure that we track along the branching process, and the running time is determined by how the measure drops in each branch. For both our algorithms, our measure will be the number of vertices that are not in the partial cut: so when we extend the partial matching cut by committing some vertices from outside $(A \cup B)$ to one of A or B, our measure decreases by as much as the number of vertices that were added to the partial matching cut.

While the improvement we achieve in the running time for general graphs compared to the state of the art is incremental, we propose a conceptually simpler branching routine. For the case of graphs with maximum degree 3 and girth at least six, we strengthen the base case and maintain a stronger invariant (namely that there are no isolated vertices in $G[A \cup B]$) to achieve a more substantial improvement in the running time.

---

[1] For example, if a vertex $v \in A$ has a neighbor in B, then all of its other neighbors must belong to A in any perfect matching cut that extends (A, B).

[2] For instance, we may have a vertex $v \notin A \cup B$, but with two neighbors $a \in A$ and $b \in B$: notice that any perfect matching cut that extends (A, B) either has the edge (v, a) or the edge (v, b). Thus, we may "branch" on the partial cuts $(A \cup \{v\}, B)$ and $(A, B \cup \{v\})$ to explore these two possibilities.

[3] A typical scenario of this kind would be when there is a vertex $v \notin A \cup B$ that has more than one neighbor in both A and B.

2 Preliminaries and Notation

We follow standard graph-theoretic terminology unless mentioned otherwise. Throughout, $G = (V, E)$ will denote a simple undirected graph, and we use $V(G)$ and $E(G)$ to denote the vertex and edge sets of G, respectively.

A *cut* is a partition $V(G) = X \cup Y$ of the vertex set into disjoint, non-empty sets X and Y. The set of all edges in G having an endvertex in X and the other endvertex in Y, written $E(X, Y)$, is called the edge cut of the cut (X, Y). A *matching cut* is an edge cut that is a (possibly empty) matching, while a *perfect matching cut* is an edge cut that is a perfect matching. Equivalently, we have the following definitions:

- a cut (X, Y) is a matching cut if and only if each vertex in X has **at most** one neighbor in Y and each vertex in Y has **at most** one neighbor in X; and
- is a perfect matching cut if and only if each vertex in X has **exactly** one neighbor in Y and each vertex in Y has **exactly** one neighbor in X.

Further, we will typically use the notation in our discussions.

- (A, B) denotes a *partial matching cut*, where $A, B \subseteq V(G)$, and while A and B are disjoint, $A \cup B$ need not be $V(G)$, and each vertex in A has at most one neighbor in B and each vertex in B has at most one neighbor in A. Note that $G[A \cup B]$ need not be a perfect matching cut.
- Given a partial matching cut (A, B), we use $F_{(A,B)}$ to denote $V \setminus (A \cup B)$. When the context is clear, we drop the subscript and just use F. Further, we use $\partial(F)$ to denote the subset of vertices in F that have at least one neighbor in $A \cup B$, i.e.:
$$\partial(F) := \{v \in F \mid N(v) \cap (A \cup B) \neq \emptyset\}.$$
- For a partial matching cut (A, B), we define:
$$A^\circ := \{v \in A \mid N(v) \cap B = \emptyset\} \text{ and } B^\circ := \{v \in B \mid N(v) \cap A = \emptyset\}$$
and
$$A^\star := A \setminus A^\circ \text{ and } B^\star := B \setminus B^\circ.$$
The vertices of A° and B° are called *unsaturated* in (A, B) while vertices of $A^\star$ and $B^\star$ are said to be *saturated* in (A, B).
- We refer to the vertices in F as *undetermined.*
- We call the vertices in $A \cup B$ *committed.*

Given a partial matching cut (A, B), a perfect matching cut (X, Y) is said to *extend* (A, B) if $A \subseteq X$ and $B \subseteq Y$.

We define an instance of the PMC-Extension problem as follows: given a graph G with a partial matching cut (A, B), determine if G admit a perfect matching cut that extends (A, B). Note that this generalizes the Perfect Matching Cut problem, which is the special case when $A = B = \emptyset$.

3 Polynomial-Time Reduction Rules

Let $(G = (V, E), A, B)$ be an instance of the PMC-Extension problem where (A, B) is a partial matching cut of G. Given $P, Q \subseteq F_{(A,B)}$ where P and Q are disjoint, we use the notation:

$$G_{AB}[P \rightsquigarrow A; Q \rightsquigarrow B] := (G, A \cup P, B \cup Q).$$

We slightly abuse this notation in a few ways: if (A, B) is clear from the context we drop the subscript, if $P = \emptyset$, we use the shorter notation $G[Q \rightsquigarrow B]$, if $P = v$ and $Q = \emptyset$, we will use $G[v \rightsquigarrow A]$, and so on. We now describe some polynomial-time reduction rules (we borrow and slightly adapt these from Le and Telle (2022)). These will be used to preprocess instances of PMC-Extension.

Reduction Rule 0. [Maintain Valid Cuts] If there is a vertex $u \in A$ (resp, B) that has more than one neighbor in B (repsectively, A), then say No.

Reduction Rule 1. [Neigbors of Saturated Vertices are Determined]
If $u \in A^\star$ (resp, $B^\star$), $v \in F$, and (uv) is an edge; then return $G[v \rightsquigarrow A]$ (resp, $G[v \rightsquigarrow B]$).

Reduction Rule 2. [Last Chance for an Unsaturated Vertex]
If $u \in A^\circ$ (resp, B°) and $N(u) \cap F = v$, then return $G[v \rightsquigarrow B]$ (resp, $G[v \rightsquigarrow A]$).

Reduction Rule 3. [No Chance for an Unsaturated Vertex]
If $u \in A^\circ \cup B^\circ$ and $N(u) \cap F = \emptyset$, then say No.

Reduction Rule 4. [Too Many Neighbors in $A \cup B$]

- If $u \in F$ with $|N(u) \cap A| \geqslant 2$ and $|N(u) \cap B| \geqslant 2$ then say No[4].
- If $u \in F$ and $|N(u) \cap A| \geqslant 2$ (resp, $|N(u) \cap B| \geqslant 2$) then return $G[u \rightsquigarrow A]$ (resp, $G[u \rightsquigarrow B]$).

Reduction Rule 5. [Degree One in F]
If $u \in \partial(F)$, $d(u) = 2$, u has one neighbor in A (resp, B), u has one neighbor v in F, and v has a neighbor in $A \cup B$; then return $G[v \rightsquigarrow B]$ (resp, $G[v \rightsquigarrow A]$).

Reduction Rule 6. [Degree One in F Extended]
If $u \in \partial(F)$, $d(u) = 2$, u has one neighbor in A (resp, B), u has one neighbor v in F, and $d(v) = 1$; then return $G[u \rightsquigarrow A, v \rightsquigarrow B]$ (resp, $G[v \rightsquigarrow A, u \rightsquigarrow B]$).

Reduction Rule 7. [Degree Zero in F]
If $u \in \partial(F)$ such that $d(u) \in \{1, 3\}$ and $d_F(u) = 0$, and if u has exactly one neighbor in A (resp, B), return $G[u \rightsquigarrow B]$ (resp, $G[u \rightsquigarrow A]$).

[4] Note that this scenario does not arise if the input graph has maximum degree at most three.

4 Maximum Degree Three and Large Girth

Let $(G = (V, E), A, B)$ be an instance of PMC-EXTENSION, where G is a graph with maximum degree three and girth at least six[5]. Throughout our discussion, F is used to denote $F_{(A,B)}$.

We will describe a recursive branching algorithm denoted by $\mathcal{A}(\cdot)$ for the problem. We use $\mathcal{P}(\cdot)$ to denote the algorithm that returns an instance after the exhaustive application of reduction rules 1—7.

We say that an instance (G, A, B) is *reduced* if $\mathcal{P}((G, A, B)) = (G, A, B)$, i.e., none of the reduction rules are applicable on the instance. An instance (G, A, B) where $A = B = \emptyset$ or $G[A \cup B]$ has isolated vertices is called *degenerate*. Before we begin, we specify some useful invariants.

Key Invariants

1. $A \cup B$ is non-empty.
2. $G[A \cup B]$ has no isolated vertices.
3. Every vertex in F has at most one neighbor in A and at most one neighbor in B.
4. If (uv) is an edge and $u \in \partial(F)$ and $v \in A \cup B$, then $v \in A^\circ \cup B^\circ$
5. Every $u \in A^\circ \cup B^\circ$ has at least two neighbors in F.

We say that an instance *clean* if it satisfies all the invariants above. Notice that the reduction rules preserve the invariants, which is to say that if the input satisfies any subset of the invariants above, then so does the output. Further, notice that a reduced instance is either clean or degenerate. Our branching preserves invariants: if the input is a clean instance, then so is the output.

Note an instance G of PERFECT MATCHING CUT can be solved by guessing one edge in the final solution and applying the algorithm for PMC-EXTENSION on the guess. In particular, we return:

$$\bigvee_{e=(u,v)\in E(G)} \mathcal{A}\Big(\mathcal{P}\big((G, A = \{u\}, B = \{v\})\big)\Big)$$

to determine if G has a perfect matching cut.

We now describe our branching algorithm. We maintain that the input to the algorithm is clean and reduced with respect to Reduction Rules 1—7. Note that this holds in the first calls above. The branching rules are executed in the order in which they are described. The measure is $|F|$.

1. Suppose there is a degree-2 vertex u in $\partial(F)$ with one neighbor (say a) in A and the other neighbor (say v) in F. Since the instance is reduced, we know that v has no neighbors in $(A \cup B)$, and at least one neighbor other than u in F. We now consider two cases based on the degree of v.

[5] This implies that G has no triangles or cycles of length four or five.

(a) $d(v) = 2$.
By invariants (2) and (5), we have that a has exactly one neighbor other than u in F. Further, because of the case we are in, v has exactly one neighbor other than u in F. Let p be a's second neighbor in F, and let q be v's second neighbor in F. Note that $p \neq q$ since G has no cycles of length four. We now branch on whether u is matched to a or v:

$$\mathcal{A}\Big(\mathcal{P}(G[u \rightsquigarrow A, \{p, q, v\} \rightsquigarrow B])\Big) \bigvee \mathcal{A}\Big(\mathcal{P}(G[\{p, q\} \rightsquigarrow A, \{u, v\} \rightsquigarrow B])\Big).$$

Note that p, q, u and v are distinct. Therefore, this is a $(4, 4)$ branch.

(b) $d(v) = 3$. As in the previous case, by invariants (2) and (5), we have that a has exactly one neighbor other than u in F. Further, because of the case we are in, v has exactly two neighbors other than u in F. Let p be a's second neighbor in F, and let q and r be v's second and third neighbors in F. Note that $p \neq q$ and $p \neq r$ since G has no cycles of length five. We again branch on whether u is matched to a or v:

$$\mathcal{A}\Big(\mathcal{P}(G[u \rightsquigarrow A, \{p, q, r, v\} \rightsquigarrow B])\Big) \bigvee \mathcal{A}\Big(\mathcal{P}(G[\{p\} \rightsquigarrow A, \{u, v\} \rightsquigarrow B])\Big).$$

In the previous case, when u was matched to a, v was matched to q by lack of choice. This is no longer the situation here, since v has two neighbors in F. However, in the branch where u is matched to v, we now have an additional vertex forced into $A \cup B$.

Recall that p, q, r, u and v are distinct. Therefore, this is a $(5, 3)$ branch.

The case when there is a degree 2 vertex u in $\partial(F)$ with one neighbor (say b) in B and the other neighbor (say v) in F is symmetric.

2. Suppose u in $\partial(F)$ is degree three and has one neighbor a in A, one in B (say b), and one in F (say v). Note that a and b have at least one neighbor other than v in F, by invariant (5). Let a' be a's second neighbor in F and let b' be b's second neighbor in F (c.f. Fig. 1). We branch on whether u is matched to a or b:

$$\mathcal{A}\Big(\mathcal{P}(G[\{v, u\} \rightsquigarrow A, \{a', b'\} \rightsquigarrow B])\Big) \bigvee \mathcal{A}\Big(\mathcal{P}(G[\{a', b'\} \rightsquigarrow A, \{v, u\} \rightsquigarrow B])\Big).$$

Note that we have $a' \neq v$, $b' \neq v$ since G has no triangles and $a' \neq b'$ since G has no cycles of length four.
Therefore, this is a $(4, 4)$ branch.

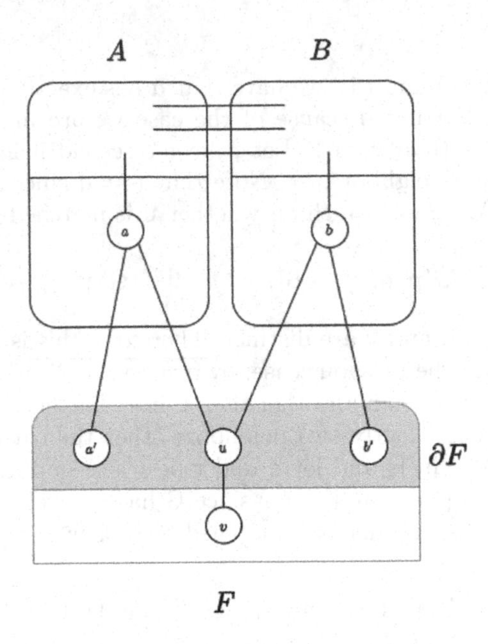

Fig. 1. A figure describing the setting of Branching Rule 2.

3. Suppose there is a degree-3 vertex v in $\partial(F)$ such that v has one neighbor a in A and two neighbors p, q in F. Note that a has exactly one neighbor other than v in F, by invariants (2) and (5), say u. Note that $u \in \partial(F)$ and:
   - If $d(u) = 1$ then the instance is not reduced, a contradiction.
   - If $d(u) = 2$ and $d_F(u) = 1$ then we would have branched on u.
   - If $d(u) = 3$ and $d_F(u) = 1$ we would have branched[6] on u.
   - If $d(u) = 3$ and $d_F(u) = 3$ then $d(u) \geqslant 4$, a contradiction.

   Therefore, we have that $d(u) = 3$ and $d_F(u) = 2$ or $d(u) = 2$ and $d_F(u) = 0$. We analyze these cases separately.
   (a) $d(u) = 3$ and $d_F(u) = 2$.
   Let the neighbors of u in F be r and s (c.f. Fig. 2). We branch on whether a is matched to u or v:

$$\mathcal{A}\Big(\mathcal{P}(G[v \rightsquigarrow A, \{u, r, s\} \rightsquigarrow B])\Big) \bigvee \mathcal{A}\Big(\mathcal{P}(G[u \rightsquigarrow A, \{v, p, q\} \rightsquigarrow B])\Big).$$

[6] Indeed, in this scenario u would then be a degree three vertex in $\partial(F)$ with one neighbor each in a A and B (notice that this is the only possibility because of our assumption that the instance is reduced).

Note that we have $\{r, s\} \cap \{p, q\} = \emptyset$ since G has no cycles of length four. Therefore, this is a $(4, 4)$ branch.

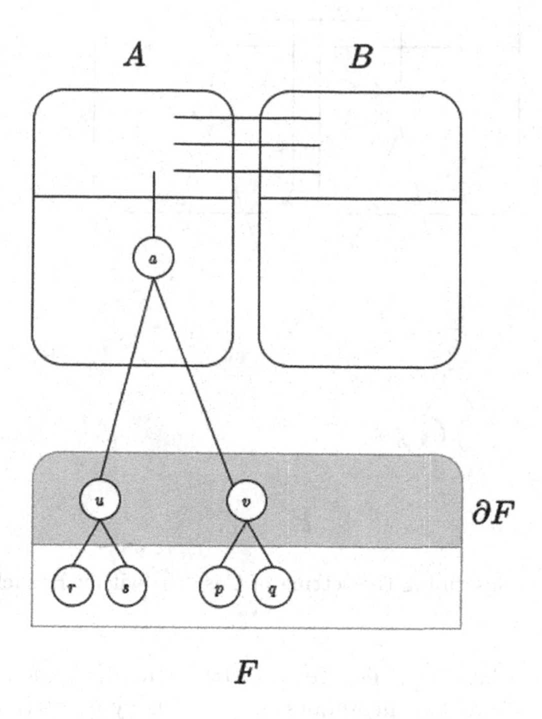

Fig. 2. A figure describing the setting of Case (a) within Branching Rule 3.

(b) $d(u) = 2$ and $d_F(u) = 0$.
By invariant (3), we know that u's second neighbor (say b) belongs to B. Note that by invariants (2) and (5), b must have exactly one more neighbor in F other than u: let this neighbor be w. Here we branch on whether u is matched to a or b:

$$\mathcal{A}\Big(\mathcal{P}(G[\{w, v\} \rightsquigarrow A, u \rightsquigarrow B])\Big) \bigvee \mathcal{A}\Big(\mathcal{P}(G[u \rightsquigarrow A, \{v, p, q, w\} \rightsquigarrow B])\Big).$$

Note that $w \neq p$ and $w \neq q$, since G has no cycles of length five. Therefore, this is a $(3, 5)$ branch.

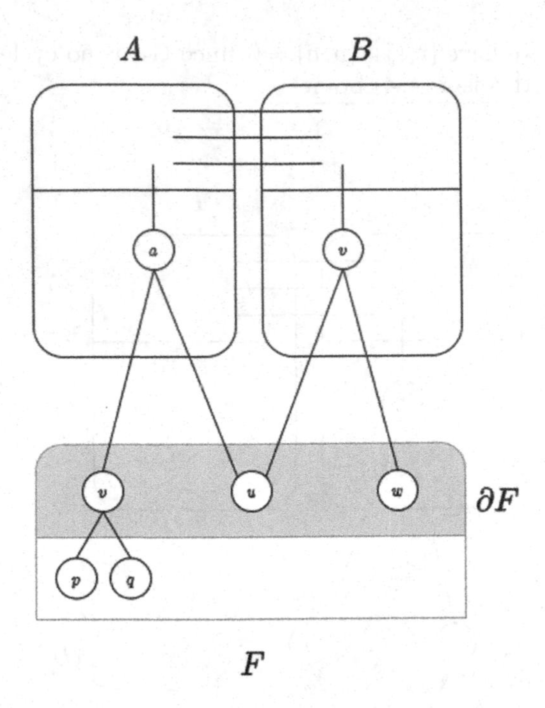

Fig. 3. A figure describing the setting of Case (b) within Branching Rule 3.

The case when there is a degree 3 vertex v in $\partial(\mathsf{F})$ such that v has one neighbor b in B and two neighbors p, q in F is symmetric and analogously handled (Fig. 3).

This completes the description of the branching rules.

Observe that if none of the branching rules are applicable on a clean and reduced instance, then note that we have the following:

1. Every vertex in $\partial(\mathsf{F})$ has degree two in G: indeed, if we have a degree three vertex in $\partial(\mathsf{F})$, then we would have branched according to branching rules 2 or 3; and if we have a degree one vertex in $\partial(\mathsf{F})$, then the instance is not reduced. Combined with invariant (3), we have that every vertex in $\partial(\mathsf{F})$ has exactly one neighbor in A and exactly one neighbor in B.
2. Every vertex in $\mathsf{A}^\circ \cup \mathsf{B}^\circ$ has exactly two neighbors in F: this follows from the assumption about the maximum degree of the graph being three, and invariants (2) and (5). Therefore, we have that every vertex has degree two in the graph $\mathsf{G}[\partial(\mathsf{F}) \cup \mathsf{A}^\circ \cup \mathsf{B}^\circ] \setminus \left(\mathsf{E}[\mathsf{G}[\mathsf{A}]] \cup \mathsf{E}[\mathsf{G}[\mathsf{B}]] \right)$, which is to say that this subgraph is a disjoint union of cycles. Further, note that $\mathsf{F} \setminus \partial(\mathsf{F}) = \emptyset$, since G is connected.

Observe that this instance can be solved in polynomial time by checking the parity of the cycles. This concludes our handling of the base case of the algorithm.

It can be verified that all the branching and reduction rules preserve the invariants, and this can be used to establish that the branching rules are exhaustive and correct. These arguments are implicit in the description of the branching rules.

Running Time Analysis. Recall that an algorithm branches on an instance of size n into r subproblems of sizes at most $n-t_1, n-t_2, \ldots, n-t_r$, then $(t_1, t_2, \ldots, t_r)$ is called the branching vector of this branching, and the unique positive root of $x^n - x^n t_1 - x^{nt_2} - \cdots - x^n t_r = 0$, denoted by $\tau(t_1, t_2, \ldots, t_r)$, is called its branching factor. The running time of a branching algorithm is $\mathcal{O}^*(\alpha^n)$, where $\alpha = \max_i \alpha_i$ and α_i is the branching factor of branching rule i, and the maximum is taken over all branching rules (c.f. the book by Fomin and Kratsch (2010)). Since $\tau(3,5) = 1.1938$ and $\tau(4,4) = 1.1892$, and therefore we have the following.

Theorem 1. *The* Perfect Matching Cut *problem admits an algorithm with running time* $\mathcal{O}^*(1.1938^n)$ *on graphs of maximum degree 3 and girth at least six.*

5 Concluding Remarks

We show improved algorithms for Perfect Matching Cut on general graphs and graphs of maximum degree 3 and girth at least six. A natural question is if we can achieve a comparable running time without the assumption of excluding short cycles. We believe this can be done with a more careful case analysis. Obtaining a sub-exponential running time for bipartite graphs with maximum degree three and girth g, matching the ETH-based lower bound from Le and Telle (2022) is also a natural open problem.

Due to lack of space, we defer our second result about a $\mathcal{O}^*(1.2599^n)$-algorithm applicable to all n-vertex graphs to a full version of this paper.

References

Bui-Xuan, B.-M., Telle, J.A., Vatshelle, M.: Fast dynamic programming for locally checkable vertex subset and vertex partitioning problems. Theoret. Comput. Sci. **511**, 66–76 (2013)

Chen, C.-Y., Hsieh, S.-Y., Le, H.-O., Le, V.B., Peng, S.-L.: Matching cut in graphs with large minimum degree. Algorithmica **83**(5), 1238–1255 (2021)

Chvátal, V.: Recognizing decomposable graphs. J. Graph Theory **8**(1), 51–53 (1984)

Feghali, C., Lucke, F., Paulusma, D., and Ries, B.: Matching cuts in graphs of high girth and h-free graphs (2023)

Fomin, F.V., Kratsch, D.: Exact Exponential Algorithms. Texts in Theoretical Computer Science. An EATCS Series. Springer, Heidelberg (2010). https://doi.org/10.1007/978-3-642-16533-7

Golovach, P.A., Komusiewicz, C., Kratsch, D., Le, V.B.: Refined notions of parameterized enumeration kernels with applications to matching cut enumeration. In: 38th International Symposium on Theoretical Aspects of Computer Science (STACS 2021), vol. 187, pp. 37:1–37:18 (2021)

Heggernes, P., Telle, J.A.: Partitioning graphs into generalized dominating sets. Nordic J. Comput. **5**(2), 128–142 (1998)

Komusiewicz, C., Kratsch, D., Le, V.B.: Matching cut: kernelization, single-exponential time FPT, and exact exponential algorithms. In: 13th International Symposium on Parameterized and Exact Computation (IPEC 2018), pp. 19:1–19:13 (2018)

Le, V.B., Telle, J.A.: The perfect matching cut problem revisited. Theor. Comput. Sci. **931**, 117–130 (2022)

Telle, J.A., Proskurowski, A.: Algorithms for vertex partitioning problems on partial k-trees. SIAM J. Discret. Math. **10**(4), 529–550 (1997)

Édouard Bonnet, Chakraborty, D., Duron, J.: Cutting Barnette graphs perfectly is hard (2023)

Connected Feedback Vertex Set on AT-Free Graphs

Joydeep Mukherjee[1] and Tamojit Saha[1,2(✉)]

[1] Ramakrishna Mission Vivekananda Educational and Research Institute, Howrah, India
joydeep.m@gm.rkmvu.ac.in, tamojitsaha1@gmail.com
[2] Institute of Advancing Intelligence, TCG CREST, Kolkata, India
http://rkmvu.ac.in/, https://www.tcgcrest.org/

Abstract. A connected feedback vertex set of a graph is a connected subgraph of the graph whose removal makes the graph cycle free. In this paper, we give an approximation algorithm that computes a connected feedback vertex set of size $(1.9091OPT + 6)$ on 2–connected AT-free graphs with running time $O(n^8m^2)$. Also, we give another approximation algorithm that computes a connected feedback vertex set of size $(2.9091OPT + 6)$ on the same graph class with more efficient running time $O(min\{m(log(n)), n^2\})$.

Keywords: Graph Algorithm · Approximation Algorithm · AT-free graph · Feedback Vertex Set · Combinatorial Optimization

1 Introduction

Feedback vertex set of an undirected graph is a subset of vertices whose removal from the graph makes the remaining graph acyclic. Connected feedback vertex set of an undirected graph is a feedback vertex set which is also connected. The minimum connected feedback vertex set problem ($MCFVS$) seeks a connected feedback vertex set of minimum cardinality.

This problem is a connected variant of the classical minimum feedback vertex set problem ($MFVS$) in which we compute a set $S \subseteq V(G)$ such that $G[V \setminus S]$ is a forest and $|S|$ is minimized. We denote a minimum feedback vertex set of a graph by *min-FVS* and minimum connected feedback vertex set by *min-CFVS*.

The feedback vertex set problem is known to be NP-complete [13]. A 2-approximation algorithm for the problem is provided by Bafna et al. [2]. The feedback vertex set problem is polynomial time solvable in several special graph classes. The feedback vertex set problem is solved in polynomial time on AT-free graphs by Kratsch at al. [15], permutation graphs by Liang [16], interval

---

The second author is a doctoral student at Ramakrishna Mission Vivekananda Educational and Research Institute (RKMVERI) and Institute of Advancing Intelligence (IAI), TCG CREST.

S.-Y. Hsieh et al. (Eds.): IWOCA 2023, LNCS 13889, pp. 319–330, 2023.
https://doi.org/10.1007/978-3-031-34347-6_27

graphs by Lu et al. [17], cocomparability graphs and convex bipartite graphs by Liang et al. [10], $(sP_1 + P_3)$–free graphs by Dabrowski et al. [9], in P_5–free graphs by Abrishami et al. [1].

However not much study has been carried out on connected feedback vertex set problem on special graph classes. A PTAS is known for $MCFVS$ on planar graphs proposed by Grigoriev et al. [11]. $MCFVS$ is solved in polynomial time on (sP_2)–free graphs by Chiarelli et al. [6], cographs and $(sP_1 + P_3)$–free graphs by Dabrowski et al. [9]. To the best of our knowledge, no other polynomial time algorithm is known for $MCFVS$.

Apart from algorithms, the ratio between the sizes of connected feedback vertex set and feedback vertex set is also studied. The price of connectivity for feedback vertex set on a graph class $\mathcal{G}$ is the maximum ratio $(\frac{|min\text{-}CFVS|}{|min\ FVS|})$ over all connected graph G, $G \in \mathcal{G}$. The price of connectivity for feedback vertex set is shown to be upper bounded by constant in H–free by Belmonte et al. [4].

In this paper we initiate the study MCFVS on 2-connected AT-free graphs from approximation algorithm point of view. In a graph $G = (V, E)$, a set of three vertices $\{u, v, w\}$ of $V(G)$ is called an *Asteroidal Triple* if these three vertices are mutually nonadjacent and for any two vertices of this set there exists a path between these two vertices which avoids the neighborhood of the third vertex. G is called Asteroidal Triple Free if it does not contain any Asteroidal Triple.

We present an approximation algorithm that computes a connected feedback vertex set of size $(1.9091OPT + 6)$ on 2–connected AT-free graphs which runs in time $O(n^8m^2)$. Note that, here OPT denotes the size of the optimal solution. We also present an approximation algorithm with approximation ratio $(2.9091OPT + 6)$ for the same graph class which runs in time $O(min\{m(log(n)), n^2\})$.

Asteroidal triple free graph class contains graph classes like permutation graphs, interval graphs, trapezoid graphs, and cocomparability graphs [8]. AT-free graphs have many desirable properties which make it amenable for designing polynomial time algorithms for many NP-complete problems. MFVS is solved in polynomial time in AT free graphs by Kratsch at al. [15]. NP-hard problems are like independent set, dominating set, total dominating set and connected dominating set respectively by Broersma et al. [5], Kratsch [14] and Balakrishnan et al. [3]. However complexity of connected feedback vertex set problem is unknown in AT-free graphs.

2 Preliminaries

For a graph $G = (V, E)$, we denote the set of vertices by $V(G)$ and the set of edges by $E(G)$. A graph $H = (V', E')$ is a subgraph of $G = (V, E)$ if $V' \subseteq V$ and $E' \subseteq E$. We denote $|V|$ by n and $|E|$ by m. A subgraph $H = (V', E')$ of G is a induced subgraph if $V' \subseteq V$ and for $u, v \in V'$, $(u, v) \in E'$ if and only if $(u, v) \in E$. The induced subgraph on any subset $S \subseteq V$ is denoted by $G[S]$.

The neighbourhood of a vertex v, denoted by $N(v)$, is the set of all vertices that are adjacent to v. Closed neighbourhood of v is denoted by $N[v] = \{v\} \cup N(v)$. The neighbourhood of a set of vertices $\{v_1, v_2, \ldots, v_k\}$ is denoted by $N(v_1, v_2, \ldots, v_k) = \cup_{i=1}^{k} N(v_i)$ and the closed neighbourhood is denoted by $N[v_1, v_2, \ldots, v_k] = \cup_{i=1}^{k} N[v_i]$.

A path is a graph, $Y = (V, E)$, such that $V = \{y_1, y_2, \ldots, y_k\}$ and $E = \{y_1y_2, y_2y_3, \ldots, y_{k-1}y_k\}$. We denote a path by the sequence of its vertices, that is $Y = y_1y_2 \ldots y_k$. Here y_1 and y_k are called endpoints of path Y. The number of vertices present in Y is denoted by $|Y|$. We denote $y_iYy_j = y_iy_{i+1} \ldots y_j$ where $1 \leq i \leq j \leq k$. A path on k vertices is denoted by Y_k and the length of the path is denoted by the number of edges present on the path that is $k-1$. The distance between two vertices in a graph is the length of the shortest path between them. A cycle is a graph, $C=(V, E)$, such that $V(C)=\{c_1, c_2, \ldots, c_l\}$ and $E(C)=\{c_1c_2, \ldots, c_{l-1}c_l, c_lc_1\}$. The shortest distance between u and v is denoted by $dist_C(u, v)$ where $u, v \in V(C)$. The number of vertices present in the cycle C is denoted by $|C|$.

A *dominating set* D of G is a subset of vertices of G such that for every v outside D, $N(v) \cap D \neq \phi$. A *dominating pair* is a pair of vertices such that any path between them is a dominating set. There is a linear time algorithm to find a dominating pair [7] in AT-free graphs. We denote a shortest path between a dominating pair by DSP.

For a minimization problem $\mathcal{P}$ an algorithm $\mathcal{A}$ is called $(\alpha \cdot OPT + \beta)$–approximation for some $\alpha \geq 1$ and $\beta > 0$, if for all instance of $\mathcal{P}$, algorithm $\mathcal{A}$ produces an output which is at most $(\alpha \cdot (optimal) + \beta)$. Here *optimal* denotes the cost of the optimal solution of the specific instance of $\mathcal{P}$.

3 Approximation Algorithm for MCFVS

This section describes the main results of this paper. The first part of this section conveys the main approach and provides an approximation algorithm with approximation ratio $(2OPT + 6)$. The second part makes some important observation towards a tighter approximation bound. The third part states an approximation algorithm with approximation ratio $(1.9091OPT + 6)$.

3.1 2-Approximation Algorithm

Let F^* be any minimum feedback vertex set of G, where G is a 2–connected AT-free graph. Recall that, since G is an AT-Free graph, it always contains a dominating pair [8]. Let p_1, p_k be a dominating pair of G. Let $P = p_1p_2...p_k$ be a shortest path between p_1 and p_k.

Our algorithm begins by choosing $F^* \cup P$ as a solution to MCFVS on 2–connected AT-free graphs. Then we modify both F^* and P and report their union as the connected feedback vertex set. We first analyze the quality of $F^* \cup P$ as a solution to MCFVS on 2–connected AT-free graphs. Next, we state the algorithm for modification of both F^* and P to obtain F_1 and P', respectively. Finally, we report $F_1 \cup P'$ as the solution of MCFVS and analyze the modified solution to achieve the approximation ratio.

We restate a well-known fact about AT-Free graphs [8].

Lemma 1. *The longest induced cycle in AT-free graphs can contain at most 5 vertices.*

Lemma 2. *The set $F^* \cup P$ is a connected feedback vertex set of G.*

Proof. As (p_1, p_k) is a dominating pair, hence every path between p_1 and p_k is a dominating set of G. Hence $F^* \cup P$ is connected. □

Lemma 3. *Let $X = uvw$ be a path of length two and F be any feedback vertex set in G. Then $F \cap N[u, v, w] \neq \phi$.*

Proof. As the graph is 2–connected there is a cycle containing any two edges of the graph [12]. Consider the shortest cycle containing X. We denote this cycle by C.

The cycle C can be induced or it may contain chords. If C is induced then from Lemma 1, it contains at most 5 vertices, hence the vertices of the cycle is a subset of $N[X]$. That guarantees that at least one vertex of $N[X]$ is in F.

Now consider the case when C is not induced. If length of C is five or less then $N[X] \cap F \neq \phi$. The following proposition holds when $|C| > 5$.

Proposition 1. *One of the endpoints of any chord of C must belong to X.*

Proof. Assume none of the endpoints is in X. The cycle containing the chord and X is a cycle which is shorter than C. Hence contradiction. □

We strengthen Proposition 1 in the following proposition.

Proposition 2. *One of the endpoints of any chord of C must be v.*

Proof. Assume for contradiction v is not an endpoint of some chord of C. Since we know from the above proposition one of the end points should belong to X, w.l.o.g let it be u. Then the cycle containing the chord and the vertices X, is a shorter cycle than C. Hence contradiction. □

Let $C_{u,v}$ be the shortest cycle in $G[C]$, that contains the edge (u, v). Similarly let $C_{v,w}$ be the shortest cycle in $G[C]$, that contains the edge (v, w). Note that, by the above proposition, each of the cycles $C_{u,v}$ and $C_{v,w}$ are chordless and exactly one of the edges of these cycles is a chord of C. In other words each of them are induced cycles and exactly one of their edges is a chords of C. Hence from Lemma 1, $|C_{u,v}| \leq 5$ and $|C_{v,w}| \leq 5$.

If one of these cycle is of length less than 5 then $N[X] \cap F \neq \phi$ holds. To see this assume w.l.o.g $C_{u,v}$ is of length less than 5. Note that $N[u, v] \cap F$ must be non empty. Hence the proposition is satisfied if $|C_{u,v}| < 5$ or $|C_{v,w}| < 5$.

The only remaining case, using Lemma 1, is when $|C_{u,v}| = 5$ and $|C_{v,w}| = 5$. We will show below that neither $C_{u,v}$ nor $C_{v,w}$ can be of length 5. Since v is the common vertex in both of these cycles, we consider the following cases.

Case 1. In the first case we consider the chord of C present in $C_{u,v}$ is different from the chord of C present in $C_{v,w}$. In this case $C_{u,v}$ and $C_{v,w}$ has only v as a common vertex. This case is depicted in Fig. 1. The edge (c, v) and the edge (v, d) are chords of C. This case is not possible as vertices $\{w, e, b\}$ forms an asteroidal triple, following we explain that.

Proposition 3. *$\{w, e, b\}$ are independent.*

Proof. Each member of $\{w, e, b\}$ is present in C and from the Proposition 2, v is an end point of every chord in C. Edge between any two vertices of $\{w, e, b\}$ will imply a chord in C for which v is not an end point, which violates the Proposition 2. Hence $\{w, e, b\}$ are independent. □

By similar argument of above Proposition 2, b and f arc not adjacent, hence we get a path from e to w namely $\{e, f, w\}$ which does not contain any neighbour of b. The w, b-path $\{w, v, u, a, b\}$ does not contain any neighbour of e, since e is not adjacent to any of $\{w, u, a, b\}$ due to Proposition 2 and e is not adjacent to v as we get a shorter cycle than $C_{v,w}$ if e and v is adjacent. Finally there is a e, b path P_c through $V(C) \setminus X$. As w is not present as a end point of any chord in C, w is not adjacent to any vertices of P_c.

Case 2. The second case is when $C_{u,v}$ and $C_{v,w}$ share a common chord in C. In this case $C_{u,v}$ and $C_{v,w}$ share exactly one edge. This case is depicted in Fig. 2. The edge (d, v) is a chord of C. This case is also not possible as $\{e, w, a\}$ is a asteroidal triple.

Hence the proof of the lemma. □

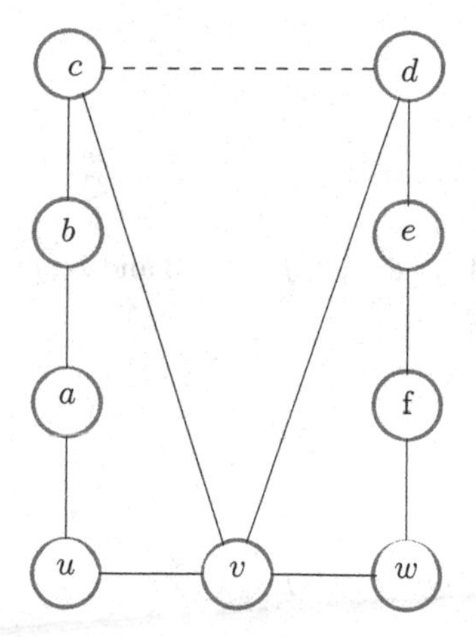

Fig. 1. Two chords of C are present in $C_{u,v}$ and $C_{v,w}$.

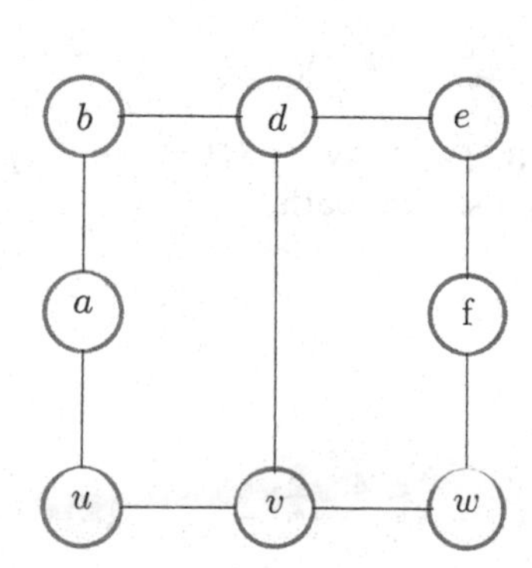

Fig. 2. One chord is common between $C_{u,v}$ and $C_{v,w}$.

We denote any arbitrary minimum connected feedback vertex set by F_c^*.

Lemma 4. *Let F_c^* be a minimum connected feedback vertex set. Then* $|P| \leq |F_c^*| + 6$.

Proof. P is a shortest path between p_1 and p_k. From Lemma 3 we know $N[p_1, p_2, p_3] \cap F_c^* \neq \phi$, similarly $N[p_{k-2}, p_{k-1}, p_k] \cap F_c^* \neq \phi$. Let $u \in N(p_3) \cap F_c^*$ and $v \in N(p_{k-2}) \cap F_c^*$. If $|P| > |F_c^*| + 6$ then we get a u, v path P^* through F_c^* which has length at most $|P| - 7$. The p_1, p_k path $p_1 p_2 p_3 P^* p_{k-2} p_{k-1} p_k$ has length strictly less than the length of P. Which contradicts minimum length property of P. □

From the above lemma we get a $(2OPT + 6)$–approximation for 2–connected AT-Free graphs in the following way. As discussed in this section $F^* \cup P$ is a solution produced by our approximation algorithm. The solution size is $|F^*| + |P|$. Using Lemma 4 solution size is upper bounded by $2|F_c^*| + 6$.

3.2 Towards Tighter Approximation

Although a minimum feedback vertex set along with a shortest path between a dominating pair gives us a 2 approximation algorithm for $MCFVS$, but modifying the solution we can achieve a tighter approximation. As a first step we modify the shortest path P according to Algorithm 1 and denote the modified path by P'. We note that the modification of the path does not increase the length of the path. Next we show that there is at least a constant fraction of vertices of P' which can be counted as part of F^*. This shows that number of non-feedback vertices in P' is upper bounded by some constant fraction of $|P'|$. This observation leads to a better approximation (Figs. 3 and 6).

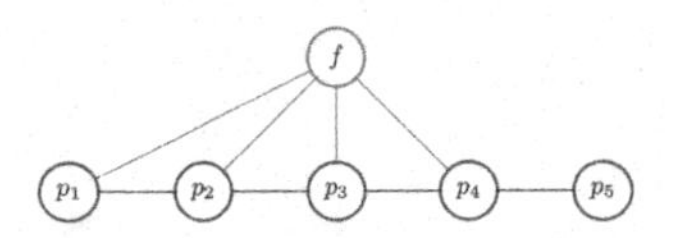

Fig. 3. $f \in F^*$, $|N(f) \cap P| > 3$. (p_1, f, p_4) gives a shorter path.

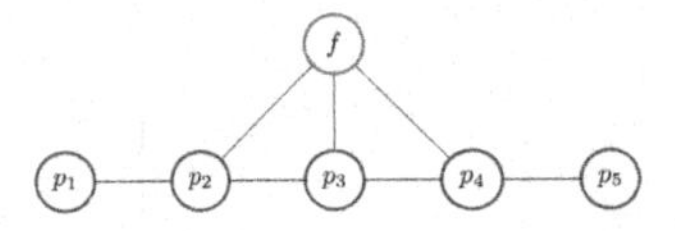

Fig. 4. $f \in F^*$, $|N(f) \cap P| = 3$ and $N(f) \cap P \cong P_3$.

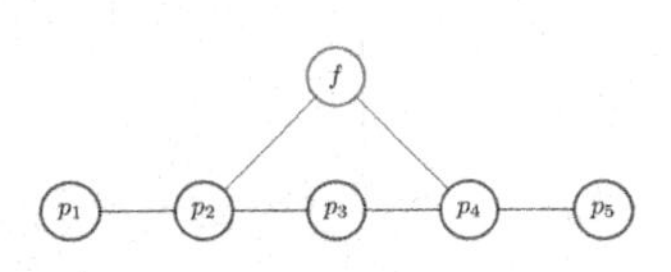

Fig. 5. $f \in F^*$, $|N(f) \cap P| = 2$ and $N(f) \cap P$ has a common neighbour in P.

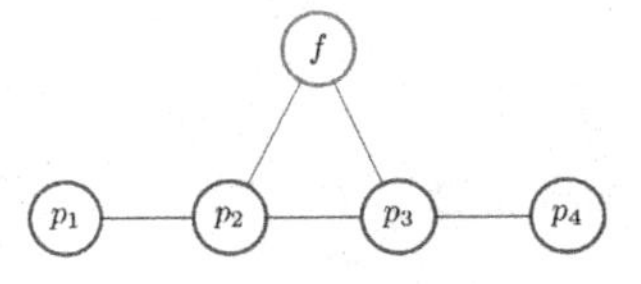

Fig. 6. $f \in F^*$, $|N(f) \cap P| = 2$ and $N(f) \cap P \cong P_2$.

Consider the following lemma.

Lemma 5. *Let $v \in F^*$. If $v \notin P$ then,*

a) $|N(v) \cap P| \leq 3$.
b) If $|N(v) \cap P| = 3$ *then* $N(v) \cap P \cong P_3$.
c) If $|N(v) \cap P| = 2$ *then either* $N(v) \cap P \cong P_2$ *or members of* $N(v) \cap P$ *has a common neighbour in* P.

Proof. Please see the proof in Appendix. □

In the light of above lemma we modify P. Suppose there exists $v \in F^*$ such that $|N(v) \cap P| = 2$ and those two vertices share a common neighbour u in P or v is adjacent to three consecutive vertices of P where u is the middle vertex, then if $u \notin F^*$ then we can replace u by the vertex v. The cases mentioned in this paragraph are depicted in Fig. 5, Fig. 4 respectively.

Based on the above paragraph we have the following algorithm.

Algorithm 1. Shortest path modification

Require: Minimum feedback vertex set F^* and shortest path P.
Ensure: A modified path P'
1: **for all vertices** $v \in F^*$ **do**
2:    **if** ($|N(v) \cap P| = 3$) or ($|N(v) \cap P| = 2$ and members of $N(v) \cap P$ has a common neighbour in P). **then**
3:       Let $p_i p_{i+1} p_{i+2}$ be the sequence of path vertices such that $p_i, p_{i+2} \in N(v) \cap P$.
4:       **if** $p_{i+1} \notin F^*$ **then**
5:          Remove the vertex p_{i+1}.
6:          Put v as the path vertex that is removed, that is $p_i v p_{i+2}$.
7:       **end if**
8:    **end if**
9: **end for**
10: Let the new path be P'.

The modified P' has the following property.

Lemma 6. *Suppose $v \in F^* \setminus P'$. Then,*

a) If v is adjacent to three consecutive vertices p_i, p_{i+1}, p_{i+2} of P' then $p_{i+1} \in P' \cap F^*$.
b) If $|N(v) \cap P'| = 2$ *then* $N(v) \cap P' \cong P_2$.
c) If v is adjacent to two vertices p_i and p_{i+2} for some i such that they share a common neighbour p_{i+1} in P', then $p_{i+1} \in P' \cap F^*$.

Proof. Please see the proof in Appendix. □

As mentioned in above paragraph we obtain a modified path P' using Algorithm 1 whose length is equal to $|P|$. This ensures that P' is a DSP. Since it is a DSP every vertex of $F^* \setminus P'$ is adjacent to at least one vertex of P'. We classify the induced cycles of G in two types. *Type-1* cycles are those induced cycles which contain at least one vertex from P'. *Type-2* cycles are those induced cycles which does not contain any vertex from P'. In the next two lemma we show that neighbourhood of a type-2 cycle is localized over P'.

Figure 7 illustrates type-1 cycle and in Fig. 9, $\{v_1, v_2, x, v_3, v_4\}$ is a type-2 cycle.

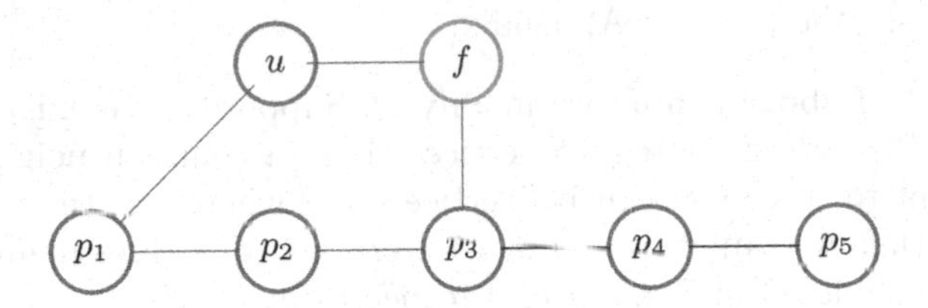

Fig. 7. (p_1, p_2, p_3, f, u) is a type-1 cycle.

Let $p_1, p_2, \ldots p_k$ denote the vertices of the path P' and for any r, s such that $1 \leq r < s \leq k$, p_r occurs before p_s in P'.

Lemma 7. *Let C be a type-2 cycle. Let $u, v \in V(C)$ such that $(u, p_i) \in E(G)$ and $(v, p_j) \in E(G)$. Then $|j - i| \leq 4$.*

Proof. Please see the proof in Appendix. □

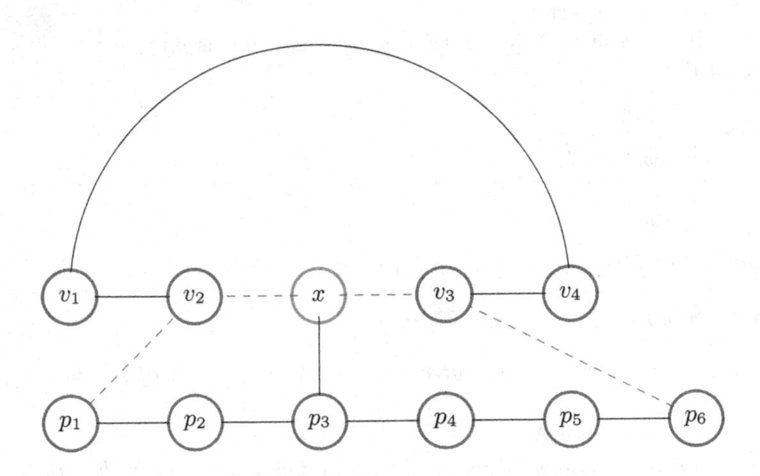

Fig. 8. Illustrating proof of Lemma 7.

As a consequence of Lemma 7, we get Lemma 8 (Fig. 8).

Lemma 8. *Let C be a type-2 cycle such that $p_i \in P'$ is adjacent to some vertex of C. Then there exist a set S consisting of at most 5 consecutive vertices of P' such that $S \ni p_i$ and $N(C) \cap P' \subseteq S$.*

Proof. Please see the proof in Appendix. □

In the next lemma we establish a condition when we can possibly remove a vertex from F^*.

Let $x \in F^* \setminus P'$ then, by Lemma 6 there are four possible cases. The first case is by the third part of Lemma 6, where x is adjacent to two vertices p_i and p_{i+2} for some i such that they share a common vertex p_{i+1} and $p_{i+1} \in P' \cap F^*$. The second case is by the second part of Lemma 6, where x is adjacent to two vertices p_j and p_{j+1} for some j such that (p_j, p_{j+1}) is a edge of P'. The third case is by the first part of Lemma 6, where x is adjacent to three consecutive vertices p_i, p_{i+1}, p_{i+2} of P' then $p_{i+1} \in P' \cap F^*$. The fourth case is when x is adjacent to exactly one vertex of P'. We will consider only the fourth case for Lemma 9, Lemma 10, Lemma 11, and Lemma 12 since the other cases will follow similarly. We will also consider C to be a type-2 cycle containing x in the above-mentioned lemma.

Lemma 9. *If $(N(C) \cap P') \cap F^* = \phi$, then $(C \setminus \{x\}) \cap F^* \neq \phi$.*

Proof. Please see the proof in Appendix. □

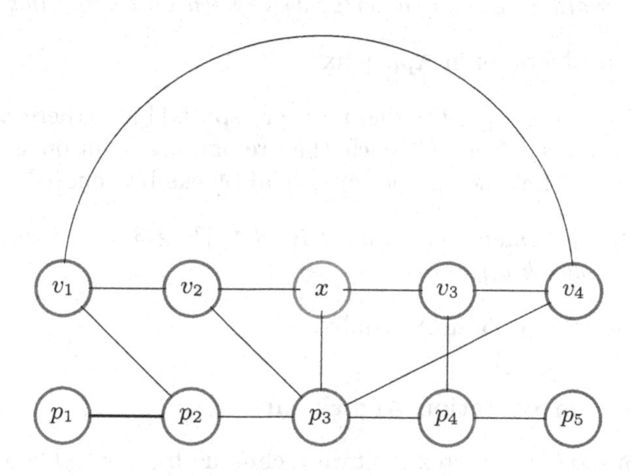

Fig. 9. Illustrating proof of Lemma 9.

We divide the path P' into set of 11 consecutive vertices each. We call each such set *block*. Let B_i denote the ith block where $i = 1, 2, \ldots, \lceil \frac{|P'|}{11} \rceil$. We note that $B_i \cap B_j = \phi$ for all $i \neq j$. The possible cases for a block B_i is either $B_i \cap F^* \neq \phi$ or $B_i \cap F^* = \phi$. We call the blocks of first type as *normal blocks* and blocks of second type as *special blocks*. Let $x \in F^* \setminus P'$ then, from Lemma 6, for every special block B only two cases are possible. First case is when x is adjacent to two vertices p_j and p_{j+1} of B for some j such that (p_j, p_{j+1}) is an edge of B. The second case is when x is adjacent to exactly one vertex of B. Here too we consider only the second case since the first case will follow similarly.

Consider B_i and B_{i+1}. We denote fifth, sixth and seventh vertices of B_i and B_{i+1} with p_5^i, p_6^i, p_7^i and $p_5^{i+1}, p_6^{i+1}, p_7^{i+1}$ respectively. From Lemma 3, $N[p_5^i, p_6^i, p_7^i] \cap F^* \neq \phi$ and $N[p_5^{i+1}, p_6^{i+1}, p_7^{i+1}] \cap F^* \neq \phi$.

The following lemma conveys a notion of disjointness among the type-2 cycles of two consecutive blocks. Note that, it is possible for some block to contain no type-2 cycle.

Lemma 10. *Let $u \in (N[p_5^i, p_6^i, p_7^i] \setminus P') \cap F^*$ and $v \in (N[p_5^{i+1}, p_6^{i+1}, p_7^{i+1}] \setminus P') \cap F^*$. Type-2 cycles containing u does not contain v and type-2 cycles containing v does not contain u.*

Proof. Please see the proof in Appendix. □

We note that the proof of Lemma 10 will also hold for blocks of size 7, but it is necessary to keep the block size 11. The reason behind keeping the block size 11 will be clear in the proof of the next lemma.

Let B_i be a special block. Hence from Lemma 3 we get a $x \in (N[p_5^i, p_6^i, p_7^i] \setminus P') \cap F^*$.

Lemma 11. *Let B_i be a special block and $x \in (N[p_5^i, p_6^i, p_7^i] \setminus P') \cap F^*$. Every type-2 cycle C containing x has a vertex $v \in C$ such that $v \neq x$ and $v \in F^*$.*

Proof. Please see the proof in Appendix. □

The following lemma specifies that for every special block there exists a vertex which can be removed from F^* such that remaining solution is a connected feedback vertex set. Let the number of special blocks be denoted by β.

Lemma 12. *If $\beta \geq 0$, then there is a set $R \subseteq F^*$, $|R| \geq \beta$ such that $(F^* \setminus R) \cup P'$ is a connected feedback vertex set.*

Proof. Please see the proof in Appendix. □

3.3 1.9091-Approximation Algorithm

Below we summarize the vertex removal technique from F^*. The removal technique consists of constructing R and then removing R from F^*. Suppose B_i is a special block. Fix y where $y \in (N[p_5^i, p_6^i, p_7^i] \setminus P') \cap F^*$. Two cases might arise.

Case 1: There is no type-2 cycle that contains y. In this case y is present in only type-1 cycles. From the definition of type-1 cycle we know it contains at least one vertex of P'. As P' is already in our solution, we can remove y from F^* without destroying connected feedback vertex set property of our solution. Hence we can include y in R.

Case 2: There are cycles containing y which has no vertex present in P'. In this case by Lemma 12, $F^* \setminus \{y\}$ is a connected feedback vertex set. Hence we can include y in R.

From the above discussion, every normal block has at least one vertex belonging to F^* and for every special block B_i there is a vertex $x \in (N[p_5^i, p_6^i, p_7^i] \setminus P') \cap F^*$

which can be removed from F^*. In the case of special blocks we remove such x and label any vertex of B_i as a vertex of F^*. We call the process of removing x and labeling a vertex of B_i as a vertex of F^* *normalization*. We perform normalization on each special block. After the process of normalization each block has at least one of its vertex labeled as F^*. Hence $|P' \setminus F^*| \leq \frac{10}{11}|P'|$. From the property of minimum feedback vertex set we get $|F^*| \leq |F_c^*|$ and from Lemma 4, we have $|P'| \leq |F_c^*| + 6$. Now,

$$|F^*| + \frac{10}{11}|P'| \leq |F_c^*| + \frac{10}{11}(|F_c^*| + 6) \leq |F_c^*| + \frac{10}{11}|F_c^*| + 6 \leq 1.9091|F_c^*| + 6$$

Hence we get a connected feedback vertex set of size $(1.9091OPT + 6)$, where OPT is the size of optimal solution.

The time complexity of the approximation algorithm for MCFVS depends on the complexity of finding a min-FVS and finding a DSP. In AT-free graphs it takes $O(n^8m^2)$ time to compute the min-FVS [15] and a DSP can be found in linear time [7]. The path modification and normalization both takes linear time. Hence complexity of our algorithm is $O(n^8m^2)$. Thus we have the following theorem.

Theorem 1. *There is a approximation algorithm for MCFVS in 2−connected AT-free graphs which produces solution of size* $(1.9091OPT+6)$ *in time* $O(n^8m^2)$.

We can achieve a better running time by compromising on the approximation factor. Instead of computing the min-FVS for an AT-free graph, we use the 2-approximation algorithm for computing min-FVS [2] on general graphs to get a better running time. Running time of the algorithm will be reduced to $O(min\{m(log(n)), n^2\})$ [2].

Before stating the following theorem, note that F_c^* denotes the minimum connected feedback vertex set and F^* denotes the minimum feedback vertex set. Consider F to be any feedback vertex set, then Lemma 6 through Lemma 12 holds for F and P' as none of this lemma depends on the minimality of the feedback vertex set. We have the following theorem based on the above mentioned fact.

Theorem 2. *Let* F *be a feedback vertex set of* 2−*connected AT-free graph* G *such that* $|F| \leq 2|F^*|$. *Then using* F, *our algorithm produces a connected feedback vertex set* F_c *such that* $|F_c| \leq (2.9091|F_c^*| + 6)$.

Proof. Please see the proof in Appendix. □

4 Conclusion

We provided a constant factor approximation algorithm for connected feedback vertex set problem on 2−connected AT-free graphs. The main open question that remains is whether the problem is exactly solvable in polynomial time or is it NP-hard on AT-free graphs.

References

1. Abrishami, T., Chudnovsky, M., Pilipczuk, M., Rzążewski, P., Seymour, P.: Induced subgraphs of bounded treewidth and the container method. In: Proceedings of the 2021 ACM-SIAM Symposium on Discrete Algorithms (SODA), pp. 1948–1964. SIAM (2021)
2. Bafna, V., Berman, P., Fujito, T.: A 2-approximation algorithm for the undirected feedback vertex set problem. SIAM J. Discret. Math. **12**(3), 289–297 (1999)
3. Balakrishnan, H., Rajaraman, A., Rangan, C.P.: Connected domination and Steiner set on asteroidal triple-free graphs. In: Dehne, F., Sack, J.-R., Santoro, N., Whitesides, S. (eds.) WADS 1993. LNCS, vol. 709, pp. 131–141. Springer, Heidelberg (1993). https://doi.org/10.1007/3-540-57155-8_242
4. Belmonte, R., van't Hof, P., Kamiński, M., Paulusma, D.: The price of connectivity for feedback vertex set. Discret. Appl. Math. **217**, 132–143 (2017)
5. Broersma, H., Kloks, T., Kratsch, D., Müller, H.: Independent sets in asteroidal triple-free graphs. SIAM J. Discret. Math. **12**(2), 276–287 (1999)
6. Chiarelli, N., Hartinger, T.R., Johnson, M., Milanič, M., Paulusma, D.: Minimum connected transversals in graphs: new hardness results and tractable cases using the price of connectivity. Theor. Comput. Sci. **705**, 75–83 (2018)
7. Corneil, D.G., Olariu, S., Stewart, L.: Computing a dominating pair in an asteroidal triple-free graph in linear time. In: Akl, S.G., Dehne, F., Sack, J.-R., Santoro, N. (eds.) WADS 1995. LNCS, vol. 955, pp. 358–368. Springer, Heidelberg (1995). https://doi.org/10.1007/3-540-60220-8_76
8. Corneil, D.G., Olariu, S., Stewart, L.: Asteroidal triple-free graphs. SIAM J. Discret. Math. **10**(3), 399–430 (1997)
9. Dabrowski, K.K., Feghali, C., Johnson, M., Paesani, G., Paulusma, D., Rzążewski, P.: On cycle transversals and their connected variants in the absence of a small linear forest. Algorithmica **82**(10), 2841–2866 (2020)
10. Daniel Liang, Y., Chang, M.-S.: Minimum feedback vertex sets in cocomparability graphs and convex bipartite graphs. Acta Informatica **34**(5) (1997)
11. Grigoriev, A., Sitters, R.: Connected feedback vertex set in planar graphs. In: Paul, C., Habib, M. (eds.) WG 2009. LNCS, vol. 5911, pp. 143–153. Springer, Heidelberg (2010). https://doi.org/10.1007/978-3-642-11409-0_13
12. Gross, J.L., Yellen, J., Anderson, M.: Graph Theory and its Applications. Chapman and Hall/CRC (2018)
13. Hartmanis, J.: Computers and intractability: a guide to the theory of np-completeness (Michael R. Garey and David S. Johnson). SIAM Rev. **24**(1), 90 (1982)
14. Kratsch, D.: Domination and total domination on asteroidal triple-free graphs. Discret. Appl. Math. **99**(1–3), 111–123 (2000)
15. Kratsch, D., Müller, H., Todinca, I.: Feedback vertex set on at-free graphs. Discret. Appl. Math. **156**(10), 1936–1947 (2008)
16. Daniel Liang, Y.: On the feedback vertex set problem in permutation graphs. Inf. Process. Lett. **52**(3), 123–129 (1994)
17. Lu, C.L., Tang, C.Y.: A linear-time algorithm for the weighted feedback vertex problem on interval graphs. Inf. Process. Lett. **61**(2), 107–111 (1997)

Reconfiguration and Enumeration of Optimal Cyclic Ladder Lotteries

Yuta Nozaki[1,2], Kunihiro Wasa[3], and Katsuhisa Yamanaka[4(✉)]

[1] Yokohama National University, Yokohama, Japan
nozaki-yuta-vn@ynu.ac.jp
[2] SKCM[2], Hiroshima University, Hiroshima, Japan
[3] Hosei University, Tokyo, Japan
wasa@hosei.ac.jp
[4] Iwate University, Morioka, Japan
yamanaka@iwate-u.ac.jp

Abstract. A ladder lottery of a permutation π of $\{1, 2, \ldots, n\}$ is a network with n vertical lines and zero or more horizontal lines each of which connects two consecutive vertical lines and corresponds to an adjacent transposition. The composition of all the adjacent transpositions coincides with π. A *cyclic ladder lottery* of π is a ladder lottery of π that is allowed to have horizontal lines between the first and last vertical lines. A cyclic ladder lottery of π is *optimal* if it has the minimum number of horizontal lines. In this paper, for optimal cyclic ladder lotteries, we consider the reconfiguration and enumeration problems. First, we investigate the two problems when a permutation π and its optimal displacement vector $\boldsymbol{x}$ are given. Then, we show that any two optimal cyclic ladder lotteries of π and $\boldsymbol{x}$ are always reachable under braid relations and one can enumerate all the optimal cyclic ladder lotteries in polynomial delay. Next, we consider the two problems for optimal displacement vectors when a permutation π is given. Then, we present a characterization of the length of a shortest reconfiguration sequence of two optimal displacement vectors and show that there exists a constant-delay algorithm that enumerates all the optimal displacement vectors of π.

Keywords: Reconfiguration · Enumeration · Cyclic ladder lottery

1 Introduction

A *ladder lottery*, known as "Amidakuji" in Japan, is a common way to decide an assignment at random. Formally, we define ladder lotteries as follows. A *network* is a sequence $\langle \ell_1, \ell_2, \ldots, \ell_n \rangle$ of n vertical lines (*lines* for short) and horizontal lines (*bars* for short) each of which connects two consecutive vertical lines. We say that ℓ_i is located on the left of ℓ_j if $i < j$ holds. The i-th line from the left is called the *line* i. We denote by $[n]$ the set $\{1, 2, \ldots, n\}$. Let $\pi = (\pi_1, \pi_2, \ldots, \pi_n)$ be a permutation of $[n]$. A *ladder lottery* of π is a network with n lines and zero or more bars such that

This work was supported by JSPS KAKENHI Grant Numbers JP18H04091, JP20H05793, JP20K14317, and JP22K17849.

S.-Y. Hsieh et al. (Eds.): IWOCA 2023, LNCS 13889, pp. 331–342, 2023.
https://doi.org/10.1007/978-3-031-34347-6_28

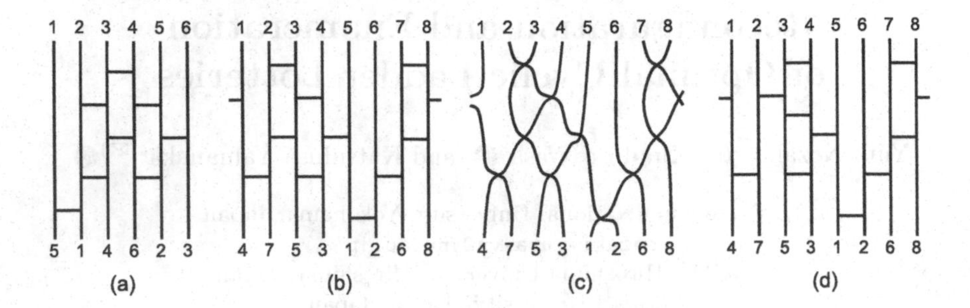

Fig. 1. (a) An optimal ladder lottery of the permutation $(5, 1, 4, 6, 2, 3)$. (b) A cyclic ladder lottery of the permutation $(4, 7, 5, 3, 1, 2, 6, 8)$ and (c) its representation as a pseudoline arrangement. (d) A cyclic ladder lottery obtained from (b) by applying a braid relation to the triple $(2, 3, 4)$.

(1) the top endpoints of lines correspond to the identity permutation,
(2) each bar exchanges two elements in $[n]$, and
(3) the bottom endpoints of lines correspond to π.

See Fig. 1(a) for an example. In each bar in a ladder lottery, two elements are swapped. We can regard a bar as an adjacent transposition of two elements in the current permutation, and the permutation always results in the given permutation π. A ladder lottery of a permutation π is *optimal* if it consists of the minimum number of bars among ladder lotteries of π. Let L be an optimal ladder lottery of π and let m be the number of bars in L. Then, we can observe that m is equal to the number of "inversions" of π, which are pairs (π_i, π_j) in π with $\pi_i > \pi_j$ and $i < j$.

The ladder lotteries are related to some objects in theoretical computer science. For instance, the optimal ladder lotteries of the reverse permutation of $[n]$ one-to-one correspond to arrangements[1] of n pseudolines[2] such that any two pseudolines intersect (see [7]). Each bar in a ladder lottery corresponds to an intersection of two pseudolines. Note that, in an optimal ladder lottery of the reverse permutation, each pair of two elements in $[n]$ is swapped exactly once on a bar. For such pseudoline arrangements, there exist several results on the bounds of the number of them. Let B_n be the number of the arrangements of n pseudolines such that any two pseudolines intersect and let $b_n = \log_2 B_n$. The best upper and lower bounds are $b_n \leq 0.6571n^2$ by Felsner and Valtr [4] and $b_n \geq 0.2053n^2$ by Dumitrecu and Mandal [3], respectively.

In this paper, we consider a variation of ladder lotteries, "cyclic" ladder lotteries. A *cyclic ladder lottery* is a ladder lottery that is allowed to have bars between the first and last lines. Now, as is the case with ladder lotteries, we introduce "optimality" to cyclic ladder lotteries. A cyclic ladder lottery of a permutation π is *optimal* if it has the minimum number of bars. It is known that

[1] An arrangement is *simple* if no three pseudolines have a common intersection point. In this paper, we consider only simple arrangements of pseudolines.

[2] A *pseudoline* in the Euclidean plane is a y-monotone curve extending from positive infinity to negative infinity.

the minimum number of bars in a cyclic ladder lottery of a permutation is equal to the cyclic analogue of inversion number and is computed in $\mathcal{O}(n^2)$ time [5].

For the optimal ladder lotteries of a permutation π, reconfiguration and enumeration problems have been solved [6,7]. The key observation is that reconfiguration graphs under braid relations are always connected, where a reconfiguration graph is a graph such that each vertex corresponds to an optimal ladder lottery of π and each edge corresponds to a braid relation between two optimal ladder lotteries. Hence, for the reconfiguration problems, the answer to a reachability problem is always yes. Moreover, Yamanaka et al. [6] characterized the length of a shortest reconfiguration sequence and proposed an algorithm that finds it. For the enumeration problem, Yamanaka et al. [7] designed an algorithm that enumerates them by traversing a spanning tree defined on a reconfiguration graph. Now, does the same observation of reconfiguration graphs hold for optimal cyclic ladder lotteries? Can we solve reconfiguration and enumeration problems for optimal cyclic ladder lotteries?

For optimal cyclic ladder lotteries, reconfiguration graphs under braid relations may be disconnected. For example, the two optimal cyclic ladder lotteries of the permutation $(4, 2, 6, 1, 5, 3)$ in Fig. 2 have no reconfiguration sequence under braid relations. Actually, it can be observed that the set of optimal cyclic ladder lotteries of a permutation π is partitioned into the sets of optimal cyclic ladder lotteries with the same "optimal displacement vectors" [5], which represent the movement direction of the each element in $[n]$ in optimal cyclic ladder lotteries. Note that applying a braid relation does not change a displacement vector. Therefore, to enumerate all the optimal cyclic ladder lotteries of a permutation π, we have to solve two enumeration problems: (1) enumerate all the optimal cyclic ladder lotteries of π with the same optimal displacement vector and (2) enumerate all the optimal displacement vectors of π. We first consider reconfiguration and enumeration problems for cyclic optimal ladder lotteries of a given permutation π and optimal displacement vector $\boldsymbol{x}$. For the reconfiguration problem, we show that any two optimal cyclic ladder lotteries of π and $\boldsymbol{x}$ are always reachable under braid relations. Then, for the enumeration problem, we design an algorithm that enumerates all the cyclic optimal ladder lotteries of π and $\boldsymbol{x}$ in polynomial delay. Next, we consider the two problems for the optimal displacement vectors of a permutation π under "max-min contractions". For the reconfiguration problem, we characterize the length of a shortest reconfiguration sequence between two optimal displacement vectors of π and show that one can compute a shortest reconfiguration sequence. For the enumeration problem, we design a constant-delay algorithm that enumerates all the optimal displacement vectors of π.

Due to space limitations, all the proofs are omitted.

2 Preliminary

Let $\pi = (\pi_1, \pi_2, \ldots, \pi_n)$ be a permutation of $[n]$. In a cyclic ladder lottery of π, each element i in $[n]$ starts at the top endpoint of the line i, and goes down

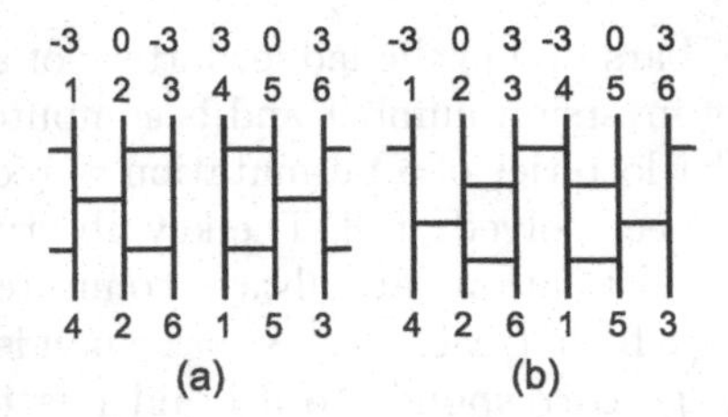

Fig. 2. (a) An optimal cyclic ladder lottery of the permutation $(4, 2, 6, 1, 5, 3)$ and its optimal displacement vector $(-3, 0, -3, 3, 0, 3)$ and (b) an optimal cyclic ladder lottery of the same permutation and its optimal displacement vector $(-3, 0, 3, -3, 0, 3)$.

along the line, then whenever i comes to an endpoint of a bar, i goes to the other endpoint and goes down again, then finally i reaches the bottom endpoint of the line j, where $i = \pi_j$. This path is called the *route* of the element i. Each bar is regarded as a cyclically adjacent transposition and the composition of all the transpositions in a cyclic ladder lottery always results in π. A cyclic ladder lottery of π is *optimal* if the ladder lottery contains the minimum number of bars. For example, the cyclic ladder lottery in Fig. 1(b) is optimal. We can observe from the figure that there exists an optimal cyclic ladder lottery L without a slit, that is, L does not come from any ladder lottery.

A cyclic ladder lottery of π is regarded as an arrangement of n pseudolines on a cylinder. The route of an element in $[n]$ corresponds to a pseudoline and a bar corresponds to an intersection of two pseudolines. Figure 1(c) shows the arrangement of pseudolines corresponding to the cyclic ladder lottery in Fig. 1(b). We use terminologies on pseudoline arrangements on a cylinder instead of the ones on cyclic ladder lotteries to clarify discussions. Let $\mathrm{pl}(L, i)$ denote the pseudoline of $i \in [n]$ in a cyclic ladder lottery L. Note that the top endpoint of $\mathrm{pl}(L, i)$ corresponds to the element i in the identity permutation and the bottom endpoint of $\mathrm{pl}(L, i)$ corresponds to π_i in π. In an optimal cyclic ladder lottery L, any two pseudolines cross at most once. From now on, we assume that any two pseudolines in L cross at most once. For two distinct elements $i, j \in [n]$, $\mathrm{cr}(i, j)$ denotes the intersection of $\mathrm{pl}(L, i)$ and $\mathrm{pl}(L, j)$ if it exists. For distinct $i, j, k \in [n]$, a triple $\{i, j, k\}$ is *tangled* if $\mathrm{pl}(L, i)$, $\mathrm{pl}(L, j)$, and $\mathrm{pl}(L, k)$ cross each other. Let $\{i, j, k\}$ be a tangled triple in L. Let M be the ladder lottery induced from L by the three pseudolines $\mathrm{pl}(L, i)$, $\mathrm{pl}(L, j)$, and $\mathrm{pl}(L, k)$, and let p, q, r be the three intersections in M. Without loss of generality, in M, we suppose that (1) p is adjacent to the two top endpoints of two pseudolines, (2) q is adjacent to the top endpoint of a pseudoline and the bottom endpoint of a pseudoline, (3) r is adjacent to the two bottom endpoints of two pseudolines. Then, $\{i, j, k\}$ is a *left tangled triple* if p, q, r appear in counterclockwise order on the contour of the region enclosed by $\mathrm{pl}(M, i)$, $\mathrm{pl}(M, j)$, and $\mathrm{pl}(M, k)$. Similarly, $\{i, j, k\}$ is a *right tangled triple* if p, q, r appear in clockwise order on the contour of the region. See Fig. 3 for examples. A tangled triple $\{i, j, k\}$ is *minimal* if the region enclosed by $\mathrm{pl}(L, i)$, $\mathrm{pl}(L, j)$, and $\mathrm{pl}(L, k)$ includes no subpseudoline in L. A *braid relation* is an operation to transform a minimal left (resp. right) tangled triple into another minimal right (resp. left) tangled triple. The cyclic ladder lottery in Fig. 1(d) is

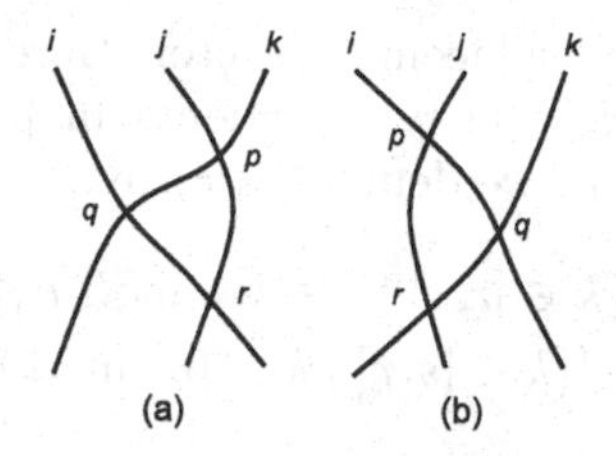

Fig. 3. Illustrations of (a) a left tangled triple and (b) a right tangled triple.

obtained from the ladder lottery in Fig. 1(b) by applying a braid relation to the triple $\{2, 3, 4\}$.

The vector $\boldsymbol{x} = (x_1, x_2, \ldots, x_n)$ is a *displacement vector* of π if $\sum_{i \in [n]} x_i = 0$ and $i + x_i \equiv \pi_i \bmod n$ for any $i \in [n]$. Let L be a cyclic ladder lottery of π. Then, a displacement vector can be defined from L and denote it by $DV(L)$. Intuitively, the element x_i in $DV(L) = (x_1, x_2, \ldots, x_n)$ represents the movement direction of the element i in $[n]$. That is, if $x_i > 0$, the element i goes right and if $x_i < 0$, the element i goes left. For instance, the displacement vector of the ladder lottery in Fig. 1(b) is $(-4, 4, 1, -3, -2, 1, 3, 0)$.

Let $\mathcal{L}(\pi)$ be the set of the cyclic ladder lotteries of π. Let $\mathcal{L}^{\mathrm{opt}}(\pi)$ be the set of the optimal cyclic ladder lotteries of π and let $\mathcal{L}^1(\pi)$ be the set of the cyclic ladder lotteries of π such that any two pseudolines cross at most once. Note that $\mathcal{L}^{\mathrm{opt}}(\pi) \subseteq \mathcal{L}^1(\pi) \subseteq \mathcal{L}(\pi)$ holds. A displacement vector $\boldsymbol{x}$ is *optimal* if there exists an optimal cyclic ladder lottery $L \in \mathcal{L}^{\mathrm{opt}}(\pi)$ such that $\boldsymbol{x} = DV(L)$ holds. We define the sets of cyclic ladder lotteries with the same displacement vector, as follows: $\mathcal{L}^{\mathrm{opt}}(\pi, \boldsymbol{x}) = \{L \in \mathcal{L}^{\mathrm{opt}}(\pi) \mid DV(L) = \boldsymbol{x}\}$ and $\mathcal{L}^1(\pi, \boldsymbol{x}) = \{L \in \mathcal{L}^1(\pi) \mid DV(L) = \boldsymbol{x}\}$. Then, we have the following lemma which shows that the set $\mathcal{L}^{\mathrm{opt}}(\pi)$ is partitioned into sets of optimal cyclic ladder lotteries with the same optimal displacement vectors.

Lemma 1. *Let π be a permutation. Then,*

$$\mathcal{L}^{\mathrm{opt}}(\pi) = \bigsqcup_{\boldsymbol{x} \in \mathcal{D}(\pi)} \mathcal{L}^{\mathrm{opt}}(\pi, \boldsymbol{x}),$$

where $\mathcal{D}(\pi)$ is the set of the optimal displacement vectors of π.

Similarly, a displacement vector $\boldsymbol{x}$ is said to be *almost optimal* if there exists a cyclic ladder lottery L of π such that $L \in \mathcal{L}^1(\pi, \boldsymbol{x})$ and $\boldsymbol{x} = DV(L)$ hold. Let L' be the ladder lottery obtained from an optimal cyclic ladder lottery L by removing $\mathrm{pl}(L, i)$ for $i \in [n]$. Note that L' is a cyclic ladder lottery of the permutation π' obtained from π by removing i and L' may be a non-optimal cyclic ladder lottery of π'.

In the study of ladder lotteries, the inversion number plays a crucial role. In [5, (3.6)], the cyclic analogue of the inversion number

$$\mathrm{inv}(\boldsymbol{x}) = \frac{1}{2} \sum_{i,j \in [n]} |c_{ij}(\boldsymbol{x})|$$

is introduced, where $\boldsymbol{x}$ is a displacement vector. Here, a crossing number $c_{ij}(\boldsymbol{x})$ is defined as follows. Let i, j be two elements in $[n]$ and let $r = i - j$ and $s = (i + x_i) - (j + x_j)$. Then, we define $c_{ij}(\boldsymbol{x})$ by

$$c_{ij}(\boldsymbol{x}) = \begin{cases} |\{k \in [r, s] \mid k \equiv 0 \mod n\}| & r \leq s \\ -|\{k \in [s, r] \mid k \equiv 0 \mod n\}| & s < r. \end{cases}$$

The number $\text{inv}(\boldsymbol{x})$ coincides with the affine inversion number for $\widetilde{\mathfrak{S}}_n$ (see [2, Section 8.3] for instance). As mentioned in [5], $\text{inv}(\boldsymbol{x})$ is equal to the number of intersections between the n pseudolines on the cylinder. Note here that [5, Lemma 3.6] corresponds to [2, Proposition 8.3.1].

3 Reconfiguration and Enumeration of Cyclic Ladder Lotteries with Optimal Displacement Vectors

Let $\pi = (\pi_1, \pi_2, \ldots, \pi_n)$ be a permutation of $[n]$, and let $\boldsymbol{x} = (x_1, x_2, \ldots, x_n)$ be an optimal displacement vector of π. In this section, we consider the problems of reconfiguration and enumeration for the set of the optimal cyclic ladder lotteries in $\mathcal{L}^{\text{opt}}(\pi, \boldsymbol{x})$.

3.1 Reconfiguration

In this subsection, we consider a reconfiguration between two optimal cyclic ladder lotteries in $\mathcal{L}^{\text{opt}}(\pi, \boldsymbol{x})$. The formal description of the problem is given below.

Problem: Reconfiguration of optimal cyclic ladder lotteries with optimal displacement vector (ReconfCLL-DV)
Instance: A permutation π, an optimal displacement vector $\boldsymbol{x}$ of π, and two optimal cyclic ladder lotteries $L, L' \in \mathcal{L}^{\text{opt}}(\pi, \boldsymbol{x})$
Question: Does there exist a reconfiguration sequence between L and L' under braid relations?

To solve ReconfCLL-DV, we consider the reconfiguration problem for the cyclic ladder lotteries in $\mathcal{L}^1(\pi, \boldsymbol{y})$, where $\boldsymbol{y}$ is an almost optimal displacement vector of π. First, we show that any two cyclic ladder lotteries in $\mathcal{L}^1(\pi, \boldsymbol{y})$ are always reachable. As a byproduct, we obtain an answer to ReconfCLL-DV.

Let L be a cyclic ladder lottery in $\mathcal{L}^1(\pi, \boldsymbol{y})$ and suppose that L contains one or more intersections. Then, in L, there exists an element $i \in [n]$ such that $\text{pl}(L, i)$ and $\text{pl}(L, (i + 1) \bmod n)$ cross, since assuming otherwise contradicts the definition of cyclic ladder lotteries.

Lemma 2. *Let π and $\boldsymbol{y}$ be a permutation in $\mathfrak{S}_n$ and an almost optimal displacement vector of π, respectively. Let L be a cyclic ladder lottery in $\mathcal{L}^1(\pi, \boldsymbol{y})$. Let $i \in [n]$ be an element such that $\text{pl}(L, i)$ and $\text{pl}(L, (i+1) \bmod n)$ cross. Then, there exists a reconfiguration sequence under braid relations between L and L', where L' is a cyclic ladder lottery in $\mathcal{L}^1(\pi, \boldsymbol{y})$ such that $\text{cr}(i, (i + 1) \bmod n)$ appears as a topmost intersection in L'.*

Using Lemma 2, we can prove the following theorem, which claims that any two cyclic ladder lotteries in $\mathcal{L}^1(\pi, \boldsymbol{y})$ with the same displacement vector are always reachable.

Theorem 1. *Let π be a permutation in $\mathfrak{S}_n$. Let L, L' be two cyclic ladder lotteries in $\mathcal{L}^1(\pi)$. Then, $DV(L) = DV(L')$ holds if and only if L and L' are reachable under braid relations.*

The following corollary is immediately from Theorem 1.

Corollary 1. *For any instance of* ReconfCLL-DV*, the answer is yes and one can construct a reconfiguration sequence between two ladder lotteries in $\mathcal{L}^{\mathrm{opt}}(\pi)$.*

3.2 Enumeration

In this subsection, we consider the problem of enumerating all the optimal cyclic ladder lotteries in $\mathcal{L}^{\mathrm{opt}}(\pi, \boldsymbol{x})$. The formal description of the problem is as follows.

Problem: Enumeration of optimal cyclic ladder lottery with optimal displacement vector (EnumCLL-DV)
Instance: A permutation π and an optimal displacement vector $\boldsymbol{x}$ of π.
Output: All the cyclic ladder lotteries in $\mathcal{L}^{\mathrm{opt}}(\pi, \boldsymbol{x})$ without duplication.

As in the previous subsection, we consider the enumeration problem for $\mathcal{L}^1(\pi, \boldsymbol{y})$, where $\boldsymbol{y}$ is an almost optimal displacement vector of π and propose an enumeration algorithm for $\mathcal{L}^1(\pi, \boldsymbol{y})$, since the algorithm can be applied to EnumCLL-DV. From Theorem 1, the reconfiguration graph of $\mathcal{L}^1(\pi, \boldsymbol{y})$ under braid relations is connected. This implies that the reverse search [1] can be applied for enumerating them.

Let $\pi \in \mathfrak{S}_n$ and let $\boldsymbol{y}$ be an almost optimal displacement vector of π. We denote by $LT(L)$ the set of the left tangled triples in a cyclic ladder lottery L. A cyclic ladder lottery L in $\mathcal{L}^1(\pi, \boldsymbol{y})$ is a *root* of $\mathcal{L}^1(\pi, \boldsymbol{y})$ if $LT(L) = \emptyset$ holds. If a cyclic ladder lottery in $\mathcal{L}^1(\pi, \boldsymbol{y})$ has no tangled triple, then $\mathcal{L}^1(\pi, \boldsymbol{y})$ contains only one ladder lottery. This is trivial from Theorem 1. For convenience, in such case, we define the ladder lottery as a root.

Lemma 3. *Let $\pi \in \mathfrak{S}_n$ and let $\boldsymbol{y}$ be an almost optimal displacement vector of π. Suppose that a cyclic ladder lottery in $\mathcal{L}^1(\pi, \boldsymbol{y})$ contains one or more tangled triples. Then, any $L \in \{L \in \mathcal{L}^1(\pi, \boldsymbol{y}) \mid LT(L) \neq \emptyset\}$ has a minimal left tangled triple.*

Let L be a cyclic ladder lottery in $\mathcal{L}^1(\pi, \boldsymbol{y})$. Let $\{i, j, k\}$ and $\{i', j', k'\}$ be two distinct left tangled triples in L, and suppose that $i < j < k$ and $i' < j' < k'$ hold. We say that $\{i, j, k\}$ is *smaller* than $\{i', j', k'\}$ if either $i < i'$ holds, $i = i'$ and $j < j'$ hold, or $i = i'$, $j = j'$, and $k < k'$ hold. The *parent*, denoted by $\mathrm{par}(L)$, of $L \in \{L \in \mathcal{L}^1(\pi, \boldsymbol{y}) \mid LT(L) \neq \emptyset\}$ is the cyclic ladder lottery obtained from L by applying a braid relation to the smallest minimal left tangled triple in L. We say that L is a *child* of $\mathrm{par}(L)$. Note that Lemma 3 implies that the parent is always defined for any $L \in \{L \in \mathcal{L}^1(\pi, \boldsymbol{y}) \mid LT(L) \neq \emptyset\}$. Moreover, the parent is unique from its definition.

Lemma 4. *Let $\pi \in \mathfrak{S}_n$ and let $\boldsymbol{y}$ be an almost optimal displacement vector of π. Let L be a cyclic ladder lottery in $\mathcal{L}^1(\pi, \boldsymbol{y})$. By repeatedly finding the parent from L, we have a root L_0 of $\mathcal{L}^1(\pi, \boldsymbol{y})$.*

Lemma 4 implies that there exists a root of $\mathcal{L}^1(\pi, \boldsymbol{y})$. Now, below, we show the uniqueness of a root of $\mathcal{L}^1(\pi, \boldsymbol{y})$.

Lemma 5. *Let $\pi \in \mathfrak{S}_n$ and let $\boldsymbol{y}$ be an almost optimal displacement vector of π. For two cyclic ladder lotteries $L, L' \in \mathcal{L}^1(\pi, \boldsymbol{y})$, if $LT(L) = LT(L')$, $L = L'$ holds.*

From Lemma 5, we have the following corollary.

Corollary 2. *Let $\pi \in \mathfrak{S}_n$ and let $\boldsymbol{y}$ be an almost optimal displacement vector of π. Then, the root L_0 of $\mathcal{L}^1(\pi, \boldsymbol{y})$ is unique.*

From Lemma 4, by repeatedly finding the parent from L, we finally obtain the root of $\mathcal{L}^1(\pi, \boldsymbol{y})$. The *parent sequence* of $L \in \mathcal{L}^1(\pi, \boldsymbol{y})$ is the sequence $\langle L_1, L_2, \ldots, L_p \rangle$ such that

(1) L_1 is L itself,
(2) $L_i = \text{par}(L_{i-1})$ for $i = 2, 3, \ldots, p$, and
(3) L_p is the root L_0 of $\mathcal{L}^1(\pi, \boldsymbol{y})$.

Note that the parent sequence of the root is $\langle L_0 \rangle$. The *family tree* of $\mathcal{L}^1(\pi, \boldsymbol{y})$ is the tree structure obtained by merging the parent sequences of all the cyclic ladder lotteries in $\mathcal{L}^1(\pi, \boldsymbol{y})$. In the family tree of $\mathcal{L}^1(\pi, \boldsymbol{y})$, the root node is the root L_0 of $\mathcal{L}^1(\pi, \boldsymbol{y})$, each node is a cyclic ladder lottery in $\mathcal{L}^1(\pi, \boldsymbol{y})$, and each edge is a parent-child relationship of two ladder lotteries in $\mathcal{L}^1(\pi, \boldsymbol{y})$.

Now, we design an enumeration algorithm of all the cyclic ladder lotteries in $\mathcal{L}^1(\pi, \boldsymbol{y})$. The algorithm enumerates them by traversing the family tree of $\mathcal{L}^1(\pi, \boldsymbol{y})$ starting from the root L_0. To traverse the family tree, we design the following two algorithms: (1) an algorithm that constructs the root L_0 of $\mathcal{L}^1(\pi, \boldsymbol{y})$ and (2) an algorithm that enumerates all the children of a given cyclic ladder lottery in $\mathcal{L}^1(\pi, \boldsymbol{y})$. Note that, if we have the above two algorithms, starting from the root, we can traverse the family tree by recursively applying the child-enumeration algorithm.

The outline of how to construct the root is as follows. First, we construct a cyclic ladder lottery from π and $\boldsymbol{y}$, which may not be the root in $\mathcal{L}^1(\pi, \boldsymbol{y})$. Next, from the constructed cyclic ladder lottery, by repeatedly finding parents, we obtain the root.

Lemma 6. *Let π and $\boldsymbol{y}$ be a permutation in $\mathfrak{S}_n$ and an almost optimal displacement vector of π, respectively. One can construct the root L_0 of $\mathcal{L}(\pi, \boldsymbol{y})$ in $\mathcal{O}(n + (\text{inv}(\boldsymbol{y}))^3)$ time.*

Let L be a cyclic ladder lottery in $\mathcal{L}^1(\pi, \boldsymbol{x})$ and let $t = \{i, j, k\}$ be a minimal right tangled triple in L. We denote by $L(t)$ the cyclic ladder lottery obtained from L by applying braid relation to t. We can observe that $L(t)$ is a child of L if

Algorithm 1: ENUM-CLL-CHILDREN(L)

```
1 Output L
2 foreach minimal right tangled triple t in L do
3 |   if t is the smallest triple in L(t) then
4 |   |_ ENUM-CLL-CHILDREN(L(t))
```

and only if t is the smallest minimal left tangled triple in $L(t)$. This observation gives the child-enumeration algorithm shown in Algorithm 1.

First, we construct the root L_0 of $\mathcal{L}^1(\pi, \boldsymbol{x})$ and call Algorithm 1 with the argument L_0. The algorithm outputs the current cyclic ladder lottery L, which is the argument of the current recursive call. Next, for every minimal right tangled triple t in L, if $L(t)$ is a child of L, the algorithm calls itself with the argument $L(t)$. Algorithm 1 traverses the family tree of $\mathcal{L}^1(\pi, \boldsymbol{y})$ and hence enumerates all the cyclic ladder lotteries in $\mathcal{L}^1(\pi, \boldsymbol{y})$. Each recursive call lists all the minimal right tangled triples. To do that, we take $\mathcal{O}(\text{inv}(\boldsymbol{y}))$ time. For each minimal right tangled triple t, we check whether or not $L(t)$ is a child of L, as follows. First, we construct $L(t)$ from L. Next, in $L(t)$, we list all the minimal right tangled triples. Finally, we check t is the smallest one in the listed triples. The answer is true implies that $L(t)$ is a child of L. This takes $\mathcal{O}(\text{inv}(\boldsymbol{y}))$ time. Therefore, a recursive call of Algorithm 1 takes $\mathcal{O}((\text{inv}(\boldsymbol{y}))^2)$ time.

Theorem 2. *Let $\pi \in \mathfrak{S}_n$ and let $\boldsymbol{y}$ be an almost optimal displacement vector of π. After constructing the root in $\mathcal{O}(n + (\text{inv}(\boldsymbol{y}))^3)$ time, one can enumerate all the cyclic ladder lotteries in $\mathcal{L}^1(\pi, \boldsymbol{y})$ in $\mathcal{O}((\text{inv}(\boldsymbol{y}))^2)$ delay.*

The discussion to derive Theorem 2 can be applied to the set $\mathcal{L}^{\text{opt}}(\pi, \boldsymbol{x})$ for an optimal displacement vector $\boldsymbol{x}$ of π. Hence, we have the following corollary.

Corollary 3. *Let $\pi \in \mathfrak{S}_n$ and let $\boldsymbol{x}$ be an optimal displacement vector of π. After constructing the root in $\mathcal{O}(n + (\text{inv}(\boldsymbol{x}))^3)$ time, one can enumerate all the cyclic ladder lotteries in $\mathcal{L}^{\text{opt}}(\pi, \boldsymbol{x})$ in $\mathcal{O}((\text{inv}(\boldsymbol{x}))^2)$ delay.*

4 Reconfiguration and Enumeration of Optimal Cyclic Ladder Lotteries

In this section, we consider the problem of enumerating all the optimal cyclic ladder lotteries in $\mathcal{L}^{\text{opt}}(\pi)$, where $\pi \in \mathfrak{S}_n$. That is, a displacement vector is not given as an input and only a permutation is given. The formal description of the problem is shown below.

Problem: Enumeration of optimal cyclic ladder lotteries (ENUMCLL)
Instance: A permutation $\pi \in \mathfrak{S}_n$.
Output: All the optimal cyclic ladder lotteries in $\mathcal{L}^{\text{opt}}(\pi)$ without duplication.

From Lemma 1, we have the following outline of an enumeration algorithm to solve ENUMCLL. First, we enumerate all the optimal displacement vectors of a given permutation. Next, for each optimal displacement vector, we enumerate all the optimal cyclic ladder lotteries using the algorithm in the previous section.

Therefore, in this section, we first consider the reconfiguration problem for the optimal displacement vectors to investigate the connectivity of the reconfiguration graph of them. Utilizing the knowledge of the reconfiguration graph, we design an enumeration algorithm that enumerates all the optimal displacement vectors of a given permutation.

4.1 Reconfiguration

Let $\boldsymbol{x} = (x_1, x_2, \ldots, x_n)$ be an optimal displacement vector of a permutation π. We denote the maximum and minimum elements in $\boldsymbol{x}$ by $\max(\boldsymbol{x})$ and $\min(\boldsymbol{x})$. Let i and j be two indices such that $x_i = \max(\boldsymbol{x})$, $x_j = \min(\boldsymbol{x})$, and $x_i - x_j = n$. If $\boldsymbol{x}$ includes two or more maximum (and minimum) values, the index i (and j) is chosen arbitrarily. Then, a *max-min contraction*[3] T_{ij} of $\boldsymbol{x}$ is a function $T_{ij} \colon \mathbb{Z}^n \to \mathbb{Z}^n$ such that $T_{ij}(\boldsymbol{x}) = (z_1, z_2, \ldots, z_n)$, where

$$z_k = \begin{cases} x_k - n & \text{if } k = i \\ x_k + n & \text{if } k = j \\ x_k & \text{otherwise.} \end{cases}$$

We consider the following reconfiguration problem under the max-min contractions.

Problem: Reconfiguration of optimal displacement vectors (RECONFDV)
Instance: A permutation $\pi \in \mathfrak{S}_n$ and two optimal displacement vectors $\boldsymbol{x}$ and $\boldsymbol{x}'$ of π.
Question: Does there exist a reconfiguration sequence between $\boldsymbol{x}$ and $\boldsymbol{x}'$ under max-min contractions?

Jerrum [5] showed the following theorem.

Theorem 3 ([5]). *Any instance of* RECONFDV *is a yes-instance.*

In the remaining part of this subsection, we consider the shortest version of RECONFDV. Let $\boldsymbol{x} = (x_1, x_2, \ldots, x_n)$ and $\boldsymbol{x}' = (x'_1, x'_2, \ldots, x'_n)$ be two optimal displacement vectors of π. We denote the length of a shortest reconfiguration sequence between $\boldsymbol{x}$ and $\boldsymbol{x}'$ under max-min contractions by $OPT_{DV}(\boldsymbol{x}, \boldsymbol{x}')$. For two optimal displacement vectors $\boldsymbol{x}$ and $\boldsymbol{x}'$, we define $\boldsymbol{x} \triangle \boldsymbol{x}'$ by

$$\boldsymbol{x} \triangle \boldsymbol{x}' = \sum_{i \in [n]} 1 - \delta_{x_i, x'_i}.$$

One can characterize the length of a shortest reconfiguration sequence using the symmetric difference, as stated in the following theorem.

[3] The contraction is originally proposed by Jerrum [5].

Theorem 4. *Let $\boldsymbol{x}$ and $\boldsymbol{x}'$ be two optimal displacement vectors of a permutation $\pi \in \mathfrak{S}_n$. Then $OPT_{DV}(\boldsymbol{x}, \boldsymbol{x}') = \frac{\boldsymbol{x} \triangle \boldsymbol{x}'}{2}$ holds. Moreover, one can compute a reconfiguration sequence of length $OPT_{DV}(\boldsymbol{x}, \boldsymbol{x}')$ in $\mathcal{O}(n + OPT_{DV}(\boldsymbol{x}, \boldsymbol{x}'))$ time.*

4.2 Enumeration

In this subsection, we consider the following enumeration problem.

Problem: Enumeration of optimal displacement vectors (ENUMDV)
Instance: A permutation $\pi \in \mathfrak{S}_n$.
Output: All the optimal displacement vectors of π without duplication.

Theorem 3 implies that the reconfiguration graph of the optimal displacement vectors of a permutation under max-min contractions is connected. Therefore, we may use the reverse search technique to enumerate them.

Let $\pi \in \mathfrak{S}_n$. Recall that $\mathcal{D}(\pi)$ denotes the set of all the optimal displacement vectors of π. Let $\boldsymbol{x} = (x_1, x_2, \ldots, x_n)$ and $\boldsymbol{x}' = (x'_1, x'_2, \ldots, x'_n)$ be two distinct optimal displacement vectors in $\mathcal{D}(\pi)$. The vector $\boldsymbol{x}$ is *larger than* $\boldsymbol{x}'$ if $x_i = x'_i$ for $i = 1, 2, \ldots, j$ and $x_{j+1} > x'_{j+1}$.

The *root* of $\mathcal{D}(\pi)$, denoted by $\boldsymbol{x_0}$, is the largest displacement vector among $\mathcal{D}(\pi)$. Intuitively, the root includes the maximum values in early indices. Note that $\boldsymbol{x_0}$ is unique in $\mathcal{D}(\pi)$. Let $\boldsymbol{x}$ be an optimal displacement vector in $\mathcal{D}(\pi) \setminus \{\boldsymbol{x_0}\}$. Let $\alpha(\boldsymbol{x})$ be the minimum index of $\boldsymbol{x}$ such that $\boldsymbol{x}_{\alpha(\boldsymbol{x})} = \min(\boldsymbol{x})$. Let $\beta(\boldsymbol{x})$ be the minimum index of $\boldsymbol{x}$ such that $\alpha(\boldsymbol{x}) < \beta(\boldsymbol{x})$ and $\boldsymbol{x}_{\beta(\boldsymbol{x})} = \max(\boldsymbol{x})$ hold. Then, we define the *parent* of $\boldsymbol{x}$ by $\text{par}(\boldsymbol{x}) = T_{\beta(\boldsymbol{x})\alpha(\boldsymbol{x})}(\boldsymbol{x})$. Note that $\text{par}(\boldsymbol{x})$ is larger than $\boldsymbol{x}$ and always exists for $\boldsymbol{x} \neq \boldsymbol{x_0}$. We say that $\boldsymbol{x}$ is a *child* of $\text{par}(\boldsymbol{x})$. The *parent sequence* $\langle \boldsymbol{x_1}, \boldsymbol{x_2}, \ldots, \boldsymbol{x_k} \rangle$ of $\boldsymbol{x}$ is a sequence of optimal displacement vectors in $\mathcal{D}(\pi)$ such that

(1) $\boldsymbol{x_1} = \boldsymbol{x}$,
(2) $\boldsymbol{x_i} = \text{par}(\boldsymbol{x_{i-1}})$ for each $i = 2, 3, \ldots, m$, and
(3) $\boldsymbol{x_k} = \boldsymbol{x_0}$.

Note that one can observe that, by repeatedly finding the parents from any optimal displacement vector in $\mathcal{D}(\pi)$, the root $\boldsymbol{x_0}$ is always obtained. Hence, by merging the parent sequence of every vector in $\boldsymbol{x} \in \mathcal{D}(\pi) \setminus \{\boldsymbol{x_0}\}$, we have the tree structure rooted at $\boldsymbol{x_0}$. We call the tree the *family tree* of $\mathcal{D}(\pi)$. Note that the family tree is a spanning tree of the reconfiguration graph of $\mathcal{D}(\pi)$ under max-min contractions. Therefore, to enumerate all the optimal displacement vectors in $\mathcal{D}(\pi)$, we traverse the family tree of $\mathcal{D}(\pi)$. To traverse the family tree, we design an algorithm to enumerate all the children of an optimal displacement vector in $\mathcal{D}(\pi)$.

Let $\boldsymbol{x} = (x_1, x_2, \ldots, x_n)$ be an optimal displacement vector in $\mathcal{D}(\pi)$. The *max-min index sequence*, denoted by $m(\boldsymbol{x}) = \langle m_1, m_2, \ldots, m_\ell \rangle$, of $\boldsymbol{x}$ is a sequence of indices of $\boldsymbol{x}$ such that either $x_{m_i} = \max(\boldsymbol{x})$ or $x_{m_i} = \min(\boldsymbol{x})$ for $i = 1, 2, \ldots, \ell$ and $m_i < m_{i+1}$ for each $i = 1, 2, \ldots, \ell - 1$. It can be observed that if $x_{m_1} = \min(\boldsymbol{x})$, $\boldsymbol{x}$ has no child from the definition of the parent. Hence, we assume

Algorithm 2: ENUM-DV-CHILDREN($\boldsymbol{x}$)

1 Output $\boldsymbol{x}$
2 Let $m(\boldsymbol{x}) = \langle m_1, m_2, \ldots, m_\ell \rangle$ be the max-min index sequence of $\boldsymbol{x}$
3 Let $m_p = \alpha(\boldsymbol{x})$ and $m_q = \beta(\boldsymbol{x})$
4 **foreach** $j = p, p+1, \ldots, q-1$ **do**
5 ⌊ ENUM-DV-CHILDREN($T_{m_{p-1}m_j}(\boldsymbol{x})$)

that $x_{m_1} = \max(\boldsymbol{x})$, below. Now, we enumerate all the children of $\boldsymbol{x}$ as follows. Suppose that $m_p = \alpha(\boldsymbol{x})$ and $m_q = \beta(\boldsymbol{x})$. (For the root $\boldsymbol{x_0}$, $\beta(\boldsymbol{x})$ is not defined. Hence, for convenience, we define $\beta(\boldsymbol{x}) = \ell + 1$ for the root.)

Lemma 7. *Let $\boldsymbol{x}$ be an optimal displacement vector of $\pi \in \mathfrak{S}_n$. Let $m(\boldsymbol{x}) = \langle m_1, m_2, \ldots, m_\ell \rangle$ be the max-min index sequence of $\boldsymbol{x}$. Then, $T_{m_i m_j}(\boldsymbol{x})$ is a child of $\boldsymbol{x}$ if and only if $i = p - 1$ and $j = p, p+1, \ldots, q-1$ hold.*

From Lemma 7, we have the child-enumeration algorithm shown in Algorithm 2. We first construct the root $\boldsymbol{x_0}$ and call the algorithm with the argument $\boldsymbol{x_0}$. By recursively calling Algorithm 2, one can traverse the family tree.

Theorem 5. *Let $\pi \in \mathfrak{S}_n$. After $\mathcal{O}(n)$-time preprocessing, one can enumerate all the optimal displacement vectors in $\mathcal{D}(\pi)$ in $\mathcal{O}(1)$ delay.*

References

1. Avis, D., Fukuda, K.: Reverse search for enumeration. Discret. Appl. Math. **65**(1–3), 21–46 (1996)
2. Björner, A., Brenti, F.: Combinatorics of Coxeter Groups. Graduate Texts in Mathematics, vol. 231. Springer, New York (2005)
3. Dumitrescu, A., Mandal, R.: New lower bounds for the number of pseudoline arrangements. J. Comput. Geom. **11**(1), 60–92 (2020). https://doi.org/10.20382/jocg.v11i1a3
4. Felsner, S., Valtr, P.: Coding and counting arrangements of pseudolines. Discret. Comput. Geom. **46**, 405–416 (2011)
5. Jerrum, M.R.: The complexity of finding minimum-length generator sequence. Theor. Comput. Sci. **36**, 265–289 (1985)
6. Yamanaka, K., Horiyama, T., Wasa, K.: Optimal reconfiguration of optimal ladder lotteries. Theor. Comput. Sci. **859**, 57–69 (2021). https://doi.org/10.1016/j.tcs.2021.01.009
7. Yamanaka, K., Nakano, S., Matsui, Y., Uehara, R., Nakada, K.: Efficient enumeration of all ladder lotteries and its application. Theor. Comput. Sci. **411**, 1714–1722 (2010)

Improved Analysis of Two Algorithms for Min-Weighted Sum Bin Packing

Guillaume Sagnol(✉)

Institut für Mathematik, Technische Universität Berlin,
Straße des 17. Juni 136, 10623 Berlin, Germany
sagnol@math.tu-berlin.de

Abstract. We study the Min-Weighted Sum Bin Packing problem, a variant of the classical Bin Packing problem in which items have a weight, and each item induces a cost equal to its weight multiplied by the index of the bin in which it is packed. This is in fact equivalent to a batch scheduling problem that arises in many fields of applications such as appointment scheduling or warehouse logistics. We give improved lower and upper bounds on the approximation ratio of two simple algorithms for this problem. In particular, we show that the knapsack-batching algorithm, which iteratively solves knapsack problems over the set of remaining items to pack the maximal weight in the current bin, has an approximation ratio of at most 17/10.

Keywords: Bin Packing · Batch Scheduling · Approximation Algorithms

Full version: arXiv:2304.02498

1 Introduction

Bin Packing is a fundamental problem in computer science, in which a set of items must be packed into the smallest possible number of identical bins, and has applications in fields as varied as logistics, data storage or cloud computing. A property of the bin packing objective is that all bins are treated as "equally good", which is not always true in applications with a temporal component. Consider, e.g., the problem of allocating a set of n patients to days for a medical appointment with a physician. Each patient is characterized by a service time and a weight indicating the severity of his health condition. The total time required to examine all patients assigned to a given day should not exceed a fixed threshold. The days thus correspond to bins indexed by $1, 2, \ldots$, and bins with a small index are to be favored, especially for patients with a high weight, because they yield a smaller waiting time for the patients.

In the Min-Weighted Sum Bin Packing Problem (MWSBP), which was formally introduced in [5], the input consists of a set of n items with size $s_i \in (0, 1]$ and weight $w_i > 0$[1]. The goal is to find a feasible allocation of

[1] In [5] the weights are assumed to be $w_i \geq 1$, but we can reduce to this case by scaling; Our lower bounds on approximation ratios are not affected by this operation.

S.-Y. Hsieh et al. (Eds.): IWOCA 2023, LNCS 13889, pp. 343–355, 2023.
https://doi.org/10.1007/978-3-031-34347-6_29

minimum cost of the set of items to bins, i.e., a partition of $[n] := \{1, \dots, n\}$ into $B_1, \dots, B_p$ such that $\sum_{i \in B_k} s_i \leq 1$ holds for all $k \in [p]$, where the cost of putting item i into B_k is given by $k \cdot w_i$. In other words, if we use the notation $x(S) := \sum_{i \in S} x_i$ for a vector $x \in \mathbb{R}^n$ and a subset $S \subseteq [n]$, the cost of a feasible allocation is

$$\Phi(B_1, \dots, B_p) := \sum_{k=1}^{p} k \cdot w(B_k) = \sum_{k=1}^{p} \sum_{j=k}^{p} w(B_j). \tag{1}$$

Another interpretation of this problem is that we have a set of jobs with unit processing times, and want to schedule them on a batch processing machine capable of simultaneously processing a batch of jobs of total size at most 1, with the objective to minimize the weighted sum of completion times. In the three-field notation introduced by Graham et al. [9], MWSBP is thus equivalent to $1|p\text{-}batch, s_j, p_j = 1|\sum w_j C_j$; we refer to [7] for a recent review giving more background on parallel batch scheduling. Broadly speaking, we see that MWSBP captures the main challenge of many real-world problems in which items must be packed over time, such as the appointment scheduling problem mentioned above, or the problem of scheduling batches of orders in an order-picking warehouse.

For a given instance of MWSBP and an algorithm ALG, we denote by OPT the cost of an optimal solution, and the cost of the solution returned by ALG is denoted by ALG as well. Recall that the approximation ratio $\mathcal{R}(\texttt{ALG})$ of an algorithm ALG is the smallest constant $\rho \geq 1$ such that, for all instances of the problem, it holds $\texttt{ALG} \leq \rho \cdot \texttt{OPT}$.

Related Work. The complexity of MWSBP is well understood, as the problem is NP-hard in the strong sense and a polynomial-time approximation scheme (PTAS) exists [5]. This paper also gives a simple algorithm based on `Next-Fit` which has an approximation ratio of 2. Prior to this work, several heuristics have been proposed for a generalization of the problem with incompatible families of items associated with different batch times [1], and it was shown that a variant of `First-Fit` has an approximation ratio of 2 as well.

The unweighted version of the problem, `Min-Sum Bin Packing` (MSBP), in which each item has weight $w_i = 1$, also attracted some attention. A simpler PTAS is described in [4] for this special case. The authors also analyze the *asymptotic approximation ratio* of several algorithms based on `First-Fit` or `Next-Fit`, i.e., the limit of the approximation ratio when one restricts attention to instances with $\texttt{OPT} \to \infty$. In particular, they give an algorithm whose asymptotic approximation ratio is at most 1.5604. In addition, it is known that the variant of `First-Fit` considering items in nondecreasing order of their sizes is a $(1 + \sqrt{2}/2)$-approximation algorithm for MSBP [10].

Another related problem is MIN-WEIGHTED SUM SET COVER (MWSSC), in which a collection of subsets $S_1, \dots, S_m$ of a ground set $[n]$ of items is given in the input, together with weights $w_1, \dots, w_n > 0$. As in MWSBP, a solution is a partition $B_1, B_2, \dots, B_p$ of $[n]$ and the cost of these batches is given by (1), but the difference is that each batch $B_k \subseteq [n]$ must be the subset of some S_j, $j \in [m]$. In other words, the difference with MWSBP is that the feasible batches

Table 1. Summary of previous and new bounds on the approximation ratio of simple algorithms for MWSBP. The results of this paper and indicated in boldface, together with the reference of the proposition (Pr.) or theorem (Th.) where they are proved.

Algorithm	lower bounds		upper bounds		upper bounds for special cases of MWSBP
	previous	new	previous	new	
`WNFI`	2 [1]		2 [1]		1.618 for $w_i = 1$, $\texttt{OPT} \to \infty$ [4]
`WFFI`	2 [1]		2 [1]		1.707 for $w_i = 1$ [10]
`WNFD-R`	2 [5]		2 [5]		
`WFFI-R`	1.354 [4]	**1.557** (Pr. 6)	2 [1]		**1.636** for $w_i = s_i$ (Th. 7)
`KB`	1.354 [4]	**1.433** (Pr. 5)	4 [6]	**1.7** (Th. 3)	

are described *explicitly* by means of a collection of maximal batches rather than *implicitly* using a knapsack constraint $s(B_k) \leq 1$. Any instance of MWSBP can thus be cast as an instance of MWSSC, although this requires an input of exponential size (enumeration of all maximal subsets of items of total size at most 1). The unweighted version of MWSSC was introduced in [6]. The authors show that a natural greedy algorithm is a 4-approximation algorithm, and that this performance guarantee is the best possible unless $P = NP$.

Contribution and Outline. Given the practical relevance of MWSBP for real-world applications, we feel that it is important to understand the performance of simple algorithms for this problem, even though a PTAS exists. Indeed, the PTAS of [5] has an astronomical running time, which prevents its usage in applications. This paper gives improved lower and upper bounds on the approximation ratio of two simple algorithms for MWSBP; In particular, we obtain the first *simple algorithm* with an approximation algorithm strictly below 2, see Table 1 for a summary of previous and new results.

The two analyzed algorithms, called `Knapsack-Batching` (`KB`) and `Weighted First-Fit Increasing Reordered` (`WFFI-R`), are introduced in Sect. 2 alongside with more background on algorithms for Bin Packing and MWSBP. In Sect. 3 we show that $\mathcal{R}(\texttt{KB}) \in (1.433, 1.7]$ and in Sect. 4 we show that $\mathcal{R}(\texttt{WFFI-R}) > 1.557$. Further, all the best known lower bounds are achieved for instances in which $w_i = s_i$ for all items, a situation reminiscent of scheduling problems where we minimize the weighted sum of completion times, and where the worst instances have jobs with equal *smith ratios*; see, e.g. [12]. It is therefore natural to focus on this regime, and we prove that `WFFI-R` is a $(7+\sqrt{37})/8$-approximation algorithm on instances with $w_i = s_i$ for all $i \in I$.

2 Preliminaries

Many well known heuristics have been proposed for the Bin Packing problem. For example, `Next-Fit` (`NF`) and `First-Fit` (`FF`) consider the list of items in an arbitrary given order, and assign them sequentially to bins. The two algorithms differ in their approach to select the bin where the current item is placed. To this end, NF keeps track of a unique opened bin; the current item is placed in this bin if it fits into it, otherwise another bin is opened and the item is placed into it. In contrast, FF first checks whether the current item fits in any of the bins used so far. If it does, the item is put into the bin where it fits that has been opened first, and otherwise a new bin is opened. It is well known that FF packs all items in at most $\lfloor 17/10 \cdot OPT \rfloor$ bins, where OPT denotes the minimal number of required bins, a result obtained by Dósa and Sgall [2] after a series of papers that improved an additive term in the performance guarantee.

In MWSBP, items not only have a size but also a weight. It is thus natural to consider the weighted variants WFFI, WFFD, WNFI, WNFI of FF and NF, respectively, where W stands for *weighted*, and the letter I (resp. D) stands for *increasing* (resp. *decreasing*) and indicates that the items are considered in nondecreasing (resp. nonincreasing) order of the ratio s_i/w_i. Using a simple exchange argument, we can see that every algorithm can be enhanced by reordering the bins it outputs by nonincreasing order of their weights. We denote the resulting algorithms by adding the suffix -R to their acronym. While WNFD, WFFD and WFFD-R do not have a bounded approximation ratio, even for items of unit weights [4], it is shown in [1] that the approximation ratio of WNFI and WFFI is exactly 2, and the same holds for WNFD-R [5]. This paper gives improved bounds for WFFI-R in Sect. 4.

Another natural algorithm for MWSBP is the Knapsack-Batching (KB) algorithm, which was introduced in [1] and can be described as follows: At iteration $k = 1, 2, \ldots$, solve a knapsack problem to find the subset of remaining items $B_k \subseteq [n] \setminus (B_1 \cup \ldots \cup B_{k-1})$ of maximum weight that fits into a bin (i.e., $s(B_k) \leq 1$). In fact, [5] argues that KB is the direct translation of the greedy algorithm for the Min-Sum Set Cover Problem mentioned in the introduction, so its approximation ratio is at most 4. In practice one can use a *fully polynomial-time approximation scheme* (FPTAS) to obtain a near-optimal solution of the knapsack problem in polynomial-time at each iteration, which yields a $(4+\varepsilon)$-approximation algorithm for MWSBP. In the next section, we show an improved upper bound of 1.7 for the KB algorithm (or $1.7+\varepsilon$ if an FPTAS is used for solving the knapsack problems in polynomial-time).

3 The Knapsack-Batching Algorithm

In this section, we study the Knapsack-Batching (KB) algorithm. Throughout this section, we denote by $B_1, \ldots, B_p$ the set of bins returned by KB, and by $O_1, \ldots, O_q$ the optimal bins, for an arbitrary instance of MWSBP. For notational convenience, we also define $B_{p'} = \emptyset$ for all $p' \geq p+1$ and $O_{q'} = \emptyset$ for all $q' \geq q+1$. Recall that KB solves a knapsack over the remaining subset of items at each iteration, so that for all k, $w(B_k) \geq w(B)$ holds for all $B \subseteq [n] \setminus (B_1 \cup \ldots \cup B_{k-1})$ such that $s(B) \leq 1$.

We first prove the following proposition, which shows that a performance guarantee of α can be proved if we show that KB packs at least as much weight in the first αk bins as OPT does in only k bins. The proof relies on expressing OPT

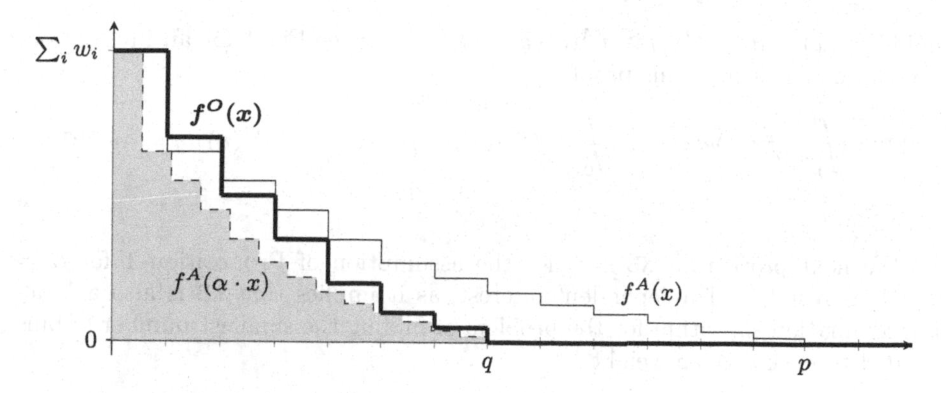

Fig. 1. Illustration of the proof of Proposition 1. The area below the thick curve of $x \mapsto f^O(x)$ is OPT, and the area under the thin curve of $x \mapsto f^A(x)$ is KB. Shrinking this area horizontally by a factor α produces the shaded area under $x \mapsto f^A(\alpha \cdot x)$, which must be smaller than OPT.

and KB as the area below a certain curve. Then, we show that shrinking the curve corresponding to KB by a factor α yields an underestimator for the OPT-curve; see Fig. 1. A similar idea was used in [6] to bound the approximation ratio of the greedy algorithm for Min-Sum Set Cover, but their bound of 4 results from shrinking the curve by a factor 2 along both the x-axis and the y-axis.

Proposition 1. *Let $\alpha \geq 1$. If for all $k \in [q]$ it holds*

$$w(B_1) + w(B_2) + \ldots + w(B_{\lfloor \alpha k \rfloor}) \geq w(O_1) + w(O_2) + \ldots + w(O_k),$$

then $KB \leq \alpha \cdot OPT$.

Proof. For all $x \geq 0$, let $f^O(x) := \sum_{j=\lfloor x \rfloor+1}^{\infty} w(O_j)$. Note that f^O is piecewise constant and satisfies $f^O(x) = \sum_{j=k}^{q} w(O_j)$ for all $x \in [k-1, k), k \in [q]$ and $f^O(x) = 0$ for all $x \geq q$. As a result, using the second expression in (1), we can express OPT as the area below the curve of $f^O(x)$:

$$\mathtt{OPT} = \sum_{k=1}^{q} \sum_{j=k}^{q} w(O_j) = \int_0^{\infty} f^O(x)\, dx.$$

Similarly, we have $\mathtt{KB} = \int_0^{\infty} f^A(x)\, dx$, where for all $x \geq 0$ we define $f^A(x) := \sum_{j=\lfloor x \rfloor+1}^{\infty} w(B_j)$. The assumption of the lemma can be rewritten as $f^A(\alpha \cdot k) \leq f^O(k)$, for all $k \in [q]$, and we note that this inequality also holds for $k = 0$, as $f^A(0) = f^O(0) = \sum_{i=1}^{n} w_i$.

Now, we argue that this implies $f^A(\alpha \cdot x) \leq f^O(x)$, for all $x \geq 0$. If x lies in an interval of the form $x \in [k, k+1)$ for some $k \in \{0, \ldots, q-1\}$, then we have $f^O(x) = f^O(k) \geq f^A(\alpha \cdot k) \geq f^A(\alpha \cdot x)$, where the last inequality follows from $k \leq x$ and the fact that f^A is nonincreasing. Otherwise, we have $x \geq q$, so it holds $f^O(x) = 0 = f^O(q) \geq f^A(\alpha \cdot q) \geq f^A(\alpha \cdot x)$; see Fig. 1 for an illustration. We can now conclude this proof:

$$\mathtt{KB} = \int_0^{\infty} f^A(x)\,dx = \alpha \cdot \int_0^{\infty} f^A(\alpha \cdot y)\,dy \leq \alpha \cdot \int_0^{\infty} f^O(y)\,dy = \alpha \cdot \mathtt{OPT}.$$

∎

We next prove that KB satisfies the assumption of Proposition 1 for $\alpha = \frac{17}{10}$. This result is of independent interest, as it implies that KB is also a $17/10$-approximation algorithm for the problem of finding the smallest number of bins needed to pack a given weight.

Proposition 2. *For all instances of MWSBP in which the bins $O_1, \ldots, O_q$ are optimal and KB outputs the bins $B_1, \ldots, B_p$, and for all $k \in [q]$ it holds*

$$w(B_1) + w(B_2) + \ldots + w(B_{\lfloor \frac{17}{10} k \rfloor}) \geq w(O_1) + w(O_2) + \ldots + w(O_k).$$

Proof. Let $k \in [q]$. We define the sets $\overline{B} := B_1 \cup \ldots \cup B_{\lfloor 1.7k \rfloor}$, $\overline{O} := O_1 \cup \ldots \cup O_k$. For all $i \in \overline{B}$, denote by $\beta(i) \in [1.7\,k]$ the index of the KB–bin that contains i, i.e., $i \in B_{\beta(i)}$. Now, we order the items in $\overline{O}$ in such a way that we first consider the items of $\overline{O} \cap \overline{B}$ in nondecreasing order of $\beta(i)$, and then all remaining items in $\overline{O} \setminus \overline{B}$ in an arbitrary order. Now, we construct a new packing $H_1, \ldots, H_{q'}$ of the items in $\overline{O}$, by applying `First-Fit` to the list of items in $\overline{O}$, considered in the aforementioned order. For all $i \in \overline{O} \cap \overline{B}$, let $\beta'(i)$ denote the index such that $i \in H_{\beta'(i)}$. Clearly, our order on $\overline{O}$ implies $\beta'(i) \leq \beta(i)$ for all $i \in \overline{O} \cap \overline{B}$.

It follows from [2] that $q' \leq \lfloor 1.7 \cdot k \rfloor$. So we define $H_j := \emptyset$ for $j = q'+1, q'+2, \ldots, \lfloor 1.7 \cdot k \rfloor$, and it holds $\overline{O} = H_1 \cup \ldots \cup H_{\lfloor 1.7k \rfloor}$. Now, we will show that $w(H_j) \leq w(B_j)$ holds for all $j = 1, \ldots, \lfloor 1.7k \rfloor$. To this end, using the greedy property of KB, it suffices to show that all elements of H_j remain when the knapsack problem of the jth iteration is solved, i.e., $H_j \subseteq [n] \setminus (B_1 \cup \ldots \cup B_{j-1})$. So, let $i \in H_j$. If $i \notin \overline{B}$, then $i \notin (B_1 \cup \ldots \cup B_{j-1})$ is trivial. Otherwise, it is $i \in \overline{O} \cap \overline{B}$, so we have $j = \beta'(i) \leq \beta(i)$, which implies that i does not belong to any B_ℓ with $\ell < j$. We can now conclude the proof:

$$w(\overline{O}) = \sum_{i=1}^{\lfloor 1.7 \cdot k \rfloor} w(H_i) \leq \sum_{i=1}^{\lfloor 1.7 \cdot k \rfloor} w(B_i) = w(\overline{B}).$$

∎

Proposition 1 and Proposition 2 yield the main result of this section:

Theorem 3. $\mathcal{R}(\mathit{KB}) \leq \frac{17}{10}$.

Remark. *It is straightforward to adapt the above proof to show that if we use an FPTAS for obtaining a $(1 + \frac{\varepsilon}{n})$-approximation of the knapsack problems in place of an optimal knapsack at each iteration of the KB algorithm, we obtain a polynomial-time $(\frac{17}{10} + \varepsilon)$-approximation algorithm for MWSBP.*

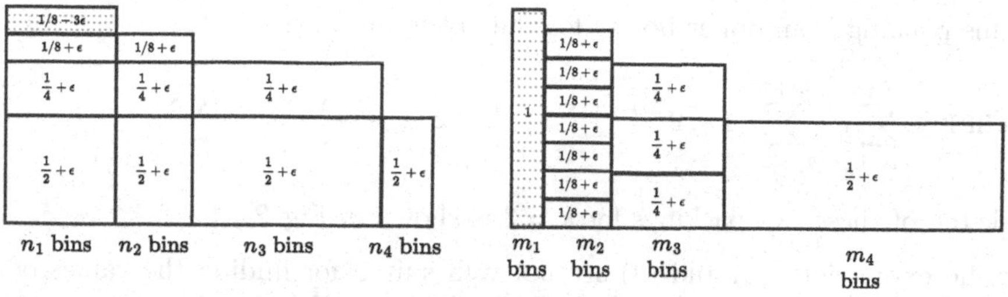

Fig. 2. Sketch of the "good" packing (left) and the packing returned by KB (right) for the instance defined before Lemma 4, for $k = 3$. The dotted blocks represent bunches of tiny items. The m_i's are chosen so that the number of items in each class is the same in both packings, i.e., $m_1 = n_1 \cdot (1/8 - 3\epsilon)$, $m_2 = (n_1+n_2)/7$, $m_3 = (n_1+n_2+n_3)/3$ and $m_4 = n_1 + n_2 + n_3 + n_4$.

We next show a lower bound on the approximation ratio of KB. For some integers s and k with $s \geq 2^k$, let $\epsilon = \frac{1}{2^k \cdot s}$. Given an integer vector $\boldsymbol{n} \in \mathbb{Z}_{\geq 0}^{k+1}$, we construct the following instance: The items are partitioned in $k+1$ classes, i.e., $[n] = C_1 \cup C_2 \cup \ldots \cup C_{k+1}$. For all $j \in [k]$, the class C_j consists of $N_j = n_1 + \ldots + n_{k+2-j}$ items of class j, with $s_i = w_i = \frac{1}{2^j} + \epsilon$, $\forall i \in C_j$. In addition, the class C_{k+1} contains $N_{k+1} := n_1 \cdot (s - k)$ tiny items with $s_i = w_i = \epsilon$. We further assume that $m_1 := n_1 \cdot (2^{-k} - k\epsilon) = \frac{n_1(s-k)}{2^k \cdot s}$ is an integer, and for all $j \in [k]$, $\frac{N_j}{2^j - 1}$ is an integer. Then, for $j = 2, \ldots, k+1$ we let $m_j := \frac{N_{k+2-j}}{2^{k+2-j}-1} \in \mathbb{Z}$.

On this instance, KB could start by packing m_1 bins of total weight 1, containing $2^k \cdot s$ tiny items each. After this, there only remains items from the classes $C_1, \ldots, C_k$, and the solution of the knapsack problem is to put $2^k - 1$ items of the class C_k in a bin. Therefore, KB adds $m_2 = N_k/(2^k - 1)$ such bins of weight $(2^k - 1) \cdot (2^{-k} + \epsilon) = 1 - 2^{-k} + (2^k - 1) \cdot \epsilon$ into the solution. Continuing this reasoning, we see that for each $j = 2, \ldots, k+1$, when there only remains items from the classes $C_1, \ldots, C_{k+2-j}$, KB adds a group of m_j bins that contain $(2^{k+2-j} - 1)$ items of class C_j, with a weight of $1 - 2^{-(k+2-j)} + (2^{k+2-j} - 1) \cdot \epsilon$ each. This gives:

$$\text{KB} = \sum_{i=1}^{m_1} i + \sum_{j=2}^{k+1} \left(1 - 2^{-(k+2-j)} + (2^{k+2-j} - 1) \cdot \epsilon\right) \cdot \sum_{i=m_1+\ldots+m_{j-1}+1}^{m_1+\ldots+m_j} i. \tag{2}$$

On the other hand, we can construct the following packing for this instance: The first n_1 bins contain one item of each class $C_1, \ldots, C_k$, plus $(s-k)$ tiny items; their total weight is thus $\sum_{j=1}^{k} (2^{-j} + \epsilon) + (s-k)\epsilon = 1 - 2^{-k} + s \cdot \epsilon = 1$. Then, for each $j = 2, \ldots, k+1$, we add a group of n_j bins, each containing one item of each class $C_1, C_2, \ldots, C_{k+2-j}$. The bins in the jth group thus contain a weight of $\sum_{i=1}^{k+2-j} (2^{-i} + \epsilon) = 1 - 2^{-(k+2-j)} + (k+2-j) \cdot \epsilon$. Obviously, the cost of this packing is an upper bound for the optimum:

$$\texttt{OPT} \leq \sum_{i=1}^{n_1} i + \sum_{j=2}^{k+1} \left(1 - 2^{-(k+2-j)} + (k+2-j)\cdot\epsilon\right) \cdot \sum_{i=n_1+\ldots+n_{j-1}+1}^{n_1+\ldots+n_j} i. \quad (3)$$

A sketch of these two packings for $k = 3$ is shown in Fig. 2.

The expressions (2) and (3) are not well suited for finding the values of n_i producing the best lower bound on $\mathcal{R}(\texttt{KB})$. The next lemma shows how we can instead focus on maximizing a more amenable ratio of two quadratics.

Lemma 4. *Let $\boldsymbol{L}$ be the $(k+1) \times (k+1)$ lower triangular matrix with all elements equal to 1 in the lower triangle, $\boldsymbol{u} := [\frac{1}{2^{k+1}}, \frac{1}{2^{k+1}}, \frac{1}{2^k}, \ldots, \frac{1}{2^2}] \in \mathbb{R}^{k+1}$, $\boldsymbol{v} := [\frac{1}{2^k}, \frac{1}{2^k-1}, \frac{1}{2^{k-1}-1}, \ldots, \frac{1}{2^2-1}, \frac{1}{2-1}] \in \mathbb{R}^{k+1}$, and let $\boldsymbol{U} :=$ Diag$(\boldsymbol{u})$ and $\boldsymbol{V} :=$ Diag$(\boldsymbol{v})$. Then, for all $\boldsymbol{x} \in \mathbb{R}^{k+1}_{>0}$, it holds*

$$\mathcal{R}(\mathit{KB}) \geq R(\boldsymbol{x}) := \frac{\boldsymbol{x}^T \boldsymbol{L}^T \boldsymbol{V} \boldsymbol{L}^T \boldsymbol{U} \boldsymbol{L} \boldsymbol{V} \boldsymbol{L} \boldsymbol{x}}{\boldsymbol{x} \boldsymbol{L}^T \boldsymbol{U} \boldsymbol{L} \boldsymbol{x}}.$$

The proof of this lemma is ommited by lack of space and can be found in the full version [11]; The basic idea is to use (2)–(3) for an instance where n_i is proportional to $\lfloor t x_i \rfloor$ for some $t > 0$ and to let $t \to \infty$.

Proposition 5. $\mathcal{R}(\mathit{KB}) > 1.4334$.

Proof. This follows from applying Lemma 4 to the vector $\boldsymbol{x} = [0.97, 0.01, 0.01, 0.01, 0.03, 0.07, 0.15, 0.38] \in \mathbb{R}^8_{>0}$. This vector is in fact the optimal solution of the problem of maximizing $R(\boldsymbol{x})$ over $\mathbb{R}^8$, rounded after 2 decimal places for the sake of readability. To find this vector, we reduced the problem of maximizing $R(\boldsymbol{x})$ to an eigenvalue problem, and recovered $\boldsymbol{x}$ by applying a reverse transformation to the corresponding eigenvector. Note that optimizing for $k = 50$ instead of $k = 7$ only improves the bound by $3 \cdot 10^{-5}$. ■

4 First-Fit-Increasing Reordered

In this section, we analyze the `WFFI-R` algorithm. First recall that `WFFI` (without bins reordering) has an approximation ratio of 2. To see that this bound is tight, consider the instance with 2 items such that $w_1 = s_1 = \epsilon$ and $w_2 = s_2 = 1$. Obviously, the optimal packing puts the large item in the first bin and the small item in the second bin, so that $\texttt{OPT} = 1 + 2\epsilon$. On the other hand, since the two items have the same ratio s_i/w_i, `WFFI` could put the small item in the first bin, which yields $\texttt{WFFI} = \epsilon + 2$. Therefore, $\texttt{WFFI}/\texttt{OPT} \to 2$ as $\epsilon \to 0$. In this instance however, we see that reordering the bins by decreasing weight solves the issue.

It is easy to find an instance in which `WFFI-R/OPT` approaches $3/2$, though: Let $w_1 = s_1 = \epsilon$ and $w_2 = w_3 = s_2 = s_3 = \frac{1}{2}$. Now, the optimal packing puts items 2 and 3 in the first bin, which gives $\texttt{OPT} = 1 + 2\epsilon$, while `WFFI-R` could return the bins $B_1 = \{1, 2\}, B_2 = \{3\}$, so that $\texttt{WFFI-R} = \frac{1}{2} + \epsilon + 2 \cdot \frac{1}{2} = \frac{3}{2} + \epsilon$.

We next show how to obtain a stronger bound on $\mathcal{R}(\texttt{WFFI-R})$. To this end, we recall that for all $k \in \mathbb{N}$ and $\epsilon > 0$ sufficiently small, there exists an instance of Bin Packing with the following properties [2,8]: There are $n = 30k$ items, which can be partitioned in three classes. The items $i = 1, \dots, 10k$ are *small* and have size $s_i = 1/6 + \delta_i\epsilon$; then there are $10k$ *medium* items of size $s_i = 1/3 + \delta_i\epsilon$ ($i = 10k+1 \dots, 20k$), and $10k$ *large* items of size $s_i = 1/2+\epsilon$ ($i = 20k+1, \dots, 30k$). The constants $\delta_i \in \mathbb{R}$ can be positive or negative and are chosen in a way that, if FF considers the items in the order $1, \dots, n$, it produces a packing $B_1, \dots B_{17k}$ in which the first $2k$ bins contain 5 small items, the next $5k$ bins contain 2 medium items and the last $10k$ bins contain a single large item. On the other hand, there exists a packing of all items into $10k + 1$ bins, in which $10k - 1$ bins consist of a small, a medium and a large item and have size $1 - O(\epsilon)$, and the two remaining bins have size $1/2 + O(\epsilon)$ (a small with a medium item, and a large item alone).

We can transform this Bin Packing instance into a MWSBP instance, by letting $w_i = s_i$, for all i. This ensures that all items have the same ratio s_i/w_i, so we can assume that WFFI-R considers the items in any order we want. In addition, we consider two integers u and v, and we increase the number of medium items from $10k$ to $10k + 2u$ (so the medium items are $i = 10k + 1, \dots, 20k + 2u$) and the number of large items from $10k$ to $10k + 2u + v$ (so the large items are $i = 20k+2u+1, \dots, 30k+4u+v$). The δ_i's are unchanged for $i = 1, \dots 20k$, and we let $\delta_i = 1$ for all additional medium items ($i = 20k + 1, \dots, 20k + 2u$). Then, assuming that WFFI-R considers the items in the order $1, 2, \dots$, the algorithm packs $2k$ bins with 5 small items, $5k + u$ bins with 2 medium items and the last $10k + 2u + v$ bins with a single large item. On the other hand, we can use the optimal packing of the original instance, and pack the additional items into $2u$ bins of size $5/6 + 2\epsilon$ containing a medium and a large item, and v bins with a single large item. This amounts to a total of $10k - 1$ bins of size $1 - O(\epsilon)$, $2u$ bins of size $5/6 + O(\epsilon)$ and $v + 2$ bins of size $1/2 + O(\epsilon)$. In the limit $\epsilon \to 0$, we get

$$\begin{aligned}\texttt{WFFI-R} &= \frac{5}{6}\sum_{i=1}^{2k} i + \frac{2}{3}\sum_{i=2k+1}^{7k+u} i + \frac{1}{2}\sum_{i=7k+u+1}^{17k+3u+v} i\\ &= \frac{5}{6}k(2k+1) + \frac{1}{3}(5k+u)(9k+u+1) + \frac{1}{4}(10k+2u+v)(24k+4u+v+1)\end{aligned}$$

and

$$\begin{aligned}\texttt{OPT} &\geq \sum_{i=1}^{10k-1} i + \frac{5}{6}\sum_{i=10k}^{10k-1+2u} i + \frac{1}{2}\sum_{i=10k+2u}^{10k+2u+v+1} i\\ &= 5k(10k-1) + \frac{5}{6}u(20k+2u-1) + \frac{1}{4}(v+2)(20k+4u+v+1).\end{aligned}$$

Proposition 6. $\mathcal{R}(\textit{WFFI-R}) > 1.5576$.

Proof. The bound is obtained by substituting $k = 10350, u = 11250, v = 24000$ in the above expressions. As in the previous section, these values can be obtained

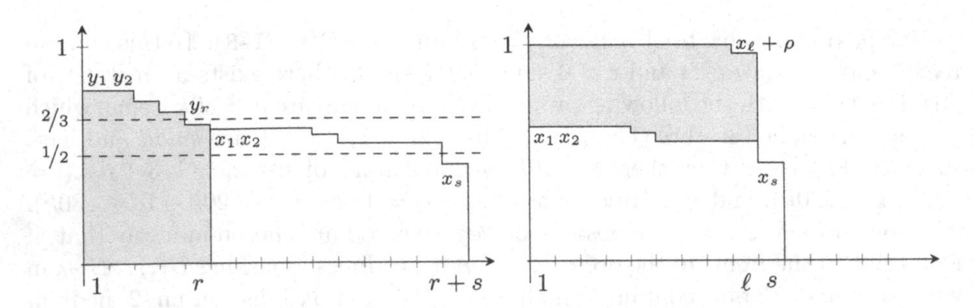

Fig. 3. Sketch of a packing defining $A(\boldsymbol{y}, \boldsymbol{x})$ (left), and corresponding pseudo-packing defining $B(\boldsymbol{y}, \boldsymbol{x})$ (right). The values of ℓ and ρ ensure equality of the two shaded areas.

by reducing the problem of finding the best values of k, u, v to an eigenvalue problem and by scaling-up and rounding the obtained eigenvector. ■

All bad examples for WFFI-R (and even for all algorithms listed in Table 1) have the property that all items have the same ratio s_i/w_i. This makes sense intuitively, as the algorithm does not benefit from sorting the items anymore, i.e., items can be presented to the algorithm in an adversarial order. While the only known upper bound on the approximation ratio of WFFI-R is 2 (as WFFI-R can only do better than WFFI), we believe that $\mathcal{R}$(WFFI-R) is in fact much closer to our lower bound from Proposition 6. To support this claim, we next show an upper bound of approx. 1.6354 for instances with $s_i = w_i$.

Theorem 7. *For all MWSBP instances with $s_i = w_i$ for all $i \in [n]$, it holds*

$$\texttt{WFFI-R} \leq \frac{7+\sqrt{37}}{8} \cdot \texttt{OPT}.$$

Proof. Consider an instance of MWSBP with $w_i = s_i$ for all $i \in [n]$ and denote by $W_1 \geq \ldots \geq W_p$ the weight (or equivalently, the size) of the bins $B_1, \ldots, B_p$ returned by WFFI-R. We first handle the case in which there is at most one bin with weight $\leq \frac{2}{3}$, in which case we obtain a bound of $\frac{3}{2}$:

Claim 1. *If $W_{p-1} > \frac{2}{3}$, then $\texttt{WFFI-R} \leq \frac{3}{2}\texttt{OPT}$.*

This claim is proved in the appendix of the full version [11]. We can thus assume w.l.o.g. that there are at least 2 bins with weight $\leq \frac{2}{3}$. Let $r \in [p-1]$ denote the index of the first bin such that $W_r \leq 2/3$ and define $s := p - r \in [p-1]$. We define the vectors $\boldsymbol{y} \in \mathbb{R}^r$ and $\boldsymbol{x} \in \mathbb{R}^s$ such that $y_i = W_i$ ($\forall i \in [r]$) and $x_i = W_{r+i}$ ($\forall i \in [s]$). By construction, we have $1 \geq y_1 \geq \ldots \geq y_{r-1} > 2/3 \geq y_r \geq x_1 \geq \ldots \geq x_s$. We also define $x_0 := y_r$ and $x_i := 0$ for all $i > s$ for the sake of simplicity. Note that $W_i + W_j > 1$ must hold for all $i \neq j$, as otherwise the First-Fit algorithm would have put the items of bins i and j into a single bin. This implies $x_{s-1} + x_s > 1$, hence $x_{s-1} = \max(x_{s-1}, x_s) > 1/2$. With this notation, we have:

$$\texttt{WFFI-R} = A(\boldsymbol{y}, \boldsymbol{x}) := \sum_{i=1}^{r} i \cdot y_i + \sum_{i=1}^{s} (r+i) \cdot x_i.$$

Next, we prove a lower bound on OPT depending on $\boldsymbol{x}$ an $\boldsymbol{y}$. Observe that among the WFFI-R-bins $B_r, B_{r+1}, \ldots, B_{r+s}$ with weight$\leq 2/3$, at most one of them can contain two items or more (the first of these bins that was opened, as items placed in later bins have a size –and hence a weight– strictly larger than $1/3$). Thus, s out of these $s+1$ bins must contain a *single item*, and there is no pair of such single items fitting together in a bin. Therefore, the s single items must be placed in distinct bins of the optimal packing. This implies

$$w(O_i) \geq w(B_{r+i}) = x_i, \quad \text{for all } i \in [s]. \tag{4}$$

Now, let ℓ be the unique index such that $\sum_{i=1}^{\ell-1}(1-x_i) < \sum_{i=1}^{r} y_i \leq \sum_{i=1}^{\ell}(1-x_i)$, and define $\rho := \sum_{i=1}^{r} y_i - \sum_{i=1}^{\ell-1}(1-x_i) \in (0, 1-x_\ell]$. We claim that

$$\texttt{OPT} \geq B(\boldsymbol{y}, \boldsymbol{x}) := \frac{1}{2}\ell(\ell-1) + (x_\ell + \rho)\ell + \sum_{i=\ell+1}^{s} i \cdot x_i,$$

where the last sum is 0 if $\ell \geq s$, which corresponds to a "pseudo-packing" with weight 1 in bins $1, \ldots, \ell-1$, weight x_i in the bins $\ell+1, \ldots, s$ and the weight of bin ℓ is adjusted to $x_\ell + \rho$ so that the total weight equals $\sum_i x_i + \sum_i y_i$, see Fig. 3 for an illustration. This clearly gives a lower bound on OPT, as transferring weight from late bins to early bins only improves the solution, and the pseudo-packing defining $B(\boldsymbol{x}, \boldsymbol{y})$ packs the whole weight earlier than any packing $O_1, \ldots, O_q$ satisfying (4). We extend the definition of the function $B(\cdot, \cdot)$ to any pair of vectors $(\boldsymbol{y}', \boldsymbol{x}') \in \mathbb{R}^r_{>0} \times [0,1]^t$ with $t \leq s$, by setting $x'_i = 0$ for all $i > t$.

The above inequalities implies that WFFI-R/OPT is bounded from above by $R(\boldsymbol{y}, \boldsymbol{x}) := A(\boldsymbol{y}, \boldsymbol{x})/B(\boldsymbol{y}, \boldsymbol{x})$. We next give a series of technical claims (proved in the appendix of [11]) which allows us to compute the maximum of $R(\boldsymbol{y}, \boldsymbol{x})$ when $\boldsymbol{x}$ and $\boldsymbol{y}$ are as above. The first claim shows that we obtain an upper bound for some vectors $\boldsymbol{y}'$ and $\boldsymbol{x}'$ of the form $\boldsymbol{y}' = [\frac{2}{3}, \ldots, \frac{2}{3}, \alpha] \in \mathbb{R}^r$ and $\boldsymbol{x}' = [\alpha, \ldots, \alpha] \in \mathbb{R}^t$ for some $\alpha \in [\frac{1}{2}, \frac{2}{3}]$ and $t \leq s$. Its proof relies on averaging some coordinates of the vectors $\boldsymbol{y}$ and $\boldsymbol{x}$, and using a sequential rounding procedure to obtain equal coordinates in the vector $\boldsymbol{x}$ and to decrease $y_1, \ldots, y_{r-1}$ to $\frac{2}{3}$.

Claim 2. *There exists an $\alpha \in [\frac{1}{2}, \frac{2}{3}]$ and an integer t such that $R(\boldsymbol{y}, \boldsymbol{x}) \leq R(\boldsymbol{y}', \boldsymbol{x}')$ holds for the vectors $\boldsymbol{y}' = [\frac{2}{3}, \ldots, \frac{2}{3}, \alpha] \in \mathbb{R}^r$ and $\boldsymbol{x}' = [\alpha, \ldots, \alpha] \in \mathbb{R}^t$.*

Next, we give a handy upper bound for $R(\boldsymbol{y}', \boldsymbol{x}')$ when $\boldsymbol{y}'$ and $\boldsymbol{x}'$ are in the form obtained in the previous claim.

Claim 3. *Let $\boldsymbol{y}' := [\frac{2}{3}, \ldots, \frac{2}{3}, \alpha] \in \mathbb{R}^r$ and $\boldsymbol{x}' := [\alpha, \ldots, \alpha] \in \mathbb{R}^t$. If $\frac{2}{3}(r-1) + \alpha \leq (1-\alpha)t$, then it holds*

$$R(\boldsymbol{y}', \boldsymbol{x}') \leq H_1(\alpha, r, t) := \frac{3(1-\alpha)(3\alpha(t+1)(2r+t) + 2r(r-1))}{3\alpha + 4r^2 + (6\alpha - 2)r + 9(1-\alpha)\alpha t(t+1) - 2}.$$

Otherwise (if $\frac{2}{3}(r-1)+\alpha > (1-\alpha)t$), the following bound is valid:

$$R(\boldsymbol{y}',\boldsymbol{x}') \leq H_2(\alpha,r,t) := \frac{6r(r-1)+9\alpha(t+1)(2r+t)}{(2r+3\alpha(t+1)-2)(2r+3\alpha(t+1)+1)}.$$

We start by bounding the second expression. In that case, we obtain a bound equal to $\frac{13}{8}=1.625$.

Claim 4. *For all $\alpha \in [\frac{1}{2},\frac{2}{3}]$, $r \geq 1$ and $0 \leq t \leq \frac{2/3(r-1)+\alpha}{1-\alpha}$, $H_2(\alpha,r,t) \leq \frac{13}{8}$.*

Then, we bound H_1. It turns out that the derivative of $H(\alpha,r,t)$ with respect to α is nonpositive over the domain of α, so we obtain an upper bound by setting $\alpha=\frac{1}{2}$.

Claim 5. *For all $\alpha \in [\frac{1}{2},\frac{2}{3}]$, $r \geq 1$ and $t \geq 0$, it holds*

$$H_1(\alpha,r,t) \leq H_1(\frac{1}{2},r,t) = \frac{3(4r^2+r(6t+2)+3t(t+1))}{16r^2+4r+9t^2+9t-2}.$$

Finally, we obtain our upper bound by maximizing the above expression over the domain $r \geq 1, t \geq 0$. Let $u := r-1 \geq 0$. This allows us to rewrite the previous upper bound as

$$H_1(\frac{1}{2},r,t) = R_1(u,t) := \frac{3(6+3t^2+10u+4u^2+9t+6ut)}{18+9t+9t^2+36u+16u^2}.$$

for some nonnegative variables u and t. Rather than relying on complicated differential calculus to maximize R_1, we give a short proof based on a certificate that some matrix is *copositive* (see, e.g. [3]), that was found by solving a semidefinite program. Observe that $R_1(u,t) = \boldsymbol{z}^T A\boldsymbol{z}/\boldsymbol{z}^T B\boldsymbol{z}$, with

$$\boldsymbol{z} = \begin{pmatrix} u \\ t \\ 1 \end{pmatrix}, \quad A = \begin{pmatrix} 12 & 9 & 15 \\ 9 & 9 & 27/2 \\ 15 & 27/2 & 18 \end{pmatrix}, \quad \text{and} \quad B = \begin{pmatrix} 16 & 0 & 18 \\ 0 & 9 & 9/2 \\ 18 & 9/2 & 18 \end{pmatrix}.$$

Let $\tau := \frac{7+\sqrt{37}}{8}$, $X := \begin{pmatrix} 0 & 0 & 3 \\ 0 & 0 & 9/8 \\ 3 & 9/8 & 0 \end{pmatrix}$. The reader can verify that the matrix

$$Z = \tau B - A - X = \begin{pmatrix} 2(1+\sqrt{37}) & -9 & 9/4\cdot(\sqrt{37}-1) \\ -9 & 9/8\cdot(\sqrt{37}-1) & 9/16\cdot(\sqrt{37}-19) \\ 9/4\cdot(\sqrt{37}-1) & 9/16\cdot(\sqrt{37}-19) & 9/4\cdot(\sqrt{37}-1) \end{pmatrix}$$

is positive semidefinite. As a result, $\boldsymbol{z}^T(\tau B - A)\boldsymbol{z} \geq \boldsymbol{z}^T X\boldsymbol{z} = 6u + 9/4\cdot t$ holds for all $u,t \in \mathbb{R}$, and this quantity is nonnegative because $u \geq 0$ and $t \geq 0$. This proves $\tau\cdot \boldsymbol{z}^T B\boldsymbol{z} - \boldsymbol{z}^T A\boldsymbol{z} \geq 0$, and thus $R_1(u,t) \leq \tau$. We note that this bound is tight, as $R_1(u_i,t_i) \to \tau$ for sequences of integers $\{(t_i,u_i)\}_{i\in\mathbb{N}}$ such that t_i/u_i converges to $\frac{2}{9}(1+\sqrt{37})$. By Claims 2–5, we have thus shown that $R(\boldsymbol{y},\boldsymbol{x}) \leq \max(\frac{13}{8},\frac{7+\sqrt{37}}{8}) = \frac{7+\sqrt{37}}{8}$, which concludes this proof. ∎

Acknowledgements. The author thanks a group of colleagues, in particular Rico Raber and Martin Knaack, for discussions during a group retreat that contributed to this paper. The author also thanks anonymous referees, whose comments helped to improve the quality of this manuscript. This work was supported by the Deutsche Forschungsgemeinschaft (DFG, German Research Foundation) under Germany's Excellence Strategy — The Berlin Mathematics Research Center MATH+ (EXC-2046/1, project ID: 390685689).

References

1. Dobson, G., Nambimadom, R.S.: The batch loading and scheduling problem. Oper. Res. **49**(1), 52–65 (2001)
2. Dósa, G., Sgall, J.: First fit bin packing: a tight analysis. In: 30th International Symposium on Theoretical Aspects of Computer Science, STACS 2013. LIPIcs, vol. 20, pp. 538–549 (2013)
3. Dür, M.: Copositive programming–a survey. In: Diehl, M., Glineur, F., Jarlebring, E., Michiels, W. (eds.) Recent Advances in Optimization and Its Applications in Engineering, pp. 3–20. Springer, Heidelberg (2010). https://doi.org/10.1007/978-3-642-12598-0_1
4. Epstein, L., Johnson, D.S., Levin, A.: Min-sum bin packing. J. Comb. Optim. **36**(2), 508–531 (2018)
5. Epstein, L., Levin, A.: Minimum weighted sum bin packing. In: Kaklamanis, C., Skutella, M. (eds.) WAOA 2007. LNCS, vol. 4927, pp. 218–231. Springer, Heidelberg (2008). https://doi.org/10.1007/978-3-540-77918-6_18
6. Feige, U., Lovász, L., Tetali, P.: Approximating min sum set cover. Algorithmica **40**(4), 219–234 (2004)
7. Fowler, J.W., Monch, L.: A survey of scheduling with parallel batch (p-batch) processing. Eur. J. Oper. Res. **298**(1), 1–24 (2022)
8. Garey, M.R., Graham, R.L., Ullman, J.D.: Worst-case analysis of memory allocation algorithms. In: Proceedings of the 4th Annual ACM Symposium on Theory of Computing, STOC 1972, pp. 143–150 (1972)
9. Graham, R., Lawler, E., Lenstra, J., Kan, A.R.: Optimization and approximation in deterministic sequencing and scheduling: a survey. In: Discrete Optimization II, Annals of Discrete Mathematics, vol. 5, pp. 287–326 (1979)
10. Li, R., Tan, Z., Zhu, Q.: Batch scheduling of nonidentical job sizes with minsum criteria. J. Comb. Optim. **42**(3), 543–564 (2021)
11. Sagnol, G.: Improved analysis of two algorithms for min-weighted sum bin packing (2023). https://doi.org/10.48550/arXiv.2304.02498
12. Schwiegelshohn, U.: An alternative proof of the Kawaguchi-Kyan bound for the largest-ratio-first rule. Oper. Res. Lett. **39**(4), 255–259 (2011)

Sorting and Ranking of Self-Delimiting Numbers with Applications to Tree Isomorphism

Frank Kammer, Johannes Meintrup(✉), and Andrej Sajenko

THM, University of Applied Sciences Mittelhessen, Giessen, Germany
{frank.kammer,johannes.meintrup,andrej.sajenko}@mni.thm.de

Abstract. Assume that an N-bit sequence S of k numbers encoded as Elias gamma codes is given as input. We present space-efficient algorithms for sorting, dense ranking and competitive ranking on S in the word RAM model with word size $\Omega(\log N)$ bits. Our algorithms run in $O(k + \frac{N}{\log N})$ time and use $O(N)$ bits. The sorting algorithm returns the given numbers in sorted order, stored within a bit-vector of N bits, whereas our ranking algorithms construct data structures that allow us subsequently to return the dense/competitive rank of each number x in S in constant time. For numbers $x \in N$ with $x > N$ we require the position p_x of x as the input for our dense-/competitive-rank data structure. As an application of our algorithms above we give an algorithm for tree isomorphism, which runs in $O(n)$ time and uses $O(n)$ bits on n-node trees. The previous best linear-time algorithm for tree isomorphism uses $\Theta(n \log n)$ bits.

Keywords: space efficient · sorting · rank · dense rank · tree isomorphism

1 Introduction

Due to the rapid growth of the input data sizes in recent years, fast algorithms that also handle space efficiently are increasingly gaining importance. To prevent cache-faults and space deficiencies we focus on *space-efficient* algorithms, i.e., algorithms that run (almost) as fast as standard solutions for the problem under consideration with decreased utilization of space.

Graphs are often used to encode structural information in many fields, e.g., in chemistry [23] or electronic design automation [4].

Model of Computation. Our model of computation is the word RAM, where we assume to have the standard operations to read, write as well as arithmetic operations (addition, subtraction, multiplication, modulo, bit shift, AND and OR) take constant time on words of size $w = \Omega(\log N)$ bits where N is the input size in bits. (In our paper log is the binary logarithm $\log_2$.) The model has three types of memory. A read-only *input memory* where the input is stored. A read-write *working memory* that an algorithm uses to compute its result and a write-only *output memory* that the algorithm uses to output its result. The space bounds stated for our results are in the working memory.

Johannes Meintrup and Andrej Sajenko: Funded by the DFG — 379157101.

S.-Y. Hsieh et al. (Eds.): IWOCA 2023, LNCS 13889, pp. 356–367, 2023.
https://doi.org/10.1007/978-3-031-34347-6_30

Sorting. Sorting is one of the most essential algorithms in computer sciences [2,8,10,16,17,24] for over 60 years. Usually sorting problems are classified into different categories. In *comparison sorting* two elements of an input sequence must be compared against each other in order to decide which one precedes the other. Pagter and Rauhe [32] gave a comparison-based algorithm that runs on input sequences of k elements in $O(k^2/s)$ time by using $O(s)$ bits for every given s with $\log k \leq s \leq k/\log k$. Let $[0, x] = \{0, \ldots, x\}$ and $[0, x) = \{0, \ldots, x-1\}$ for any natural number x. *Integer sorting* considers a sequence of k integers, each in the range $[0, m)$, which has to be sorted. It is known that, for $m = k^{O(1)}$, integer sorting can be done in linear time: consider the numbers as k-ary numbers, sort the digits of the numbers in rounds (radix sort) and count the occurrences of a digit by exploiting indirect addressing (counting sort). Han showed that *real sorting* (the given sequence consists of real numbers) can be converted in $O(k\sqrt{\log k})$ time into integers and then can be sorted in $O(k\sqrt{\log\log k})$ time [21].

These algorithms above all assume that the numbers of the input are represented with the same amount of bits. We consider a special case of integer sorting that appears in the field of space-efficient algorithms where numbers are often represented as so-called *self-delimiting numbers* to lower their total memory usage. A self-delimiting number can be represented in several ways. We use the following straightforward representation, also known as *Elias gamma code.* To encode an integer $x > 0$ write $\ell = \lfloor \log x \rfloor$ zero bits, followed by the binary representation of x (without leading zeros). When needed, the encoding can be extended to allow encoding of integers $x \geq 0$ by prepending each encoded $x > 0$ with a single bit set to 1, and encoding 0 with a single bit set to 0. E.g., the self-delimiting numbers of $0, 1, 2, 3, 4$ are $0, 11, 1010, 1011, 100100$, respectively. Throughout this paper, we assume all self-delimiting numbers are given as Elias gamma codes. Our results can be adapted to other types of self-delimiting numbers, for example, Elias delta and Elias omega codes. The property we require is the following: let $x_1 < x_2$ be two integers encoded as self-delimiting numbers e_1, e_2, respectively. Then it holds that e_1 uses at most as many bits as e_2, and if they use the same number of bits, then e_1 is lexicographically smaller than e_2.

Assume that k self-delimiting numbers in the range $[0, m)$ with $m \leq 2^N$ are stored in an N-bit sequence. If the memory is unbounded, then we can simply transform the numbers into integers, sort them, and transform the sorted numbers back into self-delimiting numbers. However, this approach uses $\Omega(k \log m)$ bits. For $k \approx N \approx m$, this is too large to be considered space-efficient. We present a sorting algorithm for self-delimiting numbers that runs in $O(k + \frac{N}{\log N})$ time and uses $O(N)$ bits.

Ranking. Let S be a sequence of numbers. The *competitive rank* of each $x \in S$ is the number of elements in S that are smaller than x. The *dense rank* of each $x \in S$ is the number of distinct elements in S that are smaller than x. I.e., competitive rank counts duplicate elements and dense rank does not.

Raman et al. [33] presented a data structure for a given a set $S \subseteq [0, m)$ of k numbers that uses $\Theta(\log \binom{m}{k}) = \Omega(k \log(m/k))$ bits to answer dense rank (and other operations) in constant time. In some sense, this space bound is "optimal" due to the entropy bound, assuming we treat all numbers in the same way. However, the representation of the self-delimiting numbers differs in their size. E.g., we have a bit vector of N bits storing self-delimiting numbers such that the vector consists of $\Theta(N)$ numbers where one number is $2^{\Theta(N)}$ and all other numbers are 1. Then, the space bound above is $\Omega(N \log(2^{\Theta(N)}/N)) = \Omega(N^2)$.

We present an algorithm to compute the dense/competitive rank on a sequence S of length N consisting of k self-delimiting numbers in $O(k + \frac{N}{\log N})$ time using $O(N)$ bits and subsequently answer dense/competitive rank queries of a number $x \in S$ in constant time. For numbers x of size $> N$ we require the position p_x of x in S as the input to the respective query.

Tree Isomorphism. In the last decade, several space-efficient graph algorithms have been published. Depth-first search and breadth-first search are the first problems that were considered [3,5,14,19]. Further papers with focus on space-efficient algorithms discuss graph interfaces [6,20,28], connectivity problems [11, 19], separators and treewidth [25,27,29]. We continue this research and present a space-efficient isomorphism algorithm for trees, based on an algorithm described in the textbook of Aho, Hopcroft and Ullman [1], which uses $\Omega(n \log n)$ bits. We improve the space-bound to $O(n)$ bits while maintaining the linear running time. We present an $O(n)$-time and $O(n)$-bit tree isomorphism algorithm that decides if two given unrooted unlabeled n-node trees are isomorphic.

Outline. We continue our paper with our results on sorting and dense/competitive ranking in Sect. 2. Afterwards we introduce definitions and notations in Sect. 3 as a preparation for our result on space-efficient tree isomorphism. Finally, we present our space-efficient tree-isomorphism algorithm for unrooted trees. Our proofs can be found in a full version [31].

2 Sorting and Ranking

In this section we consider sorting and ranking of k self-delimiting numbers, stored within an N-bit sequence S. We make use of *lookup tables*, which are precomputed tables storing the answer for every possible state of a finite universe, typically of small size. We use such tables to quickly solve sub-problems that occur in our algorithms. For the rest of this section we assume that a parameter τ with $\log N \leq \tau \leq w$ is given to our algorithms, which we use as a parameter to construct lookup tables for binary sequences of length $\lceil \tau/2 \rceil$. To give an intuitive example of such lookup tables, the universe might consist of all integers of size at most $2^{\lceil \tau/2 \rceil}$ and answer queries if the given number is prime. Such a table has $2^{\lceil \tau/2 \rceil}$ entries, and each index into the table requires $\lceil \tau/2 \rceil$ bits. Note that larger values of τ result in faster runtimes at the cost of increased space-usage.

In our application for sorting, S is a sequence of N bits containing self-delimiting numbers. By this, each number in S is of size $m = O(2^N)$. Let $q = 2^{N/\tau}$ and call $x \in S$ *big* if $q < x \leq 2^N$, otherwise call x *small*. We have to handle small and big numbers differently. To divide the problem we scan through S and write each small number of S into a sequence $S_{\leq q}$ and each big number into a sequence $S_{>q}$. On the word-RAM model, scanning through an N-bit sequence S and reporting all k numbers takes $O(k + N/\tau)$ time, which is the time bound of all our sorting algorithms. After sorting both sequences we write the sorted numbers of $S_{\leq q}$ and then of $S_{>q}$ into a sequence S' of N bits.

We first consider the sequence $S_{\leq q}$. Our idea is to run first an adaptation of stable counting sort to *presort* the numbers in several *areas* such that an area A_i consists of all numbers that require exactly i bits as self-delimiting number. By doing so we roughly sort the sequence $S_{\leq q}$ as all numbers of area A_i are smaller than any number of area A_j for all $i < j$. We then sort each area independently by another stable counting-sort algorithm.

Lemma 1. *Given an N-bit sequence S consisting of k self-delimiting numbers, each in the range $[0, m)$ $(m \leq 2^{N/\tau})$ and a parameter τ with $\log N \leq \tau \leq w$, there is an $O(k + N/\tau)$-time $(O(N) + o(2^\tau))$-bit stable-sorting algorithm computing a bit sequence of N bits that stores the given self-delimiting numbers in sorted order.*

Now, let us look at the sequence $S_{>q}$. The bound on the size of each number implies that $S_{>q}$ cannot consist of more than $O(N/\log q) = O(\tau)$ numbers and the biggest number may occupy $O(N)$ bits. The idea is to interpret each number in $S_{>q}$ as a string using the alphabet $\Sigma = \{0, \ldots, 2^{\epsilon\tau}\}$.

Similarly as in the previous lemma, we first sort the strings by their length into areas such that each area consists of all self-delimiting numbers of one length. Afterwards, we sort each area lexicographically using radix sort. Note that directly lexicographically sorting binary sequences when interpreted as strings does not result in a sorted sequence of self-delimiting numbers. For example $0, 11, 1010, 1011$ are sorted self-delimiting numbers, but directly interpreting them as strings would sort them lexicographically as $0, 1010, 1011, 11$.

Theorem 1. *Given an N-bit sequence S of k self-delimiting numbers and a parameter τ with $\log N \leq \tau \leq w$, there is an $O(k + N/\tau)$-time $O(N) + o(2^\tau)$-bit stable-sorting algorithm computing a bit sequence of N bits that stores the given self-delimiting numbers in sorted order.*

We now consider dense/competitive ranking of k self-delimiting numbers. A standard approach to compute the dense/competitive rank is to first sort S and then to use an array P of m entries, each of $\lceil \log k \rceil$ bits, to store a prefix sum over the occurrences of (different) numbers $x \in S$, i.e., in a first step for competitive rank set $P[x] = P[x] + 1$ (for dense rank, set $P[x] = 1$) for each $x \in S$. In a second step compute the prefix sums on P, i.e., for each $i = 1, \ldots, m - 2$, set $P[i] = P[i - 1] + P[i]$. The dense/competitive rank of a number x is then $P[x]$. However, the array P requires $\Theta(m \log k)$ bits. To compute the dense rank with less space, we can use a bit-vector B of m bits and set $B[x] = 1$ for each $x \in S$. Then, using a rank-select data structure on B, the dense rank of x is $\texttt{rank}_B(x)$. This approach uses $O(m)$ bits and takes $\Theta(m/w)$ time due to the initialization of a rank-select data structure on B [7]. Note that this solution does not handle duplicates, due to the use of a bit-vector as the underlying data structure.

Our novel approach provides a dense/competitive rank solution that does not rely on the universe size m in both the runtime (for initialization) and bits required, but only on N. Moreover, for our use-case in Sect. 4 we use dense rank with universe size $m = O(2^N)$ for which the approaches outlined previously require $\Omega(N \log N)$ bits, while we aim for $O(N)$ bits. Due to this specific requirement we require a novel approach, which works similar to the realization of dense rank in [7,13,26].

We first discuss dense rank. Similar to our solution for sorting, we handle small and large numbers differently. Let S be an N-bit sequence of k self-delimiting numbers. Denote with $S_{\leq N}$ the subsequence of S that contains numbers that are at most size N, and with $S_{>N}$ all numbers of S that are larger than N (and at most size 2^N). We first discuss the techniques used for enabling dense rank for all small numbers $S_{\leq N}$ of S for which we build our own data structure. Denote with m' the size of the universe of $S_{\leq N}$ and k' the number of self-delimiting numbers contained in $S_{\leq N}$. We construct the dense rank structure not on a given bit-vector, but on a given sequence consisting of k' self-delimiting numbers that correspond to the ones in the bit-vector. For this, we construct a bit vector B, which we partition into $O(m'/\tau)$ frames of $\lceil \tau/2 \rceil$ bits and create an array P that contains the prefix sum of the frames up to the ith frame ($i = 0, \ldots, \lceil m'/\lceil \tau/2 \rceil \rceil$). Subsequently, we use a lookup table that allows to determine the number of ones in the binary representation of each frame.

It remains to show a solution for big numbers $S_{>N}$. Note that the dense rank of any number cannot be bigger than N and thus use more than $\log N$ bits. On the other hand, $S_{>N}$ contains $\leq N/\log N$ numbers. Thus, we can use an N-bit vector Q consisting of $\leq N/\log N$ entries, each of $\log N$ bits, and if (intuitively speaking) drawing the vector below $S_{>N}$, we can write the dense rank of every number $x \in S_{>N}$ with $N < x \leq 2^N$ into Q below x. By requiring that the access to the dense rank of $x \in S_{>N}$ has to be done by providing the position p_x of the first bit of x in S as the input (instead of the number x itself), we can easily return the dense rank of x in constant time. Note that we need the position p_x since the binary representation of p_x can be always be written with $\log N \leq w$ bits, but not x. This allows constant time operations.

Theorem 2. *Given an N-bit sequence S of k self-delimiting numbers, a parameter τ with $\log N \leq r \leq w$ we can compute a data structure realizing dense rank. The data structure can be constructed in $O(k+N/\tau)+o(2^\tau) = O(k+N/\log N)+o(2^\tau)$ time and uses $O(N) + o(2^\tau)$ bits. For numbers x of size $> N$ the position p_x of x in S is required as the input.*

To compute the competitive rank we require the information of how many times an element appears in the given sequence. We change our approach of the previous lemma as follows: recall that τ is a parameter with $\log N \leq \tau \leq w$. We sort S to get a sorted sequence S'. Next, we partition S' into *regions* such that the ith region $\mathcal{R}_i = S' \cap [i\lceil\tau/2\rceil, \ldots, (i+1)\lceil\tau/2\rceil]$ for all $i = 0, \ldots, 2\lceil m/\lceil\tau/2\rceil\rceil - 1$. In detail, we go through S' and store for each non-empty region $\mathcal{R}_i$ a pointer $F[i]$ to a sequence A_i of occurrences of each number $x \in S'$ written as self-delimiting numbers. Similar to the usage of B for dense rank, we solve competitive rank by partitioning A_i into frames of $\lceil\tau/2\rceil$ bits and computing an array P_i storing the prefix-sums. Inside a single frame we use a lookup table to compute the prefix sum. More exactly, $P_i[j]$ stores the prefix-sum over all self-delimiting numbers in S' up to the jth frame in A_i. Figure 1 sketches an example. We so obtain the next theorem where we again require the position p_x in S of all $x > N$ as the input to our competitive rank query.

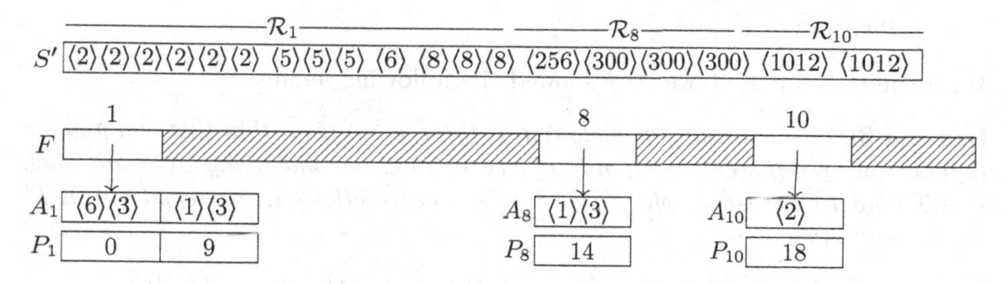

Fig. 1. A sketch of our storage schema to realize competitive rank. For each region $\mathcal{R}_i$, that contains numbers out of S', a pointer $F[i]$ points to a data structure storing the amount of occurrences for each of the numbers. In addition, P_i stores the prefix-sum over the frames up to A_i. For numbers x of size $> N$ the position p_x of x in S is required as the input.

Theorem 3. *Given an N-bit sequence S of k self-delimiting numbers, a parameter τ with $\log N \leq \tau \leq w$ we can compute a data structure realizing competitive rank. The data structure can be constructed in $O(k + N/\tau) + o(2^\tau) = O(k + N/\log N) + o(2^\tau)$ time and uses $O(N) + o(2^\tau)$ bits.*

3 Preliminaries for Tree Isomorphism

In this paper we use basic graph and tree terminology as given in [12]. By designating a node of a tree as the root, the tree becomes a *rooted* tree. If the nodes of a tree have labels, the tree is *labeled*, otherwise, the tree is called *unlabeled*. The *parent* of a node in a rooted tree is the neighbor of u on the shortest path from u to the root. The root has no parent. The *children* of a node u are the neighbors of u that are not its parent. A *leaf* is a node with no children. Two nodes that share the same parent are *siblings*. A *descendant* of a node is either a child of the node or a child of some descendant of the node. By fixing the order of the children of each node a rooted tree becomes an *ordinal* tree. The *right sibling* (*left sibling*) of a node u is the sibling of u that comes after (before) u in the aforementioned order, if it exists. We denote by $\mathtt{deg}(v)$ the degree of a node v, i.e., the number of neighbors of v, and by $\mathtt{desc}(u)$ the number of descendants of a node u in a tree. The *height* of u is defined as the number of edges between u and the longest path to a descendant leaf. The *depth* of u is defined as the number of edges on the path between u and the root.

Lemma 2. *(rank-select [7]) Given access to a sequence $B = (b_1, \ldots, b_n) = \{0,1\}^n$ ($n \in \mathbb{N}$) of n bits there is an $o(n)$-bit data structure that, after an initialization of $O(n/w)$ time, supports two constant-time operations:*

- $\mathtt{rank}_B(j) = \sum_{i=1}^{j} b_i$ *($j \in [1, n]$) that returns the number of ones in $(b_1, \ldots, b_j)$ in $O(1)$ time, and*
- $\mathtt{select}_B(k) = \min\{j \in [1, n] : \mathtt{rank}_B(j) = k\}$ *that returns the position of the kth one in B.*

With the techniques above they showed the following lemma.

Lemma 3. *Given a rooted n-node tree T there is an algorithm that computes a data structure representing an ordinal tree T' in $O(n)$ time using $O(n)$ bits such that T and T' are isomorphic. The data structure allows tree navigation on T' in constant time.*

We manage sets of nodes using *(uncolored) choice dictionaries* [18,30].

Definition 1. *((uncolored) choice dictionary) Initialized with some parameter n there is a data structure that stores a subset U' out of a universe $U = [0, n)$ and supports the standard dictionary operations* `add`, `remove` *and* `contains`. *Moreover, it provides an operation* `choice` *that returns an arbitrary element of U'. Initialization and all other operation run in $O(1)$ time.*

4 Tree Isomorphism

We start this section by giving an introduction to tree isomorphism [9].

Definition 2. *(rooted tree isomorphism) By induction two rooted trees T and T' are isomorphic if and only if*

(a) T and T' consist of only one node, or
(b) the roots r and r' of T and T', respectively, have the same number m of children, and there is some ordering $T_1, \ldots, T_m$ of the maximal subtrees below the children of r and some ordering $T'_1, \ldots, T'_m$ of the maximal subtrees below the children of r' such that T_i and T'_i are isomorphic for all $1 \leq i \leq m$.

We start to describe a folklore algorithm for tree isomorphism that requires $\Theta(n \log n)$ bits on n-node trees. Let T_1 and T_2 be two rooted trees. The algorithm processes the nodes of each tree in rounds. In each round, all nodes of depth $d = \max, \ldots, 0$ are processed. Within a round, the goal is to compute a *classification number* for every node u of depth d, i.e., a number out of $\{0, \dots, n\}$ that represents the structure of the maximal subtree below u. The correctness of the algorithm is shown in [1] and follows from the invariant that two subtrees in the trees T_1 and T_2 get the same classification number exactly if they are isomorphic.

Since we later want to modify the algorithm, we now describe it in greater detail. In an initial process assign the classification number 0 to every leaf in each tree. Then, starting with the maximal depth do the following: start by computing (for each tree) the *classification vector* of each non-leaf v of depth d consisting of the classification numbers of v's children, sorted lexicographically. After doing this in each tree, compute the classification number for the non-leafs as follows: for each tree T_1, T_2 put the classification vectors of depth d into a single sequence $\mathcal{S}_1$ and $\mathcal{S}_2$, respectively. Sort each of these sequences lexicographically by interpreting each classification vector in the sequence as a number. Then assign classification numbers $1, 2, 3$, etc. to the vectors in the (now sorted) sequences $\mathcal{S}_1, \mathcal{S}_2$ such that two vectors get the same number exactly if they are equal (among both sequences). By induction the invariant holds for all new classification numbers. Repeat the whole procedure iteratively for the remaining depths until reaching the root. By the invariant above, both trees are isomorphic exactly if the roots of both trees have the same classification number.

The algorithm above traverses the nodes in order of their depth, starting from the largest and moving to the smallest, until it reaches the root. One key modification we make to achieve our goal of making the aforementioned algorithm space efficient, is that we traverse the nodes in order of their height starting from height 0 (first round) until reaching the root with the largest height (last round), i.e., in *increasing height*. As mentioned, the standard algorithm requires the nodes of the tree to be output in *shrinking depth*. While there is a succinct data structure due to He et al. [22] that provides us with all necessary operations to implement such a shrinking depth tree traversal, the construction step of this data structure requires $O(n \log n)$ bits due to (among other things) the usage of an algorithm of Farzan and Munro [15, Theorem 1] that partitions the input tree into smaller subtrees. The aforementioned algorithm uses a stack of size $O(n)$, with each element on the stack using $\Theta(\log n)$ bits. In addition, the decomposition itself is stored temporarily. As we aim for a space usage of $O(n)$ bits we use a different approach. Note that the tree data structure outlined in Sect. 3 only allows to output the nodes in order of their *increasing depth*. While this implies a simple algorithm for shrinking depth, by storing the result of the traversal and then outputting it in reverse order, such a solution requires $\Theta(n \log n)$ bits. We therefore modify the standard isomorphism algorithm such that traversal by *increasing height* suffices, which can be implemented using the tree navigation provided by the data structure outlined in Sect. 3.

The difference to the standard approach is that we get classification vectors consisting of classification numbers, which were computed in different rounds. To avoid a non-injective mapping of the subtrees, our classification numbers consist of tuples (h_u, q_u) for each node u where h_u is the height of u and q_u is a number representing the subtree induced by u and its descendants. Intuitively, q_u is the classification number from the folklore algorithm above.

The same invariant as for the standard algorithm easily shows the correctness of our modified algorithm whose space consumption is determined by

(A) the space for traversing nodes in order of their height,
(B) the space for storing the classification vectors and the classification numbers,
(C) the space needed by an algorithm to assign new classification numbers .

We now describe $O(n)$-bit solutions for (A)–(C).

(A) Iterator Returning Vertices in Increasing Height. The idea of the first iteration round is to determine all nodes of height $h = 0$ (i.e., all leaves) of the given tree in linear time and to store them in an $O(n)$-bit choice dictionary C. While iterating over the nodes of height h in C, our goal is to determine the nodes of height $h + 1$ and store them in a choice dictionary C'. The details are described in the next paragraph. If the iteration of the nodes in C is finished, swap the meaning of C and C' and repeat this process iteratively with $h + 1$ as the new height h until, finally, the root is reached.

A node is selected for C' at the moment when we have processed all of its children. To compute this information, our idea is to give every unprocessed node u a token that is initially positioned at its leftmost child. Intuitively, the goal is to pass that token over every child of u from left to right until reaching the rightmost child of u, at which point we mark u as processed and store it in C'. More precisely, we iterate over the children in the deterministic order given by the choice dictionary, which can be arbitrary. Initially, no node is marked as processed. Informally speaking, we run a relay race where the children of a node are the runners. Before runners can start their run, they must be marked as processed. The initiative to pass the token is driven by the children of u. Whenever a child v of u is processed, we check if either v is the leftmost child of u or v's left sibling has u's token. If so, we move the token to the right sibling v' of v and then to the right sibling of v', etc., as long as the sibling is already marked. If all children of u are processed, u becomes marked and part of C'.

Lemma 4. *Given an unlabeled rooted n-node tree T there is an iteration over all nodes of the tree in order of their height that runs in linear time and uses $O(n)$ bits. The iteration is realized by the following methods:*

- `init(T)`: *Initializes the iterator and sets height $h = 0$.*
- `hasNext`: *Returns* TRUE *exactly if nodes of height h exist.*
- `next`: *Returns a choice dictionary containing nodes of height h and increments h by 1.*

(B) Storing the Classification Numbers. We now describe an algorithm to store our classification numbers in an $O(n)$-bit storage schema. Recall that a classification vector of a node u consists of the classification numbers of its children. Our idea is to use self-delimiting numbers to store the classification numbers and to choose the classification numbers such that their size is bounded by $\mathtt{min}(c_1 \cdot \log n,\ c_2 \cdot \mathtt{desc}(u))$ for some constants c_1, c_2. By the next lemma we can store the classification numbers and vectors.

Lemma 5. *Given an unlabeled rooted n-node tree T and an integer $c > 0$, there is a data structure using $O(n)$ bits that initializes in $O(n)$ time and, for each node u of T, provides operations* `read`*(u) and* `write`*(u) in constant time and* `vector`*(u) in $O(deg(u))$ time.*

- `read`*(u) (u node of T): If a number x is stored for u, then x is returned. Otherwise, the result is undefined.*
- `write`*(u, x) (u node of T, $0 \leq x \leq \min\{2^{2c\mathtt{desc}(u)}, \mathrm{poly}(n)\}$): Store number x for node u and delete all stored numbers of the descendants of u.*
- `vector`*(u) (u node of T): Returns the bit-vector of length $\leq 2c\mathtt{desc}(u)$ consisting of the concatenation of the self-delimiting numbers stored for the children of u.*

(C) Computing Classification Numbers. Let D be our data structure of Lemma 5 where we store all classification vectors. Our next goal is to replace the classification vector $D.\mathtt{vector}(u)$ of all processed subtrees with root u and height h by a classification number (h, q_u) with $|q_u| = O(\min\{\mathtt{desc}(u), \log n\})$ such that the componentwise-sorted classification vectors are equal exactly if they get the same classification number.

Our idea is to sort $D.\mathtt{vector}(u)$ by using Theorem 1 to obtain a componentwise sorted classification vector and turn this vector into a self-delimiting number for further operation on it. We subsequently compute the dense rank to replace the self-delimiting number in $D.\mathtt{vector}(u)$ by the tuple (height, dense rank). To make it work we transform each vector into a self-delimiting number by considering the bit-sequence of the vector as a number (i.e., assign the prefix $1^\ell 0$ to each vector where ℓ is the length of the vector in bits). We can store all these vectors as self-delimiting numbers in a bit-vector Z_{h+1} of $O(n)$ bits. Then we can use Theorem 2 applied to Z_{h+1} to compute the dense ranks, which allows us to determine the classification numbers for all subtrees of height $h+1$.

Our Algorithm. We now combine the solutions for (A)–(C).

Lemma 6. *Given two rooted n-node trees T_1 and T_2 there is an algorithm that recognizes if T_1 and T_2 are isomorphic in linear-time using $O(n)$ bits.*

We generalize Lemma 6 to unrooted trees by first determining the center of a tree space efficiently, which is a set of constant size that is maximal distance to a leaf. Using each vertex in the center as a (possible) root, we then run our rooted isomorphism algorithm.

Theorem 4. *Given two (unrooted) trees T_1 and T_2 there is an algorithm that outputs if T_1 and T_2 are isomorphic in linear-time using $O(n)$ bits.*

References

1. Aho, A.V., Hopcroft, J.E., Ullman, J.D.: The Design and Analysis of Computer Algorithms. Addison-Wesley, Boston (1974)
2. Asano, T., Elmasry, A., Katajainen, J.: Priority queues and sorting for read-only data. In: Chan, T.-H.H., Lau, L.C., Trevisan, L. (eds.) TAMC 2013. LNCS, vol. 7876, pp. 32–41. Springer, Heidelberg (2013). https://doi.org/10.1007/978-3-642-38236-9_4
3. Asano, T., et al.: Depth-first search using $O(n)$ bits. In: Ahn, H.-K., Shin, C.-S. (eds.) ISAAC 2014. LNCS, vol. 8889, pp. 553–564. Springer, Cham (2014). https://doi.org/10.1007/978-3-319-13075-0_44
4. Baird, H.S., Cho, Y.E.: An artwork design verification system. In: Proc. 12th Conference on Design (DAC 1975), pp. 414–420. IEEE (1975)
5. Banerjee, N., Chakraborty, S., Raman, V., Satti, S.R.: Space efficient linear time algorithms for BFS, DFS and applications. Theory Comput. Syst. **62**(8), 1736–1762 (2018). https://doi.org/10.1007/s00224-017-9841-2
6. Barbay, J., Aleardi, L.C., He, M., Munro, J.I.: Succinct representation of labeled graphs. Algorithmica **62**(1), 224–257 (2012). https://doi.org/10.1007/s00453-010-9452-7
7. Baumann, T., Hagerup, T.: Rank-select indices without tears. In: Friggstad, Z., Sack, J.-R., Salavatipour, M.R. (eds.) WADS 2019. LNCS, vol. 11646, pp. 85–98. Springer, Cham (2019). https://doi.org/10.1007/978-3-030-24766-9_7
8. Borodin, A., Cook, S.A.: A time-space tradeoff for sorting on a general sequential model of computation. SIAM J. Comput. **11**(2), 287–297 (1982). https://doi.org/10.1137/0211022
9. Buss, S.R.: Alogtime algorithms for tree isomorphism, comparison, and canonization. In: Gottlob, G., Leitsch, A., Mundici, D. (eds.) KGC 1997. LNCS, vol. 1289, pp. 18–33. Springer, Heidelberg (1997). https://doi.org/10.1007/3-540-63385-5_30
10. Chan, T.M.: Comparison-based time-space lower bounds for selection. ACM Trans. Algorithms **6**(2), 1–16 (2010). https://doi.org/10.1145/1721837.1721842
11. Choudhari, J., Gupta, M., Sharma, S.: Nearly optimal space efficient algorithm for depth first search. arXiv preprint arXiv:1810.07259 (2018)
12. Cormen, T.H., Leiserson, C.E., Rivest, R.L., Stein, C.: Introduction to Algorithms, 3rd edn. MIT Press, Cambridge (2009)
13. Elias, P.: Efficient storage and retrieval by content and address of static files. J. ACM **21**(2), 246–260 (1974). https://doi.org/10.1145/321812.321820
14. Elmasry, A., Hagerup, T., Kammer, F.: Space-efficient basic graph algorithms. In: Proc. 32nd International Symposium on Theoretical Aspects of Computer Science (STACS 2015). LIPIcs, vol. 30, pp. 288–301. Schloss Dagstuhl - Leibniz-Zentrum für Informatik (2015). https://doi.org/10.4230/LIPIcs.STACS.2015.288
15. Farzan, A., Munro, J.I.: A uniform paradigm to succinctly encode various families of trees. Algorithmica **68**(1), 16–40 (2012). https://doi.org/10.1007/s00453-012-9664-0
16. Frederickson, G.N.: Upper bounds for time-space trade-offs in sorting and selection. J. Comput. Syst. Sci. **34**(1), 19–26 (1987). https://doi.org/10.1016/0022-0000(87)90002-X
17. Hagerup, T.: Sorting and searching on the word RAM. In: Morvan, M., Meinel, C., Krob, D. (eds.) STACS 1998. LNCS, vol. 1373, pp. 366–398. Springer, Heidelberg (1998). https://doi.org/10.1007/BFb0028575

18. Hagerup, T.: A constant-time colored choice dictionary with almost robust iteration. In: Proc. 44th International Symposium on Mathematical Foundations of Computer Science (MFCS 2019). LIPIcs, vol. 138, pp. 64:1–64:14. Schloss Dagstuhl - Leibniz-Zentrum für Informatik, Dagstuhl, Germany (2019). https://doi.org/10.4230/LIPIcs.MFCS.2019.64
19. Hagerup, T.: Space-efficient DFS and applications to connectivity problems: simpler, leaner, faster. Algorithmica **82**(4), 1033–1056 (2019). https://doi.org/10.1007/s00453-019-00629-x
20. Hagerup, T., Kammer, F., Laudahn, M.: Space-efficient Euler partition and bipartite edge coloring. Theor. Comput. Sci. **754**, 16–34 (2019). https://doi.org/10.1016/j.tcs.2018.01.008
21. Han, Y.: Sorting real numbers in $O(n\sqrt{\log n})$ time and linear space. Algorithmica **82**(4), 966–978 (2019). https://doi.org/10.1007/s00453-019-00626-0
22. He, M., Munro, J.I., Nekrich, Y., Wild, S., Wu, K.: Distance Oracles for interval graphs via breadth-first rank/select in succinct trees. In: 31st International Symposium on Algorithms and Computation (ISAAC 2020). Leibniz International Proceedings in Informatics (LIPIcs), vol. 181, pp. 25:1–25:18. Schloss Dagstuhl-Leibniz-Zentrum für Informatik, Dagstuhl, Germany (2020). https://doi.org/10.4230/LIPIcs.ISAAC.2020.25
23. Irniger, C., Bunke, H.: Decision tree structures for graph database filtering. In: Fred, A., Caelli, T.M., Duin, R.P.W., Campilho, A.C., de Ridder, D. (eds.) SSPR /SPR 2004. LNCS, vol. 3138, pp. 66–75. Springer, Heidelberg (2004). https://doi.org/10.1007/978-3-540-27868-9_6
24. Isaac, E.J., Singleton, R.C.: Sorting by address calculation. J. ACM **3**(3), 169–174 (1956). https://doi.org/10.1145/320831.320834
25. Izumi, T., Otachi, Y.: Sublinear-space lexicographic depth-first search for bounded treewidth graphs and planar graphs. In: 47th International Colloquium on Automata, Languages, and Programming (ICALP 2020), vol. 168, pp. 67:1–67:17 (2020). https://doi.org/10.4230/LIPIcs.ICALP.2020.67
26. Jacobson, G.: Space-efficient static trees and graphs. In: Proc. 30th Annual Symposium on Foundations of Computer Science (FOCS 1989), pp. 549–554. IEEE Computer Society (1989). https://doi.org/10.1109/SFCS.1989.63533
27. Kammer, F., Meintrup, J.: Space-efficient graph coarsening with applications to succinct planar encodings. In: 33rd International Symposium on Algorithms and Computation (ISAAC 2022), vol. 248, pp. 62:1–62:15 (2022). https://doi.org/10.4230/LIPIcs.ISAAC.2022.62
28. Kammer, F., Meintrup, J.: Succinct planar encoding with minor operations (2023)
29. Kammer, F., Meintrup, J., Sajenko, A.: Space-efficient vertex separators for treewidth. Algorithmica **84**(9), 2414–2461 (2022). https://doi.org/10.1007/s00453-022-00967-3
30. Kammer, F., Sajenko, A.: Simple 2^f-color choice dictionaries. In: Proc. 29th International Symposium on Algorithms and Computation (ISAAC 2018). LIPIcs, vol. 123, pp. 66:1–66:12. Schloss Dagstuhl - Leibniz-Zentrum für Informatik (2018). https://doi.org/10.4230/LIPIcs.ISAAC.2018.66
31. Kammer, F., Sajenko, A.: Sorting and ranking of self-delimiting numbers with applications to tree isomorphism. arXiv preprint arXiv:2002.07287 (2020)
32. Pagter, J., Rauhe, T.: Optimal time-space trade-offs for sorting. In: Proc. 39th Annual Symposium on Foundations of Computer Science (FOCS 1998), pp. 264–268. IEEE Computer Society (1998). https://doi.org/10.1109/SFCS.1998.743455
33. Raman, R., Raman, V., Rao, S.S.: Succinct indexable dictionaries with applications to encoding k-ary trees and multisets. In: Proc. 13th Annual ACM-SIAM Symposium on Discrete Algorithms (SODA 2002), pp. 233–242. ACM/SIAM (2002)

A Linear Delay Algorithm for Enumeration of 2-Edge/Vertex-Connected Induced Subgraphs

Takumi Tada(✉) and Kazuya Haraguchi

Graduate School of Informatics, Kyoto University, Kyoto, Japan
{tada,haraguchi}@amp.i.kyoto-u.ac.jp

Abstract. In this paper, we present the first linear delay algorithms to enumerate all 2-edge-connected induced subgraphs and to enumerate all 2-vertex-connected induced subgraphs for a given simple undirected graph. We treat these subgraph enumeration problems in a more general framework based on set systems. For an element set V, $(V, \mathcal{C} \subseteq 2^V)$ is called a *set system*, where we call $C \in \mathcal{C}$ a *component*. A nonempty subset $Y \subseteq C$ is a *removable set of* C if $C \setminus Y$ is a component and Y is a *minimal removable set* (*MRS*) *of* C if it is a removable set and no proper nonempty subset $Z \subsetneq Y$ is a removable set of C. We say that a set system has *subset-disjoint* (*SD*) property if, for every two components $C, C' \in \mathcal{C}$ with $C' \subsetneq C$, every MRS Y of C satisfies either $Y \subseteq C'$ or $Y \cap C' = \emptyset$. We assume that a set system with SD property is implicitly given by an oracle that returns an MRS of a component which is given as a query. We provide an algorithm that, given a component C, enumerates all components that are subsets of C in linear time/space with respect to $|V|$ and oracle running time/space. We then show that, given a simple undirected graph G, the pair of the vertex set $V = V(G)$ and the family of vertex subsets that induce 2-edge-connected (or 2-vertex-connected) subgraphs of G has SD property, where an MRS in a 2-edge-connected (or 2-vertex-connected) induced subgraph corresponds to either an ear or a single vertex with degree greater than two.

Keywords: Enumeration of subgraphs · 2-edge-connectivity · 2-vertex-connectivity · Binary partition · Linear delay

1 Introduction

Given a graph, subgraph enumeration asks to list all subgraphs that satisfy required conditions. It could find interesting substructures in network analysis. Enumeration of cliques is among such problems [2,3], where a clique is a subgraph such that every two vertices are adjacent to each other and thus may represent a group of so-called SNS users that are pairwise friends. Pursuing further applications, there have been studied enumeration of subgraphs that satisfy weaker connectivity conditions, e.g., pseudo-cliques [12].

The work is partially supported by JSPS KAKENHI Grant Number 20K04978 and 22H00532. A preprint of this paper appeared as [8].

S.-Y. Hsieh et al. (Eds.): IWOCA 2023, LNCS 13889, pp. 368–379, 2023.
https://doi.org/10.1007/978-3-031-34347-6_31

In this paper, we consider enumeration of subgraphs that satisfy fundamental connectivity conditions; 2-edge-connectivity and 2-vertex-connectivity. For a graph G, let $V(G)$ and $E(G)$ denote the set of vertices of G and the set of edges of G, respectively. Let $n := |V(G)|$ and $m := |E(G)|$. An enumeration algorithm in general outputs many solutions, and its *delay* refers to computation time between the start of the algorithm and the first output; between any consecutive two outputs; and between the last output and the halt of the algorithm. The algorithm attains *polynomial delay* (resp., *linear delay*) if the delay is bounded by a polynomial (resp., a linear function) with respect to the input size.

The main results of the paper are summarized in the following two theorems.

Theorem 1. *For a simple undirected graph G, all 2-edge-connected induced subgraphs of G can be enumerated in $O(n+m)$ delay and space.*

Theorem 2. *For a simple undirected graph G, all 2-vertex-connected induced subgraphs of G can be enumerated in $O(n+m)$ delay and space.*

We achieve the first linear delay algorithms for enumerating 2-edge/vertex connected induced subgraphs. Ito et al. [7] made the first study on enumeration of 2-edge-connected induced subgraphs, presenting a polynomial delay algorithm based on reverse search [1] such that the delay is $O(n^3m)$. For an element set V, $(V, \mathcal{C} \subseteq 2^V)$ is called a *confluent set system* if, for every three components $X, Y, Z \in \mathcal{C}$, $Z \subseteq X \cap Y$ implies $X \cup Y \in \mathcal{C}$. Haraguchi and Nagamochi [5] studied an enumeration problem in a confluent set system that includes enumeration of k-edge-connected (resp., k-vertex-connected) induced subgraphs as special cases, which yields $O(\min\{k+1, n\}n^5m)$ (resp., $O(\min\{k+1, n^{1/2}\}n^{k+4}m)$) delay algorithms. Wen et al. [13] proposed an algorithm for enumerating maximal vertex subsets that induce k-vertex-connected subgraphs such that the total time complexity is $O(\min\{n^{1/2}, k\}m(n+\delta(G)^2)n)$, where $\delta(G)$ denotes the minimum degree over the graph G.

We deal with the two subgraph enumeration problems in a more general framework. For a set V of elements, let $\mathcal{C} \subseteq 2^V$ be a family of subsets of V. A pair $(V, \mathcal{C})$ is called a *set system* and a subset $C \subseteq V$ is called a *component* if $C \in \mathcal{C}$. A nonempty subset $Y \subseteq C$ of a component C is a *removable set of C* if $C \setminus Y \in \mathcal{C}$. Further, a removable set Y of C is minimal, or a *minimal removable set (MRS)*, if there is no $Z \subsetneq Y$ that is a removable set of C. We denote by $\textsc{Mrs}_{\mathcal{C}}(C)$ the family of all MRSs of C. Let us introduce the notion of SD property of set system as follows.

Definition 1. A set system $(V, \mathcal{C})$ has *subset-disjoint (SD) property* if, for any two components $C, C' \in \mathcal{C}$ such that $C \supsetneq C'$, either $Y \subseteq C'$ or $Y \cap C' = \emptyset$ holds for every MRS Y of C.

We consider the problem of enumerating all components that are subsets of a given component in a set system with SD property. We assume that the set system is implicitly given by an oracle such that, for a component $C \in \mathcal{C}$ and a subset $X \subseteq C$, the oracle returns an MRS Y of C that is disjoint with X

if exists; and NIL otherwise. We denote the time and space complexity of the oracle by θ_t and θ_s, respectively. We show the following theorem that is a key for proving Theorems 1 and 2.

Theorem 3. *Let $(V, \mathcal{C})$ be a set system with SD property, $C \in \mathcal{C}$ be a component and $n := |C|$. All components that are subsets of C can be enumerated in $O(n+\theta_t)$ delay and $O(n + \theta_s)$ space.*

The paper is organized as follows. After making preparations in Sect. 2, we present an algorithm that enumerates all components that are subsets of a given component in a set system with SD property, along with complexity analyses in Sect. 3, as a proof for Theorem 3. Then in Sect. 4, we provide proofs for Theorems 1 and 2. There are two core parts in the proofs. In the first part, given a 2-edge-connected (resp., 2-vertex-connected) graph G, we show that a set system $(V, \mathcal{C})$ has SD property if $V = V(G)$ and $\mathcal{C}$ is the family of all vertex subsets that induce 2-edge-connected (resp., 2-vertex-connected) subgraphs. This means that 2-edge/vertex-connected induced subgraphs can be enumerated by using the algorithm developed for Theorem 3. Then in the second part, we explain how we design the oracle to achieve linear delay and space.

For some lemmas, we omit the proofs due to space limitation. The omitted proofs are found in the preprint of this paper [8].

2 Preliminaries

Let $\mathbb{Z}$ and $\mathbb{Z}_+$ denote the set of integers and the set of nonnegative integers, respectively. For two integers $i, j \in \mathbb{Z}$ $(i \leq j)$, let us denote $[i, j] := \{i, i + 1, \ldots, j\}$.

For any sets P, Q of elements, when $P \cap Q = \emptyset$, we may denote by $P \sqcup Q$ the disjoint union of P and Q in order to emphasize that they are disjoint.

Set Systems. Let $(V, \mathcal{C})$ be a set system which does not necessarily have SD property. For any two subsets $U, L \subseteq V$, we denote $\mathcal{C}(U, L) := \{C \in \mathcal{C} \mid L \subseteq C \subseteq U\}$. Recall that, for a component $C \in \mathcal{C}$, we denote by $\mathrm{MRS}_{\mathcal{C}}(C)$ the family of all MRSs of C. Further, for $X \subseteq C$, we denote $\mathrm{MRS}_{\mathcal{C}}(C, X) := \{Y \in \mathrm{MRS}_{\mathcal{C}}(C) \mid Y \cap X = \emptyset\}$. For any two components $C, C' \in \mathcal{C}$ with $C \supsetneq C'$, let $Y_1, Y_2, \ldots, Y_\ell \subseteq C \setminus C'$ be subsets such that $Y_i \cap Y_j = \emptyset$, $1 \leq i < j \leq \ell$; and $Y_1 \sqcup Y_2 \sqcup \cdots \sqcup Y_\ell = C \setminus C'$ (i.e., $\{Y_1, Y_2, \ldots, Y_\ell\}$ is a partition of $C \setminus C'$). Then $(Y_1, Y_2, \ldots, Y_\ell)$ is an *MRS-sequence* (*between C and C'*) if

- $C' \sqcup Y_1 \sqcup \cdots \sqcup Y_i \in \mathcal{C}$, $i \in [1, \ell]$; and
- $Y_i \in \mathrm{MRS}_{\mathcal{C}}(C' \sqcup Y_1 \sqcup \cdots \sqcup Y_i)$, $i \in [1, \ell]$.

One easily sees that there exists an MRS-sequence for every $C, C' \in \mathcal{C}$ such that $C \supsetneq C'$. The following lemma holds regardless of SD property.

Lemma 1. *For any set system $(V, \mathcal{C})$, let $C \in \mathcal{C}$ and $X \subseteq C$. It holds that* $\mathrm{MRS}_{\mathcal{C}}(C, X) = \emptyset \iff \mathcal{C}(C, X) = \{C\}$.

As we described in Sect. 1, we assume that a set system $(V, \mathcal{C})$ with SD property is given implicitly by an oracle. We denote by $\textsc{ComputeMrs}_{\mathcal{C}}$ the oracle. Given a component $C \in \mathcal{C}$ and a subset $X \subseteq C$ as a query to the oracle, $\textsc{ComputeMrs}_{\mathcal{C}}(C, X)$ returns one MRS in $\textsc{Mrs}_{\mathcal{C}}(C, X)$ if $\textsc{Mrs}_{\mathcal{C}}(C, X) \neq \emptyset$, and $\textsc{Nil}$ otherwise, where we denote by θ_t and θ_s the time and space complexity, respectively.

The following lemma states a necessary condition of SD property which is not sufficient.

Lemma 2. *Suppose that a set system $(V, \mathcal{C})$ with SD property is given. For every component $C \in \mathcal{C}$, the minimal removable sets in $\textsc{Mrs}_{\mathcal{C}}(C)$ are pairwise disjoint.*

Graphs. Let G be a simple undirected graph. For a vertex $v \in V(G)$, we denote by $\deg_G(v)$ the degree of v in the graph G. We let $\delta(G) := \min_{v \in V(G)} \deg_G(v)$. Let $S \subseteq V(G)$ be a subset of vertices. A *subgraph induced by S* is a subgraph G' of G such that $V(G') = S$ and $E(G') = \{uv \in E(G) \mid u, v \in S\}$ and denoted by $G[S]$. For simplicity, we write the induced subgraph $G[V(G) \setminus S]$ as $G - S$. Similarly, for $F \subseteq E(G)$, we write as $G - F$ the subgraph whose vertex set is $V(G)$ and edge set is $E(G) \setminus F$.

A *cut-set of* G is a subset $F \subseteq E(G)$ such that $G - F$ is disconnected. In particular, we call an edge that constitutes a cut-set of size 1 a *bridge*. We define the *edge-connectivity* $\lambda(G)$ *of* G to be the cardinality of the minimum cut-set of G unless $|V(G)| = 1$. If $|V(G)| = 1$, then $\lambda(G)$ is defined to be ∞. G is called *k-edge-connected* if $\lambda(G) \geq k$. A *vertex cut* of G is a subset $S \subseteq V(G)$ such that $G - S$ is disconnected. In particular, we call a vertex cut whose size is two a *cut point pair* and a vertex that constitutes a singleton vertex cut an *articulation point*. For a subset $S \subseteq V(G)$, let $\textsc{Art}(S)$ denote the set of all articulation points in $G[S]$. We define the *vertex-connectivity* $\kappa(G)$ *of* G to be the cardinality of the minimum vertex cut of G unless G is a complete graph. If G is complete, then $\kappa(G)$ is defined to be $|V(G)| - 1$. G is called *k-vertex-connected* if $|V(G)| > k$ and $\kappa(G) \geq k$. Obviously, G is 2-edge-connected (resp., 2-vertex-connected) if and only if there is no bridge (resp., no articulation point) in G.

Proposition 1 ([15]). *Suppose that we are given a simple undirected graph G. If $|V(G)| \geq 2$, then it holds that $\kappa(G) \leq \lambda(G) \leq \delta(G)$.*

3 Enumerating Components in Set System with SD Property

In this section, we propose an algorithm that enumerates all components that are subsets of a given component in a set system $(V, \mathcal{C})$ with SD property and conduct complexity analyses, as a proof for Theorem 3.

Let us introduce mathematical foundations for a set system with SD property that are necessary for designing our enumeration algorithm.

Algorithm 1. An algorithm to enumerate all components in $\mathcal{C}(C, I)$, where $C \in \mathcal{C}$ is a component in a set system $(V, \mathcal{C})$ with SD property and I is a subset of C

Input: A component $C \in \mathcal{C}$ and a subset $I \subseteq C$
Output: All components in $\mathcal{C}(C, I)$

```
1: procedure LIST(C, I)
2:     Output C;
3:     X ← I;
4:     while COMPUTEMRS_C(C, X) ≠ NIL do
5:         Y ← COMPUTEMRS_C(C, X);
6:         LIST(C \ Y, X);
7:         X ← X ⊔ Y
8:     end while
9: end procedure
```

Lemma 3. *For a set system $(V, \mathcal{C})$ with SD property, let $C \in \mathcal{C}$ be a component and $I \subseteq C$ be a subset of C. For any $Y \in \mathrm{MRS}_{\mathcal{C}}(C, I)$, it holds that $\mathcal{C}(C, I) = \mathcal{C}(C \setminus Y, I) \sqcup \mathcal{C}(C, I \sqcup Y)$.*

Lemma 4. *For a set system $(V, \mathcal{C})$ with SD property, let $C \in \mathcal{C}$ be a component, $I \subseteq C$ be a subset of C, and $\mathrm{MRS}_{\mathcal{C}}(C, I) := \{Y_1, Y_2, \ldots, Y_k\}$. It holds that*

$$\mathcal{C}(C, I) = \{C\} \sqcup \Big(\bigsqcup_{i=1}^{k} \mathcal{C}(C \setminus Y_i, I \sqcup Y_1 \sqcup \cdots \sqcup Y_{i-1})\Big).$$

Algorithm. Let $C \in \mathcal{C}$, $I \subseteq C$ and $\mathrm{MRS}_{\mathcal{C}}(C, I) := \{Y_1, Y_2, \ldots, Y_k\}$. Lemma 4 describes a partition of $\mathcal{C}(C, I)$ such that there exists a similar partition for $\mathcal{C}(C_i, I_i)$, where $C_i = C \setminus Y_i$ and $I_i = I \sqcup Y_1 \sqcup \cdots \sqcup Y_{i-1}$, $i \in [1, k]$. Then we have an algorithm that enumerates all components in $\mathcal{C}(C, I)$ by outputting C and then outputting all components in $\mathcal{C}(C_i, I_i)$ for each $i \in [1, k]$ recursively.

For $C \in \mathcal{C}$ and $I \subseteq C$, Algorithm 1 summarizes a procedure to enumerate all components in $\mathcal{C}(C, I)$. Procedure LIST in Algorithm 1 outputs C in line 2 and computes $\mathcal{C}(C_i, I_i)$, $i \in [1, k]$ recursively in line 6. For our purpose, it suffices to invoke LIST$(C, \emptyset)$ to enumerate all components in $\mathcal{C}(C, \emptyset)$.

To analyze the complexities and prove Theorem 3, we introduce a detailed version of the algorithm in Algorithm 2. We mainly use a stack to store data, where we can add a given element (push), peek the element that is most recently added (last), remove the last element (pop), and shrink to a given size by removing elements that are most recently added (shrinkTo) in constant time, respectively, by using an array and managing an index to the last element.

In Algorithm 2, the set X of Algorithm 1 is realized by a stack. Note that X in Algorithm 2 however contains a subset of elements in each container. In addition, a stack *Seq* stores MRSs, an array *Siz* stores the number of the MRSs of a given component, and an integer *depth* represents the depth of recursion in Algorithm 1. We can see that the algorithm enumerates all components that are subsets of C in $\mathcal{C}$ by Lemma 4. To bound the time complexity by that for processing a node in the search tree, we apply the *alternative method* to our algorithm in line 4–6 and 12–14 and reduce the delay [11]. We next discuss the space complexity. We see that *Seq* stores MRS-sequence between the current component C' and C since *Seq* pops the MRS after traversing all children of C', and so it consumes $O(n)$ space. The stack X uses $O(n)$ space by Lemma 2 and the definition of $\textsc{ComputeMrs}_{\mathcal{C}}$, and moreover *Siz* uses only $O(n)$ space since the maximum depth is n.

Proof for Theorem 3. We can see that Algorithm 2 surely enumerates all components in $\mathcal{C}(C, \emptyset)$ by Lemma 4. We first prove that Algorithm 2 works in $O(n + \theta_t)$ delay. If depth is odd, then the current component C' is outputted. If depth is even and $\textsc{Mrs}_{\mathcal{C}}(C', X) = \emptyset$, then C' is outputted. If depth is even and $\textsc{Mrs}_{\mathcal{C}}(C', X) \neq \emptyset$, then $C' \setminus Y$ will be outputted for Y in line 8 in the next iteration. Then it suffices to show that operations from line 4 to 19 can be done in $O(n + \theta_t)$ time. A component can be outputted in $O(n)$ time. The difference can be traced by subtracting and adding an MRS before and after the depth changes, thus it takes $O(n)$ time. In addition, $\textsc{ComputeMrs}_{\mathcal{C}}$ works in θ_t time by definition and another operations are adding and subtracting an MRS, where computation time is $O(n)$.

We next discuss the space complexity of Algorithm 2. The maximum size of the depth is n since the size of the component C' is monotonically decreasing while the depth increases and the termination condition that $\textsc{Mrs}_{\mathcal{C}}(C', X)$ is initially empty is satisfied at most n depth. The rest to show is that the space for *Seq*, X, and *Siz* are $O(n)$. For a component $C' \in \mathcal{C}(C, \emptyset)$, we obtain that the *Seq* is equivalent to the MRS-sequence between C' and C since *Seq* store a new MRS before the depth increases and discards before the depth decreases. X can be hold in $O(n)$ space since for any subset $I \subseteq C'$, I and $\textsc{Mrs}_{\mathcal{C}}(C', I)$ are pairwise disjoint. It is obvious that *Siz* uses $O(n)$ space since the maximum depth is n. We then obtain that whole space is $O(n + \theta_s)$ space. □

4 Enumerating 2-Edge/Vertex-Connected Induced Subgraphs

In this section, we provide proofs for Theorems 1 and 2. Suppose that we are given a simple undirected graph G that is 2-edge-connected (resp., 2-vertex-connected). In Sect. 4.1, we show that a set system $(V, \mathcal{C})$ has SD property if $V = V(G)$ and $\mathcal{C}$ is the family of all vertex subsets that induce 2-edge connected subgraphs (resp., 2-vertex-connected subgraphs). This indicates that all 2-edge/vertex-connected induced subgraphs can be enumerated by the algorithm in the last section. Then in Sect. 4.2, we show how to design the oracle for generating an MRS so that the required computational complexity is achieved.

Algorithm 2. An algorithm to enumerate all components that are subsets of $C \in \mathcal{C}$ in $(V, \mathcal{C})$ with SD property

Input: A set system $(V, \mathcal{C})$ with SD property and a component $C \in \mathcal{C}$
Output: All components that are subsets of C in $\mathcal{C}$

```
1: Seq, X ← empty stack; Siz ← an array of length n, filled by 0;
2: C' ← C; depth ← 1;
3: while depth ≠ 0 do
4:     if depth is odd then
5:         Output C'
6:     end if;
7:     if ComputeMrs_C(C', X) ≠ Nil then              ▷ emulate recursive call
8:         Y ← ComputeMrs_C(C', X);
9:         Seq.push(Y); C' ← C' \ Y; Siz[depth] ← Siz[depth]+1;
10:        depth ← depth + 1
11:    else                                              ▷ trace back
12:        if depth is even then
13:            Output C'
14:        end if;
15:        C' ← C' ∪ Seq.last();
16:        X.shrinkTo(X.length() − Siz[depth]); Siz[depth] ← 0;
17:        X.push(Seq.last());
18:        Seq.pop(); depth ← depth − 1
19:    end if
20: end while
```

4.1 Constructing Set Systems with SD Property

We define $\mathcal{C}_e \triangleq \{C \subseteq V(G) \mid G[C] \text{ is 2-edge-connected and } |C| > 1\}$ and $\mathcal{C}_v \triangleq \{C \subseteq V(G) \mid G[C] \text{ is 2-vertex-connected}\}$. For $S \subseteq V(G)$, we call S an *e-component* (resp., a *v-component*) if $S \in \mathcal{C}_e$ (resp., $S \in \mathcal{C}_v$). We deal with a set system $(V, \mathcal{C})$ such that $V = V(G)$ and $\mathcal{C} \in \{\mathcal{C}_e, \mathcal{C}_v\}$. We use the notations and terminologies for set systems that were introduced in Sect. 2 to discuss the problem of enumerating all e-components or v-components. Although a singleton is a 2-edge-connected component by definition, we do not regard it as an e-component since otherwise the system $(V, \mathcal{C}_e)$ would not have SD property.

The following lemma is immediate by Proposition 1.

Lemma 5. *For a simple undirected graph G, it holds that $\mathcal{C}_v \subseteq \mathcal{C}_e$.*

By Lemma 5, every v-component is an e-component. Let $\textsc{Max}_e(S)$ (resp., $\textsc{Max}_v(S)$) denote the family of all maximal e-components (resp., v-components) among the subsets of S. For an e-component $C \in \mathcal{C}_e$ (resp., a v-component $C \in \mathcal{C}_v$), we write the family $\textsc{Mrs}_{\mathcal{C}_e}(C)$ (resp., $\textsc{Mrs}_{\mathcal{C}_v}(C)$) of all minimal removable sets of C as $\textsc{Mrs}_e(C)$ (resp., $\textsc{Mrs}_v(C)$) for simplicity. We call an MRS in $\textsc{Mrs}_e(C)$ (resp., $\textsc{Mrs}_v(C)$) an *e-MRS of C* (resp., a *v-MRS of C*). A minimal e-component $C \in \mathcal{C}_e$ induces a cycle (i.e., $G[C]$ is a cycle) since no singleton is contained in $\mathcal{C}_e$, and in this case, it holds that $\textsc{Mrs}_e(C) = \emptyset$.

A *block* of a simple undirected graph G is a maximal connected subgraph that has no articulation point. Every block of G is an isolated vertex; a cut-edge (i.e., an edge whose removal increases the number of connected components); or a maximal v-component; e.g., see Remark 4.1.18 in [14]. The following lemma is immediate.

Lemma 6. *For a given simple undirected graph G and an e-component $C \in \mathcal{C}_e$, it holds that $C = \bigcup_{H \in \text{MAX}_v(C)} H$.*

For $P \subseteq S \subseteq V(G)$, P is called a *two-deg path in* $G[S]$ (or *in* S for short) if $\deg_{G[S]}(u) = 2$ holds for every $u \in P$. In particular, P is a *maximal two-deg path in* S if there is no two-deg path P' in S such that $P \subsetneq P'$. It is possible that a maximal two-deg path consists of just one vertex. For an e-component $C \in \mathcal{C}_e$, we denote by $\text{CAN}_{=2}(C)$ the family of all maximal two-deg paths in C. We also denote $\text{CAN}_{>2}(C) := \{\{v\} \mid v \in C, \deg_{G[C]}(v) > 2\}$ and define $\text{CAN}(C) \triangleq \text{CAN}_{=2}(C) \sqcup \text{CAN}_{>2}(C)$. It is clear that every vertex in C belongs to either a maximal two-deg path in $\text{CAN}_{=2}(C)$ or a singleton in $\text{CAN}_{>2}(C)$, where there is no vertex $v \in C$ such that $\deg_{G[C]}(v) \leq 1$ since $G[C]$ is 2-edge-connected.

The following Lemma 7 states that an e-MRS of an e-component C is either a maximal two-deg path in C or a single vertex whose degree in $G[C]$ is more than two.

Lemma 7 (Observation 3 in [7]). *For a simple undirected graph G, let $C \in \mathcal{C}_e$. It holds that* $\text{MRS}_e(C) = \{Y \in \text{CAN}(C) \mid C \setminus Y \in \mathcal{C}_e\}$.

If G is 2-edge-connected, then the set system $(V(G), \mathcal{C}_e)$ has SD property, as shown in the following Lemma 8.

Lemma 8. *For a simple undirected 2-edge-connected graph G, the set system $(V(G), \mathcal{C}_e)$ has SD property.*

Proof. We see that $V(G) \in \mathcal{C}_e$ since G is 2-edge-connected. Let $C, C' \in \mathcal{C}_e$ be e-components such that $C \supsetneq C'$ and $Y \in \text{MRS}_e(C)$ be an e-MRS of C. We show that either $Y \subseteq C'$ or $Y \cap C' = \emptyset$ holds. The case of $|Y| = 1$ is obvious. Suppose $|Y| > 1$. By Lemma 7, Y induces a maximal two-deg path in $G[C]$ such that for any $u \in Y$ it holds $\deg_{G[C]}(u) = 2$. If $Y \not\subseteq C'$ and $Y \cap C' \neq \emptyset$, then there would be two adjacent vertices $v, v' \in Y$ such that $v \in C \setminus C'$ and $v' \in C'$, where we see that $\deg_{G[C']}(v') \leq 1$ holds. The C' is an e-component and thus $|C'| \geq 2$. By Proposition 1, we obtain $1 \geq \delta(G[C']) \geq \lambda(G[C'])$, which contradicts that $C' \in \mathcal{C}_e$. □

We can derive analogous results for $(V(G), \mathcal{C}_v)$. In Lemma 9, we show that a v-MRS of a v-component C is either a maximal two-deg path in C or a single vertex whose degree in $G[C]$ is more than two. Then in Lemma 10, we show that $(V(G), \mathcal{C}_c)$ has SD property when G is 2-vertex-connected.

Lemma 9. *For a simple undirected graph G, let $C \in \mathcal{C}_v$. It holds that* $\text{MRS}_v(C) = \{Y \in \text{CAN}(C) \mid C \setminus Y \in \mathcal{C}_v\}$.

Lemma 10. *For a simple undirected 2-vertex-connected graph G, the set system $(V(G), \mathcal{C}_v)$ has SD property.*

4.2 Computing MRSs in Linear Time and Space

Let G be a simple undirected graph. We describe how we compute an e-MRS of an e-component in linear time and space. Specifically, for a given e-component $C \in \mathcal{C}_e$ and subset $X \subseteq C$, we design the oracle $\text{COMPUTEMRS}_{\mathcal{C}_e}(C, X)$ so that it outputs *one* e-MRS $Y \in \text{MRS}_e(C, X)$ if $\text{MRS}_e(C, X) \neq \emptyset$, and NIL otherwise, in linear time and space. In what follows, we derive a stronger result that *all* e-MRSs in $\text{MRS}_e(C, X)$ can be enumerated in linear delay and space.

The scenario of the proof is as follows.

(1) We show that, to enumerate e-MRSs in $\text{MRS}_e(C, X)$, it suffices to examine $\text{MRS}_e(S, X)$ for each $S \in \text{MAX}_v(C)$ respectively. This indicates that we may assume C to be a v-component. It is summarized as Corollary 1, followed by Lemma 11.

(2) Using a certain auxiliary graph, we show that it is possible to output in linear time and space all candidates in $\text{CAN}(C)$ that are e-MRSs of C (Lemma 13); recall that all candidates of e-MRSs are contained in CAN(C) by Lemma 7.

The case of computing a v-MRS of a v-component can be done almost analogously.

Lemma 11. *Given a simple undirected 2-edge-connected graph G that is neither a cycle nor a single vertex, let $V := V(G)$ and $Y \subsetneq V$ be any nonempty proper subset of V. Then Y is an e-MRS of V if and only if there is $S \in \text{MAX}_v(V)$ such that*

(i) *$Y \cap S' = \emptyset$ holds for every $S' \in \text{MAX}_v(V)$ such that $S' \neq S$; and*
(ii) *Y is either a path that consists of all vertices in S except one or an e-MRS of S.*

Proof. For the necessity, every v-component $S \in \text{MAX}_v(V)$ is an e-component. By the definition of SD property, either $Y \subsetneq S$ or $Y \cap S = \emptyset$ should hold. Suppose that there are two distinct v-components $S, S' \in \text{MAX}_v(V)$ such that $Y \subsetneq S$ and $Y \subsetneq S'$. This leads to $|Y| = 1$ since $1 \geq |S \cap S'| \geq |Y| \geq 1$, where the first inequality holds by the fact that two blocks share at most one vertex (e.g., Proposition 4.1.19 in [14]). Then Y is a singleton that consists of an articulation point of G, contradicting that $V \setminus Y$ is connected. There is at most one $S \in \text{MAX}_v(V)$ that contains Y as a proper subset, and such S surely exists since there is at least one v-component in $\text{MAX}_v(V)$ that intersects Y by Lemma 6, which shows (i).

To show (ii), suppose that $G[S]$ is a cycle. Then it holds that $|S \cap \text{ART}(V)| = 1$; if $|S \cap \text{ART}(V)| \geq 2$, then no singleton or path in S is an e-MRS of V, contradicting that $Y \subsetneq S$; if $|S \cap \text{ART}(V)| = 0$, then S is a connected component in G. This contradicts that G is connected since G is not a cycle and hence $S \subsetneq V$. Then Y should be the path in S that consists of all vertices except the only articulation point. Suppose that $G[S]$ is not a cycle. Let $u, v \in S$ be two distinct vertices. We claim that every path between u and v should not visit a vertex out of S; if there is such a path, then the union of v-components visited by the path would be a v-component containing S, contradicting the maximality of S. In the graph $G - Y$, there are at least two edge-disjoint paths between any two vertices $u, v \in S - Y$. These paths do not visit any vertex out of S, and thus $S - Y$ is an e-component. It is easy to see that $Y \in \mathrm{CAN}(S)$.

For the sufficiency, suppose that $G[S]$ is a cycle. There is no e-MRS of S by definition. The set Y should be a path that consists of all vertices in S except one, and by (i), the vertex that is not contained in Y should be an articulation point (which implies $|S \cap \mathrm{ART}(V)| = 1$). Suppose that $G[S]$ is not a cycle. An e-MRS of S exists since S is a non-minimal e-component. Let Y be an e-MRS of S that satisfies (i), that is, Y contains no articulation points in $\mathrm{ART}(V)$. In either case, it is easy to see that $V \setminus Y$ is an e-component and that $Y \in \mathrm{CAN}(V)$ holds, showing that Y is an e-MRS of V. □

Corollary 1. *For a given simple undirected graph G, let $C \in \mathcal{C}_e$ be an e-component. Then it holds that*

$$\mathrm{MRS}_e(C) = \Big(\bigsqcup_{S \in \mathrm{MAX}_v(C):\ G[S] \text{ is not a cycle}} \{Y \in \mathrm{MRS}_e(S) \mid Y \cap \mathrm{ART}(C) = \emptyset\} \Big)$$
$$\sqcup \Big(\bigsqcup_{S \in \mathrm{MAX}_v(C):\ G[S] \text{ is a cycle and } |S \cap \mathrm{ART}(C)|=1} (S \setminus \mathrm{ART}(C)) \Big).$$

By the corollary, to obtain e-MRSs of an e-component C, it suffices to examine all maximal v-components in $\mathrm{MAX}_v(C)$ respectively.

We observe the first family in the right hand in Corollary 1. Let C be a v-component such that $G[C]$ is not a cycle. For each path $P \in \mathrm{CAN}_{=2}(C)$, there are exactly two vertices $u, v \in \mathrm{CAN}_{>2}(C)$ such that u is adjacent to one endpoint of P and v is adjacent to the other endpoint of P. We call such u, v *boundaries of* P. We denote the pair of boundaries of P by $B(P)$, that is, $B(P) := \{u, v\}$. We define $\Lambda_{>2}(C) \triangleq \{uv \in E(G) \mid u, v \in \mathrm{CAN}_{>2}(C)\}$. Let $\Lambda(C) := \mathrm{CAN}_{=2}(C) \sqcup \Lambda_{>2}(C)$. We then define an auxiliary graph H_C so that

$$\begin{aligned} V(H_C) &:= \mathrm{CAN}_{>2}(C) \sqcup \mathrm{CAN}_{=2}(C) \sqcup \Lambda_{>2}(C) \\ &= \mathrm{CAN}_{>2}(C) \sqcup \Lambda(C) = \mathrm{CAN}(C) \sqcup \Lambda_{>2}(C), \\ E(H_C) &:= \{uP \subseteq V(H_C) \mid u \in \mathrm{CAN}_{>2}(C),\ P \in \mathrm{CAN}_{=2}(C),\ u \in B(P)\} \\ &\quad \sqcup \{ue \subseteq V(H_C) \mid u \in \mathrm{CAN}_{>2}(C),\ e \in \Lambda_{>2}(C),\ u \in e\}. \end{aligned}$$

We call a vertex in $\mathrm{CAN}_{>2}(C)$ an *ordinary* vertex, whereas we call a vertex in $\Lambda(C) = \mathrm{CAN}_{=2}(C) \sqcup \Lambda_{>2}(C)$ an *auxiliary* vertex.

For $P \in \mathrm{CAN}_{=2}(C)$, we denote by $E(P)$ the set of all edges in the path $P \cup B(P)$. For $e \in \Lambda_{>2}(C)$, we denote $E(e) := \{e\}$. We see that $E(G[C]) = \bigsqcup_{h \in \Lambda(C)} E(h)$ holds.

Lemma 12. *Given a simple undirected graph G, let $C \in \mathcal{C}_v$ be a v-component such that $G[C]$ is not a cycle and $Y \in \mathrm{CAN}(C)$. Then $Y \in \mathrm{MRS}_e(C)$ holds if and only if there is no auxiliary vertex $h \in \Lambda(C)$ such that $\{Y, h\}$ is a cut point pair of H_C.*

Proof. For the necessity, suppose that there is $h \in \Lambda(C)$ such that $\{Y, h\}$ is a cut point pair of H_C. Then h is an articulation point of $H_C - Y$. Every edge $e \in E(h)$ is a bridge in $G[C] - Y$, indicating that $G[C] - Y$ is not 2-edge-connected, and hence $Y \notin \mathrm{MRS}_e(C)$.

For the sufficiency, suppose that $Y \notin \mathrm{MRS}_e(C)$. Then $G[C] - Y$ is not 2-edge-connected but should be connected. There exists a bridge, say e, in $G[C] - Y$. Let $h \in \Lambda(C)$ be the auxiliary vertex such that $e \in E(h)$. We see that h is a cut point of $H_C - Y$, indicating that $\{Y, h\}$ is a cut point pair of H_C. □

Lemma 13. *Suppose that a simple undirected 2-edge-connected graph G is given. Let $V := V(G)$. For any subset $X \subseteq V$, all e-MRSs in $\mathrm{MRS}_e(V, X)$ can be enumerated in $O(n + m)$ time and space.*

Proof. We can complete the required task as follows. (1) We obtain $\mathrm{ART}(V)$ and decompose V into maximal v-components. For each maximal v-component C, (2) if $G[C]$ is a cycle and $|C \cap \mathrm{ART}(V)| = 1$, then output $C \setminus \mathrm{ART}(V)$ if it is disjoint with X; and (3) if $G[C]$ is not a cycle, then we construct an auxiliary graph H_C, compute all cut point pairs of H_C, and output all $Y \in \mathrm{CAN}(C)$ that are disjoint with $X \cup \mathrm{ART}(V)$ and that are not contained in any cut point pair together with an auxiliary vertex. The correctness of the algorithm follows by Corollary 1 and Lemma 12.

For the time complexity, (1) can be done in $O(n + m)$ time [9]. For each $C \in \mathrm{MAX}_v(V)$, let $n_C := |C|$ and $m_C := |E(G[C])|$. We can decide in $O(n_C + m_C)$ time whether C is in (2), (3) or neither of them. If we are in (2), then the task can be done in $O(n_C)$ time. If we are in (3), then the task can be done in $O(n_C + m_C)$ time since H_C can be constructed in linear time and all cut point pairs of a 2-vertex-connected graph H_C can be enumerated in linear time [4,6]. An articulation point v appears in at most $\deg_G(v)$ maximal v-components, and hence $\sum_{C \in \mathrm{MAX}_v(V)} O(n_C) = O(n+m)$. The number of maximal v-components is $O(n)$, and the overall time complexity over $C \in \mathrm{MAX}_v(V)$ is $O(n) + \sum_{C \in \mathrm{MAX}_v(V)} O(n_C + m_C) = O(n + m)$. The space complexity analysis is analogous. □

Proofs for Theorems 1 *and* 2. For Theorem 1, we see that $\mathcal{C}_e = \bigsqcup_{S \in \mathrm{MAX}_e(V)} \mathcal{C}_e(S, \emptyset)$. We can enumerate all maximal e-components in $\mathrm{MAX}_e(V)$ in $O(n+m)$ time and space, by removing all bridges in G [10]. All e-components in $\mathcal{C}_e(S, \emptyset)$ for each $S \in \mathrm{MAX}_e(V)$ can be enumerated in $O(n + \theta_t)$ delay and in $O(n + \theta_s)$ space by Theorem 3. We can implement $\mathrm{COMPUTEMRS}_{\mathcal{C}_e}$ so that $\theta_t = O(n + m)$ and $\theta_s = O(n + m)$ by Lemma 13. Theorem 2 is analogous. □

Concluding Remarks. The future work includes extension of our framework to k-edge/vertex-connectivity for $k > 2$; and studying relationship between SD property and set systems known in the literature (e.g., independent system, accessible system, strongly accessible system, confluent system).

References

1. Avis, D., Fukuda, K.: Reverse search for enumeration. Discret. Appl. Math. **63**(1–3), 21–46 (1996). https://doi.org/10.1016/0166-218X(95)00026-N
2. Chang, L., Yu, J.X., Qin, L.: Fast maximal cliques enumeration in sparse graphs. Algorithmica **66**(1), 173–186 (2013). https://doi.org/10.1007/s00453-012-9632-8
3. Conte, A., Grossi, R., Marino, A., Versari, L.: Sublinear-space and bounded-delay algorithms for maximal clique enumeration in graphs. Algorithmica **82**(6), 1547–1573 (2019). https://doi.org/10.1007/s00453-019-00656-8
4. Gutwenger, C., Mutzel, P.: A linear time implementation of SPQR-trees. In: Marks, J. (ed.) GD 2000. LNCS, vol. 1984, pp. 77–90. Springer, Heidelberg (2001). https://doi.org/10.1007/3-540-44541-2_8
5. Haraguchi, K., Nagamochi, H.: Enumeration of support-closed subsets in confluent systems. Algorithmica **84**, 1279–1315 (2022). https://doi.org/10.1007/s00453-022-00927-x
6. Hopcroft, J.E., Tarjan, R.E.: Dividing a graph into triconnected components. SIAM J. Comput. **2**(3), 135–158 (1973). https://doi.org/10.1137/0202012
7. Ito, Y., Sano, Y., Yamanaka, K., Hirayama, T.: A polynomial delay algorithm for enumerating 2-edge-connected induced subgraphs. IEICE Trans. Inf. Syst. **E105.D**(3), 466–473 (2022). https://doi.org/10.1587/transinf.2021FCP0005
8. Tada, T., Haraguchi, K.: A linear delay algorithm for enumeration of 2-edge/vertex-connected induced subgraphs. arXiv cs.DS, 2302.05526 (2023). https://doi.org/10.48550/arXiv.2302.05526
9. Tarjan, R.: Depth-first search and linear graph algorithms. SIAM J. Comput. **1**(2), 146–160 (1972). https://doi.org/10.1137/0201010
10. Tarjan, R.: A note on finding the bridges of a graph. Inf. Process. Lett. **2**(6), 160–161 (1974)
11. Uno, T.: Two general methods to reduce delay and change of enumeration algorithms. NII Technical Reports (2003)
12. Uno, T.: An efficient algorithm for solving pseudo clique enumeration problem. Algorithmica **56**, 3–16 (2010). https://doi.org/10.1007/s00453-008-9238-3
13. Wen, D., Qin, D., Zhang, Y., Chang, L., Chen, L.: Enumerating k-vertex connected components in large graphs. In: 2019 IEEE 35th International Conference on Data Engineering (ICDE), Macao, China, pp. 52–63 (2019). https://doi.org/10.1109/ICDE.2019.00014
14. West, D.B.: Introduction to Graph Theory, 2nd edn. Pearson Modern Classic (2018)
15. Whitney, H.: Congruent graphs and the connectivity of graphs. Amer. J. Math. **54**, 150–168 (1932)

Partial-Adaptive Submodular Maximization

Shaojie Tang[1(✉)] and Jing Yuan[2]

[1] Naveen Jindal School of Management, University of Texas at Dallas, Richardson, USA
shaojie.tang@utdallas.edu

[2] Department of Computer Science and Engineering, University of North Texas, Denton, USA
jing.yuan@unt.edu

Abstract. The goal of a typical adaptive sequential decision making problem is to design an interactive policy that selects a group of items sequentially, based on some partial observations, to maximize the expected utility. It has been shown that the utility functions of many real-world applications, including pooled-based active learning and adaptive influence maximization, satisfy the property of adaptive submodularity. However, most studies on adaptive submodular maximization focus on fully adaptive settings, which can take a long time to complete. In this paper, we propose a partial-adaptive submodular maximization approach where multiple selections can be made in a batch and observed together, reducing the waiting time for observations. We develop effective and efficient solutions for both cardinality and knapsack constraints and analyzes the batch query complexity. We are the first to explore partial-adaptive policies for non-monotone adaptive submodular maximization problems.

1 Introduction

Adaptive sequential decision making, where one adaptively makes a sequence of selections based on the stochastic observations collected from the past selections, is at the heart of many machine learning and artificial intelligence tasks. For example, in experimental design, a practitioner aims to perform a series of tests in order maximize the amount of "information" that can be obtained to yield valid and objective conclusions. It has been shown that in many real-world applications, including pool-based active learning [5], sensor selection [2], and adaptive viral marketing [9], the utility function is adaptive submodular. Adaptive submodularity [5] a notion that generalizes the notion of submodularity from sets to policies. The goal of adaptive submodular maximization is to design an interactive policy that adaptively selects a group of items, where each selection is based on the feedback from the past selections, to maximize an adaptive submodular function subject to some practical constraints. Although this problem has been extensively studied in the literature, most of existing studies focus on the fully adaptive setting where every selection must be made after observing the feedback from *all* past selections. This fully adaptive approach can take full advantage of feedback from the past to make informed decisions, however, as

S.-Y. Hsieh et al. (Eds.): IWOCA 2023, LNCS 13889, pp. 380–391, 2023.
https://doi.org/10.1007/978-3-031-34347-6_32

a tradeoff, it may take a longer time to complete the selection process as compared with the non-adaptive solution where all selections are made in advance before any observations take place. This is especially true when the process of collecting the observations from past selections is time consuming. In this paper, we study the problem of partial-adaptive submodular maximization where one is allowed to make multiple selections simultaneously and observe their realizations together. Our setting generalizes both non-adaptive setting and fully adaptive setting. As compared with the fully adaptive strategy, our approach enjoys the benefits of adaptivity while using fewer number of batches. To the best of our knowledge, no results are known for partial-adaptive policies for the non-monotone adaptive submodular maximization problem. We next summarize the main contributions made in this paper.

- We first study the partial-adaptive submodular maximization problem subject to a cardinality constraint. We develop a partial-adaptive greedy policy that achieves a α/e approximation ratio against the optimal fully adaptive policy where α is the *degree of adaptivity* of our policy. One can balance the performance/adaptivity tradeoff through adjusting the value of α. In particular, if we set $\alpha = 0$, our policy reduces to a non-adaptive policy, and if we set $\alpha = 1$, our policy reduces to a fully adaptive policy.
- For the partial-adaptive submodular maximization problem subject to a knapsack constraint, we develop a sampling based partial-adaptive policy that achieves an approximation ratio of $\frac{1}{6+4/\alpha}$ with respect to the optimal fully adaptive policy.
- We theoretically analyze the batch query complexity of our policy and show that if the utility function is weak policywise submodular (in addition to adaptive monotonicity and adaptive submodularity), then the above sampling based partial-adaptive policy takes at most $O(\log n \log \frac{B}{c_{\min}})$ number of batches to achieve a constant approximation ratio where B is the budget constraint and $c_{\min}$ is the cost of the cheapest item. It was worth noting that if we consider a cardinality constraint k, then $O(\log n \log \frac{B}{c_{\min}})$ is upper bounded by $O(\log n \log k)$ which is polylogarithmic.

Additional Related Work. While most of existing studies on adaptive submodular maximization focus on fully adaptive setting [11], there are a few results that are related to partial-adaptive submodular maximization. [3] propose a policy that selects batches of fixed size r, and they show that their policy achieves a bounded approximation ratio compare to the optimal policy which is restricted to selecting batches of fixed size r. However, their approximation ratio becomes arbitrarily bad with respect to the optimal fully adaptive policy. In the context of adaptive viral marketing, [12] develop a partial-adaptive seeding policy that achieves a bounded approximation ratio against the optimal fully adaptive seeding policy. However, their results can not be extended to solve a general adaptive submodular maximization problem. Very recently, [4] study the batch-mode monotone adaptive submodular optimization problem and they develop an efficient semi adaptive policy that achieves an almost tight $1 - 1/e - \epsilon$ approximation ratio. To the best of our knowledge, all existing studies are focusing on maximizing a monotone adaptive submodular function. We are the first to study the non-monotone partial-adaptive submodular maximization problem subject to both cardinality and knapsack constraints. We also provide a rigorous analysis of the batch query complexity of our policy.

2 Preliminaries and Problem Formulation

2.1 Items and States

The input of our problem is a ground set E of n items (e.g., tests in experimental design). Each items $e \in E$ has a random state $\Phi(e) \in O$ where O is a set of all possible states. Let $\phi(e) \in O$ denote a realization of $\Phi(e)$. Thus, a *realization* ϕ is a mapping function that maps items to states: $\phi : E \rightarrow O$. In the example of experimental design, the item e may represent a test, such as the temperature, and $\Phi(e)$ is the result of the test, such as, $38°C$. There is a known prior probability distribution $p(\phi) = \Pr(\Phi = \phi)$ over realizations ϕ. When realizations are independent, the distribution p completely factorizes. However, in many real-world applications such as active learning, the realizations of items may depend on each other. For any subset of items $S \subseteq E$, let $\psi : S \rightarrow O$ denote a *partial realization* and $\text{dom}(\psi) = S$ is called the *domain* of ψ. Consider a partial realization ψ and a realization ϕ, we say ϕ is consistent with ψ, denoted $\phi \sim \psi$, if they are equal everywhere in $\text{dom}(\psi)$. Moreover, consider two partial realizations ψ and ψ', we say that ψ is a *subrealization* of ψ', and denoted by $\psi \subseteq \psi'$, if $\text{dom}(\psi) \subseteq \text{dom}(\psi')$ and they agree everywhere in $\text{dom}(\psi)$. Let $p(\phi \mid \psi)$ represent the conditional distribution over realizations conditional on a partial realization ψ: $p(\phi \mid \psi) = \Pr[\Phi = \phi \mid \Phi \sim \psi]$. In addition, there is an additive cost function $c(S) = \sum_{e \in S} c(e)$ for any $S \subseteq E$.

2.2 Policies and Problem Formulation

A policy is a function π that maps a set of partial realizations to some distribution of E: $\pi : 2^{E \times O} \rightarrow \mathcal{P}(E)$, specifying which item to select next. By following a given policy, we can select items adaptively based on our observations made so far. We next introduce two additional notions related to policies from [5].

Definition 1 (Policy Concatenation). *Given two policies π and π', let $\pi@\pi'$ denote a policy that runs π first, and then runs π', ignoring the observation obtained from running π.*

Definition 2 (Level-t-Truncation of a Policy). *Given a policy π, we define its level-t-truncation π_t as a policy that runs π until it selects t items.*

There is a utility function $f : 2^E \times O^E \rightarrow \mathbb{R}_{\geq 0}$ which is defined over items and states. Let $E(\pi, \phi)$ denote the subset of items selected by π under realization ϕ. The expected utility $f_{avg}(\pi)$ of a policy π can be written as

$$f_{avg}(\pi) = \mathbb{E}_{\Phi \sim p(\phi), \Pi}[f(E(\pi, \Phi), \Phi)]$$

where the expectation is taken over possible realizations and the internal randomness of the policy.

In this paper, we first study the problem of partial-adaptive submodular maximization subject to a cardinality constraint k.

$$\max_{\pi}\{f_{avg}(\pi) \mid |E(\pi,\phi)| \leq k \text{ for all realizations } \phi\}.$$

Then we generalize this study to consider a knapsack constraint B.

$$\max_{\pi}\{f_{avg}(\pi) \mid c(E(\pi,\phi)) \leq B \text{ for all realizations } \phi\}.$$

2.3 Adaptive Submodularity and Adaptive Monotonicity

We next introduce some additional notations that are used in our proofs.

Definition 3 (Conditional Expected Marginal Utility of an Item). *Given a utility function $f : 2^E \times O^E \to \mathbb{R}_{\geq 0}$, the conditional expected marginal utility $\Delta(e \mid S, \psi)$ of an item e on top of a group of items $S \subseteq E$, conditioned on a partial realization ψ, is defined as follows:*

$$\Delta(e \mid S, \psi) = \mathbb{E}_{\Phi}[f(S \cup \{e\}, \Phi) - f(S, \Phi) \mid \Phi \sim \psi] \tag{1}$$

where the expectation is taken over Φ with respect to $p(\phi \mid \psi) = \Pr(\Phi = \phi \mid \Phi \sim \psi)$.

Definition 4 (Conditional Expected Marginal Utility of a Policy). *Given a utility function $f : 2^E \times O^E \to \mathbb{R}_{\geq 0}$, the conditional expected marginal utility $\Delta(\pi \mid S, \psi)$ of a policy π on top of a group of items $S \subseteq E$, conditioned on a partial realization ψ, is defined as follows:*

$$\Delta(\pi \mid S, \psi) = \mathbb{E}_{\Phi,\Pi}[f(S \cup E(\pi, \Phi), \Phi) - f(S, \Phi) \mid \Phi \sim \psi]$$

where the expectation is taken over Φ with respect to $p(\phi \mid \psi) = \Pr(\Phi = \phi \mid \Phi \sim \psi)$ and the random output of π.

Now we are ready to introduce the notations of adaptive submodularity and adaptive monotone [5]. Intuitively, adaptive submodularity is a generalization of the classic notation of submodularity from sets to policies. This condition states that the expected marginal benefit of an item never increases as we collect more observations from past selections.

Definition 5 (Adaptive Submodularity and Adaptive Monotonicity). *A function $f : 2^E \times O^E \to \mathbb{R}_{\geq 0}$ is adaptive submodular if for any two partial realizations ψ and ψ' such that $\psi \subset \psi'$, the following holds for each item $e \in E \setminus \mathrm{dom}(\psi')$:*

$$\Delta(e \mid \mathrm{dom}(\psi), \psi) \geq \Delta(e \mid \mathrm{dom}(\psi'), \psi'). \tag{2}$$

Moreover, we say a utility function $f : 2^E \times O^E \to \mathbb{R}_{\geq 0}$ is adaptive monotone *if for any partial realization ψ and any item $e \in E \setminus \mathrm{dom}(\psi)$: $\Delta(e \mid \mathrm{dom}(\psi), \psi) \geq 0$.*

3 Cardinality Constraint

We first study the problem of partial-adaptive submodular maximization subject to a cardinality constraint k. It has been shown that if the utility function is adaptive submodular, then a fully adaptive greedy policy can achieve a $1/e$ approximation ratio against the optimal fully adaptive policy [7]. However, one weakness about a fully adaptive policy is that one must wait for the observations from all past selections before making a new selection. To this end, we develop a *Partial-Adaptive Greedy Policy* π^p that allows to make multiple selections simultaneously within a single batch. We show that π^p achieves a α/e approximation ratio with respect to the optimal fully adaptive policy where $\alpha \in [0, 1]$ is called *degree of adaptivity*. One can adjust the value of α to balance the performance/adaptivity tradeoff.

Algorithm 1. Partial-Adaptive Greedy Policy π^p

1: $t = 1; b[0] = 1; \psi_0 = \emptyset; S_0 = \emptyset; \forall i \in [n], S_{[i]} = \emptyset$.
2: **while** $t \leq k$ **do**
3: let $M(S_{t-1}, \psi_{b[t-1]-1}) \leftarrow \text{argmax}_{V \subseteq E'; |V|=k} \sum_{e \in E'} \Delta(e \mid S_{t-1}, \psi_{b[t-1]-1})$;
4: **if** $\sum_{e \in M(S_{t-1}, \psi_{b[t-1]-1})} \Delta(e \mid S_{t-1}, \psi_{b[t-1]-1}) \geq \alpha \cdot \sum_{e \in M(\text{dom}(\psi_{b[t-1]-1}), \psi_{b[t-1]-1})} \Delta(e \mid \text{dom}(\psi_{b[t-1]-1}), \psi_{b[t-1]-1})$ **then**
5: {stay in the current batch}
6: $b[t] = b[t-1]$;
7: sample e_t uniformly at random from $M(S_{t-1}, \psi_{b[t]-1})$;
8: $S_t = S_{t-1} \cup \{e_t\}$; $S_{[b[t]]} = S_{[b[t]]} \cup \{e_t\}$;
9: **else**
10: {start a new batch}
11: $b[t] = b[t-1] + 1$;
12: observe $\{(e, \Phi(e)) \mid e \in S_{[b[t-1]]}\}$; $\psi_{b[t]-1} = \psi_{b[t-1]-1} \cup \{(e, \Phi(e)) \mid e \in S_{[b[t-1]]}\}$;
13: let $M(S_{t-1}, \psi_{b[t]-1}) \leftarrow \text{argmax}_{V \subseteq E'; |V|=k} \sum_{e \in E'} \Delta(e \mid S_{t-1}, \psi_{b[t]-1})$;
14: sample e_t uniformly at random from $M(S_{t-1}, \psi_{b[t]-1})$; $S_t = S_{t-1} \cup \{e_t\}$; $S_{[b[t]]} = S_{[b[t]]} \cup \{e_t\}$;
15: $t \leftarrow t + 1$;

3.1 Algorithm Design

We next explain the design of π^p (a detailed implementation of π^p is listed in Algorithm 1). We first add a set V of $2k-1$ dummy items to the ground set, such that $\Delta(e \mid \text{dom}(\psi), \psi) = 0$ for any $e \in V$ and any partial realization ψ. Let $E' = E \cup V$ denote the expanded ground set. These dummy items are added to avoid selecting any item with negative marginal utility. Note that we can safely remove all these dummy items from the solution without affecting its utility. For any partial realization ψ and any subset of items $S \subseteq \text{dom}(\psi)$, define $M(S, \psi)$

as a set of k items that have the largest marginal utility on top of S conditional on ψ, i.e.,

$$M(S,\psi) \in \underset{V\subseteq E';|V|=k}{\operatorname{argmax}} \sum_{e\in E'} \Delta(e \mid S,\psi). \tag{3}$$

The policy π^p executes k iterations, selecting exactly one item (which could be a dummy item) in each iteration. It is worth noting that multiple iterations of π^p may be performed together in a single batch. For every $t \in [k]$ of π^p, let S_t denote the first t items selected by π^p, let $b[t]$ denote the batch index of the t-th item selected by π^p, i.e., the t-th item is selected in batch $b[t]$, and let $S_{[q]}$ denote the set of selected items from batch q. Let ψ_q denote the partial realization of the first q batches of items selected by π^p, i.e., $\mathrm{dom}(\psi_q) = \cup_{i\in[1,q]} S_{[i]}$. Set the initial solution $S_0 = \emptyset$ and the initial partial realization $\psi_0 = \emptyset$.

- Starting with the first iteration $t = 1$.
- In each iteration t, we compare $\sum_{e\in M(S_{t-1},\psi_{b[t-1]-1})} \Delta(e \mid S_{t-1}, \psi_{b[t-1]-1})$ with $\alpha \cdot \sum_{e\in M(\mathrm{dom}(\psi_{b[t-1]-1}),\psi_{b[t-1]-1})} \Delta(e \mid \mathrm{dom}(\psi_{b[t-1]-1}), \psi_{b[t-1]-1})$, then decide whether to start a new batch or not based on the result of the comparison as follows.
  - If

$$\sum_{e\in M(S_{t-1},\psi_{b[t-1]-1})} \Delta(e \mid S_{t-1}, \psi_{b[t-1]-1}) \geq \alpha \cdot \sum_{e\in M(\mathrm{dom}(\psi_{b[t-1]-1}),\psi_{b[t-1]-1})} \Delta(e \mid \mathrm{dom}(\psi_{b[t-1]-1}), \psi_{b[t-1]-1}), \tag{4}$$

    then π^p chooses to stay with the current batch, i.e., $b[t] = b[t-1]$. It samples an item e_t uniformly at random from $M(S_{t-1}, \psi_{b[t]-1})$, which is identical to $M(S_{t-1}, \psi_{b[t-1]-1})$ due to $b[t] = b[t-1]$, and updates the solution S_t using $S_{t-1} \cup \{e_t\}$. Move to the next iteration $t = t+1$.
  - Otherwise, π^p starts a new batch, i.e., $b[t] = b[t-1]+1$, and observe the partial realization $\Phi(e)$ of all items e from the previous batch $S_{[b[t-1]]}$. Then it updates the observation $\psi_{b[t]-1}$ using $\psi_{b[t-1]-1} \cup \{(e, \Phi(e)) \mid e \in S_{[b[t-1]]}\}$. Note that $S_{t-1} = \mathrm{dom}(\psi_{b[t]-1})$ in this case. At last, it samples an item e_t uniformly at random from $M(S_{t-1}, \psi_{b[t]-1})$ and updates the solution S_t using $S_{t-1} \cup \{e_t\}$. Move to the next iteration $t = t+1$.
- The above process iterates until π^p selects k items (which may include some dummy items).

Note that $\sum_{e\in M(\mathrm{dom}(\psi_{b[t-1]-1}),\psi_{b[t-1]-1})} \Delta(e \mid \mathrm{dom}(\psi_{b[t-1]-1}), \psi_{b[t-1]-1})$ in (4) is an upper bound of $\sum_{e\in M(S_{t-1},\psi_{b[t-1]-1})} \Delta(e \mid S_{t-1}, \psi_{b[t-1]-1})$. This is because $\mathrm{dom}(\psi_{b[t-1]-1}) \subseteq S_{t-1}$ and $f : 2^E \times O^E \to \mathbb{R}_{\geq 0}$ is adaptive submodular. Intuitively, satisfying (4) ensures that the expected gain of each iteration is sufficiently large to achieve a constant approximation ratio. Unlike some other criteria proposed in previous studies [4,9], evaluating (4) is relatively easy since it does not involve the calculation of the expectation of the maximum of n random variables. Under our framework, one can adjust the degree of adaptivity $\alpha \in [0, 1]$ to balance the performance/adaptivity tradeoff. In particular, choosing a smaller α makes it easier to satisfy (4) and hence leads to fewer number of batches but poorer performance.

3.2 Performance Analysis

We next analyze the performance of π^p against the optimal fully adaptive strategy. The following main theorem shows that π^p with degree of adaptivity α achieves an approximation ratio of α/e. All missing proofs are moved to our technical report [10].

Theorem 1. *If $f : 2^E \times O^E \to \mathbb{R}_{\geq 0}$ is adaptive submodular, then the Partial-Adaptive Greedy Policy π^p with degree of adaptivity α achieves a α/e approximation ratio in expectation.*

The rest of this section is devoted to proving Theorem 1. We first present two technical lemmas. Recall that for any iteration $t \in [k]$, S_{t-1} represents the first $t-1$ selected items, $\psi_{b[t]-1}$ represents the partial realization of all items selected from the first $b[t]-1$ batches, and $M(S, \psi) \in \operatorname{argmax}_{V \subseteq E'; |V|=k} \sum_{e \in E'} \Delta(e \mid S, \psi)$.

Lemma 1. *For each iteration $t \in [k]$, $\sum_{e \in M(S_{t-1}, \psi_{b[t]-1})} \Delta(e \mid S_{t-1}, \psi_{b[t]-1}) \geq \alpha \cdot \sum_{e \in M(\mathrm{dom}(\psi_{b[t]-1}), \psi_{b[t]-1})} \Delta(e \mid \mathrm{dom}(\psi_{b[t]-1}), \psi_{b[t]-1})$.*

The next lemma shows that for any iteration $t \in [k]$, the sum of expected marginal benefits of all items from $M(\mathrm{dom}(\psi_{b[t]-1}), \psi_{b[t]-1})$ is sufficiently high. This will be used later to lower bound the expected gain of each iteration of our policy.

Lemma 2. *Let π^* denote an optimal fully adaptive policy. In each iteration $t \in [k]$, $\sum_{e \in M(\mathrm{dom}(\psi_{b[t]-1}), \psi_{b[t]-1})} \Delta(e \mid \mathrm{dom}(\psi_{b[t]-1}), \psi_{b[t]-1}) \geq \Delta(\pi^* \mid S_{t-1}, \psi_{b[t]-1})$.*

Now we are ready to prove the main theorem. Let the random variable $\mathcal{S}_{t-1}$ denote the first $t-1$ items selected by π^p and let the random variable $\Psi_{b[t]-1}$ denote the partial realization of the first $b[t]-1$ batches of items selected by π^p. For any $t \in [k]$, we next bound the expected marginal gain of the t-th iteration of π^p,

$$f_{avg}(\pi_t^p) - f_{avg}(\pi_{t-1}^p) = \mathbb{E}_{\mathcal{S}_{t-1},\Psi_{b[t]-1}}[\mathbb{E}_{e_t}[\Delta(e_t \mid \mathcal{S}_{t-1}, \Psi_{b[t]-1})]]$$
$$= \frac{1}{k}\mathbb{E}_{\mathcal{S}_{t-1},\Psi_{b[t]-1}}[\sum_{e\in M(\mathcal{S}_{t-1},\Psi_{b[t]-1})} \Delta(e \mid \mathcal{S}_{t-1}, \Psi_{b[t]-1})]$$
$$\geq \frac{1}{k}\mathbb{E}_{\mathcal{S}_{t-1},\Psi_{b[t]-1}}[\alpha \cdot \sum_{e\in M(\mathrm{dom}(\Psi_{b[t]-1}),\Psi_{b[t]-1})} \Delta(e \mid \mathrm{dom}(\Psi_{b[t]-1}), \Psi_{b[t]-1})]$$
$$= \frac{\alpha}{k}\mathbb{E}_{\mathcal{S}_{t-1},\Psi_{b[t]-1}}[\sum_{e\in M(\mathrm{dom}(\Psi_{b[t]-1}),\Psi_{b[t]-1})} \Delta(e \mid \mathrm{dom}(\Psi_{b[t]-1}), \Psi_{b[t]-1})]$$
$$\geq \frac{\alpha}{k}\mathbb{E}_{\mathcal{S}_{t-1},\Psi_{b[t]-1}}[\Delta(\pi^* \mid \mathcal{S}_{t-1}, \Psi_{b[t]-1})]$$
$$= \frac{\alpha}{k}(f_{avg}(\pi^* @ \pi_{t-1}^p) - f_{avg}(\pi_{t-1}^p)) \tag{5}$$
$$\geq \frac{\alpha}{k}((1 - \frac{1}{k})^{t-1} f_{avg}(\pi^*) - f_{avg}(\pi_{t-1}^p)). \tag{6}$$

The second equality is due to the fact that at each round $t \in [k]$, π^p adds an item uniformly at random from $M(S_{t-1}, \psi_{b[t]-1})$ to the solution. The first inequality is due to Lemma 1. The second inequality is due to Lemma 2. The last inequality is due to Lemma 1 in [7] where they show that $f_{avg}(\pi^* @ \pi_{t-1}^p) \geq (1-\frac{1}{k})^{t-1} f_{avg}(\pi^*)$.

We next prove

$$f_{avg}(\pi_t^p) \geq \frac{\alpha t}{k}(1 - \frac{1}{k})^{t-1} f_{avg}(\pi^*) \tag{7}$$

by induction on t. For $t = 0$, $f_{avg}(\pi_0^p) \geq 0 \geq \frac{\alpha \cdot 0}{k}(1 - \frac{1}{k})^{0-1} f_{avg}(\pi^*)$. Assume (7) is true for $t' < t$, let us prove it for t.

$$f_{avg}(\pi_t^p) \geq f_{avg}(\pi_{t-1}^p) + \frac{\alpha}{k}((1 - \frac{1}{k})^{t-1} f_{avg}(\pi^*) - f_{avg}(\pi_{t-1}^p))$$
$$= (1 - \alpha/k) f_{avg}(\pi_{t-1}^p) + \frac{\alpha(1 - \frac{1}{k})^{t-1} f_{avg}(\pi^*)}{k}$$
$$\geq (1 - \alpha/k)(\alpha(t-1)/k)(1 - 1/k)^{t-2} f_{avg}(\pi^*) + \frac{\alpha(1 - \frac{1}{k})^{t-1} f_{avg}(\pi^*)}{k}$$
$$\geq (1 - 1/k)(\alpha(t-1)/k)(1 - 1/k)^{t-2} f_{avg}(\pi^*) + \frac{\alpha(1 - \frac{1}{k})^{t-1} f_{avg}(\pi^*)}{k}$$
$$= \frac{\alpha t}{k}(1 - \frac{1}{k})^{t-1} f_{avg}(\pi^*).$$

The first inequality is due to (6), the second inequality is due to the inductive assumption. When $t = k$, we have $f_{avg}(\pi_t^p) \geq \alpha(1 - 1/k)^{k-1} \cdot f_{avg}(\pi^*) > (\alpha/e) f_{avg}(\pi^*)$. This finishes the proof of the main theorem.

Remark: When the utility function $f : 2^E \times O^E \to \mathbb{R}_{\geq 0}$ is adaptive submodular and adaptive monotone, [5] show that $f_{avg}(\pi^* @ \pi_{t-1}^p) \geq f_{avg}(\pi^*)$ for all $t \in [k]$. Thus, for all $t \in [k]$,

$$f_{avg}(\pi_t^p) - f_{avg}(\pi_{t-1}^p) \geq \frac{\alpha}{k}(f_{avg}(\pi^*@\pi_{t-1}^p) - f_{avg}(\pi_{t-1}^p))$$
$$\geq \frac{\alpha}{k}(f_{avg}(\pi^*) - f_{avg}(\pi_{t-1}^p)). \tag{8}$$

The first inequality is due to (5). Through induction on t, we have $f_{avg}(\pi^p) \geq (1 - e^{-\alpha})f_{avg}(\pi^*)$.

Theorem 2. *If $f : 2^E \times O^E \to \mathbb{R}_{\geq 0}$ is adaptive submodular and adaptive monotone, then the Partial-Adaptive Greedy Policy π^p achieves a $1 - e^{-\alpha}$ approximation ratio in expectation.*

4 Knapsack Constraint

In this section, we study our problem subject to a knapsack constraint B. In [1,8], they develop a fully adaptive policy that achieves a bounded approximation ratio against the optimal fully adaptive policy. We extend their design by developing a partial-adaptive policy which allows to select multiple items in a single batch. Our policy with degree of adaptivity α achieves an approximation ratio of $\frac{1}{6+4/\alpha}$ with respect to the optimal fully adaptive policy.

4.1 Algorithm Design

We first construct two candidate policies: the first policy always picks the best singleton o with the largest expected utility, i.e., $o \in \text{argmax}_{e \in E} \mathbb{E}_{\Phi \sim p(\phi)}[f(\{e\}, \Phi)]$ and the second candidate is a sampling based "density-greedy" policy π^k. Our final policy randomly picks one from the above two candidates such that $\{o\}$ is picked with probability $\frac{1/\alpha}{3+2/\alpha}$ and π^k is picked with probability $\frac{3+1/\alpha}{3+2/\alpha}$. In the rest of this paper, let $f(o)$ denote $\mathbb{E}_{\Phi \sim p(\phi)}[f(\{o\}, \Phi)]$ for short.

We next explain the idea of *Partial-Adaptive Density-Greedy Policy* π^k (the second candidate policy). π^k first selects a random subset R which is obtained by including each item $e \in E$ independently with probability $1/2$. Then we run a "density-greedy" algorithm only on R. We first introduce some notations. For each iteration $t \in [n]$, let $b[t]$ denote the batch index of t, i.e., we assume that the t-th item is selected in batch $b[t]$, for convenience, we define $b[0] = 1$. Let S_{t-1} denote the first $t - 1$ items selected by π^k, and $\psi_{b[t-1]-1}$ represent the partial realization of the first $b[t-1]-1$ batches of selected items. Set the initial solution $S_0 = \emptyset$ and the initial partial realization $\psi_0 = \emptyset$.

- Starting from iteration $t = 1$ and batch $b[t] = 1$.
- In each iteration t, let e' be the item that has the largest benefit-cost ratio on top of S_{t-1} conditioned on $\psi_{b[t-1]-1}$ from $R \setminus S_{t-1}$, i.e.,

$$e' \leftarrow \underset{e \in R \setminus S_{t-1}}{\text{argmax}} \frac{\Delta(e \mid S_{t-1}, \psi_{b[t-1]-1})}{c(e)}. \tag{9}$$

Let e'' be the item with the largest benefit-cost ratio on top of $\mathrm{dom}(\psi_{b[t-1]-1})$ conditional on $\psi_{b[t-1]-1}$ from $R \setminus \mathrm{dom}(\psi_{b[t-1]-1})$, i.e.,

$$e'' \leftarrow \underset{e \in R \setminus \mathrm{dom}(\psi_{b[t-1]-1})}{\mathrm{argmax}} \frac{\Delta(e \mid \mathrm{dom}(\psi_{b[t-1]-1}), \psi_{b[t-1]-1})}{c(e)}. \tag{10}$$

It will become clear later that e'' stores the first selected item, if any, from the $b[t-1]$-th batch. Note that $\mathrm{dom}(\psi_{b[t-1]-1}) \subseteq S_{t-1}$.
Compare $\frac{\Delta(e' | S_{t-1}, \psi_{b[t-1]-1})}{c(e')}$ with $\alpha \cdot \frac{\Delta(e'' | \mathrm{dom}(\psi_{b[t-1]-1}), \psi_{b[t-1]-1})}{c(e'')}$,

- if $\frac{\Delta(e' | S_{t-1}, \psi_{b[t-1]-1})}{c(e')} \geq \alpha \cdot \frac{\Delta(e'' | \mathrm{dom}(\psi_{b[t-1]-1}), \psi_{b[t-1]-1})}{c(e'')}$ and adding e' to the solution does not violate the budget constraint, then stay in the current batch, i.e., $b[t] = b[t-1]$, add e' to the solution, i.e., $S_t = S_{t-1} \cup \{e'\}$. Move to the next iteration, i.e., $t = t+1$;
- if $\frac{\Delta(e' | S_{t-1}, \psi_{b[t-1]-1})}{c(e')} \geq \alpha \cdot \frac{\Delta(e'' | \mathrm{dom}(\psi_{b[t-1]-1}), \psi_{b[t-1]-1})}{c(e'')}$ and adding e' to the solution violates the budget constraint, then terminate;
- if $\frac{\Delta(e' | S_{t-1}, \psi_{b[t-1]-1})}{c(e')} < \alpha \cdot \frac{\Delta(e'' | \mathrm{dom}(\psi_{b[t-1]-1}), \psi_{b[t-1]-1})}{c(e'')}$, then start a new batch, i.e., $b[t] = b[t-1] + 1$, observe the partial realization of all items selected so far, i.e., $\psi_{b[t]-1} = \psi_{b[t-1]-1} \cup \{(e, \Phi(e)) \mid e \in S_{[b[t-1]]}\}$. If $\max_{e \in R \setminus \mathrm{dom}(\psi_{b[t]-1})} \frac{\Delta(e | \mathrm{dom}(\psi_{b[t]-1}), \psi_{b[t]-1})}{c(e)} > 0$ and adding

$$\underset{e \in R \setminus \mathrm{dom}(\psi_{b[t]-1})}{\mathrm{argmax}} \frac{\Delta(e \mid \mathrm{dom}(\psi_{b[t]-1}), \psi_{b[t]-1})}{c(e)}$$

  to the solution does not violate the budget constraint, then add this item to the solution, and move to the next iteration, i.e., $t = t+1$; otherwise, terminate.

A detailed description of π^k is presented in our technical report [10].

4.2 Performance Analysis

For ease of analysis, we present an alternative implementation of π^k. Unlike the original implementation of π^k where R is sampled at the beginning of the algorithm, we defer this decision in our alternative implementation, that is, we toss a coin of success $1/2$ to decide whether or not to add an item to the solution each time after an item is being considered. It is easy to verify that both versions of the algorithm have identical output distributions. A detailed description of this alternative implementation can be found in our technical report [10].

We first provide some useful observations that will be used in the proof of the main theorem. Consider an arbitrary partial realization ψ, let $W(\psi) = \{e \in E \mid \Delta(e \mid \mathrm{dom}(\psi), \psi) > 0\}$ denote the set of all items whose marginal utility with respect to $\mathrm{dom}(\psi)$ conditional on ψ is positive. We number all items $e \in W(\psi)$ by decreasing ratio $\frac{\Delta(e|\mathrm{dom}(\psi), \psi)}{c(e)}$, i.e., $e(1) \in \mathrm{argmax}_{e \in W(\psi)} \frac{\Delta(e|\mathrm{dom}(\psi), \psi)}{c(e)}$. If $\sum_{e \in W(\psi)} c(e) \geq B$, let $l = \min\{i \in \mathbb{N} \mid \sum_{j=1}^{i} c(e(i)) \geq B\}$; otherwise, if

$\sum_{e\in W(\psi)} c(e) < B$, let $l = |W(\psi)|$. Define $D(\psi) = \{e(i) \in W(\psi) \mid i \in [l]\}$ as the set containing the first l items from $W(\psi)$. Intuitively, $D(\psi)$ represents a set of *best-looking* items conditional on ψ.

Consider any $e \in D(\psi)$, assuming e is the i-th item in $D(\psi)$, let

$$x(e,\psi) = \min\{1, \frac{B - \sum_{s\in\cup_{j\in[i-1]}\{e(j)\}} c(s)}{c(e)}\}$$

where $\cup_{j\in[i-1]}\{e(j)\}$ represents the first $i-1$ items in $D(\psi)$.

Define $d(\psi) = \sum_{e\in D(\psi)} x(e,\psi)\Delta(e \mid \mathrm{dom}(\psi), \psi)$. Similar to Lemma 1 in [6],

$$d(\psi) \geq \Delta(\pi^* \mid \mathrm{dom}(\psi), \psi). \tag{11}$$

We use $\lambda = (\{S_t^\lambda, \psi_{b[t]-1}^\lambda \mid t \in [z^\lambda]\}, \psi_{b[z^\lambda]}^\lambda)$ to represent a fixed run of π^k where S_t^λ denotes the first t items selected by π^k under λ, $\psi_{b[t]-1}^\lambda$ denotes the partial realization of first $b[t]-1$ batches of selected items under λ, and $\psi_{b[z^\lambda]}^\lambda$ denotes the partial realization of all selected items under λ, i.e., π^k selects z^λ items under λ. Hence, $\mathrm{dom}(\psi_{b[z^\lambda]}^\lambda) = S_{z^\lambda}^\lambda$. Define $C(\lambda)$ as those items in $D(\psi_{b[z^\lambda]}^\lambda)$ that have been considered by π^k but not added to the solution because of the coin flips. Let $U(\lambda)$ denote those items in $D(\psi_{b[z^\lambda]}^\lambda)$ that have not been considered by π^k. (11) implies that

$$d(\psi_{b[z^\lambda]}^\lambda) = \sum_{e\in D(\psi_{b[z^\lambda]}^\lambda)} x(e, \psi_{b[z^\lambda]}^\lambda)\Delta(e \mid S_{z^\lambda}^\lambda, \psi_{b[z^\lambda]}^\lambda) \tag{12}$$

$$= \sum_{e\in U(\lambda)\cup C(\lambda)} x(e, \psi_{b[z^\lambda]}^\lambda)\Delta(e \mid S_{z^\lambda}^\lambda, \psi_{b[z^\lambda]}^\lambda)$$

$$\geq \Delta(\pi^* \mid S_{z^\lambda}^\lambda, \psi_{b[z^\lambda]}^\lambda). \tag{13}$$

Before presenting the main theorem, we provide two technical lemmas.

Lemma 3. *Let the random variable Λ denote a random run of π^k,*

$$f_{avg}(\pi^k) \geq \frac{1}{2}\mathbb{E}[\sum_{e\in C(\Lambda)} \Delta(e \mid S_{z^\Lambda}^\Lambda, \psi_{b[z^\Lambda]}^\Lambda)]. \tag{14}$$

Lemma 4. *Let the random variable Λ denote a random run of π^k,*

$$f_{avg}(\pi^k) + f(o) \geq \alpha \cdot \mathbb{E}[\sum_{e\in U(\Lambda)} x(e, \psi_{b[z^\Lambda]}^\Lambda)\Delta(e \mid S_{z^\Lambda}^\Lambda, \psi_{b[z^\Lambda]}^\Lambda)]. \tag{15}$$

Now we are ready to present the main theorem.

Theorem 3. *If we randomly pick a policy from $\{o\}$ and π^k (with degree of adaptivity α) such that $\{o\}$ is picked with probability $\frac{1/\alpha}{3+2/\alpha}$ and π^k is picked with probability $\frac{3+1/\alpha}{3+2/\alpha}$, then we can achieve the expected utility of at least $\frac{1}{6+4/\alpha} f_{avg}(\pi^*)$.*

4.3 Bounding the Batch Query Complexity

If the utility function $f : 2^E \times O^E \rightarrow \mathbb{R} \geq 0$ is adaptive monotone, adaptive submodular, and weak policywise submodular [10], we demonstrate that by selecting a suitable value of α, it takes at most $\min\{O(n), O(\log n \log \frac{B}{c_{\min}})\}$ batches for our policy to achieve a constant approximation ratio. Here, $c_{\min} = \min_{e \in E} c(e)$ is the cost of the cheapest item. Notably, if we consider a cardinality constraint k, then the above complexity is upper bounded by $O(\log n \log k)$ which is polylogarithmic. We move this part to our technical report [10].

References

1. Amanatidis, G., Fusco, F., Lazos, P., Leonardi, S., Reiffenhäuser, R.: Fast adaptive non-monotone submodular maximization subject to a knapsack constraint. In: Advances in Neural Information Processing Systems (2020)
2. Asadpour, A., Nazerzadeh, H.: Maximizing stochastic monotone submodular functions. Manage. Sci. **62**(8), 2374–2391 (2016)
3. Chen, Y., Krause, A.: Near-optimal batch mode active learning and adaptive submodular optimization. ICML (1) **28**(160–168), 8–1 (2013)
4. Esfandiari, H., Karbasi, A., Mirrokni, V.: Adaptivity in adaptive submodularity. In: COLT (2021)
5. Golovin, D., Krause, A.: Adaptive submodularity: theory and applications in active learning and stochastic optimization. J. Artif. Intel. Res. **42**, 427–486 (2011)
6. Gotovos, A., Karbasi, A., Krause, A.: Non-monotone adaptive submodular maximization. In: Twenty-Fourth International Joint Conference on Artificial Intelligence (2015)
7. Tang, S.: Beyond pointwise submodularity: non-monotone adaptive submodular maximization in linear time. Theoret. Comput. Sci. **850**, 249–261 (2021)
8. Tang, S.: Beyond pointwise submodularity: non-monotone adaptive submodular maximization subject to knapsack and k-system constraints. In: Le Thi, H.A., Pham Dinh, T., Le, H.M. (eds.) MCO 2021. LNNS, vol. 363, pp. 16–27. Springer, Cham (2022). https://doi.org/10.1007/978-3-030-92666-3_2
9. Tang, S., Yuan, J.: Influence maximization with partial feedback. Oper. Res. Lett. **48**(1), 24–28 (2020)
10. Tang, S., Yuan, J.: Partial-adaptive submodular maximization. arXiv preprint arXiv:2111.00986 (2021)
11. Tang, S., Yuan, J.: Partial-monotone adaptive submodular maximization. J. Comb. Optim. **45**(1), 35 (2023). https://doi.org/10.1007/s10878-022-00965-9
12. Yuan, J., Tang, S.: No time to observe: adaptive influence maximization with partial feedback. In: Proceedings of the 26th International Joint Conference on Artificial Intelligence, pp. 3908–3914 (2017)

Budget-Constrained Cost-Covering Job Assignment for a Total Contribution-Maximizing Platform

Chi-Hao Wang[1], Chi-Jen Lu[1], Ming-Tat Ko[1], Po-An Chen[2(✉)], and Chuang-Chieh Lin[3(✉)]

[1] Institute of Information Science, Academia Sinica, 128 Academia Road, Section 2, Nankang, Taipei 11529, Taiwan
{wangch,cjlu,mtko}@iis.sinica.edu.tw

[2] Institute of Information Management, National Yang Ming Chiao Tung University, 1001 University Road, Hsinchu City 30010, Taiwan
poanchen@nycu.edu.tw

[3] Department of Computer Science and Information Engineering, Tamkang University, 151 Yingzhuan Road, Tamsui Distict, New Taipei City 251301, Taiwan
josephcclin@gms.tku.edu.tw

Abstract. We propose an optimization problem to model a situation when a platform with a limited budget wants to pay a group of workers to work on a set of jobs with possibly worker-job-dependent execution costs. The platform needs to assign workers to jobs and at the same time decides how much to pay each worker to maximize the total "contribution" from the workers by using up the limited budget. The binary effort assignment problem, in which an effort from a worker is indivisible and can only be dedicated to a single job, is reminiscent of bipartite matching problems. Yet, a matched worker and job pair neither incurs cost nor enforces a compulsory effort in a standard matching setting while we consider such cost to be covered by payment and certain level of effort to be made when a job is executed by a worker. The fractional effort assignment problem, in which generally a worker's effort can be divisible and split among multiple jobs, bears a resemblance to a labor economy or online labor platform, and the platform needs to output an arrangement of efforts and the corresponding payments.

There are six settings in total to consider by combining the conditions on payments and efforts. Intuitively, we study how to come up with the best assignment under each setting and how different these assignments under different settings can be in terms of the total contribution from workers when the information of each worker's quality of service and cost is available. NP-completeness results and approximation algorithms are given for different settings. We then compare the solution quality of some settings in the end.

S.-Y. Hsieh et al. (Eds.): IWOCA 2023, LNCS 13889, pp. 392–403, 2023.
https://doi.org/10.1007/978-3-031-34347-6_33

1 Introduction

Nowadays, new business models have changed the ways how technology companies make profits from their operations. The platform business model is a typical and new emerging one which creates value from the exchanges between two groups, such as sellers and buyers, consumers and producers, job managers and workers, etc. The crowdsourcing platform, which leverages the development of Web 2.0 [15], provides a classic case in point. The platform outsources jobs or tasks to interested workers and tries to incentivize them so as to crop valuable profits from their deliverables (e.g., refer to [8] for more discussions). Challenges come with such business models.

In this work, we first formulate the platform business model as an assignment problem while incentivization via sufficient payment is simplified and treated as a set of constraints that must be satisfied when the workers' information is known, i.e., they are "non-strategic", and then as a mechanism design problem with "strategic" workers whose information has to be reported to the mechanism. Specifically, when the workers' information is known, we propose an optimization problem to model this situation for the platform with a limited budget paying a group of workers to work on a set of jobs with possibly worker-job-dependent execution costs. The platform needs to assign workers to jobs and at the same time decides how much to pay each worker to maximize the total "contribution" from the workers by using up a limited budget. Intuitively, the platform might be struggling in how to better design a mechanism to incentivize the workers by payment as little as possible yet to exert as much as their total efforts. From its point of view, one may consider how to better assign jobs to suitable workers who may provide different quality of contribution for various jobs. This turns out to be a typical job matching problem [5] or a kind of assignment problem which in general is NP-hard [14] although a $(1 - 1/e)$ approximation ratio can be achieved [6].

The binary effort assignment problem, in which an effort from a worker is indivisible and can only be dedicated to a single job, is reminiscent of *bipartite matching* problems. Yet, a matched worker and job pair neither incurs "worker cost" nor enforces a "compulsory effort" commitment in a standard matching setting while we consider such cost to be covered by payment and certain level of effort to be made when a job is assigned to a worker so a *payment* to this executing worker becomes rather necessary as a relatively relaxed incentive (compared with standard rationality in terms of individual utility maximization), by simply covering the cost using the payment no matter how much the remaining utility is left.

With the characteristics of our binary effort assignment highlighted above, our fractional effort assignment becomes easier to follow as well. The fractional effort assignment problem, in which generally a worker's effort can be divisible and split among multiple jobs, bears a resemblance to a *labor economy or online labor platform*, and the platform needs to output an arrangement of efforts and the corresponding payments to cover costs for participating workers (how to incentivize the workers) in a way that the total contribution from all the workers is maximized for the economy or platform.

We consider three payment settings as follows. The first one is when all workers are paid the same no matter which worker is assigned to which job, the second one is when different jobs pay differently no matter which worker is assigned to it, and the last one is when a job can pay differently if it is assigned to different workers. Workers' efforts can be discussed under two types: one type is when a worker's effort is indivisible, which means that a worker can only work on one job, and the other type is when a worker's effort is divisible among jobs, which means that a worker can work on multiple jobs fractionally.

Combining the conditions on payments and efforts, there are six settings in total to consider in this paper. Intuitively, we study *how to come up with the best arrangement under the various settings, how different these assignments under different settings can be in terms of the total contribution from workers* when the information about workers' work quality and cost is available. We summarize these main results including NP-completeness and approximation algorithms in different settings along with comparisons after introducing the models and definitions in Sect. 1.1 for discussion.

1.1 Our Models, Preliminaries and Results

We formally consider a *job assignment* problem for a platform with budget V whose goal is to assign N workers to M jobs to maximize the total contribution from all the workers. Inspired by the concept of threshold badges [2], we generalize this idea as well as considering convexity to quantify the cost according to the devoted effort as follows. For job j, worker i can provide a quality of service $q_{ij} \geq 0$ and incurs a cost of $C_\delta(c_{ij}, e_{ij}) = c_{ij}e_{ij}^{\delta}$ for $c_{ij} \geq 0$ and $\delta \geq 1$, where c_{ij} is the maximum cost worker i could bear for job j and $e_{ij} \in [0, 1]$ is the "effort" that worker i would spend on job j. Assume that each worker has a limit on her/his effort so $\sum_j e_{ij} \leq 1$. Also, the contribution provided by worker i to job j is $q_{ij}e_{ij}$. The platform has to decide how much r_{ij} to *pay* worker i for a unit of contribution on job j. Thus, there is a natural constraint for worker i's willingness to spend effort e_{ij} on job j: $r_{ij}q_{ij}e_{ij} - C_\delta(c_{ij}, e_{ij}) \geq 0$, meaning that the payment from the platform has to cover the worker's cost. The platform has a budget constraint as well: $\sum_{i,j} r_{ij}q_{ij}e_{ij} \leq V$, i.e., the total payments cannot exceed budget V. In summary, this job assignment problem can be modeled as an optimization problem whose objective function to maximize is the total contribution of all the workers $Q(G) = \sum_{i,j} q_{ij}e_{ij}$, subject to the following constraints, where $[N]$ denotes $\{1, 2, \ldots, N\}$:
(1) $\sum_j e_{ij} \leq 1$ for all $i \in [N]$; (2) $r_{ij}q_{ij}e_{ij} - C_\delta(c_{ij}, e_{ij}) \geq 0$ for all $i \in [N]$ and all $j \in [M]$; (3) $\sum_{i,j} r_{ij}q_{ij}e_{ij} \leq V$; (4) $0 \leq e_{ij} \leq 1$ for all $i \in [N]$ and all $j \in [M]$; (5) $r_{ij} \geq 0$ for all $i \in [N]$ and all $j \in [M]$.

We define a feasible assignment for a platform such that $G = (R, E)$ with $R = \{r_{ij}\}$ and $E = \{e_{ij}\}$ to be one that satisfies constraints (1)–(5). In a feasible assignment G, the contribution from job j is defined as $Q_j(G) = \sum_i q_{ij}e_{ij}$ and the payment to job j is defined as $M_j(G) = \sum_i r_{ij}q_{ij}e_{ij}$. The total contribution is

$Q(G) = \sum_j Q_j(G) = \sum_{i,j} q_{ij}e_{ij}$, and the total payment is $M(G) = \sum_j M_j(G) = \sum_{i,j} r_{ij}q_{ij}e_{ij}$.

We consider the six settings formed by two types of efforts, i.e., fractional and binary, along with three types of payments. If workers' efforts satisfy constraint (4), we say that they are *fractional*; if they further satisfy constraint (4') $e_{ij} \in \{0,1\}$ for each $i \in [N]$ and each $j \in [M]$, we say that they are *binary*. If the platform only decides a payment $r_{ij} = r$ for all workers $i \in [N]$ and jobs $j \in [M]$, the payments are called *all-independent*. If the payments satisfy $r_{ij} = r_j$ for all workers $i \in [N]$, they are called *worker-independent*; otherwise, we call them *worker-dependent*. Hence, the six settings of the job assignment problem (see Table 1) are: binary efforts with all-independent assignments (BAI), binary efforts with worker-independent assignments (BI), binary efforts with worker-dependent assignments (BD), fractional efforts with all-independent assignments (FAI), fractional efforts with worker-independent assignments (FI), and fractional efforts with worker-dependent assignments (FD). We denote by OPT_{BI}, OPT_{BD}, OPT_{FI}, and OPT_{FD} the optimal solutions to the BI, BD, FI, and FD assignment problems, respectively.

We are ready to give a summary of our main results when the information about workers' quality of service q_{ij} and cost c_{ij} per unit of effort are available. We would like to output workers' efforts e_{ij} and payments r_{ij} that they get per unit of contribution.

1. The BD assignment problem is NP-Complete, and a fully polynomial time approximation scheme (FPTAS) is given; the BAI assignment problem is also NP-Complete and has a FPTAS.
2. The BI assignment problem is also NP-Complete, and a polynomial time $(1 - 1/e)$-approximation algorithm is given.
3. The FAI assignment problem can be polynomially solved. Using the interior point method, the FD assignment problem can be also solved optimally in polynomial time.
4. If $\delta = 1$, the FI assignment problem is NP-Complete, and we give an approximation algorithm with a ratio arbitrarily close to $(1 - 1/e)$ in polynomial time.
5. We compare the solution quality of $OPT_{BI}, OPT_{BD}, OPT_{FI}$, and OPT_{FD}:
   - With fractional efforts, $Q(OPT_{FD}) \geq Q(OPT_{FI})$ because worker-dependent payments induce a better total contribution than worker-independent payments, and we show a tight inequality $2Q(OPT_{FI}) \geq Q(OPT_{FD})$.
   - With binary efforts, we provide a tight inequality $3Q(OPT_{BI}) \geq Q(OPT_{BD})$. However, the optimal binary solutions can be arbitrarily away from the optimal fractional solutions in terms of the total contribution.

1.2 Related Work

It is clear that our job assignment problem is related to the problem of job matching [5]. When there is a budget constraint, it naturally a kind of knapsack

Table 1. Our results in the six settings of the job assignment problem for non-strategic workers.

effort type	all independent	worker independent	worker dependent
binary	BAI (NP-C, FPTAS)	BI (NP-C, P.-Time Approx.)	BD (NP-C, FPTAS)
fractional	FAI (P. Time)	FI (NP-C, P.-Time Approx.)	FD (P. Time)

problem dating back to [13], and more specifically, the multiple-choice knapsack problem [9]. As we have mentioned earlier, the assignment problem studied in [6, 14] is also related to our optimization problem. Such problems ask for a collection of items that satisfies the capacity constraint(s) and maximizes the collected values. While in this work, the job assignment problem is more complex and generalized. The workers do not necessarily exert all their effort and can make contribution only when there is enough incentive for doing so.

When a manager cannot do all jobs by herself, it is natural for her to consider outsourcing the jobs to the workers. Job assignment can then be considered as a kind of outsourcing policy [8]. When the manager gives payment or compensation to the workers, problems as such resembles the idea of incentivized exploration of the classic multi-armed bandits (MAB) [7,11,12] or steering workers to enhance their behaviors of exploitation in a way that they are more engaged in pursuing badges [1,2]. In such a framework, a principal tries to maximize the expected rewards by steering myopic players to explore arms other than the current empirically best one using compensation. Generally, previous works on the incentivized exploration of MAB did not consider budget constraints and focused more on a *no-regret* setting.

2 Binary Effort Assignment

In this section, since $e_{ij} \in \{0, 1\}$ for all i, j, the value of δ does not matter in the binary assignment problems. We will first show that the BD assignment problem and the multiple-choice knapsack problem (MCKP) are equivalent. *The MCKP problem is NP-Complete and has a fully polynomial time approximation scheme (FPTAS)* [3] *so the BD assignment problem has the same properties as MCKP.* Also, we will show that the BAI assignment problem is NP-Complete. Since the BAI assignment problem can be regarded as a special version of MCKP, the BAI assignment problem thereby has a FPTAS. Nevertheless, we use the fact that the knapsack problem is NP-Complete, and it can be reduced to a special case of the BI assignment problem. Thus, the BI assignment problem is NP-Complete, and we then give a polynomial-time $(1 - 1/e)$-approximation algorithm.

We have the following lemmas whose proofs are deferred to the full version.

Lemma 1. *The BD assignment problem is equivalent to MCKP.*

Lemma 2. *The BAI assignment problem is NP-Complete and has a FPTAS.*

Lemma 3. *The BI assignment problem is NP-complete.*

Next, we give an approximation algorithm for the BI assignment problem. Many works share similar high-level ideas for such an approximation guarantee but with different technical details due to the context (e.g., see [10]). We denote by $A = \{(i, j) \mid i \in [N], j \in [M]\}$ the set of all worker-job pairs and denote by $P = ((i_k, j_k) \mid (i_k, j_k) \in A, k \in [\ell])$ a sequence of ℓ worker-job pairs. We then construct a correspondence from a sequence of worker-job pairs to a "reasonable" assignment. A reasonable assignment consists of two parts: payments and efforts so we need to give corresponding payments and efforts. A natural way to do is, according to the sequence of worker-job pairs, setting the payment for job j to *the maximum cost-to-quality ratio among the workers who are paired with job j in the sequence* so any selected worker in the sequence will be willing to do it. Because every worker can appear more than once in the sequence, a worker should be assigned according to the last worker-job pair associated with it. In notations, payment $r_j(P) = \max\{c_{i_k j_k}/q_{i_k j_k} \mid j_k = j, k \in [\ell]\}$ and $\ell(P, i) = \arg\max\{k \mid (i_k = i, j_k) \in A\}^1$ denotes *the index of the last worker-job pair in which worker i was assigned in sequence P*. Then, we set effort $e_{ij} = 1$ if $j = j_{\ell(P,i)}$ and $e_{ij} = 0$ otherwise, which means that job j was not in the last worker-job pair in which worker i was assigned in sequence P. We call the correspondence that we just define as $G(P) = (\{r_j(P), \{e_{ij}(P)\}\})$. Note that function G is not an one-to-one function, because there could be two different sequences P and P' such that assignments $G(P)$ and $G(P')$ are the same.

Given a worker-job pair $(i, j) \notin P$, we let $P + (i, j)$ denote the sequence formed by appending worker-job pair (i, j) to sequence P, and we define the cost-performance (CP)[2] ratio of adding worker-job pair (i, j) to sequence P by $CP((i, j), P) = (Q(G(P + (i, j))) - Q(G(P)))/(M(G(P + (i, j))) - M(G(P)))$. The greedy algorithm works as follows.

Algorithm 1. Greedy(P, U, V)

Input: a sequence P, a collection U of work-job pairs and the budget V.
1: **repeat**
2: Select $(i_0, j_0) \leftarrow \arg\max_{(i,j)\in U}\{CP((i, j), P) : CP((i, j), P) > 0\}$
3: **if** $M(G(P + (i_0, j_0))) \le V$ **then**
4: $P \leftarrow P + (i_0, j_0)$
5: **end if**
6: $U \leftarrow U \backslash \{(i_0, j_0)\}$
7: **until** $U = \emptyset$
8: Return P.

The basic step above is to choose a pair (i, j) of the highest CP ratio and check if the payment for $P + (i, j)$ is within budget V.

[1] Let us abuse the use of notation ℓ as a function name as well.
[2] This anecdotal abbreviation can also refer to capacity-price.

Algorithm 2. AlgoBI(U, V)

Input: U: a collection of work-job pairs; V: the budget.
1: $P_1 \leftarrow \arg\max\{Q(G(P)) : P$ is a sequence of $U, |P| < 3, M(G(P)) \leq V\}$, $P_2 \leftarrow \emptyset$
2: **for all** P s.t. $|P| = 3$ and $M(G(P)) \leq V$ **do**
3: $U' \leftarrow U$ and $V' \leftarrow V - M(G(P))$
4: $P \leftarrow$ **Greedy**(P, U', V')
5: **if** $Q(G(P)) \geq Q(G(P_2))$ **then** $P_2 \leftarrow P$
6: **end for**
7: **if** $Q(G(P_1)) > Q(G(P_2))$ **then**
8: **output** $G(P_1)$
9: **else**
10: **output** $G(P_2)$
11: **end if**

A is a set that collects all worker-job pairs, V is the platform's budget, and **AlgoBI**(A,V) considers two goals: first, P_1 looks for the assignment corresponding to the sequence that contributes most among all sequences of size less than or equal to 2; second, P_2 considers the assignment contributing most and returned by the greedy algorithm initially fed with a pair sequence of size 3. Finally, **AlgoBI**(A,V) outputs the assignment contributing most in these two goals.

Analysis of the Approximation Ratio. We are analyzing how much contribution by assignment P_2 is guaranteed in the for-loop of **AlgoBI**(A,V). Let Y be the initial sequence such that $|Y| = 3$ and $M(G(Y)) \leq V$. OPT denotes the best assignment to this BI assignment problem, and P_{OPT} is a corresponding sequence so we have that $G(P_{OPT}) = OPT$. Suppose that when running the greedy algorithm, the ℓ'th step was in P_{OPT} for the first time, but it was not added into P due to exceeding budget V. Let ℓ represent the length of sequence P in the first ℓ' steps of the greedy algorithm. Note that ℓ is not necessarily equal to ℓ' because not every step is added as new pairs. We rename the pairs in sequence P: let (i_k, j_k) denote the kth worker-job pair added in P, and (i_ℓ, j_ℓ) be the worker-job pair considered by the greedy algorithm and in P_{OPT} for the first time but not added to sequence P due to the budget constraint. We let $P_0^Y = Y$, $P_k^Y = P_{k-1}^Y + (i_k, j_k)$ for $k \in [\ell]$, and then we give a contribution guaranteeing lemma and a theorem whose proof is in the full version.

Lemma 4. *Suppose the initial sequence is a subset of P_{OPT}, $Y \subset P_{OPT}$. Then,*

$$Q(G(P_\ell^Y)) \geq (1 - 1/e)Q(OPT) + Q(G(Y))/e.$$

Theorem 1. AlgoBI(A, V) *is a $(1 - 1/e)$-approximation algorithm for the BI assignment problem.*

3 Fractional Effort Assignment

In this section, we consider FAI, FD and FI assignment problems, where workers' effort $0 \leq e_{ij} \leq 1$ are real numbers for all $i \in [N]$ and $j \in [M]$. First, we consider

FAI assignment problem with $\delta \geq 1$ and $r_{ij} = r$ for all i, j, which is to maximize the contribution $\sum_{i,j} q_{ij}e_{ij}$ of all the workers such that

$$\sum_{j=1}^{M} e_{ij} \leq 1,\ \text{for each } i \in [N];\ rq_{ij}e_{ij} \geq c_{ij}e_{ij}^{\delta},\ \text{for each } i \in [N], j \in [M];$$
$$\sum_{i\in[N],j\in[M]} rq_{ij}e_{ij} \leq V;\ 0 \leq e_{ij} \leq 1,\ \text{for each } i \in [N], j \in [M];\ r \geq 0.$$

For a given r, this optimization problem is solvable because the functions above are convex. The third constraint gives an upper bound on r and suggests that the contribution $\sum_{i,j} q_{ij}e_{ij}$ is inversely proportional to the value of r. Hence we can solve FAI assignment problem with $\delta \geq 1$ in polynomial time.

Next, we consider FD assignment problem with $\delta \geq 1$. Suppose that manager arranges worker i, who uses effort e_{ij}, to do job j. The most cost-saving payment $r_{ij}$is then $r_{ij}(e_{ij}) = \frac{c_{ij}}{q_{ij}}e_{ij}^{\delta-1}$. That is, r_{ij} can be regarded as a function of e_{ij} instead of simply a variable. Now we reconsider FD assignment problem as below. Only $\{e_{ij}\}$ are variables. The total contribution $\sum_{i,j} q_{ij}e_{ij}$ is maximized such that $\sum_{j=1}^{M} e_{ij} \leq 1$, for each $i \in [N], \sum_{i\in[N],j\in[M]} c_{ij}e_{ij}^{\delta} \leq V$, $0 \leq e_{ij} \leq 1$, for each $i \in [N], j \in [M], r_{ij} \geq 0$, for each $i \in [N], j \in [M]$.

Note that $c_{ij}e_{ij}^{\delta}$ is convex since $\delta \geq 1$. We can apply the interior point methods [4] to the FD assignment problem with $\delta \geq 1$ to find out the optimal solution in polynomial time.

Now we will show that FI assignment problem with $\delta \geq 1$ is NP-Complete. We reduce the knapsack problem, which is NP-Complete, to the FI assignment with $\delta = 1$. Then, we will propose a polynomial time $(1 - 1/e)$ approximation algorithm.

First, we argue that FI assignment problem with $\delta = 1$ admits a special optimal solution such that every worker's effort is binary except one, say $e_{ij} \in (0, 1)$, which is fractional and corresponds to a payment $r_j = \max\{r_{j'} \mid e_{ij'} \neq 0, i \in [N], j \in [M]\}$. We have the following lemmas whose proofs are deferred to the full version.

Lemma 5. *There exists an optimal payment of the FI assignment problem with* $\delta = 1$ *which satisfies: at most one effort* $0 < e_{ij} < 1$*; the corresponding payment arrangement* $r_j = \max\{r_{j'} \mid e_{ij'} \neq 0, i \in [N], j' \in [M]\}$.

Next, we will show that the FI assignment problem with $\delta = 1$ is NP-Complete. First, we consider this problem in the scenario that there are only one job and two workers. Intuitively, it resembles the knapsack problem with one inseparable item.

Let us consider an illustrating example as follows. Suppose that we are given two workers and one job such that the quality and costs are $q_{11} = a$, $q_{21} = q$, $c_{11} = 0$ and $c_{21} = c$, where $q, a, c > 0$ are three arbitrary real numbers. If the manager sets payment by $r_1 = 0$, then worker 1 can contribute a without any effort. If the manager raises the payment by $r_1 = c/q$, then both worker 1 and 2 are willing to do job 1. From the above two kinds of payment, we observe that the best contribution we can get is a if the budget less than $a(c/q)$ (note that the manager does not need to pay for this case), The best contribution we can get is $a + q$ if the budget is more than $a(c/q) + c$. The best contribution we can get is $v\frac{q}{c}$ if the budget is between $a(c/q)$ and $a(c/q)+c$. If c is very small, then we can simulate the knapsack problem for whether we want an item or not.

Lemma 6. *FI assignment problem with $\delta = 1$ is NP-Complete.*

According to Lemma 5, the optimal solution consists of only one fractional effort. Hence we can exhaustively try all the NM effort assignments by setting all efforts to be binary except one which is set to be fractional, so that **AlgoBI** can be applied.

Algorithm 3. Complete($\mathbf{G}, (\mathbf{i_O}, \mathbf{j_O})$))

Input: G: an arrangement of BI assignment problem; (i_O, j_O): a worker-job pair.
1: Let $G \leftarrow (R = \{r_j\}, E = \{e_{ij}\})$ be an arrangement with $r_{j_O} \leftarrow c_{i_O j_O}/q_{i_O j_O}$ and $e_{i_O j_O} \leftarrow 0$.
2: $e_{ij} \leftarrow (V - M(G))/c_{i_O j_O}$ if $(i,j) = (i_O, j_O)$.
3: **Output** G.

We sort all possible payments $\{c_{ij}/q_{ij} \mid i \in [N], j \in [M]\}$ in the ascending order. Assume that γ is the kth smallest payment. Let ω_γ and τ_γ be the corresponding worker and job respectively. Let $A_\gamma = \{(i,j) \mid c_{ij}/q_{ij} \leq \gamma, (i,j) \neq (\omega_\gamma, \tau_\gamma)\}$ be the set collecting the worker-job pairs with payment less than or equal to γ.

Algorithm 4. AlgoFI($\mathbf{V}, \epsilon$)

Input: V: the budget; ϵ: a positive real smaller than 1.
1: Sort the values $\{c_{ij}/q_{ij}\}$ in the ascending order.
2: Let the kth smallest value in $\{c_{ij}/q_{ij}\}$ be r_k and the corresponding worker-job pair be (i_k, j_k) for $k = 1, 2, \ldots, MN$.
3: $\gamma \leftarrow \frac{1}{1-1/e}$.
4: **for** $k \leftarrow 1$ to MN **do**
5: $U \leftarrow A_{r_k}$.
6: $G \leftarrow$ **Complete**(**AlgoBI**$(A_{r_k}, V), (i_k, j_k)$).
7: $P \leftarrow \left\lfloor \frac{-(\log \epsilon - \log q_{i_k j_k})}{\log \gamma} \right\rfloor$
8: **for** $l \leftarrow 1$ to P **do**
9: $G' \leftarrow$ **Complete**(**AlgoBI**$(A_{r_k}, V - c_{ij}\epsilon\gamma^l), (i_k, j_k)$).
10: **if** $Q(G') \geq Q(G)$ **then** $G \leftarrow G'$.
11: **end for**
12: **end for**
13: **Output** G.

We have the following theorem whose proof is deferred to the full version.

Theorem 2. *We have $Q(\mathbf{AlgoFI}(\mathbf{V}, \epsilon)) \geq (1 - 1/e)Q(OPT_{FI}(V)) - \epsilon$.*

4 Comparisons

FI vs. FD. Let OPT_{FI}, OPT_{FD} be the optimal arrangements for the FI assignment problem and the FD assignment problem, respectively. It is easy to see that $Q(OPT_{FI}) \leq Q(OPT_{FD})$ since the FI assignment problem can be regarded as a special case of the FD assignment problem. In the following we will prove that $2Q(OPT_{FI}) \geq Q(OPT_{FD})$ and provide an instance to show that the bound is tight.

We have the following lemma and theorem whose proofs are in the full version.

Lemma 7. *Given N workers, one job and a manager with budget V. Let OPT_{FI} and OPT_{FD} be the optimal arrangements for the FI assignment problem and FD assignment problem, respectively. Then $2Q(OPT_{FI}) \geq Q(OPT_{FD}) \geq Q(OPT_{FI})$.*

Theorem 3. *Given N workers, M jobs and a manager with budget V. Let OPT_{FI}, OPT_{FD} be the optimal arrangements for the FI assignment problem and FD assignment problem, respectively. Then $2Q(OPT_{FI}) \geq Q(OPT_{FD}) \geq Q(OPT_{FI}) \geq Q(OPT_{FD})$.*

Example 1. Let there be only two workers and one job such that the quality is 1 while the costs are ϵ and 1 respectively. We assume that the budget of the manager is $1 + \epsilon$. In the FD assignment problem, the highest contribution is 2. In the FI assignment problem, we can only set $r_1 = 1$ so that the highest contribution is $1 + \epsilon$. The bound is tight when ϵ approaches 0.

BI vs. BD. Let OPT_{BI}, OPT_{BD} be the optimal arrangements for the BI assignment problem and the BD assignment problem, respectively. We obtain that $Q(OPT_{BI}) \leq Q(OPT_{BD})$ since BI assignment problem is a special case of the BD assignment problem. In the following we will prove that $3Q(OPT_{BI}) \geq Q(OPT_{BD})$ and provide an instance to show that the bound is tight.

We have the following lemma and theorem whose proofs are in the full version.

Lemma 8. *Given N workers, one job and a manager with budget V. Let OPT_{BI} and OPT_{BD} be the optimal arrangements for the BI assignment problem and BD assignment problem, respectively. Then $3Q(OPT_{BI}) \geq Q(OPT_{BD}) \geq Q(OPT_{BI})$.*

Theorem 4. *Given N workers, M jobs and a manager with budget V. Let OPT_{BI} and OPT_{BD} be the optimal arrangements for the BI assignment problem and BD assignment problem, respectively. Then $3Q(OPT_{BI}) \geq Q(OPT_{BD}) \geq Q(OPT_{BI})$.*

Example 2. Suppose that there are three workers and one job. The qualities of the workers are $1 + 2\epsilon$, $1 + \epsilon$, and 1, respectively. The cost of the workers are $\epsilon^2, 1, 1$, respectively. Let say that the budget of the platform is $2 + \epsilon^2$. The largest contribution under the BD assignment is $3 + 3\epsilon$, where $0 < \epsilon < 1/2$. Under the FI assignment, the largest contribution is by paying for the first worker which leads to contribution of $1 + 2\epsilon$. The bound is tight when ϵ approaches to 0.

FAI vs. FI and BAI vs. BI. In the previous subsections we have discussed how large the difference can be between the optimal contributions of the FI and the FD assignment problems. Now we proceed to discuss FAI and FI. We can utilize the discussion in Lemma 7 to show that the contribution of the FI assignment is at most twice that of the FAI assignment and show that the bound is also tight.

Corollary 1. *Given N workers, M jobs and a manager with budget V. Let OPT_{FAI} and OPT_{FI} be optimal arrangements in this case. Then* $2Q(OPT_{FAI}) \geq Q(OPT_{FI}) \geq Q(OPT_{FAI})$.

Example 4. Suppose that there are two workers and two jobs such that $c_{1,1} = c_{1,2} = 1$, $c_{2,1} = 1$, $c_{2,2} = \epsilon$ and $q_{1,1} = q_{2,2} = 1$, $q_{1,2} = q_{2,1} = 0$, for $0 < \epsilon < 1$. Now we consider an arrangement which is prejudicial to the manager. In this arrangement, job 2 is assigned to worker 1 and job 1 is assigned to worker 2. Assume that the manager has budget $1 + \epsilon$. Then we have that the maximum contribution is 2 under FI assignment, while it is $1 + \epsilon$ under FAI assignment by setting $r = 1$. Thus the bound is tight when ϵ approaches 0.

Next, we compare BAI with BI. We can make use of the discussion in Lemma 8 to show that the optimal solution of the BI assignment is at most three times of that of the BAI assignment problem.

Corollary 2. *Given N workers, M jobs and a manager with budget V. Let OPT_{BAI} and OPT_{BI} be optimal arrangements in this case. Then* $2Q(OPT_{BAI}) \geq Q(OPT_{BI}) \geq Q(OPT_{BAI})$.

Similar to previous illustrating example, we can also create three workers and three jobs to validate that the bound is tight.

Binary Efforts vs. Fractional Efforts. Suppose that there is only one worker and only one job. The worker has quality of q and cost of c for this job. The platform has budget of $c-1$. Then the difference of the total contribution between the FI and BI assignment problem can be arbitrarily large. This argument holds similarly for the FD and BD assignment problem.

5 Conclusions and Future Work

Job assignment is an extensively studied problem. We propose a new optimization model for a platform with limited budget, when facing many workers and many jobs, to arrange them to form worker-job assignments along with their corresponding payments to workers to incentivize them simply by covering their costs for doing such jobs and as a while to maximize a global objective of the total worker contribution. Our results rely on the platform's knowledge of every worker's parameter values in order to decide assignments to maximize the total worker contribution.

However, in reality the platform may not know the true parameter values of every worker. Thus, we can treat workers' parameters as bids including quality and cost as a promising direction of future work. Since we compare the differences among settings and find that they are not different by much in terms of the total contribution optimization so we can design truthful mechanisms for workers to reveal their parameters.

References

1. Anderson, A., Huttenlocher, D., Kleinberg, J., Leskovec, J.: Steering user behavior with badges. In: Proceedings of the 22nd International Conference on World Wide Web (WWW 2013), pp. 95–106 (2013)
2. Anderson, A., Huttenlocher, D., Kleinberg, J., Leskovec, J.: Engaging with massive online courses. In: Proceedings of the 23rd International Conference on World Wide Web. (WWW 2014), pp. 687–698 (2014)
3. Bansal, M., Venkaiah, V.: Improved fully polynomial time approximation scheme for the 0–1 multiple-choice knapsack problem. Technical report, International Institute of Information Technology Hyderabad, India) (2004)
4. Boyd, S., Vandenberghe, L.: Convex Optimization. Cambridge University Press, Cambridge (2004)
5. Crawford, V.P., Knoer, E.M.: Job matching with heterogeneous firms and workers. Econometrica **49**(2), 437–450 (1981)
6. Fleischer, L., Goemans, M.X., Mirrokni, V.S., Sviridenko, M.: Tight approximation algorithms for maximum general assignment problems. In: Proceedings of the 17th Annual ACM-SIAM Symposium on Discrete Algorithms (SODA 2006), pp. 611–620 (2006)
7. Frazier, P., Kempe, D., Kleinberg, J., Kleinberg, R.: Incentivizing exploration. In: Proceedings of the 15th ACM Conference on Economics and Computation (EC 2014) (2014)
8. Katmada, A., Satsiou, A., Kompatsiaris, I.: Incentive mechanisms for crowdsourcing platforms. In: Bagnoli, F., et al. (eds.) INSCI 2016. LNCS, vol. 9934, pp. 3–18. Springer, Cham (2016). https://doi.org/10.1007/978-3-319-45982-0_1
9. Kellerer, H., Pferschy, U., Pisinger, D.: Knapsack Problems. Springer, Berlin (2004)
10. Khuller, S., Moss, A., Naor, J.S.: The budgeted maximum coverage problem. Inf. Process. Lett. **70**(1), 39–45 (1999)
11. Liu, Z., Wang, H., Shen, F., Liu, K., Chen, L.: Incentivized exploration for multi-armed bandits under reward drift. In: Proceedings of the 34th AAAI Conference on Artificial Intelligence (AAAI 2020), pp. 4981–4988 (2020)
12. Mansour, Y., Slivkins, A., Syrgkanis, V.: Bayesian incentive-compatible bandit exploration. In: Proceedings of the 16th ACM Conference on Economics and Computation. (EC 2015), pp. 565–582 (2015)
13. Mathews, G.B.: On the partition of numbers. Proc. Lond. Math. Soc. **28**, 486–490 (1987)
14. Özbakir, L., Baykasoğlu, A., Tapkan, P.: Bees algorithm for generalized assignment problem. Appl. Math. Comput. **215**(11), 3782–3795 (2010)
15. Zhao, Y., Zhu, Q.: Evaluation on crowdsourcing research: current status and future direction. Inf. Syst. Front. **16**(3), 417–434 (2014)

Correction to: A Polyhedral Perspective on Tropical Convolutions

Cornelius Brand, Martin Koutecký, and Alexandra Lassota

Correction to:
Chapter 10 in: S.-Y. Hsieh et al. (eds.):
***Combinatorial Algorithms*, LNCS 13889,**
https://doi.org/10.1007/978-3-031-34347-6_10

The original version of this book was inadvertently published without the page numbers in the first reference: 4–47. This has been corrected.

The updated version of this chapter can be found at
https://doi.org/10.1007/978-3-031-34347-6_10

S.-Y. Hsieh et al. (Eds.): IWOCA 2023, LNCS 13889, p. C1, 2024.
https://doi.org/10.1007/978-3-031-34347-6_34

Author Index

[illegible] https://doi.org/10.1007/978-3-031-34346-9

Author Index

S.-Y. Hsieh et al. (Eds.): IWOCA 2023, LNCS 13889, pp. 405–406, 2023.
https://doi.org/10.1007/978-3-031-34347-6

Printed in the United States
by Baker & Taylor Publisher Services